SMRE 3191

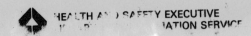

handbook of noise and vibration control

5th Edition

by R.H. Warring

TRADE & TECHNICAL PRESS LTD,
MORDEN, SURREY SM4 5EW,
ENGLAND.

ISBN 85461 093 6

© 1983 by TRADE & TECHNICAL PRESS LIMITED

All rights reserved. No part of this publication may be reproduced, stored in a retrieval system, or transmitted in any form or by any means, electronic, mechanical, photocopying, recording or otherwise, without the prior written permission of the publishers, Trade & Technical Press Limited, Crown House, Morden, Surrey.

PREFACE

The Fifth Edition of the Handbook of Noise and Vibration Control has been completely revised, up-dated and extended to keep pace with the rapid expansion in this relatively new discipline of engineering and construction. Machinery is the principal cause of noise and vibration. The constant quest for more power, more capacity, more speed, more weight loading, to meet the demands of industry today, and tomorrow, is a portend of greater noise and vibration. Yet all industrialized countries have introduced stringent laws to combat noise in the factory, the office and the environment. To sell at home and abroad, machinery manufacturers must comply with the laws appertaining to the country of origin or importation. It is certain that more man-made laws will be introduced to offset the detrimental effects of man-made machinery. A quiet machine is an acceptable machine; a noiseless machine is an efficient machine. The ultimate in machine efficiency and acceptability precludes noise and vibration completely.

This handbook is fully comprehensive and contains a wealth of technical data and information essentially of a practical nature to help machine designers, architects, public health and municipal authorities, factory managers, and all those concerned with reducing noise and vibration, wherever, whenever and however possible. Cause, effect, measurement, acceptable levels, methods of control, materials to use, working data and sources of specialized assistance and purchase are all incorporated in this one volume.

THE PUBLISHERS

ACKNOWLEDGMENTS

Bentley Nevada Corporation
E. H. Berger
G. Berry
British Gypsum R & D
P. Brown
Bruel & Kjaer (UK)
I. J. Campbell
A. Champion
Computer Engineering Ltd
Endevco UK
Entran Devices
Environmental Equipments Ltd
GEC Machines Ltd
W. T. Gracey
R. A. Heron
Hewlett Packard
M. D. Hico
ICI Acoustics
IMI Norgren Ltd
Industrial Acoustics Ltd
A. J. Jones
P. E. Jones

A. D. Liquorish
Lord Kinematics
Lucas Industries
Microdata
I. Mitchell
P. M. Nelson
C. Peacock
J. Ridings
P. P. Riley
Rolls Royce
Scientific Atlanta
I. Sharland
R. S. Skinner
Sound Research Laboratories Ltd
Transport and Road Research Laboratory
G. Tozzi
Vibro-Meter
Vitec Inc
R. H. Warring
R. V. Webb
R. F. Willmott
H. J. Wyman

We're proud of our tiny part in the space shuttle program.

Thousands of products have contributed to the successful launch and return of the space shuttle, Columbia. And our products were among the tiniest.

By design.

Entran® manufactures a broad line of ultra-miniature accelerometers and pressure transducers. These tiny, high accuracy instruments have very little mass to interfere with the system being tested, and won't distort your test scale.

Because they are so small they can be mounted almost anywhere for the ultimate in applications flexibility. And they leave plenty of room for other electronics.

Our devices have been used to obtain critical test data for the Columbia throughout its development. From the early scale-model wind tunnel tests, to main engine testing, through full-scale vibration testing of the entire shuttle and into actual flight.

Now our devices will be used to gather data for a series of intricate on-board experiments.

Our products are rugged enough to function in the toughest applications.

For more information contact Entran's exclusive agent: Thorn EMI Datatech Ltd., North Feltham Trading Estate, Feltham, Middlesex TW1W4OTD. Phone 01-890 1477. Telex 23995.

Pressure transducer.
EPI-080, only 0.080" diameter.

Accelerometer.
EGA-125D, less than 0.5 gram.

Entran Ltd.
(Our tiny products help make big projects possible.)

CONTENTS

SECTION 1
- Sound Levels 1
- Propagation of Sound 7
- Measurement of Sound 16
- Noise Scales and Noise Indices 27
- Subjective Noise Parameters 31
- Room Acoustics 37
- Acoustic Rooms 44
- Principles of Vibration 47

SECTION 2a
- Annoyance and Community Response 69
- Health and Safety (Hearing Damage) 74
- Speech Communication 78
- Hearing Conservation in Industry 82
- Hearing Protective Devices 87

SECTION 2b
- Vibration – Effect on People 97

SECTION 3a
- Noise Measuring Techniques 105
- Sound Level Meters 113
- Frequency Analysis (Spectrum Analysis) 122
- Recording and Signal Processors and Data Loggers 130
- Environmental Noise Monitoring 132
- Audiometry 137

SECTION 3b
- Vibration Measurement 147
- Vibration Transducers 154
- Dynamic Analysis of Vibration 160
- Modal Analysis 170
- Vibration Testing 177
- Machinery Health Monitoring 181

SCIENTIFIC-ATLANTA

The San Diego Group of Scientific-Atlanta Inc. (formerly Spectral Dynamics Corporation) have specialised in **Vibration Shock** and **Noise** Instrumentation for three decades.

We supply the following industries: Engineering; Aerospace; Transport; Defence; Manufacturing and Utilities with advanced products which include:-

SPECTRUM ANALYSERS – Scientific-Atlanta produce a large range of stand-alone **Digital FFT Real Time Analysers,** including a comprehensive range of accessories to perform Translation, one Third and Octave Band Analysis, Water Fall Display, Synchronous Averaging, IEEE 488 and RS232C Interfacing for operation with DEC and HP Computers, Modal Analysis, etc.

DIGITAL CONTROL SYSTEMS for **Shock Noise** and **Vibration** testing of materials and components. Scientific-Atlanta provide capabilities for Sine, Random, Classical Shock, Shock Response Spectrum and combined testing. Accessories include Multi Point Averaging, Multi Shaker Testing, System Protection, and Comprehensive Graphics.

MACHINERY MONITORING AND DIAGNOSTICS EQUIPMENT Sold under the trade name DYMAC. This instrumentation provides protection of high capital cost, high duty machinery in the Petrochemical and Electricity Generating Industries, as well as Land, Sea and Air Transport. Products include Transducers, Monitors, Balancing and Alignment Equipment for all types of machinery and applications.

Please send for a free copy of our latest product catalogue containing more details.

SCIENTIFIC-ATLANTA

USA	EUROPE	UK
San Diego Group, 4075 Ruffin Road, P.O. Box 23575, San Diego, California 92123-0575, USA. Tel: (619) 268 7000 Telex: 9103352022	Horton Manor, Stanwell Road, Horton, Slough. SL3 9PA England Tel: Colnbrook (02812) 3211 Telex: 849406	25 Bury Mead Road Hitchin, Herts. SG5 1RT England Tel: Hitchin (0462) 31101 Telex: 826087

SECTION 4a
Machines 193
Bearings 202
Internal Combustion Engines 209
Construction Site Equipment 221
Air Distribution Systems 231
Fan Noise 246
Factory Noise 257
Road Traffic Noise 267
Aircraft and Airport Noise 273
Noise in Commercial Buildings 282
Noise in Domestic Buildings 292
Auditoria 301
Noise in Ships 308

SECTION 5a
Sound Insulation and Absorption 315
Acoustic Materials 334
Acoustic Enclosures 345
Sound Barriers 354
Acoustic Treatment of Floors and Ceilings 361
Acoustic Glazing 369
Acoustic Doors 374
Fan and Air Duct Silencers 378
Industrial Silencers 391
Silencing Gas Turbines 396

SECTION 5b
Machine Balance 403
Vibration Isolation 415
Anti-Vibration Mounts 422
Damping Techniques 437
Resilient Mounting of Structures 451

SECTION 6
Legislation 463

SECTION 7
Buyers' Guide 467
Editorial Index 486
Advertisers' Index 497

SECTION 1

Sound Levels

SOUND IS a form of energy transmission, the intensity of which or *sound intensity level* (SIL or IL) is measurable in terms of watts passing through or arriving at unit area in unit time. Subjectively the sensation of sound is proportional to the log of the SIL. Where 'sound' becomes 'noise' is arguable on several counts, and is also largely dependent on the frequency content of the sound. Thus *noise* is broadly defined as any undesired sounds, usually of different frequencies, resulting in an irritating or objectionable sensation.

If sound is expressed in terms of logarithms to base 10, the unit for level is the *bel*. However, the bel is a relatively large unit, so one tenth of this is normally taken — the decibel (dB). Thus the decibel is a unit of level when the base of the logarithm is the tenth root of ten ($10^{1/10}$) and the quantities concerned are proportional to power.

Sound intensity level is thus defined as:

$$IL = 10 \log_{10} \frac{I}{I_o} \text{ dB re } I_o$$

where

I is the sound intensity at the point under consideration and
I_o is the reference level = 10^{-12} watts/m²

This gives a range of 120 dB between the threshold of audibility and that of feeling (which can be extended further upwards if necessary); (see also Table I). It also establishes that a doubling of sound intensity corresponds to a change of 3 dB.

Sound Power Level

The complement of SIL is *sound power level* (PWL) or the acoustic power in watts radiated by a noise source.

Sound power level is thus defined as:

$$PWL = 10 \log_{10} \left(\frac{W}{W_o}\right) \text{ dB re } W_o$$

where

W is the acoustic power output in watts
W_o = reference level = 10^{-12} watts

TABLE I – EXAMPLES OF ACOUSTIC POWERS AND POWER LEVELS

Acoustic Power Watts	Power Level x 10^{-12} W	Source Examples
25–50 x 10^6	190–200	Saturn booster rocket
100 000	170	Ramjet, turbojet with reheat
10 000	160	Turbojet, 10 000 lb thrust
1 000	150	Foot-propeller piston engined airliner
100	140	Turbo-prop airliner
10	130	Noisy single-engined aircraft
1*	120	Large chipping hammer, plane peak sounds
0.1	110	Public address system, large ventilating fan
0.01	100	Car on motorway
0.001	90	Shouting voice
0.0001	80	Small ventilating fan
0.00001	70	⎫
0.000001	60	⎬ Conversational voice
0.000 000 1	50	⎭
0.000 000 01	40	
0.000 000 001	30	Very soft whisper
0.000 000 000 1	20	
0.000 000 000 01	10	
0.000 000 000 001	1	Threshold of hearing (reference level)

*Threshold of feeling

Sound Pressure Level

The source of energy transmission in a sound wave is sound pressure and it is generally more suitable to deal in this term since it is easier to measure sound pressure than sound intensity. The basic relationship involved is that sound intensity is proportional to the *square* of the associated sound pressure (P), so that for a comparison of *sound pressures* the relationship becomes:

$$20 \log_{10} \frac{P_1}{P_2} \text{ dB}$$

It follows that the *sound pressure level* (SPL) of a particular sound is given by:

$$SPL = 20 \log_{10} \frac{P_1}{P_0}$$

where

P_0 is the reference level, taken as the threshold of hearing at 1 000 Hz for a sensitive ear, which is 0.0002 microbars (2 x 10^{-4} μbar)

The actual *pressure* produced by sound waves 10^{-10} x $10^{dB/20}$ bar.

Note that sound intensity (IL) sound power (PWL) and sound pressure (SPL) are all expressed as *levels* consistent with the level of a quantity being a logarithmic function, with the common

SOUND LEVELS

unit decibel (dB). Also, in the case of SPL, a doubling of sound *pressure* level is equivalent to a difference of 6 dB.

Averaging Sound Pressure Levels

To find the average of a number of separate sound pressure levels in decibels the following formula applies:

$$\text{dB (average)} = 10 \log \frac{1}{N} \left(\text{antilog} \frac{dB_1}{10} + \text{antilog} \frac{dB_2}{10} \ldots + \text{antilog} \frac{dB_n}{10} \right)$$

where

$dB_1, dB_2 \ldots dB_n$ are the separate sound pressure level readings

N = number of separate readings to be averaged

Adding Sound Levels

When two or more wave sources are present the combined sound level is determined as the arithmetical sum of each of the *mean square pressure ratios*. (Simple addition of sound *pressure* levels gives a meaningless result).

$$\text{The mean square pressure ratio} = \text{antilog} \frac{SPL}{10}$$

where

SPL = measured sound pressure level

Thus the combined sound pressure level of two or more sources is:

$$\text{Combined SPL} = \text{antilog} \frac{SPL_1}{10} + \text{antilog} \frac{SPL_2}{10} \text{ antilog} \frac{SPL_3}{10} + \text{antilog} \frac{SPL_n}{10}$$

It is generally more convenient to use a graph for rapid solutions — see Fig 1.

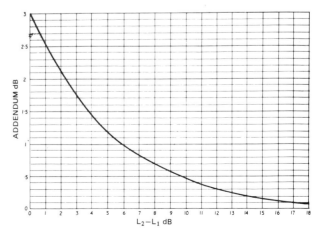

Fig 1

In either case, where the difference in sound pressure levels from different sources is greater than 6 dB, the resultant sound pressure level is substantially the same as that of the loudest sound and so the effect of the lower sound can be ignored.

Subtracting Sound Levels

It is often necessary to subtract one sound level from another, *eg* to be able to subtract background noise from a total noise measurement in order to obtain the actual noise produced by a single source, such as a machine.

The procedure to be adopted is:

(i) Measure the total noise in dB with both sources operating. Call this SPL_T

(ii) Measure the background noise only in dB with the single source shut off. Call this SPL_B

(iii) Calculate the quantity $10^{\frac{SPL_T}{10}} - 10^{\frac{SPL_B}{10}}$ Call this D

(iv) Find the numerical value of $SPL_T - 10 \log_{10} D$.

(v) Subtract this value from the total noise measurement (SPL_T) to give the value of the single noise source

Again, graphical solution is preferred for speed and simplicity — see Fig 2.

Correction for background noise is only applicable where the source to background noise ratio is less than about 20 dB. In other words, for higher source to background noise ratios the effect of background noise will be negligible and so the total noise measurement will be virtually the same as that of the single source.

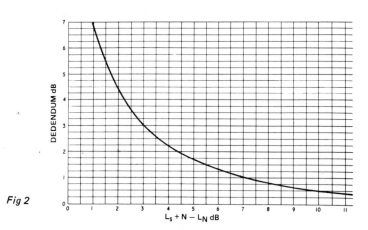

Fig 2

Noise from Identical Sources

If a number of separate noise sources are present each with an identical sound pressure level at the point of measurement, the total pressure level can be calculated from the sound pressure level of one source.

SOUND LEVELS

$$\text{dB (total)} = dB_1 + 10 \log n$$

where
    dB_1 is the sound pressure level of one source
    n is the number of identical sources

Percentage Change in Sound Level

To express an increase or decrease in sound power as a percentage instead of a difference in decibels the following relationship applies:

$$\text{percentage increase} = \text{antilog}\frac{dB2 - dB1}{10} - 1$$

$$\text{percentage reduction} = 1 - \frac{1}{\text{antilog}\frac{dB1 - dB2}{10}}$$

Solutions can be read directly from the graph given in Fig 3.

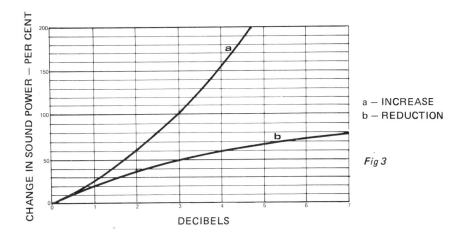

a – INCREASE
b – REDUCTION

Fig 3

Inter-relationship

Rapid estimates of level or sound pressure changes may be made simply by knowing that, on the SPL scale, 6 dB represents a pressure change of 2:1, in the case of a simple acoustic signal. In a complex acoustic field, the sum of two sources generating equal sound pressure results in a 3 dB change.

If a pressure source of 2×10^{-4} μbar is enclosed within a sphere whose surface area is one square metre, it can be shown that the resulting acoustic intensity is very nearly 10^{-12} watts/m². Although the standard reference of $I_0 = 10^{-12}$ watts/m² is not the exact intensity corresponding to 0.0002 bar, the difference is so slight as to be negligible. The exact relation between SPL and IL is:

$$SPL = IL + 0.2 \text{ dB re } 0.0002 \text{ μbar}$$

Equivalents are thus:

dB	40	50	60	70	80	90	100	110
$\dfrac{I}{I_o}$	1×10^4	1×10^5	1×10^6	1×10^7	1×10^8	1×10^9	1×10^{10}	1×10^{11}

Note that an increase in dB of 10 also gives an increase in sound intensity of 10. Intermediate values are given by the antilog of the dB value, *ie:*

dB	1	2	3	4	5	6	7	8	9	10
$\dfrac{I}{I_o}$	1.259	1.585	1.995	2.512	3.162	3.981	5.012	6.310	7.762	10.000

For example 65 dB = an intensity of 3.162×10^6 watts/m².

The relationship between SPL and PWL can be determined in the same manner. Since the intensity at the surface of the above sphere (whose surface area is one square metre) is known, the total power enclosed by the sphere can be found. Of course, for the case where the sphere has a surface area of 1 m², the acoustic power is numerically equal to the intensity. Therefore, the reference quantity for PWL is 10^{-12} watts. The relationship between SPL and PWL is:

$$SPL = PWL - 10 \log_{10} \frac{S}{S_o} + 0.5 \text{ dB re } 0.0002 \text{ bar}$$

where

S_o = reference area of 1 m²
S = area through which sound is radiated in m²

The aforementioned relationships hold only for standard conditions. In many actual applications, these simple relationships must be modified by other factors peculiar to the system under investigation.

Sound Intensity

Sound intensity as distinct from sound intensity *level*, is a vector quantity which describes both the magnitude and direction of the flow of sound energy at a given point. This is in contrast to sound pressure level which has magnitude but no direction. It is a particularly useful quantity to measure for, besides strong directivity, it distinguishes between net flow of sound, energy and any reactive part of the sound field where measurement is being made. Equally, sound intensity measurements can be used as a basis for calculating the sound power emitted by a noise source, independent of background noise.

Sound intensity is basically defined as being power per unit area, measurable by time averaging the product of the instantaneous sound pressure level and the instantaneous particle velocity (or velocity an imaginary particle would have if it was subjected to the sound field). It is only recently that suitable instruments have been evolved for the practical measurement of sound intensity — see chapter on *Measurement of Sound*.

Propagation of Sound

THE VELOCITY of sound in any medium is given by:

$$V = k\sqrt{\frac{E}{D}}$$

where E = modulus of elasticity of the medium in pascals
       D = density of medium in kg/m^3
       k = 0.843

For air at normal atmospheric pressure (1 bar)

$$V = (331.4 + 0.6T) \text{ metres/sec}$$

where T = air temperature in °C

An alternative approximate formula is

$$V = 20\sqrt{273 + T} \text{ metres/sec}$$

Thus at 20°C (68°F)

$$V = 20\sqrt{293}$$
$$= 342 \text{ metres/sec or } 1122 \text{ ft/sec}$$

The velocity of sound in non-gaseous media is appreciably higher, typical values are given in Table I.

The velocity of sound is also related to the wavelength and frequency of the sound wave:

$$V = f \times \lambda$$

where V is in metres/sec
       f is the frequency in Hz
       λ is the wavelengths in metres

TABLE I – VELOCITY OF SOUND IN NON-GASEOUS MEDIA

Medium	Velocity of Sound		Velocity compared with that of sound in air (approx)
	metres/sec	ft/sec	
lead	1 220	4 000	x 3.5
water	1 370	4 500	x 4
wood	3 350 – 4 000	11.000 – 13 000	x 10
aluminium	4 880	16 000	x 14
steel	5 000	16 500	x 15

TABLE II – FREQUENCY AND WAVELENGTH

Frequency Hz	Wavelength feet	Wavelength metres	Frequency Hz	Wavelength feet	Wavelength metres
10	117.7	34	200	5.60	1.700
20	53.9	17	250	4.50	1.368
30	37.2	11.3	500	2.25	0.684
40	27.9	8.5	1 000	1.12	0.342
50	22.4	6.8	2 000	0.56 (6.72")	0.171
63	17.8	5.429	4 000	0.28 (3.36")	0.0855
100	11.2	3.400	8 000	0.14 (1.68")	0.0428
125	8.97	2.736	16 000	0.07 (0.84")	0.0224

Table II shows the relationship between frequency (f) and wavelength (λ) for sound waves in air.

Sound Intensity

The simplest concept of a source of sound is a point without bulk. Sound is then radiated equally in all directions from this apparent centre. At any distance R from that centre the sound will be uniformly distributed over a hypothetical spherical surface of radius R. At any specific point on this surface, the *sound intensity* will be proportional to $1/r^2$, ie the sound intensity decreases as the square of the distance from the source (inverse square law).

It follows that doubling the distance from the source reduces the sound intensity to ¼, or, expressed in decibels:

$$\text{Result of doubling distance} = 10 \log \frac{1}{1/4}$$
$$= 10 \log 4$$
$$= 6 \text{ dB}$$

In other words, each doubling of the distance is equivalent to a decrease in sound pressure level of 6 dB.

In specific terms, taking a distance of one unit as a standard:

sound reduction in dB = 20 log R

where   R is the distance from the source

Thus at a distance of 10 metres from the source, the sound level would be 20 log 10 = 10 dB less than at a distance of 1 metre from the source.

Sound Power Level

The total sound power radiated by the point source, or sound power level (PWL) can be computed on a similar basis, related to the sound pressure level (SPL), as measured at a distance R:

PWL = SPL + 20 log R + 0.5 dB

where   R is the distance

Strictly speaking, this applies only at normal temperatures and pressures, and further correction should be made for differences in ambient air conditions. Such corrections are, however, small enough to be ignored for most practical purposes over small differences in temperature and pressure.

The actual acoustic power in watts can also be determined from the relationship:

$$\text{PWL (re } 10^{-12} \text{ watts)} = 10 \log \frac{W}{10^{-12}}$$

$$= 10 \log W + 120$$

The quantity 10 log W is the number of decibels corresponding to the numerical value of W watts, and can conveniently be found from tables.

Knowing the acoustic power of a point source in watts, the power level (PWL) in decibels re 10^{-12} watts can readily be determined, and from this the sound pressure level in dB at any distance from the source.

The concept of a point source is an idealized one, and the related characteristics are also dependent on the source being in free air with no physical boundaries which could modify the concentric spherical radiation pattern. This is generally referred to as *free field* conditions. Under such conditions, and with a true point source, a single measurement is sufficient to determine all the characteristics of the sound field.

Absorption by Air

The actual alteration of sound by air is due to absorption of sound energy by air molecules, converting this sound energy into heat. The degree of attenuation is small but the sound reduction in decibels is as follows:

Attenuation (dB) = 0.1f + log d

where   d is the distance (in metres) of the sound source from the point under consideration;

f is a factor dependent on the frequency of the sound, *viz:*
- 1 at 500 Hz
- 1.5 at 1 000 Hz
- 3 at 2 000 Hz
- 7.5 at 4 000 Hz

Thus theoretical values for reduction in sound due to air absorption are:-

'average' sound frequency 1 000 Hz — 0.3 dB per 100 metres
'average' sound frequency 2 000 Hz — 0.6 dB per 100 metres
'average' sound frequency 4 000 Hz — 1.5 dB per 100 metres

These sample calculations also emphasize that low frequency sounds 'travel farther' through air than high frequency sounds.

Practical Noise Sources

Noise sources are never simple point sources although they may approximate to point source characteristics for general measurement requirements. Where the sound source has definite bulk, however, the initial radiation pattern will be modified by its shape.

Assuming, for simplicity, that sound is propagated more or less uniformly from this shape, then it will be found that over an initial distance R_n, the sound intensity will decrease at a lower rate than that defined by the inverse square law. Beyond this particular distance, however, the inverse square law of propagation will apply — Fig 1.

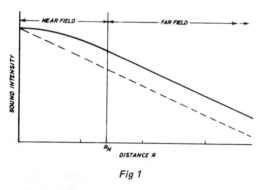

Fig 1

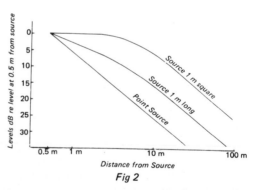

Fig 2

This characteristic of a non-point source gives rise to the conception of a *near field* in which the propagation is 'non square law' or non-spherical, and a *far field* in which propagation follows the inverse square law or is spherical. It follows that, provided free field conditions apply, a single measurement in the far field may satisfactorily define the sound field in that region, since the behaviour is then essentially independent of the shape and size of the source. In the near field, the propagation pattern is strictly dependent on the size and shape of the source.

Idealized decay rate curves are shown in Fig 2 for a point source, long cylindrical source and square or cubic source; the last two have appreciable linear dimensions (L). The near field extends up to about 5 x L in these two cases, indicating that any sound measurement taken at a distance less than 5L will be unique to the form of the source. At a greater distance, measurements are capable of extrapolation in accordance with the inverse square law, within the range of the far field. Put in simple language, the characteristics of sound propagation will be 'spherical' (or follow the inverse square law) and independent of the size of the source at a distance of several times that of the largest dimension of the source. The actual distance involved may be 5L or less — eg 3L to 4L is commonly quoted as a 'typical' figure.

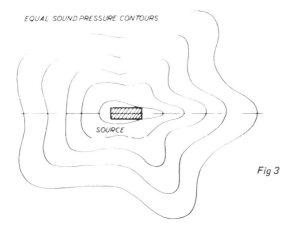

Fig 3

Directivity

Most practical sound sources are *directional*, in that they radiate more sound in one direction than another. The actual pattern of variation may be quite complex, but can be determined for any particular plane by taking measurements in that plane and plotting contours of equal sound pressure level. Fig 3 shows a simplified diagram of this type.

Although such a diagram will appear essentially non-spherical, near and far field characteristics will be apparent. Thus, as shown, in the far field, although the contours may be 'lobed' for each doubling of distance in this field the sound pressure level will be decreased by 6 dB. In other words, each lobe or similar convolution will follow a spherical pattern and will maintain a similarity of shape (*ie* the contours developed in the far field are no longer modified by the size and shape of the source).

Another method of plotting the directivity of a source is by polar diagrams. A polar diagram is plotted at a fixed distance from the source but at different angles.

The directional properties of a source may be specifically defined in terms of a *directivity factor* (Q), which is a function of direction and frequency. It may be defined as the ratio of the mean square sound pressure at a specific distance and direction (P_d^2) to the mean square sound pressure at the same distance averaged over all directions from the source (P_{av}^2)

$$Q = \frac{P_d^2}{P_{av}^2}$$

In effect the use of the directivity factor 'corrects' the performance of a point source calculation to take into account directional effects of the source, so that a similar equation for sound pressure level can apply:

SPL = PWL + 10 log Q − 20 log R − 0.5 dB

Thus knowing the power level (PWL) and the directivity factor (Q) for a specific direction, the SPL at any point R in the far field can be calculated, provided free field conditions apply.

Alternatively, the directivity index (DI) may be quoted, defined by:

$$DI = SPL_\theta - SPL_S$$

where SPL_θ is the sound pressure level in direction θ at the specific distance from the source and the SPL_S is the sound pressure level which would be produced by a point source radiating the same power at the same distance.

The directively index derives directly as:

$$Q = \text{antilog}\ \frac{DI}{10}$$

In general, sources of low frequency noise tend to be non-directional (*ie* uniformly radiated), particularly when the source itself is small in comparison with the wavelength of the sound being generated. Directional effects tend to become more apparent as the sound increases, particularly when the source of the sound becomes large in comparison with its wavelength.

Effect of Wind

It is well known that sound is 'carried' by wind, *ie* sound levels are higher downwind of a source than upwind. The effect of wind on an idealized sound radiation pattern is shown in Fig 4, providing an 'upward bending' effect on sound travelling downwind.

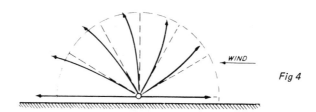

Fig 4

The effect is similar with more complex radiation patterns, but rather less predictable. The main practical point is that still air conditions are needed for accurate sound measurements.

The effect of wind can be estimated as:

Decrease/increase in sound level in dB = $3 + \log V/5$

where

V is the wind speed in metres/sec

This represents a decrease in a dead upwind direction or an increase in a dead downwind direction. The effect of intermediate wind directions can be estimated by using the approximate vector value.

This formula has the distinct limitation that wind near ground level is never constant in velocity and is generally turbulent, so it is impossible to measure and difficult to estimate the effective wind speed. As a general rule, however, a wind speed of 5 metres/sec can either decrease or increase the sound level by up to 3 dB, depending on its direction.

PROPAGATION OF SOUND

Atmospheric changes can also produce similar modification of sound propagation. A temperature inversion will tend to increase the sound energy by 'downward bending', and in certain circumstances can produce audibility of sound at distances far greater than the point at which normal attenuation renders the sound inaudible. Equally, of course, it can make a 'normal' sound louder at a particular point. Normal temperature and pressure changes have the effect shown in Fig 5.

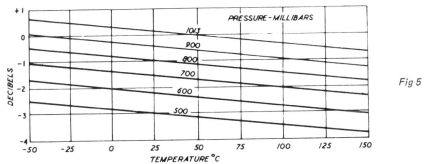

Fig 5

TABLE III – ATMOSPHERIC ATTENUATION

Temperature	Relative Humidity %	Attenuation dB per metre at		
		2 000 Hz	6 000 Hz	10 000 Hz
20°C	10	0.070	0.15	0.20
	20	0.040	0.15	0.27
	40	0.015	0.08	0.17
	50	0.015	0.07	0.15
10°C	10	0.070	0.08	0.09
	20	0.070	0.17	0.20
	40	0.030	0.13	0.25
	50	0.020	0.11	0.22

Atmospheric moisture will have an attenuating effect, which increases with the relative humidity — see Table III. The effect is generally negligible for low frequencies, but increases with increasing frequency.

Ground Absorption

Ground absorption is negligible where the surface is hard — *eg* concrete, paving, metalled roads *etc*. It can, however, be appreciable where the surface has marked absorptive rather than reflective properties — *eg* grass. In this case, the likely reduction in sound is given by:

Reduction in sound level (dB) = $4 a \log d$

where a is the absorptivity of the ground, expressed as a dimensionless factor
d is the distance which the sound travels over this surface

Values of a are essentially empirical, a figure of 0.6 being commonly quoted for grass. On this basis a reduction in sound level of slightly less than 5 dB per 100 metres distance of grass can be anticipated.

Sound Reflection

Sound is reflected by any plane surface within the sound field on which sound waves can be incident. The proportion of sound energy reflected from the surface will depend on the sound absorption properties of the surface and may range from high (low absorption) to low (high absorption).

The effect of reflection is to yield an image from which the reflected rays appear to come — Fig 6. At point R, therefore, sound is received from two sources which, with a perfectly reflecting surface, will effectively double the number of sound waves and cause an average increase in sound level of 3 dB. The direct and reflected rays, however, travel different paths and thus arrive at R out of phase. The extent of out-of-phase can either amplify or attenuate the final sound. If ideal amplification conditions are achieved (the peaks of the two waves being coincident at R), then the sound level will be increased by 6 dB. If peak and trough coincide, then effectively the sound will cancel out.

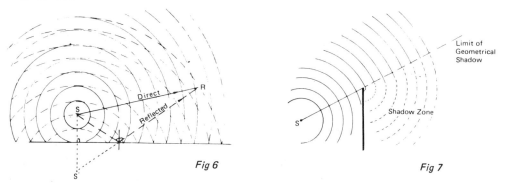

Fig 6   Fig 7

In practice — and with less than a perfectly reflecting surface — this will mean that at varying distances from the source there will be points at which the sound level is boosted by reflection, with intermediate points where the sound level is reduced. This is most likely to be noticeable near to the source. Once the distance becomes large, both with regard to the wavelength of the sound and the distance of the reflecting surface from the source, the increase in sound level due to reflection will tend to a uniform level somewhat less than 3 dB (depending on the reflective value of the plane surface).

Barriers

Any object or surface representing a barrier to the sound waves will produce a shadowing effect. That is to say, the sound level will be reduced in the shadow of the barrier — Fig 7.

The magnitude of this reduction depends primarily on the dimensions of the barrier and the wavelength of the sound. If the barrier is large compared with the wavelength, then the sound reduction in the shadow zone will be large. It follows that a barrier of specific size will be more effective in reducing high frequency sound than lower frequencies.

The shadow zone is not properly defined by optic principles. Considerable refraction can occur, causing sound energy to penetrate down into the shadow zone, with a preference for the lower frequencies. The shadow zone may also be 'fed' by sound transmitted through the barrier.

The presence of a small opening in the barrier can also greatly increase the sound level present behind the barrier, since diffraction can spread this directly fed sound into the true shadow area — see Fig 8.

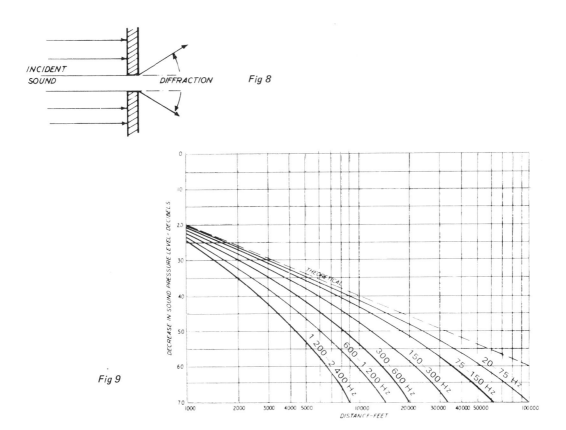

Fig 8

Fig 9

This latter effect can severely limit the usefulness of natural barriers such as trees and shrubs, although these can provide attenuation by the filter-type absorption yielded by leaves. Very dense planting and considerable depth is usually necessary to produce any significant reduction in noise level, and considerable height is necessary to offset diffraction.

Effect of Frequency

Even under free field conditions, the inverse square law characteristics of sound propagation are modified by the frequency of the sound waves. This is because attenuation is not the same for all frequencies, the decrease in sound pressure level being greater for higher frequencies. This is shown in Fig 9.

Such effects, however, are most noticeable at great distances and are commonly ignored for short distances.

See also chapter on *Room Acoustics*.

Measurement of Sound

SOUND MEASUREMENTS may be taken under the following conditions:
  (i)   *Free field* in a completely open space where there are no sound reflections or other modifying factors present.
  (ii) *Reverberant field* where the sound energy at any point is the sum of that directly radiated from the source and sound levels reflected from adjacent surfaces. In a fully reverberant field all the sound energy striking the bounding surfaces is reflected without loss.
  (iii) *Semi-reverberant field* where the prevailing conditions may be anywhere between free field and reverberant field conditions.
  (iv) *Anechoic field* where all the sound measured comes directly from the source, equivalent to taking measurements in an acoustically dead room where all incident sound energy striking the walls is fully absorbed.

Advantages and disadvantages of these different conditions are summarized in Table I.

As a general rule measurements taken *outdoors* can be considered to approximate to free field conditions and show reasonable agreement with anechoic measurement, provided there are no sources of strong reflection nearby.

Measurements taken *indoors* can be considered as approximating to diffuse field conditions and show reasonable agreement with reverberant field measurement. Average rooms, without sound deadening can approximate closely to diffuse field conditions where the flow of sound energy in all directions is more or less equal. On the other hand, those with sound deadening treatment or with inherent sound deadening characteristics could approximate more closely to anechoic or free field conditions. Reverberant field measurements may provide a suitable approximation for general indoor use when the directional characteristics of the sound source are not important.

With both outdoor and indoor measurements considerable difference may be experienced between measured levels and predicted levels (assessed on the basis of anechoic or reverberant field measurements respectively), and between individual measurements taken under different site or enclosure conditions. This makes it difficult — or even impossible — to quote a specific noise level for any particular source as such a figure is unlikely to be valid when assessing the same source under a variety of different sets of practical conditions. Direct measurement under the actual conditions involved is the only reliable method of obtaining a true objective reading in semi-reverberant field. Measurements of the same source under free field conditions are not valueless, however, as these can serve as a very useful indication of sound reduction possibilities. Rendered in terms of the sound power level of the source, such data can be used to predict average sound pressure levels in a semi-reverberant field.

TABLE I — COMPARISON OF MEASUREMENT CONDITIONS

Condition	Advantages	Disadvantages
Free field	simplest technique simple distance relationship in far-field zone provides a basic reference level (for comparison or correction) can contain compound and directivity information	measurement can be dependent on position measurement is frequency dependent in near-field zone measurements may be modified by atmosphere true free field conditions may not hold in practice even in far-field zone
Reverberant field	good general standard for indoor measurement with hard (reflective) walls, floor and ceiling	readings may be modified by standing waves no directivity information
Semi-reverberant field	good general standard for measurement in rooms or buildings	room characteristics must be known if correction to free field conditions is required. sound pressure has no simple relationship to distance from source
Anechoic field	true free field measurement	requires the use of a special anechoic chamber or anechoic room

Noise levels can be predicted with a fair degree of accuracy for semi-reverberant field conditions from the sound power level of the source (determined under anechoic or reverberant field conditions), provided the decay rate for the semi-reverberant field can be measured or estimated.

The approximate formula is:

$$SPL = PWL - 10 \log V - 10 \log D + 47.8 \text{ dB}$$

where

    SPL is the average sound pressure level in dB
    V is the room volume
    D is the distance

Only the average sound pressure level can be predicted in this manner and there may be considerable variations. In particular, the sound pressure level will usually be higher than that predicted in the vicinity of a hard surface (*ie* within one quarter of a wavelength distance). Considerable variation will also be experienced if the source is located near to a hard reflecting surface.

Quantitative Measurement

The parameter adopted for quantitative measurement of sound is sound pressure level in dB. Of necessity this has to be an average since the actual sound pressure measured by the microphone of a sound level meter alternates from positive to negative levels at frequencies from about 10–20 Hz. Signals are thus squared, so that both positive and negative parts of the square then give a positive result, although varying in amplitude at high frequency. To obtain a steady level 'average' reading the signal is put through a *time constant*. The square root of the average signal is then usually extracted to give a root mean square (RMS) average reading (exceptions being special instruments designed to indicate peak amplitudes instead of average amplitudes).

For such level measurements it is necessary to rate continually changing levels for fairly short periods, so the averaging time constants chosen are typically of the order of 0.9 seconds (slow) and 0.125 seconds (fast). Changes in the sound level occurring at lower rates can then be detected. The speed is normally selected according to the nature of the noise being measured — 'fast' for measuring a steady continuous noise, 'slow' for measuring a fluctuating noise — but preferably readings should be made at both speeds in the latter case. If, at fast speed, the fluctuations cover less than 6 dB range, the average reading should agree with the 'slow speed' readings.

A-weighting Scale

Since the human ear is frequency sensitive, subjective response differs from objective measurement of overall sound pressure levels, *ie* high frequency sounds sound much 'louder' than lower frequencies. To provide meter response characteristics which approximate to those of the human ear a filtering system is incorporated to 'correct' such differences by applying weightings to produce a relative response in the form shown in Fig 1.

Meter readings obtained are then correctly designated dB(A).

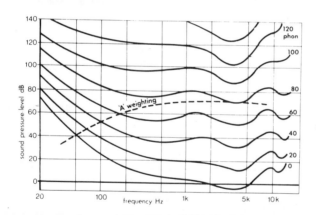

Fig 1 A-weighting compared with the ear's response.

TABLE II – A, B, C AND D WEIGHTINGS COMPARED
(weightings in decibels)

Weighting	Frequency Hz									
	16	31.5	63	125	250	500	1000	2000	4000	8000
A	−56.7	−39.4	−26.2	−16.1	−8.6	−3.2	0	1.2	1.0	−1.1
B	−28.5	−17.1	−9.3	−4.2	−1.3	−0.3	0	−0.1	−0.7	−2.9
C	−8.5	−3.0	−0.8	−0.2	0	0	0	−0.2	−0.8	−3.0
D	−22.5	−16.5	−11.0	−6.0	−2.0	0	0	8.0	11.0	−6.0

MEASUREMENT OF SOUND

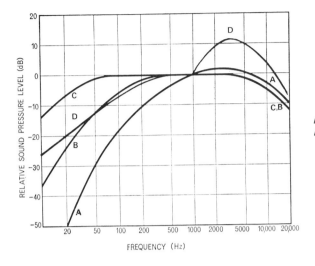

Fig 2 International standard A, B, C and D weighting curves for sound level meters.

Other Weighting Scales (See Fig 2 and Table II)

Strictly speaking A-weightings are only substantially 'correct' for sound power levels below **55 dB**. Other standard weightings designed to extend 'correction' to higher levels are B-weighting (dB(B)) for measurements between 55 and 85 dB and C-weighting (dB(C)) for measurements above **85 dB**. The D-weighting (dB(D)) standard was originally evolved specifically for aircraft noise measurement, rather than general use.

Comparatively little use is now made of these weightings (except for measurement of impulse sounds), all present standards for simple single number measurement being based on dB(A). Single number dB(A) ratings are generally in close agreement with subjective ratings, but have obvious limitations when the noise source is complex and/or the source has directional properties. They are also only meaningful when associated with a specific distance at which measurement is or has to be taken. The advantages of single-meter measurement are obvious, (*eg* simple instrument design and a single reading for comparative purposes), but more complex analysis is necessary for correctly reading *sound spectra*.

Note: Unweighted measurement values are readily corrected to weighted values by combining with the weighting figure applicable at each individual frequency. The following is an example based on octave analysis, converting dB to dB(A).

Frequency Hz	31.5	63	125	250	500	1000	2000	4000
Measured dB	81	76	74	68	62	70	64	58
Weighting dB(A)	−39.4 42	−26.2 50	−16.1 58	−8.6 59	−3.2 59	0 70	+1.2 65	+1 59

Although the weightings involve fractional numbers, converted values should be rounded off to whole numbers.

In a similar manner dB(A) values could be converted to unweighted dB values (*eg* to obtain the objective rather than the subjective spectrum); or dB(A) values could be converted to equivalent dB(B), dB(C) or dB(D) values by applying the difference between weightings to each frequency value.

Infrasound

Very low frequency sounds bordering on or below the normal audio frequency limit (*ie* below 50 Hz) are difficult, or impossible, to measure accurately with a meter, largely because measurements at these frequencies — known as *infrasound* — are easily influenced by wind, not readily correctable by windscreens for the microphone. Thus special meters are employed for measuring infrasound.

Typical sources of infrasound are:
- (i) Combustion noise, *eg* in boilers and burners.
- (ii) Vibrating objects, *eg* vibrating sieves and conveyors
- (iii) Compression waves, *eg* generated by compressors, diesel generators, aircraft, automobiles and ships.
- (iv) Fluid flow, *eg* waterfalls, waves hitting breakwaters, water falling on sand embankments and water released from dams.
- (v) Natural phenomena, *eg* wind, lightning, magnetic storms, ocean waves, volcanic erruptions and earthquakes.

Free Field Measurements

If free field conditions apply or can be simulated, *eg* by an anechoic room or chamber, the power level and directivity of the noise source can be determined from sound pressure levels measured at a number of different points. From such data reasonably accurate predictions can be made of the sound power level likely to be achieved by the same source in a reverberant field.

Measurement points are determined by the surface of a hypothetical sphere, of sufficient radius for 'square law' conditions to apply. Ideally, sound pressure levels should be measured all over the surface of such a sphere, but for practical purposes a number of measuring points are taken, distributed over the surface.

Calculation is further simplified if these points are uniformly distributed in a regular geometric pattern, *eg*:

8 points forming regular spherical triangles
12 points forming regular spherical pentagons
20 points forming regular spherical polygons

Sound Power and Directivity Index

From free field measurements under spherical conditions, it is possible to estimate the *sound power* radiated from the noise source as well as its *directivity index*.

The *sound power* can be calculated from the formula:

$$SPL = 10 \log_{10} \left(\frac{P}{P_o} \right) = 20 \log_{10} \left(\frac{P_m}{P_o} \right) + 10 \log_{10} \left(\frac{2\pi r^2}{S_o} \right)$$

MEASUREMENT OF SOUND

where
> P is the estimated sound power of the machine in watts
> P_o is the general sound power reference level, $P_o = 10^{-12}$ W
> p_m is the mean sound pressure being measured
> p_o is the general sound pressure reference, $p_o = 2 \times 10^{-5}$ N/m²
>   (2×10^{-4} μbar)
> $2\pi r^2$ is the surface of the test hemisphere of radius r
> S_o is a reference surface, $S_o = 1$ m²

According to standard recommendation the quantity p_m^2 is the mean value in space of the squares of the RMS sound pressures recorded at the measurement stations. When the measuring points are chosen so that all the measurement stations are associated with equal areas on the hypothetical hemisphere, and there are n measurement points then:

$$p_m^2 = \frac{1}{n} \sum_n^o p_1^2$$

An alternative formula, which might be easier to use in practice is:

$$\frac{p_m}{p_o} = \frac{1}{n} \sum_n^o \left(\frac{p_1}{p_o}\right)^2$$

where:

$$\sum_n^o \left(\frac{p_1}{p_o}\right)^2 = \left(\frac{p_1}{p_o}\right)^2 + \left(\frac{p_2}{p_o}\right)^2 + \left(\frac{p_3}{p_o}\right)^2 + \ldots \left(\frac{p_n}{p_o}\right)^2$$

$p_1; p_2; p_3 \ldots p_n$ are the sound pressures measured at the various measurement stations.

When the mean sound pressure is determined as described previously, it is possible to determine also a *directivity index, DI*. This may be of considerable interest in practice as most noise sources

Fig 3 Directivity as indicated by equal sound pressure contours from a typical source.

do not radiate the sound equally in all directions, Fig 3. The *directivity index* in the required direction can be calculated from the following formula:

$$DI = 20 \log_{10}\left(\frac{p_1}{p_0}\right) - 20 \log_{10}\left(\frac{p_m}{p_0}\right) + 3 \text{ dB}$$

Hemispherical Measurement

Where the source is normally mounted on a horizontal plane surface, the corresponding free field is a hemisphere, and thus sound pressure measurements are made at points on the surface of the hypothetical hemisphere. Exactly the same considerations as those just mentioned apply, but require only one half the number of measuring points. The power level can be determined in a similar manner except that 3 dB is subtracted from the final result because one half of a full spherical surface is involved.

The *directivity index* (DI) can be determined from the average sound pressure level, determined as above, and the sound pressure level measured at a particular point corresponding to a particular direction (SPL_0).

$$DI = SPL_0 - SPL \text{ dB}$$

The directivity factor (Q) can be obtained from the directivity index (DI) by converting the dB value of DI into a power ratio.

Reverberant Field Measurement

For accurate reverberant field measurements in a room, the room should be suitably proportioned. No two dimensions of the room should be alike, the 'ideal' proportions normally recommended being $1 : \sqrt[3]{2} : \sqrt[3]{4}$ for the height, width and length, respectively. The smallest dimensions should also be at least equal to the wavelength of the centre frequency of the lowest octave band of interest if octave band analysis is to be attempted.

The room boundaries should be good reflectors, *ie* smooth and hard, with the absorption small enough so that the decay rate is less than about 50 dB per second for a room of 1 000 cubic feet. Smaller decay rates are acceptable for larger rooms, but for accurate measurement at lower frequencies, the decay rate should be doubled.

The position of the noise source is also important — it should be mounted on a suitable surface, or suspended, as appropriate, at least a quarter of a wavelength from any of the boundary surfaces.

Where the source is directional, it should be oriented so that the major 'rays' cannot strike a boundary surface at right angles.

Only two readings are required, with the microphone at a suitable point:

(i) a reading of the steady value of the average sound pressure level.*

(ii) a plot (*eg* by a graphic level recorder) of the sound decay when the source is abruptly turned off.

*Note: A single reading for (i) only applies in the case where the field is truly diffuse. To obtain a representative average value it may be necessary to take sound pressure levels at several points, preferably at least half a wavelength apart, and from these data calculate the representative average on a power level basis, as previously described for free field measurement.

MEASUREMENT OF SOUND

The rate of decay (D) is then calculated as the initial slope of the decay curve, in decibels per second. The sound power level of the source is then determined from the formula:

$$PWL = SPL + 10 \log V + 10 \log D - 47.8$$

where
V is the volume of the room in cubic feet

Semi-Reverberant Field Measurement

The only requirement of the room in which such measurements are to be made is that it should be large enough to allow the microphone to be situated away from the near field of the noise source. The main factor in the measurements is that a *noise source of known sound power,* and preferably with radiation and frequency characteristics similar to those produced by the machine to be tested, is available. What is actually accomplished by the measurements is thus a comparison between the sound power produced by the known noise source and that produced by the machine being investigated. If a source of known sound power is available which produces a sufficiently uniform power spectrum this may also be used as a reference source.

Measurements are taken at a number of measurement positions, preferably situated over a hypothetical hemisphere with the machine at its centre in the same manner as discussed earlier for free field measurements. The reference source is then substituted for the machine and the measurements repeated at the same measurement stations.

From the sound power level of the reference source and the mean sound pressure levels determined from the measurements, the sound power level of the machine under test can be calculated:

$$SPL = 10 \log_{10}\left(\frac{p}{p_o}\right) = 10 \log_{10}\left(\frac{p_r}{p_o}\right) + 20 \log_{10}\left(\frac{p_m}{p_o}\right) - 20 \log_{10}\left(\frac{p_{m_r}}{p_o}\right)$$

In the above formula the same notation has been used as in earlier formulas given in this section. However, p_r is the sound power of the reference source, and p_{m_r} is the mean sound pressure produced by the reference source at the surface of the hypothetical hemisphere.

Of course, only limited information on the directivity index of the machine can be obtained from measurements under these conditions.

It might be worth mentioning that if a suitable reference sound source is not available, a prototype of the machine to be tested may be taken to a well equipped acoustical laboratory where its sound power and directivity index can be established under more ideal conditions. This prototype can then later be used as a reference source.

If the machine cannot be moved it might be possible to find an acoustically equivalent location for the reference source. In such cases, however, it may be more advantageous to choose measurement positions in the reverberant field of the room than on a hypothetical hemisphere surrounding the test object.

When comparison measurements are made it might not be necessary to perform a complete frequency analysis of the sound pressure level at each measurement station, and use may then be made of a portable precision sound level meter switched either for linear operation or with the A-weighting network inserted.

The use of A-weighting is preferable from a practical point of view because disturbing low frequency sounds are then automatically attenuated. (Care should be taken, however, that the main part of the sound radiation from the machine under test is not also heavily attenuated).

With regard to background noise the same considerations as described previously should be applied.

Finally, a method which might allow useful comparison measurements to be made in a standardized manner should be mentioned. This method is based on *near field sound pressure level measurements* and makes use of a so called 'prescribed surface'.

The 'prescribed surface' should be as simple as possible and its area should be easy to calculate. It should conform approximately to the external casing of the machine being tested and should be marked out around the machine at a well defined average distance (to be specified in the test report or test code). The number and disposition of the measuring stations depend upon the irregularity of the acoustic field and on the size of the machine. An example of the dispostion of

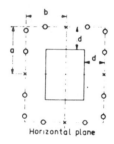
Horizontal plane

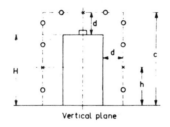
Vertical plane

Fig 4 Disposition of measuring points on a 'prescribed' surface.

l (Metres)	d (Metres)
⩾ 0·25	1
< 0·25	4l ⩽ d ⩽ 1 d > 0·25

L = Maximum linear dimension of machine

$h = \dfrac{H}{2}$ but not less than 0.25 metre

x = Key measuring points
o = Other measuring points marked off at intervals of 1 m from key points.

measuring points on a 'prescribed surface' is shown in Fig 4.

When the sound pressure at the various measurement stations has been determined, a mean sound pressure, p_m, can be calculated as described earlier. Furthermore, if the area of the 'prescribed surface' is set equal to the surface-area of an 'equivalent' hemisphere the radius of this hemisphere can be found from:

$$r = \sqrt{\dfrac{S}{2\pi}}$$

where

S is the area of the 'prescribed surface'

From the mean sound pressure, p_m, and the radius of the 'equivalent' hemisphere, r, it is recommended that the sound pressure level be calculated at a *reference distance,* d, from the machine according to the following formula:

$$20 \log_{10}\left(\dfrac{p_d}{p_0}\right) = 20 \log_{10}\left(\dfrac{p_m}{p_0}\right) - 20 \log_{10} \dfrac{d}{r}$$

MEASUREMENT OF SOUND

The reference distance, d, should preferably be chosen to equal 1, 3 or 10 m, depending on the size of the machine.

As the method of near field sound pressure level measurements is, basically, intended for comparison purposes, use may be made of A-weighting instead of octave (or 1/3 octave) band measurements. In this case, however, proper correction for background noise must be made before determining p_m, and sound reflections from walls or nearby objects should have no significant influence on the measurements.

Noise Averages

Although the sound level meter gives averaged readings, these are averages only for the short time constants involved (0.5 seconds or less). It is now generally recognized that both the subjective assessment of nuisance and the risk of noise-induced deafness are dependent on much longer term noise averages which may range from seconds to hours or even days. The parameter most favoured to express this relationship is equivalent continuous noise level (L_{eq}). Whilst this can be obtained from summation of steady level measurements taken over specific time intervals or spot samples of noise level, such data do not necessarily give a true measure of the actual sound energy average concerned.

Ideally, to measure long-term noise averages, the instrument should have:

(i) A wide enough dynamic range to ensure that large impulses are not 'clipped' and this sound energy under-recorded.

(ii) A non-sampling continuous integration mode of operation so that all the sound energy is recorded, and thus a true average obtained.

Impulsive Sound

Impulse sound pressure level is given by the formula:

$$L_{p_1} = 20 \log \frac{P_1}{P_0}$$

where
P_1 = sound pressure
P_0 = reference sound pressure

Impulse sound may be measured by an unweighted meter; or with A-, B- or C-weightings. In the latter case, weighted measurements are designated as follows:

$$L_{A_1} = 20 \log \frac{P_A}{P_0} \quad \text{dB(AI)}$$

$$L_{B_1} = 20 \log \frac{P_B}{P_0} \quad \text{dB(BI)}$$

$$L_{C_1} = 20 \log \frac{P_C}{P_0} \quad \text{dB(CI)}$$

Relationship to Octave Band Analysis

Octave band analysis involves the measurement of sound pressure levels at specific (preferred) frequencies, from which a spectrum of the complete sound can be drawn (*ie* showing the actual frequency content of the sound as well as individual sound pressure levels in dB at the frequencies measured). The resulting *total sound pressure level* or single-number dB equivalent can be found from the following relationship:

$$dB = 10 \log \frac{I_t}{I_o}$$

where I_t is the total intensity of the sound, or

$$I_t = I_1 + I_2 + I_3 + \ldots \ldots \text{etc}$$

where $I_1, I_2,$ *etc* are the sound intensities of the individual frequencies analyzed

and I_o is the reference level = 10^{-12} in/m²

The following worked out example illustrates the method taking the measured (dB) levels in I/I octave analysis (8 separate readings).

Frequency Hz	Level dB	$I/I_o \times 10^6$
31.5	55	0.316
63	60	1.000
125	80	100.000
250	82.5	177.800
500	75	31.620
1000	72	15.850
2000	68	6.310
4000	60	1.000
	Total (I_t) =	333.9

Thus total SPL = $10 \log_{10} 333.9 \times 10^6$
 = $10 \log_{10} 3.339 \times 10^8$
 = 85.236
 say 85 dB

An alternative method is to add pairs of individual sound levels successively (using the rules for 'addition of decibels'). Following is the same example as previously worked in this manner.

Frequency Hz	Level dB		Progressive Additions		Sum dB
31.5	55 ⎫	61.5 ⎫			
63	60 ⎭		84.5 ⎫		
125	80 ⎫	84.5 ⎭			
250	82.5 ⎭				85
500	75 ⎫	76.75 ⎫			
1000	72 ⎭		77 ⎭		
2000	68 ⎫	68.6 ⎭			
4000	60 ⎭				

See also chapters on *Sound Levels, Noise Measuring Techniques, Sound Level Meters,* and *Frequency Analysis.*

Noise Scales and Noise Indices

UNLIKE A single figure sound pressure reading or dB(A), which measures a sound *level* with an averaging period of either 250 ms or 2 000 ms, a *noise scale* gives a composite measure of noise over a period of time, the numerical values of such a scale corresponding with the annoyance potential of a noise. A *noise index* is also a composite measure of noise devolved over a period of time, derived from a noise scale to allow for additional factors which are relevant to rating or assessment for purposes of planning and regulation.

Both concepts have led to a multiplicity of scales and indices being devised, some for general use and others for specific applications. To rationalize the situation the Noise Advisory Council recommended that these should all be replaced by a single measure of environmental noise, the Equivalent Continuous Sound Level, L_{eq}. Thus in the UK, L_{eq} has largely become the main parameter for all long term average noise levels, as distinct from single number dB(A) measurement of instantaneous levels.

This average level is defined mathematically by:

$$L_{eq} = 10 \log \frac{t_1 \text{ antilog } \frac{L1}{10} + t_2 \text{ antilog } \frac{L2}{10} + \ldots t_n \text{ antilog } \frac{Ln}{10}}{t_1 + t_2 + t_n}$$

where
t_1 is the duration of noise level L1
t_2 is the duration of noise level L2
etc.

Fig 1 is a nomogram facilitating calculation of L_{eq} from universal measurements.

The total time involved $(t_1 + t_2 + t_n)$ may range up to 24 hours or more, depending on the survey requirements. The principal limitation of the formula is that it becomes cumbersome to use when a large number of noise levels are involved in the time period. Also the accuracy of the result will depend both on the amount of variation which may occur in any noise level and on the number of results summed. In general, therefore, the shorter the time periods selected, the more accurate the result is likely to be. These objectives are largely met by direct measurement of L_{eq} on specially developed sound level meters.

Single Event Noise Exposure Level (L_{AX})

The *single event noise exposure,* L_{AX}, is the level which, if maintained constant for a period of one second, would cause the same A-weighted sound energy to be received as is actually received

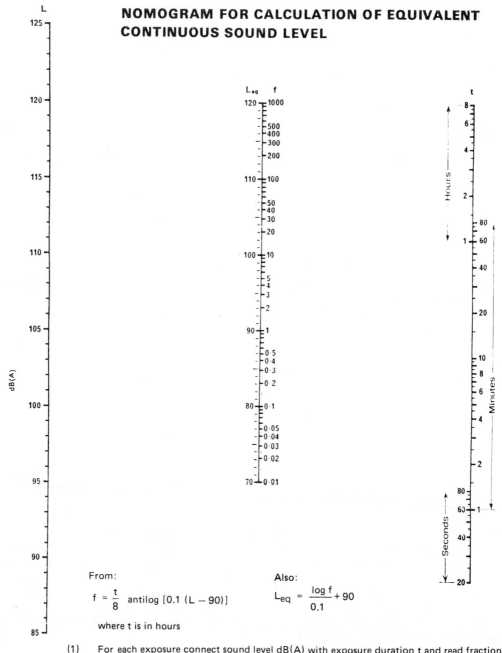

Fig 1

NOMOGRAM FOR CALCULATION OF EQUIVALENT CONTINUOUS SOUND LEVEL

From:

$$f = \frac{t}{8} \text{ antilog } [0.1(L - 90)]$$

where t is in hours

Also:

$$L_{eq} = \frac{\log f}{0.1} + 90$$

(1) For each exposure connect sound level dB(A) with exposure duration t and read fractional exposure f on centre scale.
(2) Add together values of f received during one day to obtain total value of f.
(3) Read equivalent continuous sound level L_{eq} opposite total value of f.

from a given noise event. L_{AX} values for contributing noise sources can be considered as individual building blocks being used in the construction of a calculated value of L_{eq} for the total noise.

L_{AX} is defined methematically thus:

$$L_{AX} = 10 \log_{10} \int_0^\infty \frac{P_A^2(t)}{P_{ref}^2} \, dt/\text{ref dB}$$

where ref = 1 second. As with L_{eq} there is an equivalent form:

$$L_{AX} = 10 \log_{10} \int_0^\infty 10^{L_A(t)/10} \, dt/\text{ref dB}$$

L10 Noise Index

The L10 Noise Index is a standard for road noise, and in particular is used to implement planning and remedial measures aimed at reducing the effect of road traffic noise on people.

L10 (18 hour) is the arithmetic average hourly value of the level of noise in dB(A) at one metre from the facade of a dwelling, except for just 10% of the time between the hours of 6 am and midnight on any normal weekday.

Summary of Evaluation Parameters

Wind — all predictions of L10 assume an adverse wind component from the road to the dwelling of about 10 km/hr.

Determination — either by measurement at site or by calculation.

Adjustment — measured (or calculated) noise levels should take into account any foreseeable increase in traffic flow on the road — *eg* particularly the proportion of heavy vehicles.

Gradients — an allowance of 1 dB(A) should be added for moderate gradients (between 1:25 and 1:50) and 2 dB(A) for steeper gradients.

Ground Absorption — corrections may be made for the acoustically attenuating effect of soft ground surfaces.

Other parameters, now largely obsolete but still quoted in literature, include the following:

Noise and Number Index (NNI)

The Noise and Number Index (NNI) was a single number evaluation based on perceived noise levels and developed specifically as a rating for aircraft noise. It was based on measurement of the peak perceived noise level which occurs with the passage of each aircraft. An average level is calculated, and a value added for the actual number of aircraft heard in a specific period. From this is deducted 80, as representing the PNdB at which the subjective rating of noise is zero (*ie* the basic criterion used in the maximum peak level in PNdB during the passing of each aircraft making more than PNdB in noise).

The mathematical relationship is:

$$NNI = 10 \log_{10} \left(\frac{1}{N} \int_1^N 10^{PNdB/10} \right) + 15 \log_{10} N - 80$$

where N = number of aircraft heard in a specific period

The approximate number of NNI can be derived directly from dB(A) measurement, in which case the basic formula is:

$$\text{NNI (approx)} = \text{average peak dB(A)} + 15 \log_{10} N - 67$$

Noise Pollution Level (LNP or NPL)

Measured in dB, the Noise Pollution Level is a rating similar to the NNI in concept, developed for the assessment of annoyance caused by aircraft noise, traffic noise, *etc* in specific environmental areas. It is defined as follows:

$$\text{LNP} = L_{eq} + 2.56\sigma$$

where L_{eq} = equivalent (sone) level
σ = standard deviation value

Traffic Noise Index (TNI)

The Traffic Noise Index (TNI) was originated by the Building Research Station (UK) as a measure of the social nuisance caused by traffic noise, and is based on the percentage time of noise at different levels. It is still in current use. The basic relationship is:

$$\text{TNI} = 4X \,(10\% \text{ level} - 90\% \text{ level}) + 90\% \text{ level} - 30$$

Noise exposure level L_{AX} relates to single event noise and is defined as:

$$L_{AX} = 10 \log_{10} \int_{t_1}^{t_2} 10^{L_A(t)/10} \, dt$$

where $L_A(t)$ is the time function of the noise level in dB(A)
t_1 and t_2 represent the time interval during which
L_A is within 10 dB of the maximum noise level during the event

Noise Exposure Forecast (NEF)

The Noise Exposure Forecast (NEF) was introduced in America in 1964 for commercial jet aircraft as an extension of Composite Noise Ratings (CNR) for civil aircraft noise. It is a somewhat complicated method based on the Single Event Perceived Noise Level (EPNL) in dB related to the class of aircraft and flight path and an addendum taking into account the number of flight operations. The complete formula is:

$$\text{NEF}_{ij} = \text{EPNL}_{ij} + 10 \log \left(\frac{\eta D^{(ij)}}{20} + \frac{\eta N^{(ij)}}{1.2} \right) - 75$$

where the subscripts i and j define the class and flight path respectively;
ηD is the number of daytime flights and ηN the number of night-time flights.

See also chapter on *Subjective Noise Parameters*.

Subjective Noise Parameters

MEASUREMENT OF sound pressure levels with A-weighting (*ie* dB(A)) is the universally accepted standard for single reading determination of instantaneous noise levels. The apparent loudness of sound, however, is basically a feature of hearing sensation rather than actual sound pressure measurements but the resulting data are still objective rather than subjective, and the two have still to be related on a 'response' basis. As a consequence many attempts have been made to establish subjective loudness scales for sound, a number of which have been widely quoted in the past and still continue to be used in certain applications. These are based on *subjective* units of sound, *ie*

phon — a measure of 'loudness level'.

sone — a measure of 'subjective loudness'.

noys — a measure of 'perceived noise'.

The *phon* is a measurement of 'loudness level'. Loudness levels are derived as contours of sound pressure levels of simple tones over a range of frequencies, each contour representing a constant loudness. Each contour is designated by a number, representing the loudness level in *phons*. At a specific frequency (1 000 Hz) this number is defined by the band sound pressure level in dB. The contours express the frequency dependency of subjective response.

Since these data are essentially subjective different contours can be derived from different techniques or from different sampling methods, although the ISO Normal equal loudness contours are generally accepted.

Loudness levels may be determined from octave band analysis when values may be expressed as phons (OD); or phons (GF) for loudness levels in a free field, and phons (GD) for loudness levels in a diffuse field.

Loudness Scale

A loudness scale can be derived as a transfer function from equal loudness contours, the unit for such a scale being the *sone*. As an arbitrary starting point a loudness of 1 sone is equivalent to a loudness level of *40 phons*. The complete relationship is then defined as:

$$\log_{10} L_S = 0.03 L_L - 1.2$$

where

L_S = loudness in sones

L_L = loudness in phons or the sound pressure level of a 1 000 Hz pure tone (which is assessed to be equal in loudness to the sound under comparison).

The Sone

The *sone* as a unit has the attractive feature that numerical values are substantially linear in expressing loudness levels, *eg* a reduction in sound level to one half of its original value is equivalent to halving the number of sones. Similarly, well separated components of sound are additive on the sone scale, *eg* a loudness of 51 sones from one source combined with a loudness of 52 sones from a second source gives a combined loudness of 51 + 52. In the case of two sounds of equal loudness 'S', the combined effect will be equal to 2 x S. By comparison, with loudness *level* two sources of equal loudness will give a theoretical increase of *10 phons.*

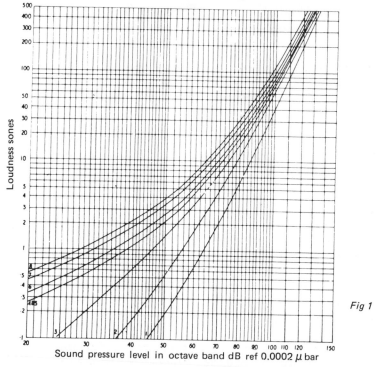

Fig 1

Loudness scale values in sones can be obtained directly from octave-band analysis by reference to tables or charts (*eg* see Fig 1). In the case of charts, the procedure is as follows:

(i) Tabulate loudness values (sone equivalent) for the measured sound pressure levels in each band.
(ii) Find the highest value given in this tabulation.
(iii) Multiply the loudness value of all *other* bands by 0.3.
(iv) Determine the sum of (ii) and all the other corrected values (iii).

This will give the total loudness of the noise in sones.

(v) The total loudness in sones can then be converted to Loudness Level (LL).

A similar procedure is adopted in the case of one-third band analysis, the appropriate charts being shown in Fig 2. The only difference (apart from the greater number of bands) is the correction factor value used in step (iii). This is 0.15.

SUBJECTIVE NOISE PARAMETERS

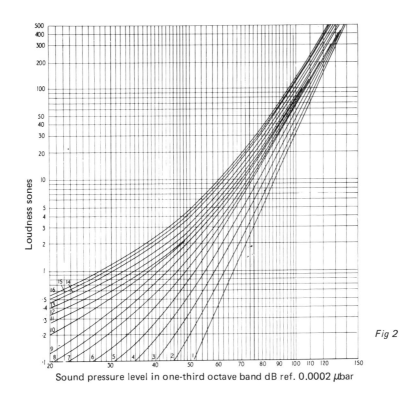

Fig 2

Sound pressure level in one-third octave band dB ref. 0.0002 μbar

Determination of loudness in this manner is applicable only to a diffuse sound field and is accurate only for steady noise with a broad band spectrum.

Loudness scale values derived from octave band analysis may be referred to as sones (OD) and sones (GF) for loudness in a free field or sones (GD) for loudness in a diffuse field.

Loudness Index (LI)

The *loudness index* (LI) is an alternative method of expressing loudness levels in the form of a series of curves established with a slope of −3 dB per octave through the 1 000 Hz reference frequency. The curves are thus rendered in straight line form; although above 9 000 Hz the slope changes to a constant 12 dB per octave and below a certain frequency each curve changes to a slope of −6 dB per octave — see Fig 3. The diagonal line which defines this lower frequency change point passes through the 10 dB pressure level at 1 000 Hz.

Again loudness index can be determined from octave band analysis. The procedure is:

(i) Determine the LI for each band.

(ii) Calculate the total loudness (in terms of LI value) from the formula—

$$LI \text{ (total)} = LI_{max} + 0.3 (LI_{sum} - LI_{max})$$

where

LI_{max} is the greatest numerical value of LI found in (i)

LI_{sum} is the sum of all the values found in (i)

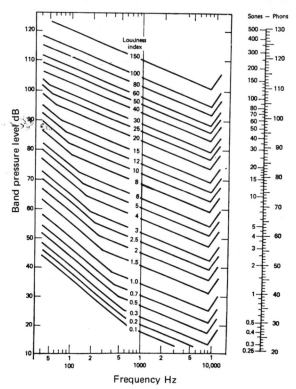
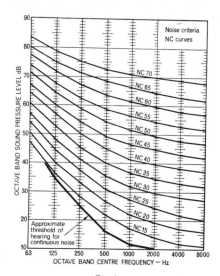

Fig 4

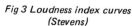

Fig 3 Loudness index curves (Stevens)

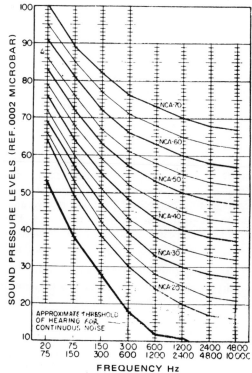

Fig 5 NCA curves

SUBJECTIVE NOISE PARAMETERS

Noise Criteria (NC) Curves

NC curves are octave band contours, each curve being given a number numerically equal to the sound pressure level in the 1 200 – 2 400 Hz octave – Fig 4. Each curve has a loudness level in phons of 22 units greater than speech interference level. Basically they are 'smoothed' equal loudness contours, originated in America in the early 1950's.

NCA curves are another similar family with the main difference that they designate a loudness level in phons of 30 units above speech interference level (see Fig 5). They are somewhat more realistic than NC curves where there is a predominance of low frequency content in the sound spectrum.

Noise Rating Curves (N or NR)

Noise rating curves are subjective data comprising a series of octave contours similar to equal loudness curves but 'smoothed'. The number given to each curve is the sound pressure level at 1 000 Hz – see Fig 6.

Noise Rating (NR) curves are similar in concept to NC curves, the numerical value of each curve being read as the sound pressure level at 1 000 Hz.

The noise rating of a particular sound is established as the next number above the *highest* NR equivalent of the sound pressure level measurement in dB in any octave band. That is, after converting all band dB measurements into equivalent NR numbers, the highest value of NR appearing in the tabulation is taken and 1 added to it to determine the final single figure noise rating in NR.

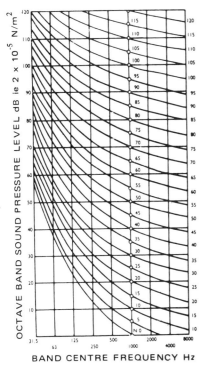

Fig 6 Noise rating curves.

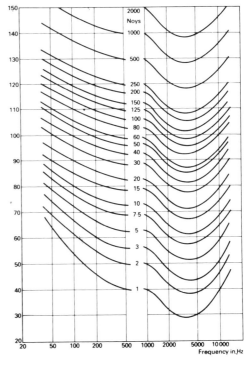

Fig 7

Noise rating numbers have the advantage of being simply related to dB(A) values, for general observation, *viz*:

$$N = dB(A) - 5\,dB$$

Perceived Noise Level (PNL)

Perceived noise is an assessment of subjective response to the 'noise' of sounds rather than its 'loudness'. Again it can be rendered in the form of contours related to sound pressure levels, the basic reference being the sound pressure level of a band of noise around 1 000 Hz that sounds as 'noisy' as the sound index comparison — see Fig 7. Perceived noise is expressed in units called *noys*.

Perceived noise is normally referred to as Perceived Noise Level (PNL) with values rendered in dB rather than noys, *ie* the perceived noise level is determined in noys units but then corrected to dB. In this case the more correct designation is PN dB.

PNL or PN dB can be derived directly from octave band measurement in a similar manner to that for loudness, *viz*:

$$PNL\,(noys) = N_m + K(N - N_m)$$

where

N_m = maximum level of noys found in the bands
N = sum of the noys values of all the bands
K = 0.3 for octave band summation
   = 0.15 for one-third-octave band summation

Effective Perceived Noise Level (EPNL)

Effective Perceived Noise Level (ENPL) is basically the PNL to which correction has been made to take into account the 'annoying' effects of narrow band noise and/or distinct tones present in the measured spectrum. The duration of the noise is also measured and again correction is made if the noise being analyzed is longer or shorter than reference duration.

Preferred Noise Criterion (PNC)

Preferred Noise Criterion (PNC) were developed on the lines of providing more pleasant sound spectra than NR or NC curves, with a particular application for specifying minimum acceptable noise levels in buildings designed for multiple occupancy.

Room Acoustics

IN A room the walls, ceiling and floor all form reflecting surfaces. Assuming a point source of sound, the sound field within the room comprises two components:

(i) the direct sound S_D between the source (S) and receiver (R) (Fig 1). This is the same as free field conditions and is not modified by the presence of the enclosure.

(ii) the reflected sound S_R, which reaches the receiver after one or more reflections from the enclosure surfaces. The value of this is determined by the power of the source and the reflecting properties of the enclosure surfaces.

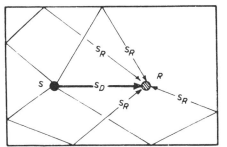

Fig 1

With perfect reflection, the value of S_R would rise to infinity, regardless of the value of the source. In practice this is impossible, but nevertheless the combined values of S_D and S_R can give rise to very high levels at any point R with highly reflective enclosure surfaces. Marked variations in sound pressure level at different points of R will also be experienced due to interference effects between the multi-directional waves. This will be further modified by a non-point source of sound.

Assuming that the size of the source is small relative to the dimensions of the room, there will be a definite distinction between the effects of the direct sound and the combination of direct and reflected sound. Near to the source, direct sound will predominate and the effect of reflected sound components will tend to be negligible. At some greater distances reflected sound will predominate, the region in which this occurs being known as the *reverberant field*. The inner boundaries of this field will be determined by the size of the room, the absorption characteristics of the reflecting surfaces in the room and the acoustic power, size and directivity characteristics of their source.

Fluctuations in sound pressure levels in the reverberant field due to 'interference' effects yield patterns which are known as *standing waves*. In general, these are most marked when the source has frequency components corresponding to, or close to, any of the possible resonances of the room. The spacing of such standing waves tends to be slightly greater than one half of the wavelength concerned.

Semi-Reverberant Rooms

Most rooms are semi-reverberant by nature, *ie* walls, ceilings and floor are neither completely reflecting nor completely absorbent. If sound measurements are made in such a room in the same way as in a free field, correction is necessary to obtain equivalent free field sound pressure level readings. This necessitates evaluation of the room characteristics or room constant. Rooms with a large degree of absorption have a large room constant and approach free field conditions. Rooms with a large degree of reflection and only a small degree of absorption have a low room constant and approach reverberant field conditions. Equally the nature of the reverberant field has a marked effect on the sound climate as experienced by an observer in the room.

Quantitatively the room content can be expressed in terms of *reverberation time* or the time in seconds required for the sound pressure level to fall to 60 dB when the source is shut off. The decay rate then follows as:

$$D = \frac{60}{\text{reverberation time}}$$

The reverberation time can also be calculated directly from formulas.

Once the decay rate has been established, or estimated, the average sound level in that part of the room where the reverberant field applies can be estimated from:

$$SPL = PWL - 10 \log V - 10 \log D + 48 \text{ dB}$$

where

PWL is the source power level
V is the volume of the room
D is the decay rate, as defined above

It should be noted that this formula gives approximate answers only. Also correction may be needed to compensate for changes in atmospheric pressure, temperature and relative humidity.

More usefully, the following formula gives the maximum sound level generated by a continuous sound in a reverberant room:

$$P = \frac{7.31 \times 10^{-2} tW}{V} \text{ W}_s/\text{m}^3$$

where

t is the reverberation time in seconds
W_s is the sound source power in watts
V is the volume of the room in m^3

The decay rate is then defined by:

$W_t = W_i - 0.04\, T/t$

where W_t is sound power level after an elapsed time T seconds from cessation of sound

t is the room reverberation times, as above

W_i is the initial sound power level

Optimum reverberation times for a room depend both on the volume of the room and the purpose. For good speech communication in small rooms of under 1 000 ft³ volume reverberation times of the order of 0.5 seconds or less are desirable — see Fig 2. For larger rooms and other purposes, much higher reverberation times are acceptable, or desirable (see chapter on *Auditoria*).

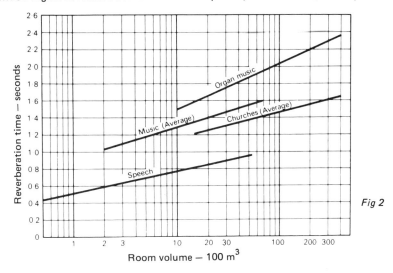

Fig 2

An empirical formula which may be used for simple calculation is:

$t = f\,(0.012\, V^{133} + 0.1070)$

where   t is in seconds and V is in m³ and

f = 4 for speech

5 for orchestral music

6 for singing voices (*eg* choir)

Room Constants

The acoustic power of the reverberant sound field in a room is related to the absorption characteristics of the boundary surfaces. The overall effect can be expressed in terms of the *room constant* (R), defined as:

$R = \dfrac{\alpha S}{1-\bar{\alpha}}$

where   S is the total surface area of the room

$\bar{\alpha}$ is the average absorption coefficient of these surfaces

Specifically, therefore, the room constant is a function of the size of the room (increasing with room size). Also, since the absorption coefficient varies with sound frequency, a particular value of R applies for only one frequency. Thus room constants need to be calculated for each individual frequency of interest. The relative sound pressure level can then be determined for each band. The value of this is that the relative sound pressure at any distance from the acoustic centre of a non-directional source can be determined — see Fig 3.

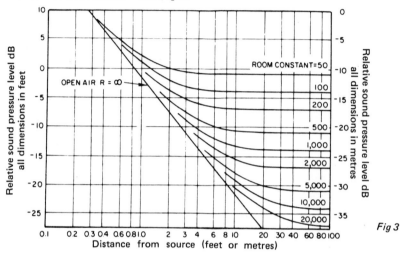

Fig 3

Analysis on this basis can also take into account the directivity factor (Q), when the sound pressure level in a room or enclosure is given by:

$$\text{SPL (watts)} = \text{PWL} + 10 \log_{10} \left(\frac{Q}{4\pi r^2} + \frac{4}{R_T} \right)$$

where

$\text{PWL} = 10 \log_{10} W + 130$ dB re 10^{-12} watts

R_T = room constant, m² sabines

r = distance from source, m

Note that this formula expresses both the direct field (Q/4π r) and the reverberant field (4/R).

Absorption and Reverberation Time

Reverberation time is directly related to the absorption performance of the surfaces of the room, expressed as a total of absorption units as given by the particular surfaces involved (*ie* materials lining walls, ceiling and floor). The relationship is:

$$t = \frac{0.16 \, V}{T}$$

where

V is the room volume in m³

T = total number of absorption units in sabines (unit ft²)

Air Absorption

In small rooms and enclosures the number of reflections is large and the mean free path small. As a consequence the contribution of air absorption is usually negligible compared with boundary absorption. In very large enclosures, however, the mean free path is large and air absorption can be significant, particularly for frequencies above 1 000 Hz, effectively modifying the room constant, viz:

$$\text{Amended room constant } R_T = \frac{\bar{\alpha}_T S}{1-\bar{\alpha}_T}$$

where

$$\bar{\alpha}_T = \bar{\alpha} + \frac{4 m V}{S}$$

where

V is the room volume

m is the air attenuation constant

The reverberation time is also modified, viz:

$$T_T = 0.049 \left(\frac{V}{\bar{\alpha} + 4 m V} \right) \text{(in English units)}$$

The value of the energy attenuation constant of air (m) is mainly influenced by frequency (see Table I), but is also dependent on relative humidity and temperature.

TABLE I – ENERGY ATTENUATION CONSTANT (m) FOR AIR AT 20°C (68°F)

Frequency kHz		1	2	4	6	8	10
Relative Humidity	20%	0.001	0.002	0.006	0.011	0.014	0.019
	40%	0.0005	0.001	0.0035	0.005	0.0075	0.010
	60%	0.0002	0.0008	0.002	0.004	0.006	0.008
	80%	–	0.0005	0.002	0.0035	0.0055	0.007

Reverberation Rooms

For specific purposes rooms may be built with highly reflective surfaces resulting in high levels of reflected sound with multiple image sources. The latter can be controlled by the geometric proportions of the room, yielding the possibility of making the flow of sound energy equal in all directions. If this is achieved the field is then said to be *diffuse,* and the enclosure geometry which produces this effect is said to comprise a *reverberation room*. It has the practical use that diffuse field measurement can be used to assess the absorption characteristics of materials, as well as being suitable for sound power measurements, where the directional characteristics of the source do not require analyzing.

The general characteristics of a true reverberation room are that the sound pressure level thoughout the room does not vary greatly, except close to the source. This is quite distinct from free field conditions, where doubling of the distance from the source decreases the sound pressure level by 6 dB. The characteristics of most rooms fall between these two extremes. Sound pressure level tends to fall by a fixed amount for each doubling of the distance in the far field, but this amount is less than 6 dB. This is particularly characteristic of rooms where the ceiling height is small compared with the length and width of the room.

Anechoic Rooms

The anechoic rooms aim, by suitable design geometry and the use of highly absorbent surfaces, to reproduce conditions where reflected sound is absent, and thus provide the equivalent of free field conditions of sound propagation. They are alternatively known as 'free field' rooms. In practice, the extent to which free field conditions can be simulated within the room is normally limited to a particular region within the room. Measurements taken at this point can then give accurate readings of radiated sound power and directivity of the source.

Cut-off Frequencies

The *cut-off frequency* of a room is determined by the room diagonal length from floor to ceiling. This dimension represents half the wavelength of the *lowest frequency* which can be sustained by the room. In other words, no frequency with a greater wavelength (or lower frequency) will be heard properly — see Table II.

TABLE II — CUT-OFF FREQUENCY OF ROOMS

Cut-off frequency Hz	Room diagonal dimension feet	metres
60	9.25	2.53
50	11.2	3.40
45	12.4	3.78
40	14.0	4.25
35	15.9	4.85
30	18.5	5.06
25	22.4	6.6
20	78.0	8.5

The room diagonal dimension is obtained directly from the length (L), width (W) and height (H) of the room, if of conventional rectangular volume, *ie*:

$$\text{Room diagonal} = \sqrt{L^2 + W^2 + H^2}$$

Example: to find the lowest frequency sound which can be reproduced satisfactorily in a room 11 ft long by 10 ft wide by 8 ft high:

$$\text{Room diagonal} = \sqrt{11^2 + 10^2 + 8^2}$$
$$= \sqrt{285}$$
$$= 16.9 \text{ feet}$$

ROOM ACOUSTICS

Thus the cut-off frequency will be between 35 Hz (15.9 ft) and 30 Hz (18.5 ft) — say about 33 Hz.

Alternatively, by formula calculation:

Room diagonal = ½ wavelength of lowest frequency

Therefore wavelength of lowest frequency = 2 x 16.9 = 33.8 Hz ft

Thus cut-off frequency = $\dfrac{1117}{33.8}$

= 33.04 Hz

External Sounds

External sounds will only affect the noise levels within a room if they can penetrate the room boundary surfaces as airborne sound, or be transmitted to the room by flanking paths. Such additional sounds will have no effect if they are less than the normal background level inside the room. Equally the penetration of extreme sound into a room can be controlled by sound insulation of the surfaces through which such sound can penetrate.

See also chapters on *Acoustic Rooms* and *Acoustic Enclosures.*

Acoustic Rooms

AN ACOUSTIC room, in the general sense, is a room or enclosure lined with sound absorbent materials to eliminate sound reflections, and also isolated from external sources of noise and vibration. It thus provides 'free field' conditions within the room for sound testing, sound measurement, *etc*. For other tests such as audiometry, cardiology, *etc*, the primary requirement is a 'silent' room. Here the emphasis is on the exclusion of external sound rather than the production of a 'dead' (*ie* echo-free) chamber. This principle is further extended in the design and construction of noise barriers in factories, *etc*.

A room completely free from sound reflections is specifically known as an anechoic chamber. Where used for free field measurement it is necessary that its size be adequate to cope with the largest noise source to be tested, since an anechoic room can never be completely 'free field' in characteristics. The best that can be achieved is a free field extending nearly, but not completely, to the boundaries. As the room becomes more restricted, relative to the size of the sound source, it may still be substantially echo-free, but is more properly described as a 'dead' room.

All categories, however, can generally be described as free field rooms, where the emphasis is on sound absorption treatment of the walls, floor and ceiling. It has been common practice to design such rooms with an air gap behind the acoustic lining to improve the low frequency cut-off of the room, although this can cause 'drop-out' at particular frequencies higher in the range (typically 100–150 Hz). The other objection to an air gap is that it reduces the size of the room.

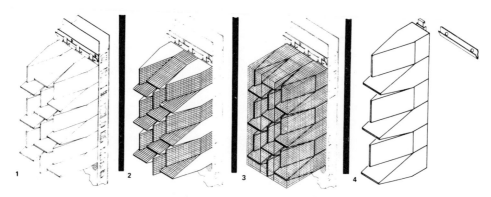

Examples of anechoic wedges.
(IAC).

Modular acoustic studio, 30 metres long by 20 metres wide by 10 metres high, formed from IAC modular acoustic panels. The acoustic door being 4.5 metres high by 3.5 metres wide providing a 55 dB(A) acoustic performance.

IAC acoustic chambers provide test rooms for the telephones of tomorrow at the Telecoms Research Establishment at Martlesham

For low level requirements, such as recording studios or production testing of audio equipment, foam linings can provide a simple and cheap method of wall treatment. Thickness can be varied between, say, 50 mm (2 inches) and 100 mm (4 inches) to vary the reverberation time. A secondary layer of profiled foam can be added if necessary, further to improve sound absorption. For higher level anechoic chambers it is usual to line the room with acoustic wedges which may range in length from 300 mm (12 inches) for a low level anechoic chamber to 1.4–2.4 metres (54–60 inches) for a full anechoic room for tests down to 50 Hz or below.

Where an anechoic chamber is designed for occupancy (*eg* a free field test room), acoustic treatment of doors, windows (if any are necessary) and the ventilation system are also of major importance. Doors need to be soundproofed and sealed and ventilation or air-conditioning must be noiseless. Run-throughs for electrical cables, *etc*, must also be hermetically sealed.

In the case of silent cabins the primary requirement is a specific degree of attenuation — *eg* 35–45 dB for general application or over 60 dB where 'deep' sound insulation is required. Maximum

CLASSIFICATION OF ACOUSTIC ROOMS

Description	Construction	Measurement Possibilities
Anechoic	Walls, floors and ceiling covered with fully effective sound absorbent materials. Room itself may also be fully isolated from external vibrations and noise	Free field measurement Direct sound of source Directivity of sound of source Actual PWL
Semi-Anechoic	Absorbent walls and ceilings, but hard sound-reflecting floor	'Hard ground' measurements (*eg* over hemispherical surface)
Reverberant Room	Walls, floor and ceiling all hard sound-reflecting surfaces	Room constant Reverberation time Average SPL Reduce PWL Diffuse field sound

attenuation is usually required over the middle (audio) frequency range. If anechoic or 'dead room' properties are also required, these can be provided by inner surface treatment.

Air-gapping may be used where 'deep' insulation is called for, or there are other special requirements. The usual principle here is the construction of two separate cabins, one floating inside the other on suitable absorbent mounts to avoid any rigid mechanical connections between the two (and to provide low frequency attenuation). This would also involve the use of double doors and double windows.

Choice of constructional materials is important as all surfaces must provide uniform, and predictable, attenuation (and sound absorption if required). Birckwork is generally unsuitable for the basic structure since its performance is largely unpredictable and can only be determined accurately by actual tests after building. Concrete provides a generally predictable performance, but steel is mainly favoured for smaller enclosures. Sound absorption can then be provided by an inside lining of absorbent material, or by a sandwich construction with the absorbent material behind an inner perforated wall. One point which may have to be considered in the case of silent cabins is the weight of the complete cabin. If too great, existing floors may have to be reinforced to bear the weight of the cabin.

See also chapters on *Room Acoustics, Acoustic Enclosures* and *Audiometry*.

Principles of Vibration

MECHANICAL VIBRATION is a form of periodic motion (oscillation) which may be simple or complex. It is generally definable in terms of frequency displacement about a reference point, velocity and acceleration. *Displacement* represents the magnitude of the motion involved. *Velocity* represents a time rate of change of displacement. *Acceleration* represents the time rate of change of velocity. A further quantity which may be of significance is the time change of acceleration. This is known as *jerk*.

Vibration may be repeated after a fixed time interval (periodic), or transient (aperiodic). Also it may be *free vibration* ('natural' vibration), *forced vibration* excited by a periodically varying disturbance, or *random vibration* excited by non-periodic disturbances. *Shock* is a related phenomenon which is aperiodic transient motion of a high order.

Simple Periodic Vibration

The simplest form of periodic motion is simple harmonic motion, displacement (X), velocity (V) and oscillation (A) are defined by the following relationship:

$$X = X_{peak} \sin \omega t$$

$$V = V_{peak} \sin (\omega t + \frac{\pi}{2})$$

$$A = A_{peak} \sin (\omega t + \pi)$$

where ω = angular velocity = 2π x frequency
t = time

The relationship between displacement, velocity and acceleration can be expressed in the form of general conversion charts as in Figs 1 and 2.

Unlike sound waves, maximum or peak-to-peak displacements can be measured directly, this being a simple form of analysis. Where vibration signals are analyzed, however, RMS or rectified average values are normally more useful. The definitions shown in Fig 3 apply to any sinusoidal wave.

There is also a trend towards using ratio values to express displacement, velocity and acceleration, particularly the last two, as levels rather than as absolute or mean values. Level values are then rendered in decibels, related to standard reference values, *eg:*

10^{-8} metres/second for velocity levels

10^{-5} metres/second^2 for acceleration levels

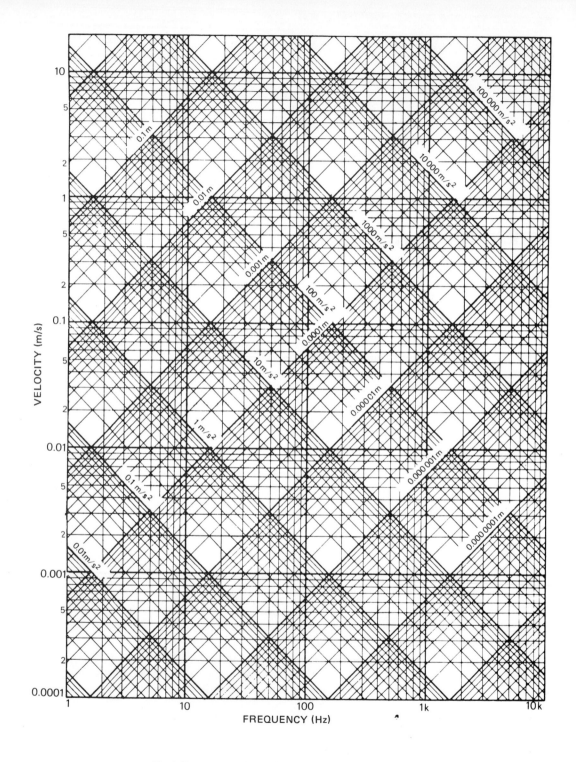

Fig 1 Chart giving relationship between velocity, acceleration and frequency for sinusoidal vibration.

PRINCIPLES OF VIBRATION

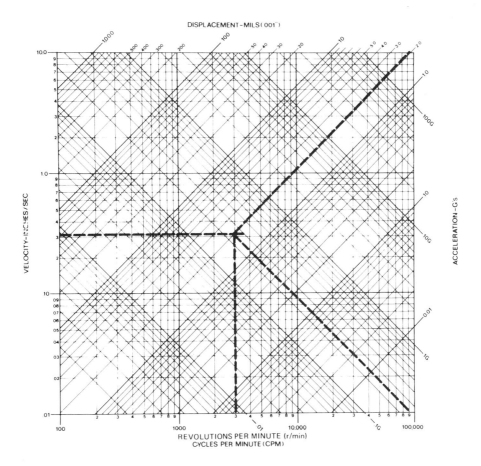

*Fig 2 Conversion chart for vibration parameters — inch units.
Example: vibration amplitude 2.00 mils at 3 000 rev/min
Velocity = 0.3 in/sec. Acceleration = 0.25 g*

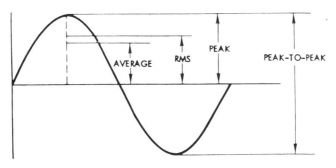

Fig 3

The energy content is represented by the absolute average of the vibration waveform, given by:

$$X_{average} = \frac{1}{t}\int_0^t X.dt$$

A more useful figure, however, is the root mean square (RMS) average, *viz:*

$$X_{RMS} = \sqrt{\frac{1}{t}\int_0^t X^2 \, t.dt}$$

which is more conveniently defined as:

$$X_{RMS} = \frac{1}{\sqrt{2}} . X_{peak}$$

Absolute average and RMS average are related by:

$$X_{RMS} = F_F.X_{average} = \frac{1}{F_C}.X_{peak}$$

where

F_F = form factor
F_C = crest factor

These factors give an indication of the waveform and for periodical sinusoidal (harmonic) motions are:

$$F_F = \frac{\pi}{2\sqrt{2}} = 1.11 \text{ or approximately 1 dB}$$

$$F_C = \sqrt{2} = 1.414 \text{ or 3 dB}$$

Although all harmonic motion is periodic, not all periodic motion is harmonic. However the Fourier relationship establishes that all periodic waves can be represented as the sum of a series of simple harmonic waves, the frequencies of which are simple multiples of the fundamental frequency of the complex wave in question. This is known as a Fourier series, represented algebraically by:

$$X = A_1 \sin(\omega t + \Phi_1) + A_2 \sin(2\omega t + \Phi_2)$$

where Φ is the phase angle (in displacement of the wave component relative to the datum starting point).

As a simple example, Fig 4 is a complex wave which can be represented by the sum of two simple harmonic waves X_1 and X_2, each with a phase angle of zero, when:

$X_1 = A_1 \sin \omega t$
$X_2 = A_2 \sin 2\omega t$

PRINCIPLES OF VIBRATION

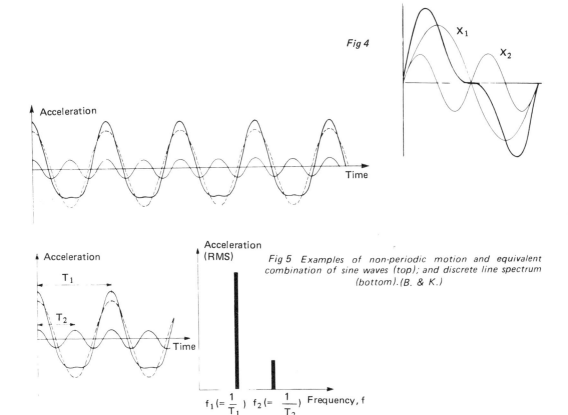

Fig 5 Examples of non-periodic motion and equivalent combination of sine waves (top); and discrete line spectrum (bottom). (B. & K.)

Thus the equation of the complex wave is:

$$X = A_1 \sin \omega t + A_2 \sin 2\omega t$$

(Note: this means that the function X can be plotted by graphical addition of the two curves, X_1 and X_2).

More conveniently, the Fourier relationship is usually expressed in differential form:

$$X = A_0 + \sum_{K=1}^{n} A_K \sin(K\omega t + \phi_K)$$

Thus, applying Fourier analysis, in the case of complex periodic motion it is possible to derive the waveform as a combination of harmonically related sine waves. The various elements in that case constitute the vibration frequency spectrum or display in the *frequency domain*. This will consist of discrete lines — Fig 5 — and is thus quite distinct from a random vibration where the spectrum is irregular and continuous. In the latter case the true history can only be established on a probability basis from analysis of a finite length of spectrum. Criteria accepted are amplitude probability distributions in terms of *probability density* and continuous vibration frequency spectra in terms of mean square spectral density or *power spectral density*.

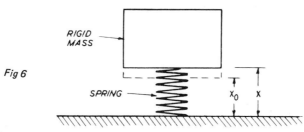

Fig 6

Steady State Vibration

Steady state vibration at a single frequency is analogous to a single-degree-of-freedom system, (or simple elastic system) represented by a rigid body, supported on a massless spring, constrained by frictionless guides, so as to be capable of movement in a vertical direction only — Fig 6. The amplitude of displacement is then represented by the height of the body above a reference plane.

The basic equation for motion is:

$$X = A \sin \sqrt{\frac{K}{m}} \, t + \phi$$

Once excited the spring-mass system thus undergoes sinusoidal vibration with an angular velocity of:

$$\omega = \sqrt{\frac{K}{m}}$$

with a period of

$$t = 2\pi \sqrt{\frac{m}{K}}$$

The body has one natural frequency given by:

$$f_n = \frac{1}{2\pi} \sqrt{\frac{Kg}{W}} \quad Hz$$

where

K = stiffness of spring
g = acceleration due to gravity
W = weight of the body
m = $\frac{W}{g}$ = mass

The natural frequency can also be expressed directly in terms of the static deflection (d) of a perfect spring under the weight of the body, *viz:*

$$f_n = 3.13 \sqrt{\frac{1}{d}} \qquad\qquad f_n = 0.62 \sqrt{\frac{1}{d}}$$

where   where
d is in inches   d is in millimetres

PRINCIPLES OF VIBRATION

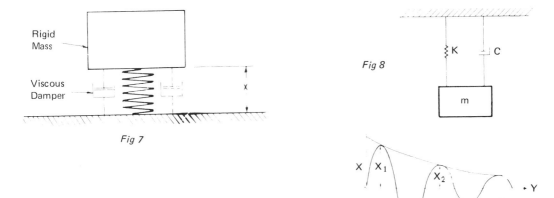

Fig 7

Fig 8

Theoretically such a system, once excited, will continue to vibrate indefinitely at its natural frequency unless some opposing force or damping is present. In practice natural damping is always present due to friction.

Damped single frequency transient vibration is analogous to the elastic system shown in Fig 7 which is similar to that of Fig 8 with the addition of a massless viscous damper. The natural frequency of the system follows as:

$$f_n = \frac{1}{2\pi} \sqrt{\frac{Kg}{W} \left(1 - \frac{c}{c_c}\right)^2}$$

where c = the damping coefficient
c_c = value of c for which damping is critical

The latter value is given by the fact that when critically damped a system returns to its equilibrium position without oscillation, *viz:*

$$f_n = \text{zero, whence } c_c = 2\sqrt{Kg/W}$$

Any other degree of damping can be considered in terms of its ratio to critical damping or *damping ratio,* normally quoted as c/c_c.

In the case of less than critical damping the sinusoidal vibration decays with diminishing amplitude as shown in Fig 9. The rate of decay of vibration decrement (A), defined as the natural log of two amplitudes can be expressed in terms of the logarithmic ratio of the successive peaks, *ie:*

$$\Delta = \log_e \frac{X_1}{X_2}$$

In terms of damping ratio

$$\Delta = \frac{2\pi \, c/c_c}{\sqrt{1 - (c/c_c)^2}}$$

In practice

$$\Delta = 2c/c_c \text{ to a close approximation}$$

Damping may be measured in terms of this damping ratio (c/c_c), using the observed pattern of decay of free vibration. In a good many cases, however, the damping is relatively light so that measurement of differences in amplitude between successive cycles is impractical. In such cases damping may be expressed in terms of the damping ratio, taking the ratio of amplitudes over a suitable number of cycles.

decrement = $Kf_n\ c/c_c$

For a value of K = 54.5 the decrement is given in dB per second.

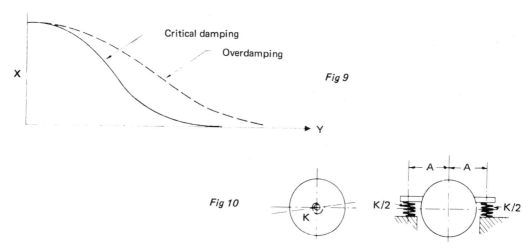

Fig 9

Fig 10

Most practical applications involve multi-degree-of-freedom systems, rather than single-degree-of-freedom systems, in which case each of the natural modes of vibration has to be considered. Various modes may be coupled (in which case one coupled mode cannot occur independent of vibration in the other), or decoupled. This depends on the geometry (and stiffness) of the elastic system, and also the mass distribution of the rigid body. Mathematical analysis thus becomes more complex, although it can be simplified to some degree, in the case of forced vibrations, by ignoring transients.

Similar analysis may be applied to simple-freedom-of-movement systems, where there is freedom of movement in rotation about an axis through the centre of gravity of the rigid body — see Fig 10. The simplest case is where elastic resistance to rotation is provided by a single torsion spring, when:

$$f_n = \frac{1}{2\pi}\sqrt{\frac{K}{I}} \text{ Hz}$$

where

K = spring constant
I = moment of inertia of the body

PRINCIPLES OF VIBRATION

If the body is elastically mounted on two identical linear springs, the corresponding formula becomes:

$$f_n = \frac{1}{2\pi} \cdot \sqrt{\frac{KA^2}{I}} \text{ Hz}$$

All these basic formulas may readily be extended to take into consideration non-linear stiffness of the elastic system.

Forced Vibration

Forced vibration, as opposed to free vibration, is excited by the application of a periodically varying disturbance. The additional energy supplied to the system causes the amplitude of vibration to be maintained. Significant parameters are:

the undamped natural frequency $\omega_n = \sqrt{\frac{Kg}{W}} = \sqrt{\frac{K}{m}}$

the static deflection of the system $d_o = \frac{F_o}{K}$

where
F_o is the deflecting force
ω = weight

The relative response or *magnification factor* (R) is then given by:

$$R = \frac{d}{d_o}$$

$$= \frac{1}{\sqrt{1 - \left(\frac{\omega}{\omega_n}\right)^2 + 2(c/c_c)\left(\frac{\omega}{\omega_n}\right)^2}}$$

Graphical analysis for single-degree-of-freedom systems is shown in Fig 11 rendered in terms of the dimensionless-parameters-damping-ratio (c/c_c) and frequency-ratio (forcing frequency to undamped natural frequency, f_n). Such curves show the displacement amplitude generated by forced vibration of the system. Similar curves can be derived to show the transmissibility, or ratio of the resultant displacement to the applied displacement. Graphical solutions are also preferred for solutions of forced vibration applied to multi-degree-of-freedom systems. In all such cases, practical control of vibration is concerned with the design of a suitable elastic system.

The phase angle is similarly defined by:

$$\tan \phi = \frac{2 c/c_c \frac{\omega}{\omega_n}}{1 - \frac{\omega}{\omega_n}}$$

This relationship is shown graphically related to frequency ratio – Fig 12.

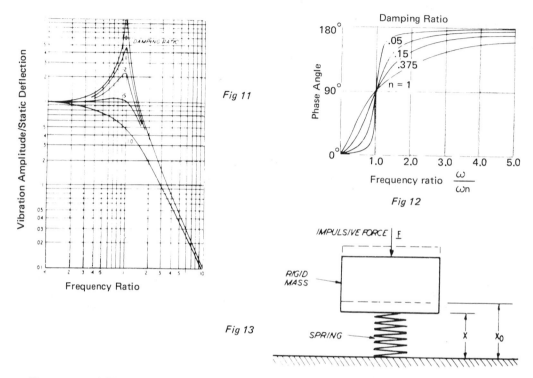

Fig 11

Fig 12

Fig 13

The curves of Figs 11 and 12 show that both the vibration amplitude and the phase angle (between the forcing function and the resultant displacement of mass) are strongly affected by the damping ratio and the frequency of vibration. In general, as damping becomes smaller, amplitude of vibration at resonance becomes larger. Small values of damping are said to produce high 'Q' resonance. (In theory, with zero damping the vibration amplitude could become infinitely large). Also, for small damping the phase angle shifts more rapidly from 0° to 180°. For lightly damped systems, one technique for finding resonance is to determine the frequency of 90° phase shift.

The problem is somewhat modified when the forcing system is of impacting rather than steady periodic nature, since transient conditions apply. Taking the simplest case of an undamped single-degree-of-freedom-of-movement elastic system (Fig 13), the unfavourable parameters are the impulsive force F and the impulse frequency f, and anything that can be done to reduce either, or both, will be beneficial in reducing the maximum amplitude induced. The single favourable parameter is the mass of the system (W/g). Control of impact vibration displacement is thus largely concerned with manipulating these three parameters. The duration of the transient vibration is, of course, directly dependent on the degree of damping present. Damping is thus the other important parameter in the control of impact-generated vibration.

Random Vibrations

The mathematics relating the excitation of linear systems to random vibration (theory of random processes) is complex and can be analyzed either in terms of the power spectral density or the autocorrelation function. The important fact to emerge from this is that the response power spectral density of a linear system at any frequency is equal to the excitation power spectral density multiplied by the square of the complex frequency response function at that frequency. If the system is non-linear, however, this simple relationship no longer applies since the power spectral density function for the response varies with excitation level.

Basically, random aperiodic motion is described mathematically in terms of statistics rather than algebraic function. Thus taking the random motion plotted in Fig 14, at any given instant (t) the (Gaussian) probability that the acceleration value is between:

a_0 and $a_0 + da$ is $p(a)da$, and for a normal process

$$p(a) = \frac{1}{\sigma\sqrt{2\pi}} \exp\left(-\frac{a^2}{2\sigma^2}\right)$$

$p(a)$ plotted as a function of a is shown in Fig 15. The probability that the instantaneous value of acceleration is between a_1 and a_2 is represented by the shaded area and given by:

$$\int_{a_1}^{a_2} p(a)da = \frac{1}{\sigma\sqrt{2\pi}} \int_{a_1}^{a_2} \exp\left(-\frac{a^2}{2\sigma^2}\right) da$$

The quantity σ is the root mean square derivation of the instantaneous acceleration value from the mean acceleration value. In the case of random accelerations the mean value is zero, hence σ is equal to the root mean square value of the instantaneous acceleration.

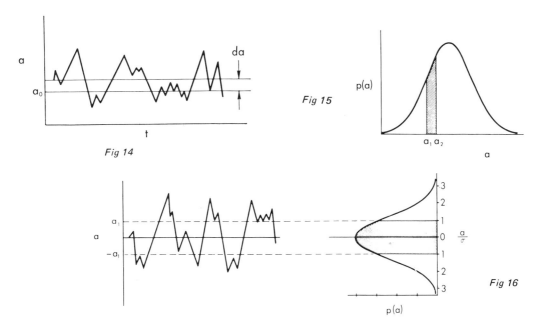

Fig 14

Fig 15

Fig 16

The probability density curve is usually normalized, *ie* the scales adjusted, so that the total area under the curve is unity (representing a probability of 1). Presented in this manner (Fig 16), the probability that the instantaneous value of acceleration is between $a \pm a_1$ is equal to the shaded area under the normalized probability density curve.

Random Amplitude Sine Wave

If a random vibration signal is passed through a narrow bandwidth filter the result will be a single frequency wave with random varying amplitude. For such a wave the probability of an acceleration/ peak having a value between a_o and $a_o + da_p$ is:

$p(a_p) \, da_p$

when

$$p(a_p) = \frac{a_p}{\sigma^2} \exp\left(\frac{-a_p^2}{2\sigma^2}\right)$$

The probability that the peak value of acceleration is between a_1 and a_2 is:

$$\int_{a_1}^{a^2} p(a_p) \, da_p = \frac{1}{\sigma^2} \int_{a_1}^{a^2} a_p \exp\left(\frac{-a_p^2}{2\sigma^2}\right) da_p$$

Again this may be related to a normalized probability density curve — Fig 17.

A single frequency component of a random vibration will vary in amplitude in a random manner and so cannot be specified by its peak value. Rather its RMS value must be used.

The RMS value of a single frequency wave can be found by dividing the filter output by its bandwidth.

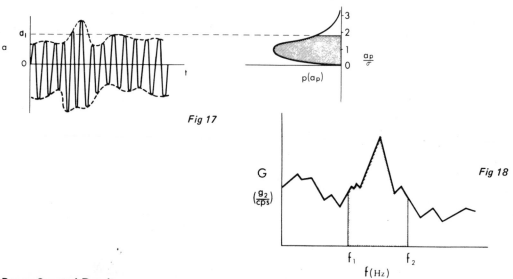

Fig 17

Fig 18

Power Spectral Density

A plot of the RMS value of single frequencies against frequency gives the power spectral density (PSD) curve. This shows the power distribution of the vibration as a function of frequency. Thus the shaded area under the curve (Fig 18), represents the mean squared acceleration between f_1 and f_2. The RMS acceleration between these frequencies is then the square root of the shaded area.

PRINCIPLES OF VIBRATION

White Noise

Random vibration which exhibits a constant acceleration density is known as *white noise*. Mathematically, in the case of white noise:

$$G_{RMS} = \sqrt{G_0 B}$$

where G_0 = constant acceleration density
B = bandwidth under consideration

Shock

Shock is defined as a transfer of kinetic energy to a system in a period of time which is short relative to the natural period of oscillation of the system. As a consequence the shock pattern is influenced by the nature and time of the shock pulse, and can involve an initial shock with maximum response, and a residual shock after the pulse has occurred. It is significant that the initial or maximum system response can be greater in value than the magnitude of the shock pulse.

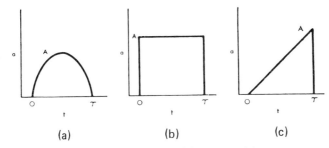

Fig 19 Simple pulse shapes: (a) half sine, (b) square wave, (c) sawtooth (terminal peak).

Shock can occur in an infinite variety of ways and can be very complex. The simplest concept is a single impulse of extremely short duration and large acceleration. This results in a 'step' velocity change (velocity shock). This is the type of shock produced by impact between very hard materials.

Simple shock analysis normally consists of considering shock as a simple pulse specified by its acceleration amplitude, time duration and pulse shape. Simple pulse shapes most commonly employed are the half sine, square wave and sawtooth — Fig 19. Mathematical expressions are:

(i) half sine pulse : $a(t) = A \sin \dfrac{\pi + t}{\tau}$ $(0 < t < \tau)$

$a(t) = 0$ $(t > \tau < 0)$

(ii) square wave : $a(t) = A$ $(0 < t < \tau)$

$a(t) = 0$ $(t > \tau < 0)$

(iii) sawtooth : $a(t) = \dfrac{A}{\tau} t$ $(0 < t < \tau)$

$a(t) = 0$ $(t > \tau, t < 0)$

Complex Shock

A complex shock may be analyzed in terms of Fourier components occurring at discrete frequencies. In the case of non-periodic transient waves, however, the Fourier spectrum is a continuous function obtained by integration, *viz*:

$$F(\omega) = \int_{-\infty}^{+\infty} a(t) \exp(-j\omega t)\, dt$$

Shock Analysis

The most useful form of analysis is the *shock response system*, normally obtained by applying the shock pulse to an undamped linear single-degree-of-freedom system and plotting the system's maximum response as a function of its natural frequency.

Suppose that several such resonators, each with a different natural frequency, are mounted on to a rigid structure and subjected to the same shock input. If the maximum response of each system is recorded and plotted as a function of resonant frequency, the result is the response spectrum for that shock input. In practice, a continuous spectrum — corresponding to a very large number of resonators differing only slightly from one another in natural frequency — is normally plotted.

In particular, a shock spectrum is usually defined as the maximum acceleration responses of a series of simple systems to the shock motion (Fig 20).

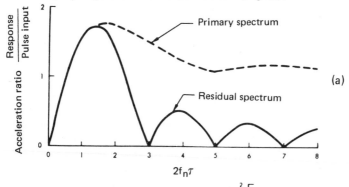

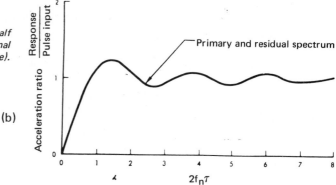

Fig 20 Shock spectrum of: (a) a half sine pulse of period T and (b) a terminal peak sawtooth of period T (after Lowe).

PRINCIPLES OF VIBRATION

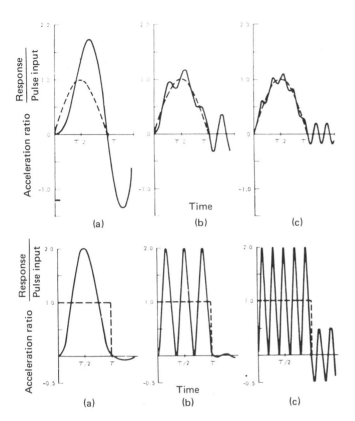

Fig 21 Acceleration response to a half sine acceleration pulse, of duration T, of an undamped single-degree-of-freedom system whose natural period is equal to: (a) 1.014 times the pulse duration, (b) 0.338 times the pulse duration, and (c) 0.203 times the pulse duration (after Levy & Kroll).

Fig 22 Acceleration response to a rectangular acceleration pulse, of duration T, of an undamped single-degree-of-freedom system whose natural period is equal to: (a) 1.014 times the pulse duration, (b) 0.338 times the pulse duration. and (c) 0.203 times the pulse duration (after Levy & Kroll).

Examination of Figs 21 and 22 shows that the response of a system during application of a shock may differ considerably from the residual motion after the shock input has ended. Because of this, it is customary to consider the primary spectrum (response during shock input) and the residual spectrum (response after shock input) separately. The primary spectrum defines peak acceleration in both directions. The residual spectrum is important not only in providing more information about dynamic loading, but because it also indicates the fatigue loading due to flexural motion. In view of this, it can be seen that a shock test that provides a flat response in both the primary and residual spectra would be highly desirable. Such a test would be equally severe in the loading it imposes on all equipments tested, regardless of their natural frequencies. Further, a spectrum that rises smoothly to about 100 Hz and is flat for all higher frequencies is a very good average of the shocks actually encountered in transporting and handling equipment. A terminal peak sawtooth acceleration pulse exhibits a spectrum of this type. In addition, the primary and residual spectra are almost equal (Fig 20b).

Shock spectra can be used to compare intensities of different shocks and are directly applicable to several dynamic design techniques, and the numbers quoted correlate better with the damage potential of the related shock. The difficult job of placing tolerances on pulse shapes is avoided and tests can be conducted that compare very well with actual·field conditions. One of its great values is that it permits useful analysis of complex shock.

Torsional Vibration

Torsional vibrations are largely associated with rotating shaft systems and are essentially a function of the mass/elastic characteristics of the system. They may be excited by inertia forces resulting from accelerations or decelerations, or from torque loading. The resulting torsional vibration takes the form of a small alternating rotational velocity, superimposed on the steady rotational velocity of the shaft. The resulting material disturbance is in the form of shear waves which, being contained within the shaft, are unlikely to generate any appreciable noise directly. However, if the frequency of vibration is resonant, severe 'shaking' can occur on coupled components which can give rise to noise — *eg* transmission gear rattle in the case of an automobile. The presence of shear waves peaking at resonance imposes additional stress on the shafting elements concerned. The natural frequency of shaft systems is thus of considerable significance as far as torsional vibrations are concerned.

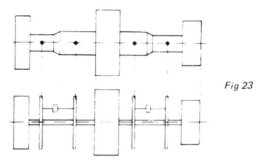

Fig 23

In the general case a shaft system can be considered as a multi-rotor system. The mass of the shaft can then be represented mathematically by a series of small rotors, each of which has a moment of inertia equivalent to the shaft segment which it represents — Fig 23. Geometrically each rotor is positioned at the centre of mass of the appropriate section. The other basic requirement is that where a stepped shaft connects two rotors in the original system, this stepped section must be replaced by a torsionally equivalent shaft of uniform diameter. This can be derived from the specific relationship:

$$L = l_1 + l_2 \left(\frac{G_1 J_1}{G_2 J_2} \right) + l_3 \left(\frac{G_1 J_1}{G_3 J_3} \right)$$

where

L = equivalent length of shaft
l_1, l_2 and l_3 are the stepped shaft lengths
G_1, G_2 and G_3 are the respective moduli of rigidity or shear
J_1, J_2 and J_3 are the respective second moments of areas

In the case of a shaft with the same material for l_1, l_2 and l_3, this reduces to:

$$L = l_1 + l_2 \left(\frac{d_1}{d_2} \right)^4 + l_3 \left(\frac{d_1}{d_3} \right)^4$$

PRINCIPLES OF VIBRATION

For a tapered length of shaft:

$$L = \frac{d \, l \, (D^3 - d^3)}{3 D^3 (D - d)}$$

where d is the diameter of the torsionally equivalent shaft. See also Fig 24.

Bearings, in theory at least, represent torsional dampers applied to the shaft system. In practice, their effect in this respect is usually negligible and is generally ignored. In certain cases the mass effect of the shaft may also be small enough to be ignored (*ie* the shaft masses are very small compared with the rotor masses).

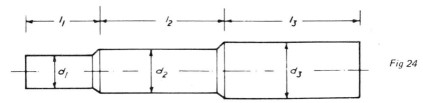

Fig 24

Mathematical Analysis

There is a direct relationship between linear and torsional vibration, and similar mathematical formulas apply. The main difference is that radius (r) and r^2 appear as factors in torsional system formulas. A direct comparison between the units of the two systems is given here:

	Linear system	**Torsional system**
Mass/Inertia	M = W/g	$I = M r_m^2 = \dfrac{W + r^2}{g}$
Spring	K = (mass/L)	K_R (mass − length/radian)
Damper	C = mass/length/sec	$C_R = C_{rc}^2 \dfrac{\text{mass} \cdot \text{length} \cdot \text{sec}}{\text{radians}}$
Acceleration	length/sec²	radians/sec²
Velocity	length/sec	radians/sec
Displacement	length	radians

Single-degree-of-freedom system,   Fig 25   *Two-inertia, one-spring system.*

Two basic cases are shown in Fig 25 — a single-degree-of-freedom torsional system and a two-inertia one-spring system. A force is imposed on inertial mass I_1 in both cases. In the case of the single-degree-of-freedom system Transmissibility is the torque ratio transmitted through the spring K_R to the support. In the case of two-inertia single-spring system transmissibility is the torque ratio transferred to the driven inertia mass I_2. Basic formulas are:

$$\text{Displacement : single-degree-of-freedom} = \frac{1}{2\pi} \sqrt{\frac{K_R}{I_1}} \text{ Hz}$$

$$\text{two-inertia one-spring} \quad \frac{1}{2\pi} \sqrt{\frac{K_R}{I_e}} \text{ Hz}$$

$$\text{where } I_e = \frac{I_1 I_2}{I_1 + I_2}$$

$$\text{Transmissibility : single-degree-of-freedom} = \frac{1}{1 - \left(\frac{f}{f_n}\right)^2}$$

$$\text{two-motion one-spring} = \frac{I_2}{I_1 + I_2} \times \frac{1}{1 - \left(\frac{f}{f_n}\right)^2}$$

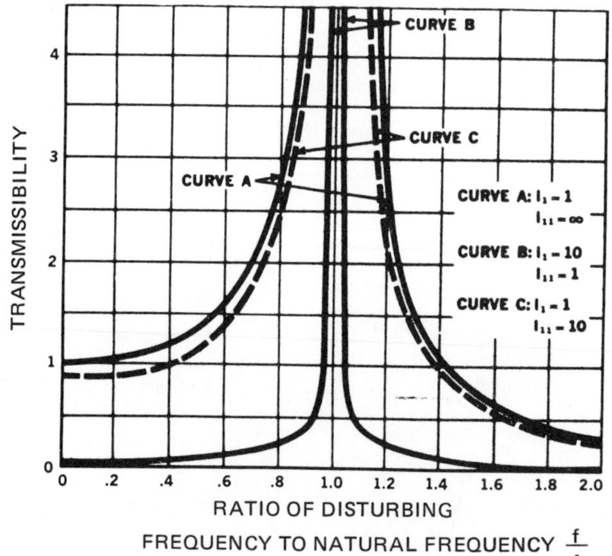

Fig 26

Comparative transmissibility curves for these two systems are shown in Fig 26. Curve A represents a single-degree-of-freedom system. It illustrates how transmissibility increases from 1:1 at very low frequencies to infinity when the disturbing frequency equals the natural frequency. Then it moves into the isolation range, where transmissibility is less than 1, when the ratio of the disturbing frequency to the natural frequency $\frac{f}{f_n}$, is more than $\sqrt{2}$ or 1.41.

Curve B illustrates the torque transmissibility in an application where the inertia of the driven part of the system is approximately 1/10th that of the driving part. In this case, the transmissibility is 0.1 at very low frequencies. It then increases in the resonance range when the disturbing frequency is equal to the natural frequency, and rapidly decreases in the isolation range.

Now refer to both the transmissibility formulas for the single-degree-of-freedom and the two-inertia, one-spring system, and to Curves A and B in Figure 26. It is immediately obvious that the transmissibility of the two-inertia system is approximately 1/10th of the transmissibility of the single-degree-of-freedom system throughout the frequency range, except at resonance.

Curve C illustrates how the two-inertia, one-spring system is approximately equal to a single-degree-of-freedom system when the driven inertia is ten or more times the magnitude of the driving inertia. In other words, when I_1 is small compared to I_2, the fraction $I_2 (I_1 + I_2)$ of the two-inertia, one-spring transmissibility formula approaches a maximum value of 1, and transmissibility approaches that of a single-degree-of-freedom system. Transmissibility based on a single-degree-of-freedom formula will always result in a conservative value.

Vibration in Structures

Structural elements such as beams and plates are, in effect, elastic masses with an infinite number of degrees of freedom, and thus an infinite number of resonances or *modes*. Thus when the frequency of an exciting force is identical to one mode of the structure a 'standard wave' will be created in the structure, to which shape it will tend to conform. Each resonance will have its own mode shape, depending on the physical form of the structure and its boundary condition.

Vibrations excited may be transient, torsional or compressional, or a combination of two or three. Mathematical analysis involves the use of differential equations of motion for the structure involved and can be extremely complex.

In the case of free transient vibrations in beams the following fourth-order partial differential equation applies:

$$\frac{\rho A}{g} \times \frac{\delta^2 z}{\delta^2 t^2} + \frac{\delta^2}{\delta \delta^2 x^2} (EI \frac{\delta^2 z}{\delta x^2}) = 0$$

where ρ = mass density of the beam
 A = cross sectional area of the beam
 E = modulus of elasticity of the beam material
 I = moment of inertia of cross section of beam
 g = acceleration of gravity

Examples of mode shapes with different boundary conditions are given in Fig 27.

In the case of free transient vibration in plates, the following fourth-order partial differential equation applies:

$$\frac{\delta^4 z}{\delta x^4} + 2 \frac{\delta^4 z}{\delta x^2 \delta y^2} + \frac{\delta^4 z}{\delta^4 y} + \frac{12\rho (1 - \mu^2)}{E b^2 g} + \frac{\delta^2 z}{\delta t^2} = 0$$

where b = thickness of plate material
 μ = Poisson's ratio for plate material

PRINCIPLES OF VIBRATION

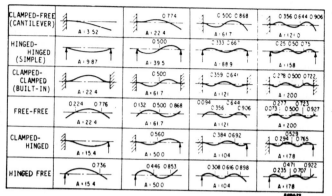

Fig 27 Boundary conditions and mode shapes for single uniform beams. (B. & K.)

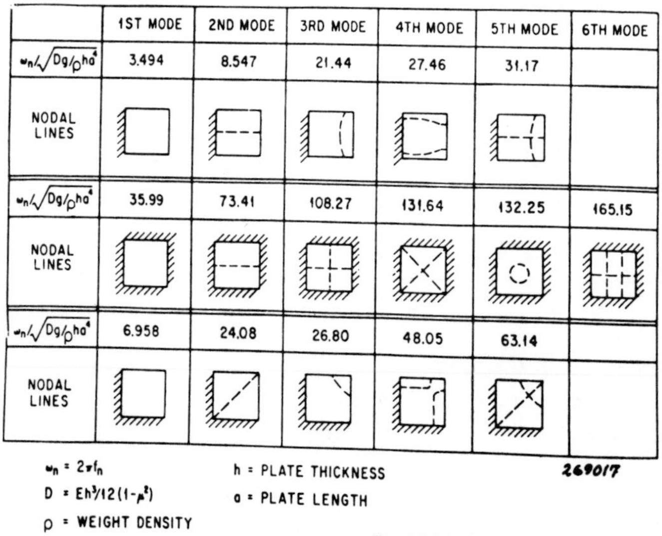

Fig 28 Nodal line configurations for square plates. (B. & K.)

Examples of mode configurations for square plates under different boundary conditions are shown in Fig 28.

See also chapters on *Dynamic Analysis of Vibration* and *Modal Analysis*.

SECTION 2a

Annoyance and Community Response

NOISE RELATED to annoyance or nuisance is essentially a subjective reaction, hence it is only to be expected that individual reactions will show considerable variations. This is further compounded because reactions to a given noise depend on the conditions under which it is heard and on attitudes to the noise source. For example, noise from a machine is often considered less annoying by the operator than by others around him; music that is considered enjoyable by the audience may be extremely annoying to neighbouring residents. Equally, a noise which may be unnoticed in a busy urban area would be very obtrusive in a quiet rural environment, and so on.

Individual variations in response mean that it is impossible to predict or assess the reaction of a particular person to a particular noise, nor is it possible to define the point at which a noise changes from being merely noticeable to being a nuisance. It is, however, possible to make a more general assessment, based on a knowledge of the noise level and of the conditions under which it is heard, to predict the typical response of the population.

People everywhere are showing increasing awareness of noise pollution as a major hazard of modern life, especially in urban areas. Opinion surveys commonly rate noise as the greatest single annoyance to people living in residential areas, whilst at the same time over the last twenty years noise levels have not only increased, but noise pollution has spread both in time (evening and night-time traffic, weekend and holiday activities) and in space (into suburbs and the countryside). These further inroads are due to continuing urbanization, higher density of population and increased traffic. The urban population in OECD countries has increased by 50% during this period for example and the number of towns with more than a million inhabitants has doubled. Large-scale urban renewal and new projects such as motorways and airports have also contributed to increasing noise pollution. As a consequence over 100 million people in OECD countries now suffer in their homes from 'unacceptable' noise levels (above 65 dB). Another 200 million live with 'uncomfortable' noise levels (above 55 dB). In a large city, such as Paris, more than half the inhabitants are exposed to 65 dB or more.

Specifically, the nuisance caused by a noise is closely related to its intrusiveness, *ie* the extent to which it stands out from the general ambient level of noise; it is also likely to be considered a nuisance when it reaches a certain (undefined) level, irrespective of the level of other ambient noise. Thus, it is possible to devise a method of rating noises based on a calculation of the extent to which they obtrude above the general background noise level.

Any site can be expected to have a more or less ambient background noise level, depending on its location. This ambient level may also vary with time of day. Superimposed on this will be fluctuating noise with differing peak noise levels due to local incidents, or peak-repetitive (*eg* due to the passage of vehicles or aircraft). The latter are referred to as *neighbourhood noise*. The difference between the normal ambient background noise and the neighbourhood noise is an indication of the 'noisiness' of the site.

The observer may be indoors or outdoors, the main difference being that indoors the observer has the benefit of attenuation of outside noise offered by the structure of the building. Noise sources themselves may also be classified as 'outdoors' or 'indoors'.

Chief generators of outdoor noise in residential areas are aircraft, road vehicles, trains, road repairs, building site work, children playing, noisy local activities and service deliveries. In industrial areas noise may also be caused by factory plant, *etc*, which, although perhaps primarily generated 'indoors', if radiated from the buildings concerned, is classed as 'outdoor' noise.

Noise sources have increased most rapidly in the field of transport. The number of *surface motor vehicles* (private cars, heavy lorries, buses and motorcycles) has trebled in the past twenty years. In fact, measurements and surveys show that the number of people exposed to and affected by road traffic noise far exceeds the number exposed to all other sound sources combined. *Air traffic* has increased tenfold in terms of passenger/km. The impact of the two modes of transport differs from country to country. In Europe and Japan, for example, an estimated twenty times more people are exposed to road traffic noise than to aircraft noise, compared with three times more in the United States. *Railways* have a smaller, but far from negligible impact: between 1–3% of the population, depending on the country, is subjected to more than 65 dB(A) by trains.

Simple Measurement

The sound level received at a point distant from a single source of noise (such as a machine or process), assuming that noise is radiated uniformly in all directions over flat open ground is given by:

$$\text{sound level dB(A)} = L_W - 20 \log_{10} R - 8$$

where
L_W = A-weighted sound power level of the source
R = distance of the source in metres

The actual sound level received may be influenced by other factors. If the point under consideration is at a building facade, for example, an additional 3 dB(A) should be added to allow for the effect of sound reflections from the facade. Further correction may be necessary to allow for ground and atmosphere attenuation, although these are generally ignored for general estimates and calculations.

Direct treatment for reduction in noise level at the point under consideration could be:

(i) *Increase the distance R.* This will have the effect of reducing the sound level received by about 6 dB(A) for each doubling of the distance (or estimate likely reduction directly from Fig 1).

(ii) *Screening.* Generally offering a reduction of 7–10 dB(A), but possibly up to 15 dB(A) as a maximum.

(iii) *Sound reduction at source.* Enclosure of the noise source could give a reduction of up to 50 dB(A) but usually less than one half this value. Silencing in the case of machines could give a reduction of from 5 to 20 dB(A). It will be appreciated from this that the most effective noise control treatment is reduction of sound at source.

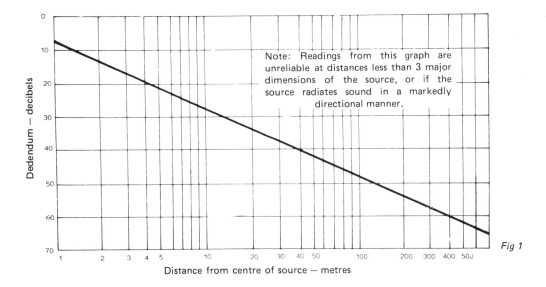

Fig 1

Distance from centre of source — metres

Corrected Noise Levels

The measurement of neighbourhood noise from the site may be complicated by the presence of ambient noise, since a measuring device cannot distinguish between noise sources. Measurements represent either the ambient noise alone (from sources other than the site) or the neighbourhood noise (from the site) and the ambient noise combined. Where the total measured sound level exceeds that of the ambient noise by 10 dB(A) or more, the total value is equal to the level of neighbourhood noise. Where the total measured level exceeds the level of ambient noise by less than 10 dB(A) the correction given in Table I should be applied to the total level in order to obtain the level of neighbourhood noise only.

TABLE I – CORRECTIONS FOR AMBIENT BACKGROUND NOISE (BS5228 : 1975)

Difference between total measured sound level and level of ambient noise dB(A)	Amount to be subtracted from total measured sound level to determine level of neighbourhood noise dB(A)
3	3
4 to 5	2
6 to 9	1
10 or more	0

More realistically, Corrected Noise Levels should incorporate corrections for noise duration and character, using the background noise level as a basis for comparison whenever it can be measured; BS4142 defines it as the noise level exceeded for 90% of the time (L_{90}), whilst ISO R1996 uses the noise level exceeded for 95% of the time (L_{95}), all measurements being made using the A-weighting scale. In practice, the difference between L_{90} and L_{95} is likely to be small.

If the Corrected Noise Level is 10 dB(A) or more above the background noise level (or derived criterion), then complaints are expected: excesses of 5 dB(A) are rated as 'marginal' by BS4142

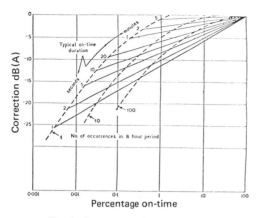

Fig 2 Corrections for duration during night-time periods as specified in BS4142.

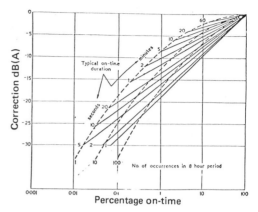

Fig 3 Correction for duration during periods other than night-time as specified in BS4142.

TABLE II — CORRECTIONS FOR DEGREE OF NOISE

Percent relevant time period	Correction dB(A)
100 – 56	0
56 – 18	–5
18 – 6	–10
6 – 1.8	–15
1.8 – 0.6	–20
0.6 – 0.2	–25
Less than 0.2	–30

and as likely to arouse 'sporadic complaints' by ISO R1996; whilst if the corrected noise level is less than the background level (or derived criterion) complaints are unlikely.

The duration correction is slightly different in BS4142 and ISO R1996; BS4142 gives two graphs, one for day-time and one for night-time (Figs 2 and 3) from which the correction is obtained, whilst ISO R1996 gives a table of corrections for various percentage 'on times' for the noise (Table II). The percentage 'on time' is taken as a percentage of the period under consideration, which may be day-time (*ie* 8 am to 6 pm) or night-time (10 pm to 7 am), or some other period depending on local conditions. BS4142 gives duration corrections for night-time which are much smaller than for day-time, to allow for the fact that noise which disturbs sleep for even a short period can be very annoying. ISO R1996 uses the same corrections for day and night, but suggests that the most unfavourable half hour of the night be taken as the period for assessment. Thus, to the measured noise level will be added the correction for character, while the duration correction will be subtracted if the noise does not occur all the time.

dB(A) Measurement

It is required to measure the level of the noise under investigation in dB(A) at, or near, the point from which complaints may arise. When the noise is reasonably steady and continuous and is significantly higher than the background noise level, it can be easily measured; if the noise fluctuates over a fairly small range (10 dB(A)), the typical level can be assessed by visual averaging of the meter reading and, if the noise consists of intermittent or regular bursts of reasonably constant level, the level of each burst can be measured and a duration correction applied according to the length of the burst. (If bursts of different level or intervening levels of relatively low noise occur, these can be separately assessed with appropriate duration corrections). Problems arise, however, when the noise is too irregular to permit a 'typical' level to be visually assessed or when the noise cannot be reliably measured above the background noise level. Here a more realistic form of measurement is L_{eq}.

Noise Exposure Levels

The Noise Advisory Council (UK) recommended the adoption of L_{eq} (equivalent continuous sound level) as a single measurement of all environmental noise (*eg* measurement of the noise of road traffic, aircraft, industrial noise, *etc*), as detailed in A Guide to Measurement and Prediction of the Continuous Sound Level L_{eq}, 1978.

L_{eq} is the level of a notional steady sound which, at a given position and over a definite period of time, would have the same A-weighted acoustic energy as the fluctuating noise. Thus L_{eq} is essentially itself a noise scale.

The single even noise exposure level L_{AX} is the level which, if maintained constant for a period of 1 second, would cause the same A-weighted sound energy to be received as is actually received from the noise event. Where there are a number of different sounds contributing to the overall noise level, L_{AX} values for each source can be used separately to produce the value of L_{eq} due to each source over the same total period. However, it is impractical to express background noise in terms of a value of L_{AX}, hence the suggested use of a notional L_{eq} due to background noise. This is then combined with the value of L_{eq} due to the noise sources as a measurement of (total) environmental noise.

Distinctive Background Noise

Another problem arises when the noise cannot be measured above the prevailing background noise, although it may be audible because of some distinctive character. One possibility is to measure the noise nearer to the source and calculate the level at the point of potential complaint. This may be quite accurate where there is propagation over open ground for moderate distances but the effects of screening and climatic effects on propagation over distances of a few hundred metres or more are difficult to calculate. An alternative is to arrange for the noise source to operate at some time when the background noise is lower and the noise can be reliably measured; this will not be possible if the intended period for assessment corresponds to the lowest background noise conditions. A reasonable assessment can, however, be made if it is assumed that when the noise is audible, but not measurable, above the background noise level, it is likely to be between 0 and 5 dB(A) below the background level and therefore likely to arouse sporadic complaints unless a (subtractive) correction for duration is appropriate. Such an assessment is not, of course, as reliable as one based on a measured noise level.

It is also possible to use the method of assessment to predict the likelihood of complaints from a future noise source by measuring the background noise level and comparing it to the calculated corrected noise level at possible sources of complaint. Providing that adequate data are available to calculate the corrected noise level, this is often easier than an assessment based on an existing noise, which may be difficult to separate from the background noise. It will be necessary to leave an adequate margin for error in the calculations if complaints are to be avoided.

See also chapters on *Road Traffic Noise, Aircraft and Airport Noise, Factory Noise, Noise in Commercial Buildings* and *Noise in Domestic Buildings.*

Health and Safety (Hearing Damage)

THE HUMAN ear (Fig 1) is a delicate mechanism which accepts sound waves along its auditory canal and through to the brain where it translates them into pitch and volume. Specifically, the auditory canal is confined to the external ear, separated from the middle ear by the tumpanic membrane (eardrum). The middle ear is a cavity in the temporal bone containing three articulated bones (ossicles) transmitting the motion of the eardrum to an oval 'window' opening into the inner ear, but covered by a membrane. The inner ear is filled with fluid and contains the cochlea or actual 'hearing' portion of the ear which is essentially a system of coiled tubes also filled with fluid and separated by membranes. The basilar membrane separates the media from another tube, the scala tympani, and contains on its surface the hair cells which are the sound receptors. Motion of the basilar membrane causes bending of the hair cells, this movement generating impulses in the cochlea nerve endings which encircle the cells. These impulses are then transmitted through a series of lower auditory structures to the brain.

Apart from disorders or disease (or congenital defects), the two main causes of hearing loss are the deteriorating effects of age and damage to the delicate hair cells by exposure to excessive noise. (Damage to hair cells can also be caused by ototoxic drugs).

Hearing Impairment

Traditionally, hearing impairment is looked upon as an inability to understand speech communication. On this basis the ability to hear major speech components can be assessed as the average

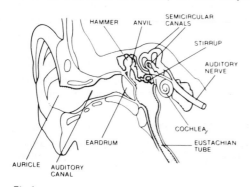

Fig 1

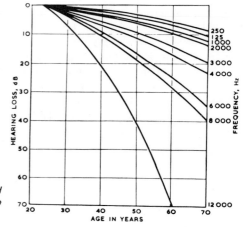

Fig 2 Presbyacusis curves for men and women with no history of exposure to excessive noise.

hearing levels at frequencies of 500, 1 000 and 2 000 Hz. An average loss of up to 25 dB represents no actual handicap, but each decibel of average loss above 25 dB represents a 1.5 per cent handicap.

More realistically, hearing handicap is assessed over a wider frequency range particularly as, when damaged by noise, the frequency affected first is 4 000 Hz (and higher frequencies). On this basis, recommended criteria representing hearing loss are:

More than 15 dB at frequencies of 500, 1 000 or 2 000 Hz.

More than 20 dB at 3 000 Hz.

More than 30 dB at 4 000 and 6 000 Hz.

Presbyacusis

This ageing effect, known as presbyacusis, normally starts at an age as early as 25 years, and with advancing age there is an ancillary loss of hearing, particularly with the higher speech frequencies. Fig 2 shows typical average patterns for healthy male and female subjects, with no history of exposure to excessive noise.

Damaging Noise Levels

Classification of noise levels related to dB measurement, is given in Table I. This is necessarily of a generalized nature, with both subjective and objective responses included. 'Interference' and 'annoyance' levels are particularly difficult to define, specifically since environment, background and other factors can all have a modifying effect, notably as regards tolerance towards a particular noise. The frequency content of the noise is also significant, for individuals are normally far less tolerant of high frequency sounds than low frequency sounds of the same or even higher sound pressure levels. A high frequency noise is normally defined as a frequency in excess of 2 000 Hz.

Noise within the sound pressure range 30 dB to 70 dB is generally accepted as 'safe'. That is, it has no adverse physiological effects on the human ear. The degree of tolerance within this range is, however, very variable. The higher figure represents what is probably the *logical* upper limit for tolerance provided this is based on A-weightings, *ie* sound pressure level measurement in dB(A). From the practical point of *achieving* such levels in typical 'noisy' atmospheres 80 dB or even 90 dB may be argued as 'acceptable'.

Continual exposure to sound pressure levels above 80 dB is now generally recognized as potentially damaging. Short term exposure to sound pressure levels up to 100 dB is likely to lead to a temporary shift in the hearing threshold, generally with complete recovery. Longer term exposure presents the hazard that recovery of hearing will not be complete and a permanent loss of hearing will result. Above 100 dB the damage risk is drastically increased and only very short term exposure can be accepted if permanent loss of hearing is to be avoided. A sound pressure level of 120 dB also represents a *discomfort level, ie* a level of sound at which extreme discomfort is felt. At around 140—150 dB, discomfort is experienced as actual pain. At sound pressure levels with peaks of 160—180 dB there is the possibility of immediate mechanical damage to the ear — a burst eardrum. At lower levels, down into the 'risk' range, the damaging effect of sound is cumulative.

Noise has three main effects on hearing — acoustic trauma, temporary threshold shift and permanent threshold shift. It is important to realize that this type of deafness is 'perceptive' and, unlike 'conductive deafness', cannot be cured by either medical or surgical intervention. Hence the serious nature of hearing damage caused in this way.

Acoustic trauma describes the immediate injury caused by exposure to intense sounds like blasts, explosions and gunfire. The perceptive deafness resulting from such sounds can be considerable and it is imperative that personnel exposed to noise levels in excess of 120 dB, no matter how short the duration, should take adequate precautions to protect their ears.

It is a general characteristic of both a temporary and a permanent threshold shift that the maximum loss of hearing occurs at around 4 kHz. In the case of permanent threshold shift the loss can be of the order of 60 dB at this frequency, with a substantial loss of hearing over the whole of the normal speech frequency range of 500 Hz to 4 kHz. It also follows that signs of permanent threshold shift are first associated with loss of acuity at the higher frequencies (*eg* the sharp consonants in speech), and progressive loss of acuity throughout the remainder of the speech frequency range. Such hearing loss is also accentuated by presbyacusis.

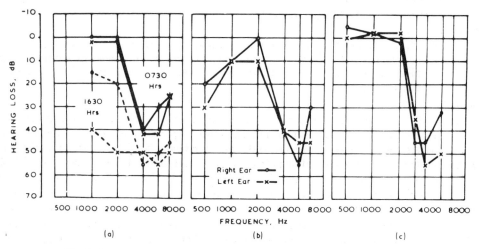

Fig 3 Pure tone audiogram showing temporary and permanent threshold shift caused by industrial noise:
(a) Temporary threshold shift after one day's exposure to a sound field of 104 dB(A).
(b) Permanent threshold shift in a 28 year old man after 8 years' exposure to the automatic hammering of copper boilers.
(c) Permanent threshold shift in a 23 year old man showing the effect of working in a rolling mill and a machine shop.

Frequency Response

The lowest frequency which the average person with normal hearing can detect is about 20 Hz. This does not alter with health or age. The highest frequency which the ear can detect is about 20 kHz but the ability to hear higher frequencies is very dependent on the individual's inherited acuity, age, health and history of exposure to noise. A typical normal limit for a younger person would be about 16–18 kHz, and for an older person could be very much less (*eg* 10–12 kHz). It is appropriate to comment here that 'Hi-Fi' performance calls for a minimum frequency range of 40 Hz to 12.5 kHz, or for complete 'Hi-Fi' capable of accommodating the full dynamic range of musical instruments, a range from 20 Hz to more than 15 kHz.

The human hearing mechanism is also more responsive to high frequency sounds and relatively insensitive to lower frequencies. In other words, listening to a high frequency sound and a low frequency sound, both at the same sound pressure level (or having the same sound energy), the high frequency sound would be judged 'louder'. This frequency content of a sound plays a large part in the psychological effect of noise at normal (below risk) levels.

HEALTH AND SAFETY (HEARING DAMAGE)

Since assessment of 'loudness' is necessarily subjective, the relative loudness of different frequencies can only be measured by sampling — *ie* by a series of subjective tests. Moreover, as the tests are subjective, age will also influence the pattern established. As a general rule age difference is only noticeable to any extent at frequencies above 2 kHz.

Specifically, for the average person, a sound at 20 Hz would need to have a sound pressure level as much as 50 dB higher than a sound at 1 kHz to be judged equally as loud at low to moderate sound pressure levels. This would appear to indicate a relative insensitivity to low frequency sounds. However at higher sound pressure levels the difference to give 'equal loudness' response decreases.

Temporary threshold shift occurs when a person has been exposed for a few hours to noise levels of about 80 dB and above. These often leave a ringing in the ears for some time afterwards. The greater part of the hearing loss occurs soon after exposure and, similarly, recovery occurs largely in the 30 minutes following removal from the noise. Persons exposed to continuous noise at a level of 100 dB(A) for an eight hour working day could show a temporary threshold shift up to 40 dB in that part of the spectrum most affected (see Fig 3). However, this could overlap a degree of hearing loss already present; an older man with 30 dB presbyacutic loss, or an individual with a similar permanent hearing loss, would not be significantly affected.

Permanent threshold shift. When the ear is subjected to high intensity noise day after day, more lasting damage will take place. It is possible that a person may not recover full hearing between exposures, and what started as a temporary threshold shift eventually becomes permanent (see Figs 3b and 3c).

In the first stages, the pure tone audiogram shows a hearing loss in either the 3 000 or 4 000 Hz range, followed by further loss in these and higher ranges. At this stage the individual is not aware of his loss, but may complain of ringing in the ears. This is because frequencies between 3 000 and 6 000 Hz contribute little to the intelligibility of speech, although they are important to music. Gradually the condition progresses, extending down the frequency scale to affect speech frequencies (500–2 000 Hz) until the sufferer is irreversibly deaf.

See also chapter on *Hearing Protection Devices.*

Speech Communication

THE MAJOR source of intelligibility in speech communication is in the audio frequency range of 1 000 — 4 000 Hz, although maximum sound energy (loudness) of speech is normally generated at around 500 Hz; and the most significant frequency range for understanding speech is 500 — 2 000 Hz. Equally, intelligibility will depend on the sound levels generated by individual noises, as well as articulation and phonetic balance of the spoken words. Typical sound levels for various bands of speech at a distance of 1 metre are given in Table I. In general a typical female voice is about 5 dB lower than a male voice, unless deliberately raised, although sharing a similar peak pressure level at different frequencies — Fig 1.

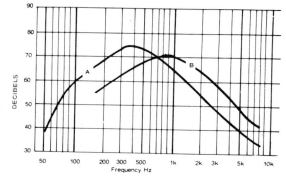

Fig 1 Generalized sound spectra of human speech. A — male. B — female.

TABLE I – SOUND PRESSURE LEVELS FOR SPEECH AT DISTANCE OF 1 METRE

Type of Speech	Male Voice		Female Voice	
	Level	Maximum Sound Energy	Level	Maximum Sound Energy
Whisper	20–30 dB(A)	At circa 400 Hz	20–25 dB(A)	At circa 900 Hz
Low voice	50–60 dB(A)		45–55 dB(A)	
Normal voice	65–68 dB(A)		60–65 dB(A)	
Raised voice	70–75 dB(A)		68–70 dB(A)	
Loud voice	75–80 dB(A)		70–75 dB(A)	At Circa 1000–1100 Hz
Shouting	85 dB(A) and over		75–85 dB(A)	

Speech Communication

Speech Interference Level (SIL)

Background noise will inevitably have a masking effect on speech and the higher the background noise level the more the voice will have to be raised for satisfactory speech communication. Intelligibility will depend on the distance between speaker and listener.

Speech Interference Levels (SIL) are a quantitative assessment of these parameters, expressing the 'masking' effect of background noise which has to be overcome to establish reasonably reliable conversation — Tables IIA and IIB. Background noise levels are given in both dB and dB(A) in these

TABLE IIA — SPEECH INTERFERENCE LEVELS

DISTANCE FROM SPEAKER	WHISPER		LOW		NORMAL		VOICE LEVELS RAISED		VERY LOUD		SHOUTING	
feet	dB	dB(A)	dB	dB(A)	dB	dB(A)	dB	dB(A)	dB	dB(A)	dB	dB(A)
0.5	46	51	60	66	71	78	77	83	83	88	89	96
1	40	45	54	60	65	72	71	77	77	83	83	90
2	34	39	48	54	59	66	65	71	71	79	77	84
3	30	35	44	50	55	62	61	68	67	73	73	80
4	—	—	42	48	53	60	59	65	65	72	71	78
5	—	—	40	46	51	58	57	63	63	70	69	76
6	—	—	38	44	49	56	55	62	61	68	67	74
8	—	—	—	—	47	54	53	59	59	66	65	72
10	—	—	—	—	45	52	51	57	57	64	63	70
12	—	—	—	—	43	50	49	56	55	62	61	68

Note: These figures are approximate

TABLE IIB — SPEECH INTERFERENCE LEVELS

DISTANCE FROM SPEAKER	WHISPER		LOW		NORMAL		VOICE LEVELS RAISED		VERY LOUD		SHOUTING	
metres	dB	dB(A)	dB	dB(A)	dB	dB(A)	dB	dB(A)	dB	dB(A)	dB	dB(A)
0.25	41	47	56	62	67	74	73	79	79	86	85	90
0.50	35	41	50	56	61	68	67	73	73	80	79	84
0.75	32	38	47	53	57	64	63	69	69	76	75	82
1.0	29	35	44	50	55	62	61	67	67	74	73	78
1.5	26	32	41	47	51	58	57	63	63	70	69	76
2	—	—	38	44	49	56	55	61	61	68	67	72
3	—	—	—	—	45	52	51	57	57	64	63	70
4	—	—	—	—	43	50	49	55	55	62	61	68
5	—	—	—	—	41	48	47	53	53	60	59	66
6	—	—	—	—	39	46	45	51	51	58	57	64

Note: These figures are approximate

IAC modular acoustic control room provides a quiet environment for operating staff at a sand refining plant in the Midlands.

tables. Original SIL levels were based on averaging the actual sound levels present (in dB) in the three octaves above 600 Hz (*ie* 600 — 1 200 Hz, 1 200 — 2 400 Hz and 2 400 — 4 800 Hz). The equivalent numerical values for dB(A) measurement, taken directly with a sound level meter with A-weighting, are also given in the table.

Examples of use:

(i) At a distance of 1.5 metres (5 feet) a speaker using a normal voice should be intelligible provided the background noise level does not exceed 58 dB(A).

(ii) If the background noise is 70 dB(A) and the distance between speaker and listener is 3 metres (10 feet), the speaker will have to shout to be heard. Alternatively, to use a normal voice the distance between speaker and listener would have to be reduced to about 0.3 metres (1 foot).

(iii) To be able to make conversation using a normal voice with a background level of 60 dB(A) the distance between speaker and listener will have to be 1.2 metres (4 feet) or less.

Specifically, SIL values are based on the loudness of an average male voice and with the listener suffering no hearing loss. They are minima for barely reliable conversation (*ie* not less than 75% of the speech heard correctly). If the listener suffers from hearing loss, (usually expressed in dB), then this figure must be added to the background noise level.

SIL limits are as follows:

(i) Below 40 dB (45 dB(A)) background noise will be negligible.

(ii) If the background level exceeds 90 dB (97 dB(A)), intelligible speech communication is virtually impossible and can only be carried out satisfactorily with a headphone communications set.

Communication Wearing HPDs

Contrary to popular belief, hearing of speech can be improved, not reduced, wearing hearing protective devices. Wearing HPDs reduces the overall sound levels which overload the ear, permitting accurate discrimination of different sounds so that the ear can operate more efficiently under the circumstances. Thus the wearing of HPDs can actually improve speech communication when sound levels are greater than about 85 dB(A). However, this only applies to individuals with normal hearing. For individuals with impaired hearing, HPDs may have the opposite effect — *ie* reduce their capacity to understand speech. But this in no way justifies their not wearing HPDs where required for hearing protection.

Primarily ear protectors are used to prevent hearing damage. The degree of attenuation required should, ideally, be matched to the noise spectrum. It is also desirable, wherever possible, to provide differential attenuation, so that the more harmful or distracting higher frequencies are stopped, whilst speech frequencies are still transmitted with the minimum of attenuation. This is not always possible, especially where a high degree of attenuation is required, or where the background noise contains all the audible frequencies in a regular distribution — *eg* 'white noise'. Speech communication is equally good with or without ear protectors for a white noise level of about 60 dB. At lower noise levels, speech communication is better without ear protectors. At white noise levels of 70 to 100 dB, there are definite gains in ease of speech communication facilities with a suitable type of ear protector offering minimum attenuation at speech frequencies.

Headsets

For satisfactory speech communication in very noisy areas, headsets may be required. Headsets comprise telephones or earphones (receiver inserts) mounted on a headband and, where these are designed to be operated under conditions of high white noise, the earphone is mounted in a suitable ear muff. Such headsets may also be used as ear protectors with the earphone replaced by a suitable plug or cushion or just plain earphone socket.

It is also possible, when using a plain socket or hollow chamber, to tune the device to a particular frequency (or series of frequencies, by using two or more resonators in series) and thus produce a flatter attenuation curve if desirable — *ie* to provide improved protection from an earmuff against high level low frequency sounds.

There may be a certain value in using earplugs in combination with a headset. Since the wearer will be required to transmit as well as receive speech, a microphone is usually combined with the headset, fitted on an adjustable and pivoted boom, so that the microphone can be swung clear of the speaking position when not in use. The microphone is normally a noise-cancelling type which, basically, discriminates between spherical and plane sound waves in favour of the former. Thus speech (spherical sound waves due to the close proximity of the mouth) is picked up, and extraneous sound sources and noise (plane sound waves) are rejected.

Telephone Speech Communication

Since one ear is still exposed to background noise level, the higher this is the more difficult satisfactory communication becomes. In terms of SIL values, average subjective assessment is:

Satisfactory communication — SIL below 45 dB (51 dB(A))

Slight difficulty — 45–60 dB (50–66 dB(A))

Difficult — 60–75 dB (66–81 dB(A))

Unsatisfactory — above 75 dB (above 81 dB(A))

There is another series of SIL values in dB based on the arithmetic average of background sound pressure levels taken at 'preferred' octaves, 50 Hz, 100 Hz and 200 Hz. These are known as Preferred Speech Interference Levels (PSIL). These are generally quoted in American literature in preference to SIL.

For normal communication, *eg* in offices, the Articulation Index is now commonly employed — see chapter on *Noise in Commercial Buildings*. SIL values are no longer widely used, though still quoted.

See also chapters on *Subjective Noise Parameters* and *Noise in Commercial Buildings*.

Hearing Conservation in Industry

APART FROM any moral obligation for an employer to protect the hearing of workers exposed to excessive noise, there is a financial incentive in avoiding claims for compensation for damage to hearing under common law, or current legislation. The latter are not clear cut, nor all-embracing, although at present they remain the only guidance available to industry on the control of noise levels. They include the limited provisions stated in the Noise Regulations in the Woodmaking Machines Regulations (1974); the Health and Safety Executive's report in Running Noise Legislation (1974); the Health and Safety Executive's report in Running Noise Legislation (1975); the Health and Safety at Work etc Act (1974); and various others in draft form.

Basically an industrial hearing conservation programme aims at controlling noise exposure of employees to levels which are not damaging. To be effective this involves four separate sections:

Measure —
  the work environment:
  measure the noise level using a sound level meter which gives an overall measurement in dB(A); it will indicate whether there is a problem but not its nature.
  the worker:
  measure his hearing to establish whether there is any hearing loss (audiometry).

Evaluate —
  the work environment:
  determine the size and nature of the problem using an octave band analyzer, which breaks noise down into frequencies and intensities.
  the worker:
  evaluate any hearing loss with more sophisticated audiometry to see whether it is conductive or sensor-neural hearing loss (this must be done by a qualified technician).

Control —
  the work environment:
  carry out good planned maintenance to reduce noise at source and along its pathways to the ears, within engineering capabilities.
  the worker:
  issue appropriate hearing protection.

Monitor —
  the work environment:
  establish a continuous programme to check on noise levels. Consider the possibility of

introducing personal dosimeters which monitor the level to which an individual is exposed.

the man:
re-test hearing on a yearly or two yearly basis (serial audiometry).

The following abbreviations are commonly used relative to these and other aspects of hearing conservation.

HSWA — Health and Safety at Work etc Act
HSC — Health and Safety Committee
HCP — Hearing Conservation Programme
HPD — Hearing Protective Devices

The HSE insists that Section 2 of the HSWA has given them a general foundation in law and one which can be used to pressure industry to impose stricter controls on noise. Indeed, the Inspectorate has already demonstrated its willingness to issue improvement notices under the HSWA to companies who fail to protect their employees from excessive noise 'so far as is reasonably practicable'. Any employer who fails to comply with the notice within the specified period lays himself open to prosecution.

At the present level, treatment should involve noise and vibration control techniques, applied to reduce noise levels at all stations to below the hazard level (not exceeding 90 dB(A) L_{eq} (8h)). This is the only satisfactory long-term solution, but even this may require further modification should future legislation demand more stringent reduction of noise levels. (See also Table I).

In many situations, however, noise control treatment can only go so far in reducing noise, and in others noise control techniques are either impractical or too costly. In all such cases personal hearing protectors need to be provided and must be worn by employees. This, essentially, is a compromise solution. The use of HPDs is not foolproof and is seldom used to best advantage, so realistic protection is very largely dependent on a satisfactory frequency of noise monitoring and regular testing of personnel concerned.

Straightforward noise monitoring can be used to ensure that noise levels are not increasing. This is apparently adequate where noise treatment has been applied originally to reduce the noise climate to below the hazard level, except for the fact that a certain percentage of exposed persons may suffer damage at lower levels than 90 dB(A). Thus it is still desirable to monitor the hearing of all personnel exposed to a continuous noise level of, say, 80 dB(A). Where ear protectors are in use, noise monitoring alone is not sufficient. The possibility of introducing monitoring audiometry must also be considered as a check on workers' hearing, and thus the effectiveness of the hearing protection provided and its proper use.

An Effective Programme

An effective hearing conservation protection programme is not assured merely by making good and properly fitting protectors available to those persons exposed to high-level noise. Management, medical, industrial hygiene and safety personnel must all be aware of the problem, as well as those habitually exposed to noise. They must all support the programme to the full if it is to be effective. In addition, the programme will be much more efficient if a responsible member of the organization is designated as co-ordinator to initiate the programme and follow it through, thereby sustaining both management's and workers' support.

TABLE I – SUMMARY OF PROPOSED STATUTORY REGULATIONS (HSC AND EUROPEAN COMMISSION)

Liability	Duty	Below 90 dB(A) (L_{eq} (8h))	Above 90 dB(A) L_{eq} (8h)	Above 105 dB(A) L_{eq} (8h)
Employers	Reduce exposure likely to be injurious to lowest reasonable practical level	X	X	X
	Reduce exposure by reduction of noise levels to lowest reasonable practical level; and then, if necessary, by other means		X	X
	Arrange for surveys		X	X
	Provide information, instruction and training		X	X
	Provide ear protectors		X	X
	Check that control measures are used		X	X
	Produce a programme of action		X	X
	Appoint a qualified adviser		X	X
	Keep records of exposure		X	X
	Arrange for audiometry			X
	Arrange for individual monitoring of exposure			X
Employees	Make full and proper use of control measures, ear protectors, etc	X	X	X
	Co-operate with employer	X	X	X

Note: Additionally manufacturers are to ensure, as far as is reasonably practical, that noise produced is not likely to be injurious to hearing and carry out testing, examination and research to this end.

The co-ordinator must determine the noise exposure patterns, both in terms of sound level and duration of exposure, for all persons under his responsibility, and hence pinpoint those areas where hearing protection is necessary. He must evaluate other existing environmental factors such as temperature, humidity, and possible communication and warning signal requirements. Also, he should ascertain the need for other personal safety equipment such as glasses, helmets and gloves in that these may affect the choice of hearing protector. With such information and by personally experimenting with different types of hearing protection in the various environments, he will be able to decide upon the most suitable types of protector required for the different situations existing in his organization.

A close relationship must be maintained by the co-ordinator with management, medical staff and employees in order to be aware of reactions to various stages of the programme. In this way any problem will become evident early on and corrective measures can be taken before any serious disorder occurs. Subjects may also have added problems when wearing hearing protectors.

In addition to these arguments there is the even greater need to protect the hearing of a person who already has a hearing loss. He has, so to speak, less hearing to lose than normally-hearing persons and therefore requires greater protection. Although a worker may have experienced many years in a noisy environment without hearing protection and thus adapted gradually to a hearing loss and the needs of his work and social life, his hearing should still be protected to prevent further increase in any social handicap he may have.

Estimating Hearing Handicap

The British Standard method of estimating the risk of hearing handicap due to noise is based on the average hearing threshold level at 1 kHz, 2 kHz and 3 kHz, this frequency combination being in conformity with Department of Health and Social Security (1973) Occupational Deafness: Report by the Industrial Enquiries Advisory Council.

The hearing of a person is deemed to be impaired sufficiently to cause a handicap if the arithmetic average of the hearing threshold levels, of the two ears combined, at 1 kHz, 2 kHz and 3 kHz is equal to or greater than 30 dB referred to the audiometric zero of BS2497: Parts 1, 2 and 3.

Handicap percentage is further defined as the percentage of persons in a population attaining or exceeding the 30 dB average hearing threshold level, where age and noise exposure are the causative factors and no pathological condition is involved. The definition assumes that the hearing levels in the two ears are substantially similar.

The procedure for calculating handicap percentage from noise exposure is detailed in BS 5330: 1976. In the case of continuous sound levels, the equivalent continuous sound level (L_{eq}) is determined from measurements. For intermittent sounds, or sound with step-wise variations, the separate levels and durations are measured and these values used to determine the L_{eq} from the formula

$$L_{eq} = 70 + 10 \log_{10} \Sigma E_1$$

where

ΣE_1 is the composite noise exposure index
= sum of partial noise exposure indices

In the case of intermittent noise or fluctuating noise a statistical analysis of the noise levels over a typical period is made with automatic recording equipment having an overall performance equivalent to that of a precision sound level meter used under similar conditions. The noise levels are grouped into classes with a width of 5 dB, 2.5 dB or 1 dB as appropriate and the total duration within a day recorded for each class.

The equivalent continuous noise level is then determined as:

$$L_{eq} = 10 \log_{10} \Sigma(t_1 \text{ antilog } L_1/10/\Sigma t_1)$$

The *handicap percentage* can be devised directly from the Noise Immission Level (NIL), which itself is derived directly from the L_{eq}, *viz*

$$NIL = L_{eq} + 10 \log_{10} (T/T_0)$$

where

T is the duration of exposure in years
T_0 is one calendar year

The *handicap percentage* is then determined by relating the NIL value to the age of the subject in years — see Table II.

See also chapters on *Audiometry* and *Hearing Protective Devices*.

TABLE II – PERCENTAGE OF PERSONS ATTAINING OR EXCEEDING A MEAN HEARING LEVEL OF 30 dB (MEAN OF 1 kHz, 2kHz and 3 kHz)

NIL	Age in years									
	20	25	30	35	40	45	50	55	60	65	70
90	0	0	0	0	0	0	0	1	1	2	4
92	0	0	0	0	0	0	1	1	2	3	5
94	0	0	0	0	0	1	1	1	2	4	7
96	0	0	0	1	1	1	1	2	3	5	9
98	1	1	1	1	1	2	2	3	5	7	11
100	1	1	1	2	2	2	3	4	7	10	14
102	2	2	2	3	3	4	5	7	9	13	19
104	3	3	4	4	5	6	7	9	13	17	23
106	5	5	6	6	7	8	10	13	17	22	29
108	8	8	8	9	10	12	14	18	22	28	35
110	11	11	12	13	14	16	19	23	28	35	42
112	15	16	16	18	19	22	25	30	35	42	49
114	21	21	22	23	25	28	32	37	43	50	57
116	27	28	29	30	32	36	40	45	51	58	65
118	35	35	36	38	40	44	48	53	59	65	72
120	43	43	44	46	49	52	56	61	67	72	78
122	51	51	52	55	57	60	64	69	74	79	84
124	59	59	60	62	65	68	72	76	80	84	88
126	67	67	68	70	72	75	78	82	85	89	92
128	74	74	75	77	79	81	84	87	89	92	94
130	80	80	81	82	84	86	88	91	93	95	96

Hearing Protective Devices

EAR DEFENDERS or Hearing Protective Devices (HPDs) as they are now called fall into three categories, ear plugs, ear muffs and ear defender head sets (although the latter are not commonly used in industry).

The choice of a suitable protector will depend on the specific situation or application, as well as the sound pressure level to which the wearer is subjected. At best an ear protector can be expected to provide an attenuation of 25–35 dB, although actual performance is frequency dependent.

Under certain circumstances, the attenuation desirable may be higher than that provided by a single type, when a combination of two types may be necessary (*eg* ear plugs together with ear muffs). However, where ear protectors are to be worn continuously, comfort and convenience may be primary factors in selecting a suitable type. Other factors which may need to be considered are wearer acceptability, hygiene, cost, durability, chemical stability and availability. The latter is much less significant these days with a wide variety of different types and makes generally available.

No ear protector working on the basis of an insertion loss between the source of sound and the ear drum can be entirely effective in excluding sound, since sound vibrations are transmitted through the skull by bone conduction as well as through the auditory canal. The wearing of an HPD modifies both the air conduction (AC) sensibility and the bone conduction (BC) sensitivity. In fact, four distinct sound paths are present — Fig 1.

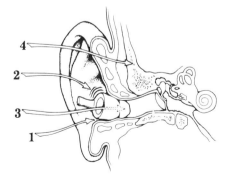

Ear plug

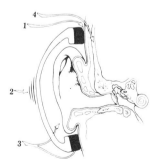

Ear muff

Fig 1 Illustrations of the four paths by which sound reaches the occluded ear. (Cabot Safety Ltd).

1. *Air Leaks.* For maximum protection the device must make a virtual airtight seal with the canal. or the side of the head. Inserts must accurately fit the contours of the ear canal and ear muff cushions must accurately fit the areas surrounding the external ear (pinna). Air leaks can typically reduce attenuation by 5—15 dB over a broad frequency range.
2. *Vibration of the HPD.* Due to the flexibility of the ear canal flesh, ear plugs can vibrate in a piston-like manner within the ear canal. This limits their low frequency attenuation. Likewise an ear muff cannot be attached to the head in a totally rigid manner. Its cup will vibrate against the head as a mass/spring system, with an effective stiffness governed by the flexibility of the muff cushion and the flesh surrounding the ear, as well as the air volume entrapped under the cup. For ear muffs, premoulded inserts and foam inserts these limits of attenuation at 125 Hz are approximately 25, 30 and 40 dB, respectively.
3. *Transmission through the Material of the HPD.* For most inserts this is generally not significant, although with lower attenuation devices such as cotton or glassdown, this path is a factor to be considered. Because of the much larger surface areas involved with ear muffs, sound transmission through the cup material and through the ear muff cushion is significant, and can limit the achievable attenuation at certain frequencies.
4. *Bone Conduction.* Since an HPD is designed effectively to reduce the AC path and not the BC path, BC may become a significant factor for the protected ear. However, the relationship between AC and BC thresholds is not dependent on sound level. Any BC advantage that muffs may have over inserts will be independent of sound level, and will be apparent in a standard threshold level attenuation test.

Field Performance

When an HPD is properly sized, fitted and adjusted for optimum performance on a laboratory subject, air leaks will be minimized and paths 2, 3 and 4 (see Fig 1) will be the primary sound transmission paths. In the work environment, this is usually not the case, and path 1, sound transmission through air leaks, often dominates. Air leaks arise when plugs do not seal properly in the ear canal or muffs do not seal uniformly against the head around the pinna. The causes of poor HPD sealing are:

Comfort. In most situations the better the fit of an HPD, the poorer the comfort. Inserts must be snugly fitted into the canal and ear muff cups must be tightly pressed against the head. This is not conducive to comfort and although some employees may adapt, many will not. This is why it is important to select several hearing protectors (generally one muff and two ear plugs) from the more comfortable available HPDs and to encourage the employee to make the final decision as to which he will use.

Utilization. Due to poor comfort, poor motivation or poor training, or user problems, ear plugs may be improperly inserted and ear muffs may be improperly adjusted.

Fit. All HPDs must be properly fitted when they are initially dispensed. For multi-sized premoulded inserts a suitably sized ear plug must also be selected during this fitting procedure. Companies must stock all available sizes of multi-sized ear plugs and must be willing to use different size plugs for an employee's two ears, this latter situation occurring in perhaps 2—10% of the population. The correct size premoulded insert will always be a compromise between a device that is too large and therefore uncomfortable, and a device that is too small and therefore provides poor protection. The appropriate compromise can sometimes be achieved, but only with care and skill.

Compatibility. Not all HPDs are equally suited for all ear canal and head shapes. Certain head contours cannot be fitted by any available muffs and some ear canals have shapes that can only be fitted with certain inserts or canal caps, if at all. Ear muffs can only work well when their cushions properly seal on the head. Spectacles, sideburns, or long or bush hair underneath cushions will prevent this and will reduce attenuation by varying amounts.

Readjustment. HPDs can work loose or be jarred out of position during the day. It must be remembered that laboratory tests require the subject carefully to adjust a device prior to testing. Under more normal conditions, wearers will eat, talk, move about and may be pumped or jostled, resulting in jaw motion and possible perspiration. These activities can cause muff cushions to break their seal with the head and cause certain inserts to work loose. Premoulded inserts tend to exhibit this problem, whereas custom moulded and expandable foam plugs tend to maintain their position more effectively in the ear canal.

Deterioration. Even when properly used, hearing protectors wear out. Some premoulded plugs shrink and/or harden when continuously exposed to ear canal wax and perspiration. This may occur in as little as three weeks. Flanges can break off and plugs may crack. Custom ear moulds may crack, or the ear canal may gradually change shape with time, so that the moulds no longer fit properly. Ear muff cushions also harden and crack or can become permanently deformed and headbands may lose their tension. Therefore it is important to inspect or reissue 'permanent' HPDs on a regular basis. This may be 2—12 times per year or more, depending upon the HPDs that are utilized.

Abuse. Employees often modify HPDs to improve comfort at the expense of protection. These techniques include springing ear muff headbands to reduce the tension, cutting flanges off of premoulded inserts, drilling holes through plugs or muffs, removing the canal portion of custom ear moulds, or deliberately obtaining undersized HPDs.

Specified Performance

A British Standard for ear muffs is to be published on April 4 1983 with the number BS6344. This will be followed by a kitemark scheme which will be granted to manufacturers conforming to the rigorous requirements of the appropriate certification mark licensing procedures; and also in 1984 by a British Standard for ear plugs, now in committee stage, and a supportive kitemark scheme.

It is recommended that on considering attenuation data, results to BS5108 be used. It will be mandatory in the aforementioned standards for manufacturers to provide results to BS5108.

Ear Plugs

Ear plugs are simple to use, inexpensive and can be (relatively) comfortable, but attenuation can vary widely with different types. Simple cotton wool plugs, for example, produce only low attenuation and are not recommended even if soaked in oil or wax.

Ear plugs can be classified in five groups:
(i) Disposable, mouldable (malleable) ear plugs.

(ii)   Re-usable, prefabricated (moulded) ear plugs.
(iii) Individually moulded ear plugs.
(iv)   Semi-inserts.

i) Disposable ear plugs are usually made from low-cost materials such as wax or glass wool. Typical of this type of disposable plugs are glass wool encapsulated in a polyethylene film wrapper that keeps the shape of the plug. Another type which is cheap enough to be disposable but which can be washed a number of times is the cylindrical shaped foam ear plug. It is rolled between the fingers to as small a diameter as possible and inserted into the ear canal, where it expands and moulds itself to fit the ear canal.

Although disposable ear plugs are sometimes more comfortable to wear than prefabricated plugs and also have the advantage of universal fit, they require a greater standard of cleanliness from the wearer. Such ear plugs should be formed and inserted with clean hands because any dirt or foreign bodies inserted into the ear may cause irritation or infection. This means that disposable and malleable ear plugs should be carefully inserted at the beginning of a work shift and not removed or re-inserted during the work period unless the hands are clean. Thus, this type of ear plug (and to a lesser extent all types of ear plug) may be a poor choice for use in dirty areas having intermittent high noise levels or in other locations where it is necessary to remove and re-insert protective devices during the work periods.

ii) Re-usable ear plugs are generally prefabricated and made from a soft flexible material such as silicone rubber or various combinations of plastics. The flanged type are designed to accommodate to the size and shape of the ear canal and maintain a tight fit.

Prefabricated ear plugs, even when supplied in three or four sizes, may not fit everyone. Emphasis should be placed on careful selection to obtain a correctly fitted earplug.

The chief advantage offered by moulded ear plugs is that the attenuation characteristics can be adjusted in the design, thus offering a more scientific approach to selective attenuation than a simple plug material. Also, of course, moulded ear plugs are permanent and washable, and thus can be used over and over again.

iii) Individually moulded ear plugs are obtained by moulding a plastic solution into a permanent shape within a person's ear canal.

This type of ear plug possesses the advantages of both prefabricated and malleable ear plugs. It is comfortable and usually provides a high degree of protection if correctly made. When fitted with small handles, it can be removed and re-inserted without getting dirty although, as with any type of ear plug, problems of hygiene may still occur unless the plugs are kept in a clean place when not being worn.

A further advantage of individually moulded ear plugs is their greater appeal to the wearer. In situations where difficulty is encountered in persuading men to wear hearing protection, the provision of an individually moulded device that will only fit the person for whom it is intended is a psychological inducement.

Semi-aural Devices

Semi-aurals are basically short ear plugs mounted on a headband. They do not, therefore, rely on 'deep' insertion into the ear to hold them in place. Quite high attenuation is possible with such designs, which can offer some of the advantages of an ear muff, as well as an ear plug. For comfort, the headband need not press directly on to the ear insert, but on to a secondary spring. Semi-inserts are usually less comfortable to wear than ear plugs, however, but are lighter and less obtrusive than ear muffs. They have the advantage that they can be readily combined with deaf-aid type ear-phones to form a light headset for telephonic speech communication.

Semi-insert protectors have the advantage that one size will fit the majority of ears, unlike prefabricated plugs. As they are captive and may be re-inserted hygienically at any time, they are suitable for industries such as the food industry, where the loss of an ear plug must be avoided, and for people who must frequently enter noisy environments for short periods, although other types of ear plug are equally appropriate in this context.

Semi-inserts are not as comfortable as ear plugs since a relatively high pressure has to be exerted over a small area to achieve satisfactory performance.

Selective Ear Plugs

Amplitude sensitive protectors are available which attempt to reduce loud sounds more than quiet sounds. Ear plugs have been developed for use against gunfire noise, which may be useful for other explosive sounds (*eg* cartridge operated tools). However evidence to date suggests these are not suitable for most industrial noises.

Ear Muffs

The ear muff provides a complete 'cushion' entirely surrounding and isolating each ear. A headband is necessary to hold the muffs in place, but the shape is far less critical than with ear plugs or semi-inserts and one size can readily be adjusted to fit virtually any person. They are essentially bulky devices, and considerably more costly than ear plugs, although not necessarily uncomofrtable to wear. The main disadvantage in the latter respect is usually overheating of the skin because of the large surface area of contact between the compliant sealing surface of the muff and the skin.

Most types of ear muff are of similar design and are made from rigid cups specially designed to cover the external ear completely. They are held against the sides of the head by a spring-loaded adjustable band and are sealed to the head with soft circumaural cushions.

Individual designs vary in detail. Ideally the muff should be of rigid material enclosing a large air volume in order to achieve high attenuation. In order to adapt to the controus of the head around the ears (and for comofrt), however, this part of the muff, at least, must be soft and compliant, which can considerably decrease the degree of attenuation achieved. Thus invidiual designs vary in detail, although the most favoured form is the use of a hard plastic shell sealed to the contours of the head with a soft foam rubber or foam plastic cushion. The enclosed air volume, which may range from as low as 25 cm^3 to 300 cm^3, can also be partially filled with absorbent foam plastic, or the elastic section be filled with a soft wax or viscous fluid. Typical performance achieved is shown in Fig 4.

For maximum attenuation of sound, protector cups should be made from a rigid, dense, non-porous material. The volume enclosed with the muff is proportional to the attenuation provided at low frequencies. Each cup should be partially filled with an absorbent material to reduce the high frequency resonances that may otherwise occur within the shell.

Ear muff seals may be liquid-filled or plastic foam-filled. Liquid-filled seals usually provide marginally better protection with only slight headband tension, all things being equal, but suffer from the additional problem of leakage of fluid if treated roughly. Modern foam-filled seals are almost as good as the liquid seals and have the additional advantage of robustness. These seals should have a small hole in the skin to allow them to distort to the shape of the side of the head. They usually require slightly higher headband pressures to provide a satisfactory seal.

In general, the degree of attenuation achieved with simple ear muffs can be expected to increase with an increase in enclosed air volume, and with the highest attenuation obtained at the higher frequencies.

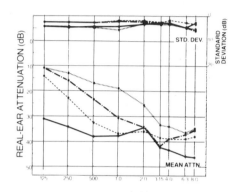

Fig 2 Comparison in performance of two ear plugs and two ear muffs. *(Cabot Safety Ltd).*

A comparison between the performance of two ear plugs and two ear muffs is given in Fig 2.

The attenuation provided by ear muffs is related to the force with which they are pressed against the sides of the head. Maintenance of the correct headband pressure is therefore important and care must be taken that this is not reduced by deliberately bending the headband so that it is more comfortable. The suspension force is usually chosen by the manufacturer as a compromise between performance and comfort and this should not be altered.

The disadvantages of ear muffs lie in their bulkiness, initial cost and the fact that they tend to make the ears hot. However, their bulk has the advantage that it can easily be seen whether they are being worn correctly. They are also usually more susceptible to damage than other forms of hearing protection.

Ear muffs are available incorporating an electronic microphone loudspeaker system which will transmit low level sound, but in which, an electronic circuit, stops high level sounds so the basic ear muff then provides the attenuation. These muffs are considerably more expensive than normal types and must be handled with more care.

Helmets

A protective helmet normally consists of a shell of reinforced glass fibre lined with sound proofing material formed into a close fitting seal around the face and neck, together with internally fitted ear pads or ear muffs. The total sound proofing is very little better than that possible with ear muffs alone, so the helmet itself would appear to offer very little attenuation. Its particular value as a protector is based on its ability to protect the skull and other more vulnerable soft areas of the head, against transmission of damaging pressure waves acting through bone conduction. The use of a full helmet, therefore, is commonly specified where ultra-high noise levels are encountered. Under such conditions a helmet may be integrated with a sound-excluding suit for additional body protection.

See also chapter on *Hearing Conservation in Industry.*

TABLE I – BRITISH STANDARD ATTENUATION DATA FOR EAR PROTECTORS*
Figures are the assumed protection *ie* the mean values minus one standard deviation
of a series of test results measured in decibels

Ear Protectors	Octave Band Centre Frequency (Hz)								Test Date
	63	125	250	500	1K	2K	4K	8K	
Ear Muffs									
Bilsom Viking 2318	9.7	9.6	16.4	27.6	29.1	31.5	31.5	26.6	1980
Bilsom Red 2135 (Foam)	6.2	5.8	13.7	20.9	29.9	29.1	32.7	29.2	1980
Bilsom Red 2135 (Fluid)	5.9	7.4	12.2	18.2	26.4	31.2	27.8	22.5	1980
Bilsom Yellow 2301	9.1	9.0	14.6	18.6	27.5	31.9	33.6	24.4	1977
Bilsom Blue 2308	5.8	4.3	10.3	17.0	24.7	26.7	31.1	24.4	1977
Bilsom Marksman 2316	7.6	7.9	9.5	17.9	27.9	28.3	33.3	38.9	1977
BAO 112000	5.8	2.9	9.4	16.4	26.9	29.4	31.1	25.3	1979
BAO 117000	4.9	2.9	6.6	14.0	23.6	26.0	26.8	20.2	1979
BAO 116751	5.7	1.1	9.7	16.9	24.0	29.3	29.5	25.4	1979
BAO 117200	6.5	4.7	11.7	18.8	27.7	29.2	27.2	23.0	1979
BAO 116750	5.9	2.0	4.8	11.9	18.2	21.3	23.6	25.5	1977
Centurion ED Type A		8.1	9.3	16.1	22.6	27.9	32.4	25.5	1977
Centurion ED Type B	5.1	4.3	10.0	17.0	23.9	27.0	26.5	24.1	1978
Johnstones "J" Muffs	5.7	3.1	10.5	18.1	24.0	26.1	31.1	27.7	1977
MSA Noise Foe MK IV	2.8	1.3	8.3	16.9	24.8	26.6	26.8	26.4	1980
MSA Radaunix	1.6	0.6	6.1	12.1	18.2	24.7	26.1	28.0	1980
MSA Comfo 500	4.7	4.0	7.5	8.9	18.1	25.8	34.5	30.3	1977
Norths Saturn	5.0	4.0	8.5	16.9	24.4	25.7	26.1	21.5	1979
Peltor FH9A	4.7	4.4	10.4	19.7	28.2	29.0	29.4	24.9	1977
Peltor FH9B	5.8	4.1	8.3	17.9	25.7	25.3	26.2	19.3	1977
Peltor FH7A	6.5	4.1	14.8	22.1	29.2	29.2	27.1	27.0	1977
Peltor FH6A	4.6	2.9	5.3	15.0	24.3	26.5	29.2	23.2	1977
Peltor FH8H	2.6	1.4	4.1	13.0	22.6	21.0	28.6	28.4	1977
Protector Safety EMU 44	4.8	4.3	13.4	19.1	27.5	26.7	28.4	26.0	1977
Protector Safety EM 62	4.8	3.3	9.5	16.3	22.9	24.5	30.4	31.2	1976
Protector Safety EMLU 47	6.0	4.3	8.5	16.6	25.1	29.0	30.5	23.3	1976
Protector Safety EMLU 60	1.9	1.7	7.6	14.3	21.6	23.8	25.8	23.3	1977
Protector Safety EMLF 48	2.1	1.0	1.7	3.5	12.5	15.0	22.6	19.1	1976
Racal Sonoguard	11.4	13.7	19.5	27.6	35.8	33.5	35.6	26.1	1977
Racal Sonomuff	12.7	14.5	16.7	23.0	31.8	31.9	35.3	31.1	1981
Racal Sonomuff (B.T.H.)	11.5	13.0	16.0	22.9	31.2	31.8	32.4	27.3	1981
Racal Auralgard 3	9.1	9.1	13.0	19.8	28.7	31.2	35.6	24.9	1977
Racal Ultramuff 2	5.3	4.7	11.2	19.4	26.4	27.2	28.5	23.5	1977
Itex 212 Mk 2	9.5	8.3	9.3	16.4	25.0	29.0	34.3	26.9	1977
Safir Standard ED/2SL (Fluid)		6.8	11.1	17.2	24.9	28.2	32.3	27.3	1976

*A Champion (Compiled for the Steel Castings & Trade Association)   cont...

TABLE I – BRITISH STANDARD ATTENUATION DATA FOR EAR PROTECTORS* (contd.)

Ear Protectors	Octave Band Centre Frequency (Hz)								Test Date
	63	125	250	500	1K	2K	4K	8K	
Ear Muffs (Contd)...									
Safir Standard ED/1SF (Foam)		3.5	9.0	17.6	24.3	27.5	32.9	28.3	1976
Safir Junior S4/JL (Fluid)		5.1	6.0	13.7	23.4	28.3	33.9	26.5	1976
Safir Junior S4/J2 (Foam)		2.9	5.0	13.3	21.4	25.9	32.7	28.5	1976
Safir Coronet C6/J2	4.8	2.7	3.7	9.2	18.4	22.5	28.8	20.2	1979
Safir Trojan ST/21F	5.0	3.2	6.0	11.2	19.8	22.8	25.8	18.4	1979
Silenta Super		16.5	21.9	23.5	33.5	31.2	33.7	29.5	1981
Silenta Universal		5.3	10.0	17.6	27.2	27.5	33.3	23.1	1976
Silenta Mil		6.9	7.5	13.1	23.0	26.0	35.7	29.8	1981
Silenta Bel		4.1	7.9	15.4	26.2	27.9	33.3	25.6	1981
Silenta Pop		8.5	12.5	17.6	27.3	29.4	35.9	29.8	1981
Windsor	5.7	3.1	10.5	18.1	24.0	26.1	30.8	26.3	1980
Ear Muffs/Helmet/Visor									
Bilsom Viking (2314)		9.5	16.3	27.1	31.9	30.8	38.3	33.9	1981
BAO 17761 BX21	0.9	1.3	3.0	11.9	22.1	23.8	22.3	19.0	1979
BAO 1720	0.3	−2.8	0.3	10.1	20.0	20.2	20.2	18.1	1979
Centurion ED Type C	4.9	2.1	6.2	14.3	21.1	24.6	27.4	22.8	1978
MSA Comfo 600/Super V Guard	5.1	3.4	8.8	13.3	16.4	23.4	33.6	26.9	1980
Peltor FH 3PB	−0.5	−1.1	−0.4	11.5	20.4	24.9	26.9	18.7	1977
Protector Safety EMCC 50	0.9	0.4	1.6	5.1	14.1	19.4	21.3	12.8	1977
Special Headset									
PEL/Seawell Musical	8.3	7.2	12.9	22.5	25.7	27.7	30.6	26.9	1979
PEL/Seawell Communicator	4.6	7.8	12.5	21.6	23.3	28.6	30.2	26.2	1980
Disposable Ear Plugs									
Bilsom Soft		9.7	12.4	15.6	17.5	26.2	31.1	26.2	1981
Bilsom Propp-o-Plast	7.5	9.3	10.6	12.0	13.3	20.2	24.9	28.6	1977
Bilsom Propp	1.6	2.6	5.6	7.9	12.0	17.3	21.2	29.0	1977
E-A-R Plugs	17.5	18.3	18.9	21.9	24.5	27.9	38.6	38.9	1977
Racal DBA	14.1	13.7	13.6	14.2	17.2	25.5	35.4	29.2	1982
Safir Deci-Damp	12.7	13.0	12.3	15.2	18.9	24.8	32.2	28.7	1978
3M Ear Plug 8773	10.4	10.9	10.5	12.1	12.6	15.6	16.3	27.7	1977
Reusable Plugs									
Hear Guard Ear Insert	14.2	15.2	14.5	13.8	16.0	22.7	25.6	23.6	1979
Norton Comfit	14.9	14.9	15.4	16.6	16.7	20.8	21.9	28.6	1977
Personally Moulded Ear Plugs									
Custom Protectors	6.3	6.6	6.6	8.4	10.1	20.8	29.5	25.2	1977

SECTION 2b

Vibration - Effect on People

THE HUMAN threshold of perception of vibration is of the order of 3–5 Hz or higher, depending on how the vibration is sensed and the amplitude of vibration present. Thus the fingertip is probably the most sensitive sensor, but relatively insensitive as regards the *frequency* of vibration felt. Particularly, the fingertips are sensitive to vibration *amplitude* and can indicate when vibration amplitude is reaching what appears to be alarming levels — see Fig 1.

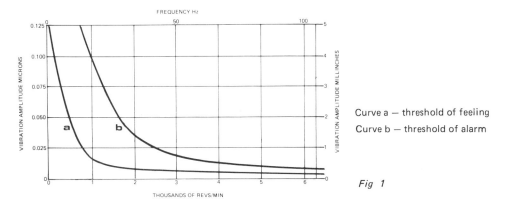

Curve a — threshold of feeling
Curve b — threshold of alarm

Fig 1

Vibrations generated at frequencies within the audio frequency range are heard as noise. Their effect may be further emphasized by the fact that vibrations are readily structural borne and thus may be transmitted for considerable distances through inter-connected (non-isolated) structures, with the distinct possibility of exciting resonance at one or more points along the transmitted path. Vibration can thus be a primary source of noise, and a source of excitation of a secondary generator of noise. The latter can be more disturbing or annoying than the original source.

Effects of Infrasound

Infrasound is very low frequency vibrations below the audio range (*eg* below about 20 Hz), although in analysis it is generally assumed to cover the frequency range 1 to 100 Hz. Individual response to infrasound varies considerably, those with acute audio sensitivity being most affected. Symptoms include nausea, irritation, headache, palpitation and ringing in the ears.

More commonly the effect of infrasound is shown up by sympathetic (resonant) vibrations set up in fixtures and fittings, including sliding doors and windows; and even loosening of roof tiles.

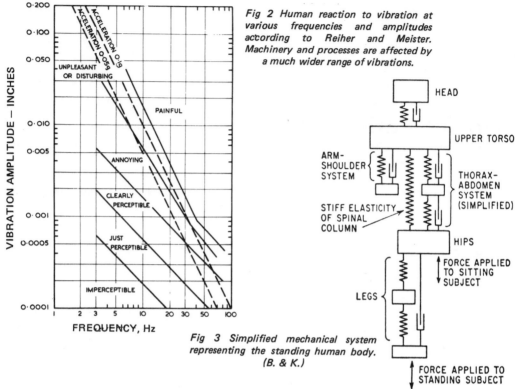

Fig 2 Human reaction to vibration at various frequencies and amplitudes according to Reiher and Meister. Machinery and processes are affected by a much wider range of vibrations.

Fig 3 Simplified mechanical system representing the standing human body. (B. & K.)

Response to Vibration

Any device which is basically a vibration generator, or any mass which is liable to be excited into forced vibration, is thus an actual or potential source of noise and its suppression may require particular treatment. Typical overall reactions are shown in Fig 2, but the subjective response to vibrations also varies greatly among individuals, where a tolerance to vibration may be built up over a period of exposure. However, continual exposure to vibration is known to be fatiguing, and even short exposure to vibratory accelerations in excess of 1g can cause severe discomfort.

Vibration frequency would appear to play an important part in the latter case, tolerance being lowest in the 5–20 Hz range, with the tolerable g force increasing with increasing frequency. This rather general statement is modified by the fact that the mechanical effect of vibration is considerably modified by the attitude or position of the subject, and also the manner and area in which he receives the vibration. There is, however, a frequency range of 40–80 Hz where vibration may directly affect vision, due, it is thought, to resonant response of the eyeball and loss of acuity.

Human Figure Models

A more complete picture is provided by the human figure model devised by Reiher and Meister. Here the human body is represented by a mechanical sprung system of masses, with damping. At low frequencies, and low vibration levels, a simplified mechanical system takes the form shown in Fig 3. The most significant part of this system is the thorax-abdomen system which develops a distinct resonance in the 3–6 Hz range and thus makes effective vibration isolation of a sitting or

VIBRATION — EFFECT ON PEOPLE

standing human very dfficult to achieve. Individual perception and tolerance levels may well differ appreciably. Also subjective response is dependent on the attitude of the subject (*eg* sitting or standing) and orientation with respect to the g force axis of the vibration, as well as the part of the body in contact with the vibration (fingertips, hands, feet).

Typical transmissibilities of vertical vibrations are shown in Figs 4 and 5 for standing and sitting subjects respectively.

Apart from the 3—6 Hz resonant range for the thorax-abdomen system, there is a further resonance in the 20—30 Hz range in the head-neck-shoulder system, eyeball resonance in the region 60—90 Hz, and lower jaw-skull resonance in the region of 100 Hz and above. At higher frequencies, analysis in terms of a simple mechanical equivalent model is no longer satisfactory and impedance modelling is to be preferred.

Another significant parameter is the ability of the human body to attenuate vibration. This has been analyzed in some detail by von Bekesey and others. The results of von Bekesey's measurements are shown in Fig 6 for the attenuation of 50 Hz vibration. Typical human body attenuation at this frequency is 30 dB from foot to head, and 40 dB from hand to head.

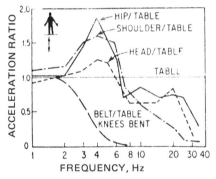

Fig 4 Transmissibility of vertical vibration from table to various parts of the body of a standing human subject as a function of frequency.
(B. & K.)

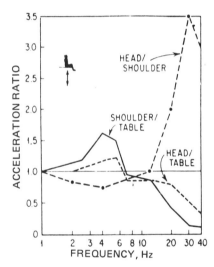

Fig 5 Transmissibility of vertical vibration from table to various parts of a seated human subject as a function of frequency.
(B. & K.)

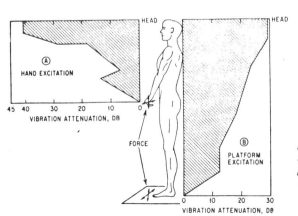

Fig 6 Attenuation of vibration at 50 Hz along human body. The attenuation is expressed in decibels below values at the point of excitation. For excitation of (A) hand, and (B) platform on which subject stands.
(After von Bekesy).

Damage from Vibration

Very little empirical information is available on the damaging, or potentially damaging effects of sub-audio frequency vibrations, except for hand damage resulting from the continual manipulation of hand-held vibration tools, which is well documented. Even here the actual cause and the exact nature of physiological damage are not clearly established — only symptoms — although it would appear that the greatest hazard arises from impacting type reaction shocks, such as are generated by heavy percussive tools, rather than exposure to steady vibrations of higher frequencies. The actual weight of a hand tool also plays a significant part. Hand damage, temporary or permanent, is far less likely to occur with the continual use of light high-speed tools.

At ultra-sonic frequencies, vibration can produce thermal as well as mechanical effects, the latter being a direct conversion of the vibrational energy absorbed. Thermal heating is generally insignificant in man, however, because the amount of energy involved is readily dissipated and can be disregarded as a hazard in the case of air-borne ultrasonic frequencies. Equally, the mechanical effects appear negligible with ultrasonic air-borne frequencies, except at power levels well beyond values obtainable in practice. The main 'damage' region, therefore, lies in the audio range of frequencies and infra-sound levels where quite high power levels may be achieved with the effects discussed elsewhere under noise — *ie* mainly affecting the ear, but also potentially representing a bodily hazard (mechanical effects) at sound pressure levels of, say, 120 dB and above.

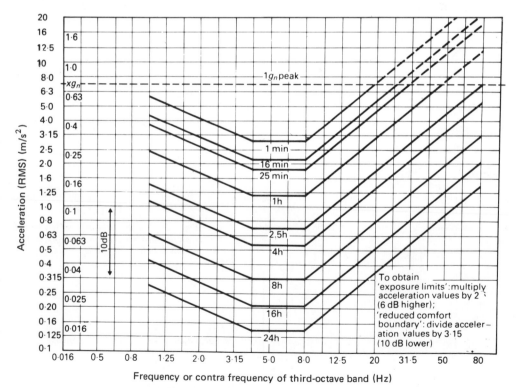

Fig 7 Vertical (head-to-toe) acceleration limits as a function of frequency and exposure time: fatigue-decreased proficiency boundary. (ISO 2631).

White Finger

The classic disease suffered by operators subject to prolonged exposure to heavy vibrations transmitted through the hands from pneumatic hammers, chipping tools, chain saws and similar vibratory tools is 'dead finger' or constitutional cold finger, known as 'vibration-induced white finger' or VWF. Additional effects may include neuritis, and damage to joints, bones or muscles. The particular problem of introducing recommendations or standards limiting vibration levels and exposure times is aggravated by the fact that the cause-effect relationship is by no means adequately investigated and established. Yet VWF can also affect all operators using hand-held tools or working at hand-held continuous mechanical processes.

Where continuous exposure does not amount to more than 2½ hours per day, then these limits can be increased by a factor of 10. If continuous exposure is less than ½ hour daily, or 2½ hours of interrupted exposure, then these limits need not apply.

Vibration exposure criteria defined in ISO 2631 are shown in Fig 7, covering the vibration frequency range 1–80 Hz. Vibration levels are given in terms of RMS acceleration levels which produce equal fatigue effects, or represent 'fatigue-decreased proficiency' boundaries. Corresponding levels considered a hazard to health would be twice as high (*ie* 6 dB up). Corresponding levels for 'reduced comfort' would be one third of these levels (10 dB lower).

SECTION 3a

SECTION 5

Noise Measuring Techniques

COMMONLY MEASUREMENTS taken outdoors using a simple hand-held sound level meter pointed at the noise source are assumed to correspond to free field conditions, which may be far from true (even body reflections may be present unless the meter is held at arm's length). Equally, wind noise errors may be ignored, the microphone diaphragm may not be clean, the battery level may be low (particularly at near-freezing temperatures), and the instrument itself may be in need of recalibration. Basically, therefore, measurement of sound levels is not a simple 'point and read' technique. However good the technical specification of the instrument, its performance is only as good as the person using it.

Outdoor measurement may also be affected by wind developing a turbulent airstream around the microphone, giving a false high reading. If necessary this can be avoided by fitting a windshield, although in a number of meter designs satisfactory shielding may be provided by the microphone nose cone itself. If the microphone used has omni-directional characteristics, the presence of wind noise can be detected by rotating the meter through 90 to 180 degrees and observing the effect on the meter reading.

In the case of measurements taken out of doors and in the absence of nearby reflecting surfaces, a single measurement taken in the 'far field' will then define the noise along that particular path of alignment with the noise source (*ie* the noise level reading will be independent of the shape and size of the source). The effect of distance along that particular path will be defined by the inverse square law. Where the source has directional characteristics, however, several individual readings will be necessary to define the noise level along different paths, or establish noise levels at different points.

In practical terms the 'free field' can be taken as existing at any point more than three times the longest dimension of the noise source away from that source. At a closer distance (*ie* in the 'near field') the actual propagation pattern of noise will be dependent on the size and shape of the source. Any readings taken in this region are valid only for that particular spot.

Whilst a single measurement taken at a point in the far field establishes the noise level in that 'line of sight' (as well as at a particular point on a site) in free field conditions, many more readings are normally required for measurement of noise in a diffuse or semi-reverberant field (*eg* indoors). This is because the noise level at any point is enhanced by reflections, the strength of which will vary at different points. Similar considerations apply for measurement of machine noise, *etc*, where the noise can be highly directional even in a free field, and further modified by reflections in a semi-reverberant field.

Various Codes and Standards, and other guidelines lay down specific positions and distances at which noise level measurements are to be taken in such cases. In the absence of such guidelines, the following general rules can be applied:

Measurement of Machine Noise — see Fig 1.

(i) Divide the space around the machine into four quadrants centred on the centre of an industrial machine, or the coupling in the case of a machine with a driver.

(ii) Sound level readings are then taken, one in each quadrant, holding the microphone at a height of approximately 1.52 m (5 ft) above the floor and a distance of approximately 1 m (3 ft) from the machine surface.

(iii) In each quadrant, traverse the area swept by the 1 m (3 ft) arc to find the highest noise level reading.

(iv) Move the microphone around within an approximate 1 ft^3 volume at this position and note the variation in noise level. If the variation is less than 3 dB, this position for the microphone is satisfactory. However, if variations are greater than 3 dB, the microphone is probably in the near field of the machine. In this case, the microphone should be moved to a distance of, say, 1.52 m (5 ft) from the machine and the position of the highest noise and re-established. If the noise level still varies significantly with position, the microphone may still be in the near field or in the reverberant field. Where large variations persist, continue to seek a microphone position where variations are less than 3 dB. Make sure that the microphone is greater than 1 m (3 ft) from reflecting surfaces, such as walls or adjacent machines which can cause deflections.

(v) Repeat the above procedure to find suitable microphone positions in quadrants 2, 3 and 4.

(vi) Sound measurements should also be taken at the machine operator stations.

The microphone should be located at the approximate position of the operator's ear.

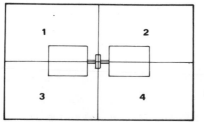

Fig 1

To check that 'far field' conditions really do exist at the measurement stations, use can be made of the inverse-distance law, *ie* by doubling the distance to the source the sound pressure level should increase by 6 dB. (This distance should actually be measured from the acoustic centre of the source for the frequency band in question. For general purposes, however, it may be sufficiently accurate to measure the distance from the surface of the source).

'Far field' conditions exist roughly at distances of more than one wavelength away from the source, or two to three times the largest linear dimensions that the machine will suffice.

From the free field measurements over a hemisphere as described previously, it is possible to estimate the sound power radiated from the noise source as well as its *directivity index*. (See chapter *Measurement of Sound*).

General Measurements in Rooms

The space in which the measurement is conducted can have a significant effect on the readings obtained. Thus the effects of the acoustical environment are measured with the sound pressure level of the noise source, unless the room is completely anechoic. The only 'artificial' object present is the observer himself (and the meter case or hand held instrument). It is important, therefore, that the observer stands in a position which offers minimum shielding of the microphone. In the case of a hand held instrument, it is generally recommended that the microphone be held well in front of the observer, with the noise source located to one side.

Most microphones used for sound measurement are omni-directional, although this may only apply as far as low frequency sounds are concerned. At frequencies high enough for their wavelength to be comparable to the size of the microphone, considerable directional effects may be present and response will vary with the direction in which the microphone is pointed. As a general rule the microphone should be positioned so that the response to incident sound is as uniform as possible.

It may also be desirable to explore the sound field fairly thoroughly before deciding on the optimum position for the microphone, unless a large number of readings is to be taken in any case, to determine the actual sound field at specific distances. The subject of sound fields is dealt with in another chapter.

Measurement of Average Noise

Although the sound level meter gives averaged readings, these are averages only for the short time constants are involved (0.5 seconds or less). It is now generally recognized that both the subjective assessment of nuisance and the risk of noise-induced deafness are dependent on much longer term noise averages which may range from seconds to hours or even days. The parameter most favoured to express this relationship is equivalent continuous noise level (L_{eq}). Whilst this can be obtained from summation of steady level measurements taken over specific time intervals or spot samples of noise level, such data do not necessarily give a true measure of the actual sound energy average concerned.

Ideally, to measure long term noise averages, the instrument used should have:

(i) A wide enough dynamic range to ensure that large impulses are not 'clipped' and thus sound energy under-recorded.

(ii) A non-sampling continuous integration mode of operation so that all the sound energy is recorded, and thus a true average obtained.

The significance of (i) is shown by a simple example. A level of 130 dB lasting for only three seconds is equivalent in energy content to one of 90 dB lasting for eight hours. Thus, if the dynamic range of the meter results in 'clipping' of levels above 90 dB, say, many serious errors could occur in the measured noise average. As an alternative to (ii), a fast sampling rate could be used to minimize errors likely to occur due to short impulses. This is open to the same objection as above if an impulse is 'missed', but a fast time constant and a sampling rate faster than 125 milliseconds will help minimize errors under impulsive sound conditions.

Average Noise Levels – Offices

Basically the aim should be to take individual readings at operative positions, and also at other points more or less equally spaced throughout the room. Thus Fig 2 shows suggested measurement positions in a typical typing pool. A reasonably accurate average level can be obtained by taking the average of the dB(A) values measured with a portable sound level meter.

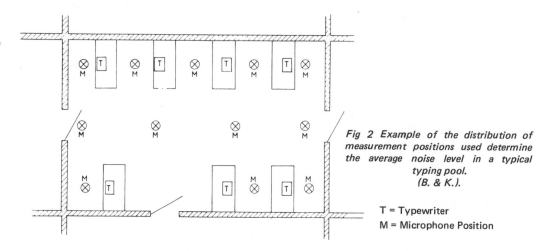

Fig 2 Example of the distribution of measurement positions used determine the average noise level in a typical typing pool. (B. & K.).

T = Typewriter
M = Microphone Position

General Measurement

As a general rule the instrument should be held at arm's length sideways, with the microphone pointed away from the noise source, *ie* with the sound impinging on the microphone at grazing incidence (90 degrees). However, this will depend on the type of microphone. Some may need to be pointed directly at the noise source for free field measurement. Requirements in this respect will be specified in the manufacturer's instructions. Errors as large as ± 6 dB may occur through bad positioning of a hand held sound level meter, due mainly to the presence of the operator. For most accurate results the meter (or separate microphone) should be clamped in position, and operator and any other instrumentation moved a short distance away.

Using hand-held noise meter fitted with microphone windshield. (GenRad Ltd).

Measurement of Impulsive Noise

Conventional sound level meters have distinct limitations for the measurement of transient noise or impulsive sounds because of inherent 'lag', although this would be apparent on a suitable readout device, such as an oscilloscope. This 'lag' may become more apparent when the meter is used with an analyzer, particularly a narrow band analyzer.

To overcome this, particularly for field use, a modified circuit can be employed, where response time may be reduced to the order of one-ten-thousandth of a second. The circuit can be designed to measure three separate characteristics — peak, instantaneous level and average level. Thus, the time decay constant of the impulse noise can be estimated with good accuracy.

Impulse sound pressure level is given by the formula:

$$L_{p_1} = 20 \log \frac{P_1}{P_0}$$

where P_1 = sound pressure
P_0 = reference sound pressure

Impulse sound may be measured by an unweighted meter; or with A—, B—, C—, or D— weightings. In the latter case, weighted measurements are designated as follows:

$$L_{A_1} = 20 \log \frac{P_A}{P_0} \quad dB(A1)$$

$$L_{B_1} = 20 \log \frac{P_B}{P_0} \quad dB(B1)$$

$$L_{C_1} = 20 \log \frac{P_C}{P_0} \quad dB(C1)$$

$$L_{D_1} = 20 \log \frac{P_D}{P_0} \quad dB(D1)$$

Sound level meters, other than special types, may have a decay time constant of the same order as, or greater than, the impulse noise decay time. In that case, if used for noise analysis in conjunction with a suitable read-out device, they may indicate the decay time of the analyzer and not of the transient noise.

One basic problem concerned with subjective assessment of impulsive noise measurement is that the time constant of the human ear is appreciably different from that of impulse sound level meters conforming to the IEC recommendations and that a conventional sound level meter with a high overload capacity and a detector capable of giving good results with high crest factor can give equally realistic results. Used in conjunction with a noise analyzer, this can enable physical peak intensities to be measured, displayed and recorded, and also provide 'quasi-peak' and 'true average' characteristics from which the form of the impulsive wave can be estimated. It is the form of the impulsive wave expressed in terms of peak level and duration which is of major significance as regards damage-risk.

Measurement of Sound Reduction Index

The sound reduction index (R) is a measure of the ability of a wall or panel to prevent sound passing through it. It is defined specifically as:

$$R = 10 \log (1/\tau) \text{ dB}$$

where τ is the transmission coefficient or the ratio of acoustic energy transmitted through the panel to the total acoustic energy incident on it.

The sound reduction index of a panel is normally determined by laboratory tests involving the use of a *source* room and *receiving* room both with high reflectance (*ie* reverberant rooms) with a

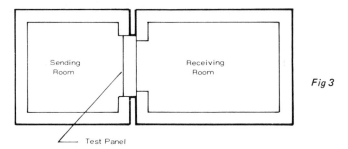

Fig 3

common cavity wall having an opening in which a test panel can be fitted — see Fig 3. Sound is generated in the source room at a measured average sound pressure level L1, and the corresponding average sound pressure level in the receiving room measured as L2. The sound reduction index is then given by:

$$R = L1 - L2 + 10 \log S - 10 \log A$$

where S = area of the test panel in m^2
A = the absorption in the receiving room in m^2 units

The sound reduction index so described can be measured as an *average* using a broad band sound source in the source room; or in terms of redundancy values measured at standing centre frequencies from 100 Hz upwards, using 1/3 octave analysis of L1 and L2. It should also be noted that the empirical sound reduction index measured in this way is specific not only to the construction of the test panel but also the way in which it is mounted.

Measurement of Absorption Coefficient

The sound absorption coefficient of a material may be measured by a standing wave tube in the laboratory, or in a reverberation room. The latter is the preferred method and the procedure for measurement is given in BS3635 : 1963.

Briefly this involves mounting a large sample (at least 10 m^2) of the material in a reverberation room and measuring the reverberation time. The random incidence absorption coefficient (α) is then given by:

$$\alpha = 0.161 \frac{V}{S} \left(\frac{1}{T1} - \frac{1}{T2} \right)$$

where V is the volume of the reverberation room, m^3
S is the area of the sample, m^2
T1 is the reverberation time with the sample in position
T2 is the reverberation time without the sample in position

See also chapter on *Sound Level Meters.*

Measurement of Sound Intensity

Sound intensity, which is a vector quantity embracing both the magnitude and direction of the flow of sound energy at a given point, can be measured in terms of sound pressure and particle velocity in a single direction. However, due to practical difficulties which arise in measuring

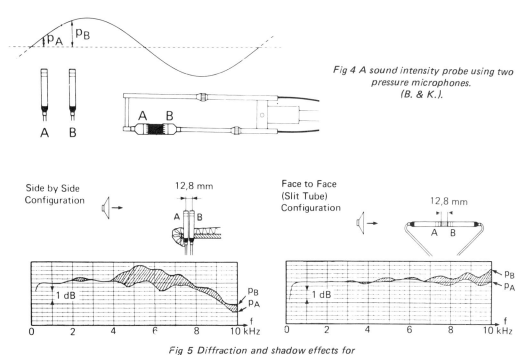

Fig 4 A sound intensity probe using two pressure microphones. (B. & K.).

Fig 5 Diffraction and shadow effects for side by side and face to face microphone configurations. (B. & K.).

particle velocity directly a suitable compromise is to derive this from two closely spaced sound pressure measurements using two pressure microphones. The pressure gradient then follows by dividing the pressure difference by the separation between the two points — Fig 4.

The configuration of the microphones is important, because each microphone will influence the sound field seen by the other. Ideally, two microphones placed in the same sound field should register the same sound pressure level. However, diffraction and shadow effects will always cause the first microphone to influence the pressure seen by the second. Fig 5 illustrates this. Two closely spaced microphones are being excited by a plane wave in an anechoic chamber. They are placed side by side with a separation of 12.8 mm, and both should register the same pressure. However, due to the influence microphone A has on the sound field seen by microphone B, differences of up to 2.5 dB occur in the responses of the two microphones. Placing the microphones face to face produces an improvement. The design is more symmetrical, and the use of a slit tube between the microphones prevents diffraction around the edges of the microphone grids. Even so, though, there is up to 1 dB difference between the pressures registered by the two microphones.

Fig 6 illustrates a further configuration, where the two microphones are used face to face but with a solid space in between them. Similar measurements to those described in the previous paragraph now indicate that the two microphones see the same sound pressure to within 0.5 dB. Also, the responses of the two microphones are almost flat.

The face to face configuration with a solid spacer also shows the best controlled effective separation of the microphones. The maximum in the response is quite broad, and it is possible to be as much as ±20% off the direction of the intensity vector, (or more at lower frequencies), before any significant errors are introduced. The mimimum in the response, on the other hand, is very sharp, and is located in the plane midway between the two microphone diaphragms and parallel to them. Because of its sharpness, it is the minimum response of the probe and the associated change in the sign of the intensity which is used in locating the direction of the sound intensity vector. This is especially important in noise source location work, where the probe can be used as an acoustic 'pointer' to indicate the direction of a noise source.

Frequency Range of Sound Intensity Measurements

Measurement of intensity using two closely spaced microphones introduces some errors not normally encountered in sound pressure measurements. These errors are important, because they effectively determine the frequency response of the probe. At high frequency, they are a function of the spacing used between the microphones, while at low frequency, they are a function of the phase mismatch between the two measurement channels and, (indirectly) the microphone spacing.

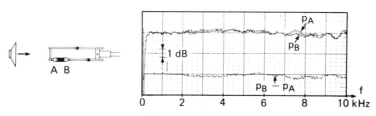

Fig 6 Diffraction and shadow effects for face to face microphone configuration with a solid spacer. (B. & K.).

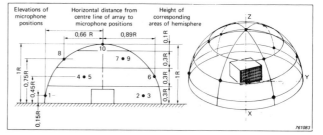

Fig 7 The measurement of sound power with a hemisphere enclosing the sound source and the microphone positions according to ISO3745.

Practical Measurement

In a practical measurement, the procedure is first to set up a suitable test surface around the object, (Fig 7). Although the shape of the surface is completely arbitrary, it is usual to make it a fairly simple geometrical shape since its surface area must be known. Also, in order to take account account of the directional characteristics of the source, the test surface will be divided up into a number of equal areas, and a measurement made in each area, (note that even for a relatively simple noise source, if the intensity is measured or scanned around the source, and the difference between the maximum and the minimum is of the order of 10 dB, then the test surface must be split into at least 16 equal areas). The intensity passing through each area is then measured at the geometric centre of each area or by sweeping the probe across the area, (note that in either case, the probe should always be kept perpendicular to the test surface), the results being averaged together to give a final figure for the average intensity, from which the sound power can be calculated. Measurement of the aforementioned nature is independent of the background noise. In practice, it has been found that the background noise can be 10–15 dB higher than the noise from the measured object, before significant errors are introduced.

Sound Level Meters

BASIC ELEMENTS of a simple sound level meter are a non-directional microphone coupled to an amplifier; some method of rectification to provide a d.c. output together with a time constant to extract a steady level for meter indication in the form of the root mean square of the 'averaged' signal. Usually a choice of two time constants is provided — Fast (125–200 ms) and Slow (500 ms). Also filters to modify the strength of signals passed by specific frequency bands to 'correct' objective measurement to subjective response, A-weighting being universal. Additionally all meters normally incorporate an attenuator to adapt the instrument for measuring noise over a wide range of levels in 'steps' so that full scale meter range is available for each step. The attenuator is usually arranged in steps of 10 dB to simplify calibration.

Such an instrument, with A-weighting, can be simple and compact, and suitable for most types of single number noise level measurements for general noise abatement purposes, *etc*, particularly in the hands of semi-skilled persons. The same form of instrument can, however, be produced in various grades to meet specific standards for performance requirements, specifically the new IEC651 standard which stipulates requirements for sound level meters and classifies them in four grades of accuracy — see Table I.

TABLE I – GRADES AND STANDARDS FOR SOUND LEVEL METERS

Grade	Type	Equivalent Standards	Remarks
0	Laboratory		most precise
1	Pressure	BS4197:DIN45633: ANSI1:AFNOR31009	original International Standard IEC179
2	General purpose	DIN45634	
3	Survey	BS3489:ANSI 2 (IEC 123)	ANSI 3 is also a US standard for survey meters
	Indicators	Open University	
	Non-classified		may satisfy certain standards such as ANSI 514 Type 52A

Circuitry

The signal from the microphone is often fed to the main sound level meter circuitry *via* a suitable impedance matching circuit. In instruments with detachable microphones this circuit will often be housed within the microphone assembly, thus allowing the use of adequate lengths of extension cable with minimal effects upon performance. The amplifier of a sound level meter usually comprises several stages and strategically placed around the early stages are the range attenuators.

Where more than one attenuator is used they will be so arranged that for any weighting network or filter in use the optimum signal-to-noise ratio and signal handling capacity will be provided. Where high rejection filters are to be used, this aspect of performance is particularly important. Whilst the level of the selected band of frequencies to be measured may be low, the wide band level could be up, say, 60 dB higher. Under these conditions the stages of the sound level meter circuit prior to the filter must be capable of operating over this wide range without any distortion at the higher levels. To achieve this, multiple attenuators are employed, thus allowing the signal levels in the early stages of the instrument to be maintained within their working limits, whilst providing the correct overall sensitivity for any given measuring range.

The weighting networks of a sound level meter almost invariably comprise close-tolerance resistor-capacitor networks. Their disposition about an instrument's circuitry will vary according to its design. In simple instruments, coupling and decoupling components may be used to provide all or part of a particular network, thus providing the maximum efficiency of component utilization. In precision instruments, however, the weighting networks are always separate elements in order that greatest accuracy is achieved.

Extraction of the root mean square of the squared input signal has already been described as the standard method of providing an input signal to the meter. In some more versatile instruments the a.c. output from the amplifier is also made available for coupling to auxiliary equipment, and sometimes a d.c. signal is also available.

Microphones

The two main types of microphones used in sound level meters are the piezoelectric and condenser types. Piezoelectric microphones utilize the effect of voltages being produced when the element, nowadays usually a ceramic material, is stressed, in this case by the receiving surface of diaphragm. Ceramic microphones offer good and reliable performance at very economic price levels and thus they are widely used in general purpose industrial grade instruments. The other main microphone type, the condenser microphone, offers the highest possible performance, but the cost and demands upon associated circuitry usually limit their application to precision grade instruments. As its name suggests, the condenser microphone has as its working element a condenser (capacitor), one plate of which is formed by the diaphragm. Movements of the diaphragm cause variations of the self capacitance which modulate a polarizing voltage applied *via* a high value series resistor. There are now available electric foil microphones which are of a capacitor type but are permanently polarized by an electrostatic charge implanted during manufacture. These microphones offer the same high performance standards as ordinary condenser types but with less demands upon associated circuitry.

Other types of microphone which may be used include moving coil or dynamic types, and special directional microphones, which may incorporate any one of the three types of element. Dynamic microphones are, however, relatively limited in application and have the basic disadvantage that comparatively large sizes are necessary to avoid distortion at low frequencies.

Calibration

Initial (manufacturer's) calibration of a simple sound level meter can usually be regarded as adequate for about a year, after which a calibration check should be made. Such calibration is normally based on free field response to noise of random incidence. The accuracy given by the calibrated meter then depends on the stability of the microphone and stability of the electrical system; whilst the validity of the measurement is dependent on the characteristics of the sound field present when measurement is taken.

SOUND LEVEL METERS

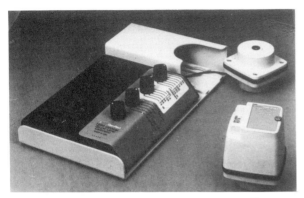

Omnical calibrator   Minical calibrator

Complete microphone calibration can be performed using GenRad's Omnical or Minical. (GenRad Ltd).

RS 104 sound level analyzer. (Reten Acoustics).

Electrical calibration is a useful check on the stability of the electrical system of the calibrator and can be carried out quite simply at regular intervals, *eg* monthly (especially where meters have built-in electrical calibration and internal calibration controls).

The use of an additional (calibrated) microphone can provide a simple comparative check as to whether the original microphone has changed in characteristics and thus needs recalibration. This method, used with electrical calibration, provides the simplest overall calibration check. Note, however, that this depends on the second microphone being used only for checking and being carefully stored.

Acoustic calibrators can provide an overall calibration check under 'field' conditions, but the accuracy of such a calibration check depends both on the suitability of the instrument and its proper method of employment (particularly as regards sealing of the microphone in the sound chamber).

An exact calibration check can only be carried out under laboratory conditions. It is generally recommended that instruments be returned to the manufacturer at suitable intervals (*eg* once a year) for such a check.

Calibration should always be checked if the instrument is subject to shock, such as being dropped, or to rough handling. Some recommendations and standards also give specific requirements for calibration checks and the frequency at which they are carried out.

Calibrators fall into two main categories:

(i) *Microphone Reciprocity Calibrators* — which can be used for the absolute calibration of standard microphones, to calibrate other microphones and perform overall calibration of systems using microphones. These determine acoustic impedance and electrical impedance, from which the microphone sensitivity can be determined at different frequency levels. The more elaborate types of reciprocity calibrators incorporate an analogue computer which automatically determines the sensitivity from the impedance data input to indicate sensitivity directly on a dial.

(ii) *Sound Level Calibrators* — for overall calibration of the system. These may range from built-in circuits to check the overall gain of the amplifiers in the electrical circuit, to true acoustic calibrators, which include the microphone as well as the electrical circuit in the calibration. Such calibrators accommodate the microphone in a sealed enclosure, which produces a fixed sound pressure level. This sound pressure can be generated in a variety of ways, both electronically and mechanically. The simplest electrical types utilize a single frequency derived from a pure tone oscillator driving a small, stable loudspeaker. Single-tone calibration is generally quite adequate for checking response up to about 1 000 Hz because of the flat response of most microphones below this frequency. For higher frequency calibration, random noise is generally to be preferred because of the more irregular microphone response at such frequencies. However, convenience of use may be a major factor in determining the type of acoustic calibrator employed. Thus mechanical random noise calibrators may only be suitable for use in comparatively quiet conditions. Also the accuracy of calibration may be affected by the manner in which the microphone is fitted in the calibrator, and the ambient temperature when so fitted.

The most popular form of calibrator used nowadays is the coupler type or *piston phone*. The microphone of the sound level meter to be calibrated is fitted into the open end of the coupler cavity which is closed at its other end by a driving transducer. This is energized by an amplitude stabilized, battery powered oscillator. Because the power required to produce the calibration level of 94 dB or 124 dB within the small volume cavity is low, the physical dimensions of the whole equipment are small, thus allowing it to be easily carried out with the sound level meter.

Small precision sound level meter with windshield fitted. (Rohde & Schwarz).

RS 440 integrating sound level meter. (Reten Acoustics).

Additional Facilities

Additional facilities may be provided on laboratory and precision grade instruments and also some general purpose meters. These may include a wider measuring range; and A-, B-, C- and D-weighting networks to provide a class of linear frequency response. Principal among the additional features is likely to be provision for the use of filters. Such filters will be either one-octave or one-third octave band types. With filters added, the sound level meter's application is extended into the field of sound analysis. It is thus not only possible to measure the overall weighted sound level, but to further measure the levels within the various frequency bands of its spectrum. Such additional information is an essential requirement in measurements for diagnostic purposes and many specifications, particularly within the engineering and building profession. Further extensions to the use of this type of instrument are often possible when a.c. and d.c. output facilities are available.

Where analytical measurements are frequently required, it is very convenient to have the sound level meter and filters in a compact and easily used form. Some modern sound level meters have self-contained filters which provide instruments with very wide-ranging facilities. In other types the module concept is adopted, with plug-in filters. This has the advantage of considerably reducing the cost and complexity of the basic instrument.

Signal Outputs

In cases where analysis of measurements is required but the environment at the measuring site is not conducive to the use of more complex equipment for fairly lengthy periods, it may be more convenient to record the noise spectra on a magnetic tape recorder. A suitable signal for recording may be obtained from the a.c. output of the sound level meter in use. It must be remembered that because a recording and replay process is involved, calibration signals must be placed on to the recording in order that the level of the recorded spectra may be established upon play back. This is easily accomplished by recording a short period of level produced by an acoustic calibrator fitted to the sound level meter before and after the recording of the noise spectra of interest.

The a.c. output of a sound meter is also useful in another application where analysis is required. The frequency resolution of one-octave band filters may not always be adequate. A typical situation is where there is a fairly high level of broad band noise and several discrete frequency components closely spaced in terms of frequency, a situation common in industry. To isolate any particular component from the overall spectrum requires analytical measurements with excellent frequency discrimination. There are available A.F. Analyzers with pass bands of only a few per cent of tuned frequency and having very high rejection rates outside of the pass band. Such an analyzer may readily be used to isolate and measure the level of a discrete component in a complex spectrum by connecting it to the a.c. output of a sound level meter. In fault diagnosis situations a pair of headphones may be added, enabling the user to hear the sound of the selected component and thus further assisting diagnosis.

Where measurements are required over long periods it is very convenient to use a chart recorder. When connected to the d.c. output of a sound level meter a chart recorder will provide a continuous record of sound level over long periods unattended. Thereafter the recorded chart provides a permanent record of such measurements. Because the chart speeds of recorders are usually closely controlled, it is a simple matter to establish the times at which various recorded events occurred.

Integrating Sound Level Meters

In many situations fluctuations of sound levels are too large, or too erratic, to allow accurate measurement even in the Slow mode of an ordinary sound level meter. To overcome this limitation integrating sound level meters have been developed which give a true energy average of the noise level over the entire measurement period.

Integrating sound level meters are capable of time-varying noise energy measurement, thus providing a reliable means of measuring fluctuating, intermittent and transient noise, even for long measurement periods. These instruments average the energy signal to provide an equivalent continuous level on L_{eq}, as a single figure direct read-out. Such a figure provides a direct assessment of potential hearing damage.

The mathematical relationship involved is:

$$L_{eq} = 10 \log_{10} \left[\frac{1}{T_m} \int_0^{T_m} \frac{P^2(t)}{P_0^2} \right] \text{ dB}$$

where T_m = measurement period
$P(t)$ = incident sound pressure
P_0 = reference sound pressure level (2×10^{-5} pascals)

See also *Sound Exposure Meters*.

Sound Exposure Meters

The principle of operation of a sound exposure meter is to provide an integrated measurement of sound pressure and time, as defined mathematically by:

$$E = \int_0^{T_h} P^2(t) \, dt$$

where E = sound exposure in (pascal)^2 hours (Pa^{2}h)
T_h = measurement duration in hours
P = A-weighted sound pressure in pascals
t = time in hours

This has the same relationship as noise risk which is proportional to the product of noise power and time, *ie*

$$\text{risk} \propto \text{noise power} \times \text{time}$$
$$\propto (\text{sound pressure})^2 \times \text{time}$$

For time-variant signals, this reduces to:

$$\text{risk } \alpha E \text{ or } \int_0^{T_h} P^2(t) \, dt$$

Equally there is a direct relationship between E and L_{eq} which, for a specific exposure time T in hours becomes:

$$L_{eqT} = 10 \log_{10} \left[\frac{E}{TP_0^2} \right]$$

where P_0 is the reference sound pressure level

Some worked out equivalents are given in Table II.

TABLE II – RELATIONSHIP BETWEEN L_{eq} AND SOUND EXPOSURE IN $(Pa)^2 h$

Eight hour equivalent continuous level L_{eq} dB	Sound Exposure $(Pa)^2 h$
64.95	0.01
74.95	0.1
84.95	1.0
87.96	2.0
90.00	3.2
90.97	4.0
91.94	5.0
94.95	10.0
104.95	100.0

Noise Dosimeters

Noise dosimeters or personal sound exposure meters are noise exposure monitoring devices which, whilst not actively measuring noise dosage specifically, integrate sound pressure level and time to give a measurement of sound exposure. Being attached to the individual a dosimeter adjusts to variations in sound level received at the position occupied by the wearer, *ie* is equally effective worn by a stationary or mobile user, traversing areas of varying sound intensity.

In the absence (until recently) of recommended design guidelines, this has led to a wide variety of instruments designed around inappropriate or irrelevant sections of sound level meter standards.

Most of these instruments display noise dose as a percentage of the legal limit of sound exposure. Such displays are open to misinterpretation as such legal limits vary from country to country and time to time. Additionally, certain manufacturers provide special units where 100% noise dose may correspond to a variety of eight hour L_{eq} measurements, with a significant risk that misinterpretation of 100% noise dose may occur.

The objective of the personal sound exposure meter is the measurement of the sound exposure in the vicinity of an individual's head that would exist in the same place in his absence. Research have demonstrated that for many practical situations, the sound exposure integrated by a user worn instrument is likely to be greater by up to 60% (corresponding to an increase in average sound level of approximately 2 dB) than measurements performed in the absence of the user. The consequence of this environmental effect on noise measurement accuracy is that recommendations no longer specify dosimeters with an accuracy equivalent to precision or Type 1 instrumentation.

Personal sound exposure meters are now defined as either Type 2 or Type 3 and both types hae the same overall linearity accuracy of ±0.5 dB from 85 dB to 115 dB, ±1.0 dB over the full dynamic range. The capability to handle transient and impulsive noise is relaxed slightly on the Type 3 instrument.

Personal monitoring should be considered for operators working in the near field of noise sources because large variations in sound pressure may occur for relatively small movements through the noise field. Operators employed in this area of activity are in potential risk situations and to provide additional warning of noise hazards, a 600 pascal peak indicator is desirable as proposed by the Health and Safety Commission.

It is currently recommended that the sound exposure of employees working in a noise environment where the equivalent continuous level for eight hours is greater than 105 dB(A) should be continuously monitored using personal sound exposure meters. This will help ensure that, where the risk is very high, there is a detailed history of exposure and that exposures higher than have been planned for can be identified. This is particularly important if operator hearing protection is provided by hearing defenders to ensure that there is sufficient safety margin in the protection products used.

See also *Integrating Sound Level Meters* and *Sound Exposure Meters*.

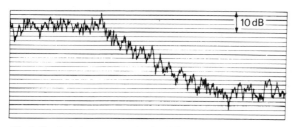

Fig 1 Typical decay trace obtained from a level recorder. The frequency was 125 Hz, and the slope represents an RT of about 4s.

RS 293 peak integrating sound level meter.

Reverberation Time Meter

Reverberation time (RT) is a fundamental quantity used in building acoustics, and is defined as the time taken for a sound to decay by 60 dB. Traditionally, RT is measured by registering the decaying sound level on a high speed level recorder and using judgment to estimate the mean slope of the resulting trace. A typical trace is shown in Fig 1. This is satisfactory if only a small number of decays are to be evaluated but if measurements are made in many frequency bands at several microphone positions the number of traces to be read becomes very large and the analysis is not only open to subjective bias but is also very time consuming.

Automatic devices originally developed for measuring RT directly formed part of a comprehensive, and expensive, measuring system. It is only recently that simple, low cost RT meters have appeared capable of yielding results accurate enough for most practical purposes.

An example is the RT meter designed by the Building Research Establishment (UK), employing an integrated circuit giving an output proportional to the log of the RMS value of the input (*ie* an RMS to dB converter). The input to the meter comes from the a.c. output of a sound level meter and filter set. The meter contains a timer which controls the output of a pink noise generator and this signal is amplified and reproduced by a loudspeaker. When the timer turns the noise generator off, the decaying signal from the sound level meter passes through the RMS to dB converter and is used to start and stop another timer when the signal is 5 and 25 dB respectively below the steady state level. As the steady signal level often fluctuates it was designed to use the last −5 dB crossing as the starting point to prevent spurious triggering.

The time taken to decay by 20 dB is assumed to be one third of the RT and so the measured value is simply trebled and then displayed. It is usual to base the determination of RT on a limited part of the decay curve because of deviations from linearity caused by the effects of background noise and differing rates of decay of the three mode types.

Impulse Sound Meters

Meters for measuring impulse sound are basically modified sound level meters with a response time reduced to the order of 1/10 ms. The circuitry may be designed to measure RMS average level,

Impulse sound level meter. (Rohde & Schwarz).

Rion infrasound level meter.

continuous peak level or peak instantaneous value; or all three. Where multiple measurement facility is provided the duration of the impulse can also be determined from the difference between peak instantaneous level and average level.

Infrasound Meters

Infrasound can be taken as lying within the frequency range of 1 to 50 Hz but the upper limit may be extended further into the audio range (*eg* up to 100 Hz) for useful measurement. An infrasound meter, therefore, is usually more correctly called a low frequency sound level meter. The chief requirements are a flat weighting from 1 to 50 Hz (or 100 Hz) and a sensitivity within −12 dB/octave.

A ceramic microphone is normally used since this type is stable and has a flat frequency response down to 1 Hz. A pistonphone is generally used for calibration, although a loudspeaker method may also be used. The indicating meter also needs to have low frequency rating in consideration of low frequency audio sensibility, with the circuit typically designed to hold the peak sound pressure level of a sound wave pattern, indicate true RMS sound pressure level, or hold the maximum sound level. Dynamic characteristics of the indicator are normally equivalent to the 'slow' setting of a conventional sound level meter.

Frequency analysis is simplified by the fact that high resolution is not required for measuring ultra low frequency sound. Thus 1/3 octave analysis is generally satisfactory.

Low frequency sound wave patterns may be recorded using a pen, discharge recorder or by oscillograph.

See also chapters on *Measurement of Sound, Noise Measuring Techniques, Frequency Analysis, Recorders, Signal Processors and Data Loggers.*

Frequency Analysis (Spectrum Analysis)

FREQUENCY ANALYSIS of a sound provides a method of determining the frequency content of the sound and thus the distribution of sound energy in the signal or the sound *spectrum*. Until some ten years ago the only measuring instruments available were those employing filters for stepping or sweeping a pass band filter over the frequency range of interest. This technique is still widely used, and incorporated in types of sound level meters which also have provision for octave band frequency analysis. Alternative systems now available are based on real time frequency analysis capable of providing parallel analysis in all the frequency bands over their entire analysis range simultaneously. At the same time they can present a virtually instantaneous graphical display of the analyzed spectra on a (CR tube) screen. Both types, however, are based on octave analysis.

Octave Analysis

An octave is a measurement of *interval*. The interval, in octaves, between any two frequencies f_1 and f_2 is defined mathematically as:

$$\text{octaves} = \log_2 \frac{f_2}{f_1} = 3.322 \log_{10} \frac{f_2}{f_1}$$

Thus the frequency ratio defining an octave interval of 1 between two sounds is 2:1

$$3.22 \log_{10} 2 = 1$$

Sound levels determined in this way enable the distribution of noise to be determined as a function of the frequency, and from this a noise spectrum can be plotted. This spectrum may be determined relatively coarsely, as with octave band analysis, or with greater resolution with narrow band analyzers.

In the case of full octave analysis (1/1 octave) each band represents an octave centred around selected frequencies, *viz* 63 Hz, 125 Hz, 250 Hz, 500 Hz, 1 kHz, 2 kHz, 8 kHz. A simple sound spectrum devised in this way is shown in Fig 1, showing peak sound levels occurring at the octaves centred in 63 Hz and 125 Hz, or in the frequency range 45–180 Hz.

The chief limitation of octave band measurement is that it gives a poor indication of steep spectrum slopes. Also it may be quite inadequate to deal with narrow band noise which may predominate in specific spectra, and in particular the separation of individual noise sources where the noise source is complex. Separation is limited to the fundamental and the second harmonic.

For rather more detailed analysis 1/3 octave band filters can be used, with centre frequencies based on preferred number series. Fig 2 shows the same noise displayed in 1/3 octave analysis. In

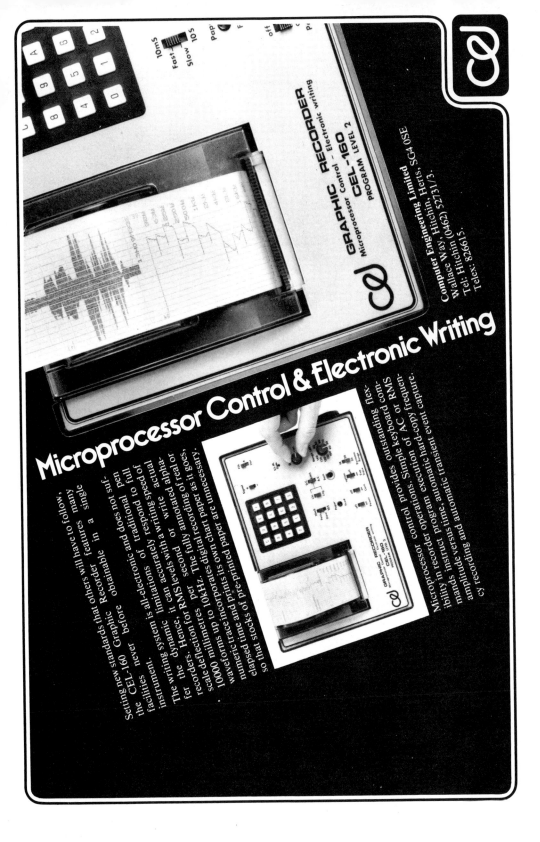

Microprocessor Control & Electronic Writing

Setting new standards that others will have to follow, many facilities the CEL-160 Graphic Recorder obtainable in a single instrument. The writing system is all-electronic and does not suffer the dynamic limitations of traditional full recorders. Hence, it can accurately respond to the speed of scale deflection per second or write alpha-10000 millimetres RMS levels as it goes, waveforms up to 10kHz. The fully annotated real or numeric trace incorporates digital recording as it goes, elapsed time and pre-printed paper are unnecessary, so that stocks of its own chart paper are unnecessary.

Microprocessor control provides outstanding flexibility in recorder program execution. Simple keyboard commands instruct time, automatic hard-copy frequency recording and automatic transient event capture.

GRAPHIC RECORDER – Electronic writing
Microprocessor Control
CEL-160
PROGRAM LEVEL 2

Computer Engineering Limited
Wallace Way, Hitchin, Herts. SG4 0SE
Tel: Hitchin (0462) 52731/3.
Telex: 826615.

FREQUENCY ANALYSIS (B)

Only Nicolet offers the Ubiquitous line of
signal processors that lets you study anything you want —
from the condition of a simple machine to the
dynamic properties of a nonexistent structure —
and now includes a new multi-channel signal processor
that has started a new era...

SINGLE-CHANNEL
446B Analyser can include built-in tape to store reference spectra; 1/3-octave, interface options.

STRUCTURAL ANALYSIS 6602 System extends modal testing to include structural modification; finite element programs optional.

DUAL-CHANNEL
660B Analyser with added power of complex math and dual-disk storage

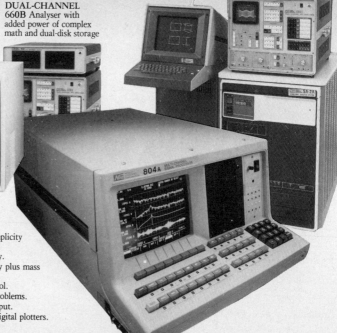

MINI-ANALYSER
100A is portable and easy to use; ideal for predictive maintenance on-site.

The 804A multi-channel signal processor.

The power of a computer with the simplicity of an instrument.
- 4 Channels processed simultaneously.
- Over 500K bytes of built-in memory plus mass storage on disc.
- User friendly through soft key control.
- Programmable for solving specific problems.
- Large built-in display with video output.
- Plots in 8 colours directly to most digital plotters.
- Standard Modal Analysis.

 Nicolet Instruments Limited

A Nicolet Instrument Subsidiary   Budbrooke Road, Warwick, England, CV34 5XH. Tel: (0926) 494111 Telex: 311135

FREQUENCY ANALYSIS (SPECTRUM ANALYSIS)

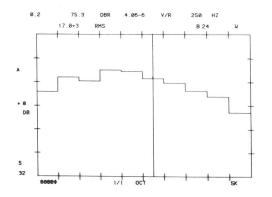

Fig 1 Acoustic analysis using octave bandwidth (flat weighting). (Nicolet Scientific Corp.).

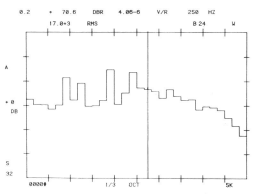

Fig 2 1/3-octave analysis shows 'discrete frequency components' (Nicolet Scientific Corp.).

this case the separate bands are narrower (one third of an octave) and there are more individual bands (three times as many as 1/1 octave analysis). More individual peaks are apparent, the main ones occurring in the bands centred at 20 Hz, 31.5 Hz, 80 Hz and 160 Hz. The most significant difference between the spectra is seen at the lower frequencies. This is a general characteristic of 1/3 octave analysis — *ie* it provides good resolution at lower frequencies but increasingly poorer resolution at higher frequencies. Also, of course, it still cannot show the presence of separate inductive frequencies which may be present in any 1/3 octave band, representing a spread (bandwidth) of 23% of the centre frequency.

For an even more detailed analysis it is necessary to go to *narrow band analysis*. This may be provided directly by constant bandwidth or constant percentage type filters. A constant bandwidth analyzer has the same narrow bandwidth, regardless of the frequency to which it is tuned. A constant percentage bandwidth analyzer has a bandwidth which is always the same percentage of the frequency to which it is tuned. Even higher resolution is attained by using analyzers which employ Fourier transforms to implement constant bandwidth filters. Such an analyzer is usually capable of several hundred live frequency resolutions to cover any one of its analysis range with the equivalent of constant bandwidth filters.

Fig 3 shows a typical narrow band analysis of the same sound displayed in Figs 1 and 2, this time using a high resolution analyzer providing 400 spectrum values covering the frequency range 1–800 Hz and providing 1.25 Hz resolution. It shows that the following frequency components are outstanding:

(i)   80 Hz or the fundamental frequency of the sound.
(ii) 161.28 Hz, 241.25 Hz, 321.25 Hz and 480 Hz as the second, third, fourth and sixth multiples of the fundamental frequency.
(iii) 120 Hz.
(iv) 18.75 Hz and 31.25 Hz.

In this particular example, the 120 Hz peak can be related to the second harmonic of mains frequency, leaving the peaks at 18.78 Hz and 31.25 Hz unexplained. Also they are not indicated on 1/1 or 1/3 octave analysis.

FREQUENCY ANALYSIS (SPECTRUM ANALYSIS)

Model 2449 — a low cost, precision analogue wave and spectrum analyzer with a frequency range of 10 Hz to 50 kHz, manually tuned or automatically scanned by an interval sweep generator.
(Spectral Dynamics)

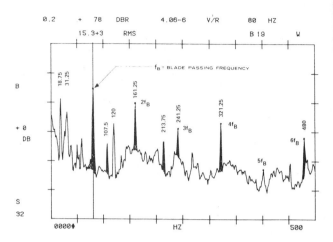

Fig 3 Narrow band analysis of acoustic signal to 500 Hz.

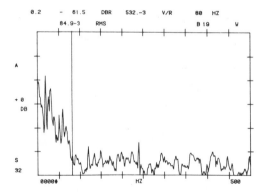

Fig 4 Ambient background spectrum.

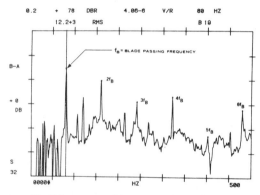

Fig 5 Spectrum of fan noise without low frequency background.

They are, in fact, background noise as a separate analysis of ambient background spectrum show (Fig 4). A connected spectrum for subject noise only is obtained by withholding the power of the ambient background spectrum from the constant spectrum, as shown in Fig 5.

These two figures serve to illustrate that background noise can be significant in high resolution analysis. Thus, no second reducing treatment applied to the noise source itself in this case would have been effective in reducing the 18.75 Hz and 31.25 Hz peaks. The cause of these would have to be sought elsewhere in the background noise.

Parallel Filters *vs* Swept Spectrum

A practical problem with parallel filters is that resolution is directly related to the number of filters provided, but the greater the number of filters employed the more expensive the analyzer becomes. Thus typical parallel filter analyzers are based on 1/3 octave analysis, demanding the use of 32 filters.

FREQUENCY ANALYSIS (SPECTRUM ANALYSIS)

One method of overcoming this particular problem is to use only one filter and sweep it slowly through the frequency range of interest. This is the basis of the swept spectrum analyzer. The particular limitation here is the response time of the filter. The speed of sweeping cannot be faster than this (*ie* the sweep time must not be less than the filter response time), and since in its pass the filter is tuned to individual frequencies in turn, it can miss an event occurring during the sweep time at a frequency to which it is not, at that moment, tuned. This does not happen with a parallel filter network, except, of course, that it will miss events occurring at a frequency not covered by an individual filter.

Summarizing, a parallel filter analyzer can be fast, but has limited resolution and tends to be expensive. A swept spectrum analyzer can be cheaper, yet have a higher resolution, but measurements take longer (especially at high resolution) so it can miss transient events.

Dynamic frequency analyzers both high speed and high resolution at competitive costs.

Time Compression Analysis

The spectrum obtained using analogue or digital filters involves passing the signal through a bank of such filters in parallel, set up to span the frequency range of interest, or by sweeping the centre frequency of a single variable filter across the range. The pass bandwidth may be constant or, more commonly, vary in proportion to the filters' centre frequency.

The multiple parallel filter equipment becomes uneconomic for finer discrimination than 1/3 octaves, whilst unfortunately, a major limitation is also found with continuously swept analogue filters. The response time of the filter network rapidly increases as the bandwidth is reduced, which severely limits the rate of frequency sweep.

It can take several hours to obtain a valid estimate of a spectrum with a 1% bandwidth, but some instruments overcome this limitation by time compression analysis. The original signal is digitized, stored in a micro-processor memory and then replayed through a filter network at a rate up to several thousand times faster. The high frequency filter can have a relatively wide bandwidth to give a rapid response time. This wider bandwidth, when the data is transformed back to the original time scale, decreases in proportion to the increase in play-back rate, restoring the narrow discrimination of the analysis.

Fourier Analysis

The Fourier technique produces a constant bandwidth analysis rather than the proportional analysis usual with analogue equipment. The problems of filter response time are overcome, the hard-wired devices producing a spectrum in a fraction of the time required to acquire the data

Nicolet 804A expendable FFT analyzer has all the single functions (power spectra, RMS spectra, spectra of averaged time) on up to four channels simultaneously, the normal cross-functions (cross-spectra, transfer function, coherence and coherent output power) and the inverse functions (correlation and impulse response). It processes to 40 kHz with high 400-line resolution increased to 2.5 mHz by 2 000:1 zoom.

from the audio-frequency range. Fourier analysis also provides phase information which opens up many possibilities for more complex analysis interrelating two or more signals.

Real Time Analysis

Narrow band DIGITAL spectrum analyzers usually transform a time domain signal into its Fourier spectrum. The input signal is considered over some finite period of time known as the 'time window, block or frame' and is stored by sampling and then digitizing the sampled levels. A discrete Fourier transform is then performed on the stored signal giving a spectral analysis of the instantaneous waveform.

If the signal is periodic and fits 'nicely' into the selected time window well and good, but this is rarely the case and the frequency content may be varying from frame to frame. If this is so our single sample is probably just as random and unrepresentative as the input signal it was derived from.

Also, sampling, digitizing, storing and subsequent Fourier transform all take time and the real time capability of a spectrum analyzer is limited to the speed at which it can analyze and process a signal, *ie* not a single datum line is lost.

Analysis time is a function of the required resolution and therefore filter bandwidth. We are limited by the fact that the narrower a filter is the longer it must be exposed to be sure it acquires the spectral energy available. The relationship between time and bandwidth will be (1 000/400) 2.5 Hz and the dwell time or window cannot be less than 0.4 sec, real time is maintained. At 1 kHz to 20 kHz, in fact many 'fall out' of real time under 5 kHz. In our example we used 400 lines as a good compromise between resolution and the optimum real time currently available.

From sampling theory, the highest sinusoidal frequency of interest must be sampled at least twice per cycle. Any components above half the sampling frequency will not be sampled correctly and will introduce errors; these higher frequencies would appear to be components of lower frequencies than they are — in other words they would take on an 'alias'.

To reduce this effect 'anti-alias' filters are introduced, but of course the perfect filter does not exist (around 100 dB/octave is typical today) and some high frequency components get through. To further minimize this (aliasing) effect, the sampling rate is often taken as 2.56 times the highest frequency rather than just two times. Taking our example of 400 lines therefore, a sample of 400 x 2.56 *ie* 1024 may well be taken, giving an actual frequency domain result of 1 024/2, *ie* 512 lines. It may well be because of 'aliasing' that the upper 112 lines are contaminated — these are thrown away and we are left with 400 lines across the selected frequency range.

The modern real time analyzer utilizing advanced design techniques can sample at 256 000 times a second *ie* 4 msec. At 2.56 x the upper frequency this gives typically an upper range of 100 kHz — not to be confused with real time operation.

Fitting the 'Window'.

Because it has no information to the contrary the finite Fourier analysis algorithm assumes the captured signal will be repeated indefinitely. In the unlikely event of a sinusoidal component that just 'fitted' the frame this assumption would be correct and a single line spectrum would result — Fig 6. However, even the rare pure tone is unlikely to match the 'window' exactly and we may have, say, 2.5 cycles captured. Ideally this should give the result shown in Fig 7, but in practice, due to the abrupt change or discontinuities at each end of the frame, we get a whole series of lines which would exist if the input signal were to reproduce itself identically before and after the frame as in Fig 8. In other words the analyzer is faithfully calculating the spectrum necessary to represent this 'unrepresentative' sample. The extra lines are known as leakage of spectral smearing

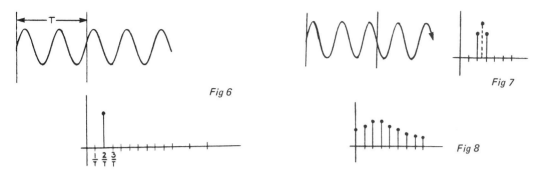

Fig 6

Fig 7

Fig 8

(Fig 9). The leakage problem can obviously be corrected by the introduction of a weighting, but like the A-weighting curve the ideal curve varies with the input signal. Various weightings have been tried and the one that is generally used is the 'Hanning function', a tapering waveform that is applied to the captured input frames by multiplication. It has a cosine bell shape (Fig 10).

Multiplying the input frame by the Hanning function essentially removes these discontinuities by tapering the ends of the frame and forcing the signal to be quasi-periodic in the frame length. This has two side effects as can be seen in Fig 11. First by attenuating the input signal at the beginning and end it reduces the energy in the frame by a factor of a half. Secondly it effectively amounts to 100% modulation of the input signal by a modulating frequency of $1/T$. Modulation of course introduces sum and difference frequencies or sidebands which in this case, being of relative frequency 1, show up mainly in the two lines immediately adjacent to each component of the input.

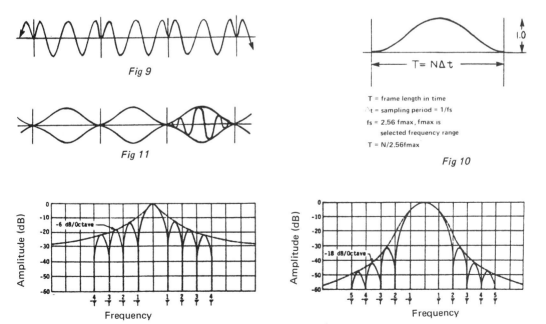

Fig 9

Fig 11

T = frame length in time
Δt = sampling period = $1/f_s$
f_s = 2.56 f_{max}, f_{max} is selected frequency range
$T = N/2.56 f_{max}$

Fig 10

Fig 12 Unweighted and Hanning-weighted filter shapes.

The Hanning weighting dramatically improves the rolloff of the filter skirt to 18 dB/octave but at the price of increasing the width of the main lobe. The Hanning equivalent noise bandwidth is 1.5 as opposed to 1 for unweighted analysis (see also Fig 12).

The reduction of the input signal is easily compensated for in the power spectrum by multiplying by a factor of 8/3 *ie* the reciprocal of energy reduction squared $(1/2)^2$ by the equivalent noise bandwidth (1.5). This factor is only truly appropriate for random signals. For deterministic signals which fall in the centre of a filter line, the Hanning weighting will reduce the apparent amplitude by 1.76 dB, corresponding to a 1:1.5 change in equivalent noise bandwidth of the filter shape. Even in this case, however, the total energy is conserved as the power is recaptured in the sidebands immediately adjacent.

In general then, Hanning should almost always be used for continuous free running inputs. For broadband noise with relatively flat spectra it may make little difference, but for spectra with peaks or notches and for deterministic signals in noise, Hanning will usually increase the selectivity or sharpness of the spectrum considerably by reducing the smearing or leakage effect of the skirts.

On the other hand, for observing transients, *etc*, Hanning will usually not be appropriate — the reason for this is that the transient may be concentrated at or near the start of the window and Hanning will multiply it by very small fractions compared with the centre window value, probably drastically affecting the spectrum.

Averaging

Averaging of a number of instantaneous spectra over a period of time will now be necessary to obtain more stable estimates for processing through the FFT programme.

Four averaging modes typically used are:

(i) Simple summation (normalized to the number of frames processed) which is equivalent to an ideal integrator.

(ii) Subtractive — useful for removing background noise when the object to be analyzed cannot be removed from a noisy environment.

(iii) Exponential — equivalent to a leaky integrator, that discounts the contribution of past spectra for varying signals.

(iv) Peak or Maximum hold — useful for analysis of machinery run-ups and shutdowns.

How much averaging must be performed? Obviously the more the better; in fact after averaging 100 successive spectra all calculated from independent input frames the result will, with high probability, be within ±10% of its true average. After averaging 1 000 it is likely to be ±3.2%. This comes from the relationship $100 \div \sqrt{}$ averages, for example 64 averaged would give 100/8 = ±12.5%.

Having processed and averaged our input windows the results are usually displayed on an annotated screen showing levels versus frequency. The frequency can often be displayed on a linear or log basis and similarly once the processing is complete any number of sums can be done on the levels to present power spectral density, energy density, volts, volts^2, dB and other parameters, and this is reflected in the price directly or indirectly, for example, a relatively inexpensive analyzer may offer a wide range of features but these might be at the expense of poor resolution or a small real time bandwidth. At the other end of the scale you can pay for facilities only rarely used or which can be done much more cheaply feeding the analyzer into one of the inexpensive desk top computers which will also give greater data storage and comparison features *etc*.

FREQUENCY ANALYSIS (SPECTRUM ANALYSIS)

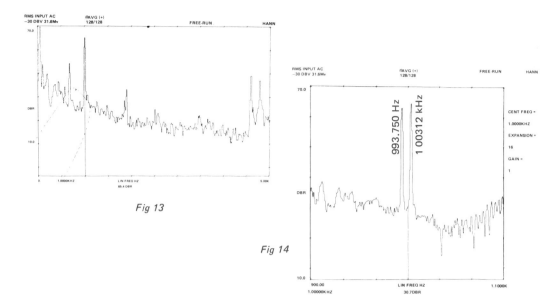

Fig 13

Fig 14

One-third octave synthesis, Zoom and Engineering is shown in Figs 13 and 14 below. In both cases we are analyzing the same input, *ie* random noise with tones, but in case 1, with a 5 kHz bandwidth, thus resolving 5 000/400, *ie* 12.5 Hz. However, using the Zoom facility to inspect the same spectrum we find two not one tones at 993.75 Hz and 1 003.12 Hz. Our resolution is now 200/260 *ie* 0.77 Hz per line. This only represents an expansion of 16:1 and expansions of 128 are common. It should be remembered that if we Zoom or reduce our bandwidth by 16 or more times our time window or averaging time will increase proportionally.

Finally, engineering units are what you might expect, the ability to read sound pressure level or mV/g being directly proportional to the sensitivity of the transducer you are using. It is usually arranged that if the conversion factor is known it can just be entered, or alternatively a calibration signal can be applied and the reading adjusted to suit.

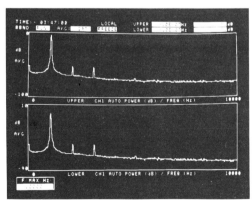
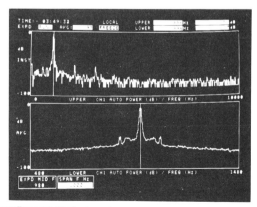

Examples of Zoom and Expand displays
(Solatron 1200 Signal Processor)

Recording and Signal Processors and Data Loggers

GRAPHIC RECORDERS are widely used for documentation of instrument measurements. They have the advantage of providing quick accurate graphic plots of measured signal levels on perforated paper with level and frequency graduations. Suitable control facilities enable them to operate synchronously with generators, filters and frequency analysis. For operation with digital equipment, the recorder takes the form of an *alphanumeric* printer.

For data storage and later playback, a tape recorder is most commonly used, taking compact cassettes or standard tapes. Here it is important that this be an *instrumentation recorder,* not a conventional audio tape recorder, *eg* with recording format compatible with 41/150 3407/ BS 5079 or DIN 66211 and 66212/ANSI X3.48. Conventional audio tape recorders do not necessarily have the required dynamic range and incorporate bias and pre-emphasis designed to give optimum performance recording speed and music (and optimum results with a particular type of tape). Instrumentation recorders, too, are also available with more than four tracks on wider (½ in and 1 in) tapes.

Data Logging

Recording measurement results digitally on magnetic tape is also the basis of *data logging*. Basically this technique samples the parameters at regular intervals and records such data for subsequent off-line input to a microprocessor, computer or other data processing equipment. Such a system is shown in block form in Fig 1, comprising a *transducer* (or trasducer) to provide the signal input, *signal conditioner* or *processor* to convert the input signal to a form acceptable to the *data logger,* the output from which is passed to the data processor or analyzer.

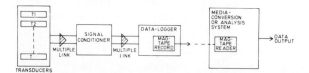

Fig 1 The general arrangement of a data logging system for multi-channel noise or vibration monitoring with provision for ancillary data inputs such as real time, temperature, pressure, etc.

Normally a separate conditioner is required for each transducer. Outputs from the conditioners are then scanned by the data logger at preset intervals (*ie* sampled sequentially), and the results recorded in digital form on magnetic tape. For fast systems, inputs may have to be sampled and held at the beginning of the scan so that recorded information for a scan all correspond to the

OmniScribe strip chart recorder. (Bausch & Lomb).

same point in time. Thus for optimum flexibility the signal conditioning system needs to be programmable.

Another major requirement of the signal conditioning system is the ability for it to have some intelligence. Many logging applications require some inputs only to be logged as background information whilst others are logged more frequently, or only to be logged when input signals exceed preset levels. With the availability of today's large scale integrated circuits it is fairly easy to design this level of intelligence into a conditioning unit accommodated on a single printed circuit board.

The final step in the data signal handling procedure is that of transferring the data from the recording medium to the appropriate analysis system, which normally takes the form of either a mainframe computer that also performs many other tasks or a dedicated minicomputer or microprocessor-controlled system. In the former case a suitable media-conversion system is usually required, capable of reading the logger's cassette or cartridge and outputting the data in suitable format for input to the mainframe. The hardware involved in such a system could well be comparable with that required for the complete dedicated analyzer but with much simpler software programming.

See also chapter on *Frequency Analysis*.

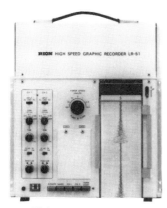

High speed graphic recorder. (Rion Co Ltd).

High speed Continuous Recorder by Scientific Atlanta interfacing directly with their SD350 analyzer to display spectra in time-grain format.

Environmental Noise Monitoring

IN DECIDING which noise rating index to use, due consideration must be given to the prime noise source in the area and the type of district. The selection can be made from a wide range of noise measuring parameters which have been proposed for various situations, but the basic requirement of the rating unit must be to reflect the subjective reaction. Another important function of the rating index is to reflect the variation of the noise level with respect to time. As it is not convenient to interpret a continuous level against time trace, the noise rating index must also perform a data reduction function. It could, therefore, range from a simple maximum dB(A) level to one of the more complex units that are designed to reflect the variation of the signal in its time domain; such as L_{eq} or an L_n value.

The basic problem which then remains is choice of a suitable system based not only on measurement requirements but duration of measurement required and cost. Thus for short duration temporary measurement a system could be constructed on a hybrid basis from standard instruments and these then enclosed in some form of all-weather enclosure. Such an approach wold prove to be the most economical and would also have the flexibility that would be necessary with semi-permanent installations. Standard enclosures are available for this purpose, some of which include facilities for mounting an all-weather microphone system.

Readout

Readout is an important consideration since for continuous monitoring an operator cannot be continuously employed to read a conventional analogue meter; hence some form of recording must be used. The new generation of instruments that employ digital maximum capture systems would be best suited to these applications because there is no risk of the held value decaying away with time. Traditional analogue hold arrangements are only suitable for holding a maximum for a few minutes, whereas we are looking for many hours and on some occasions, even days between readings. Although a maximum hold arrangement offers considerable cost advantages the information provided is very limited because it cannot be related to the time of the occurrence nor can it be used in the calculation of time related noise rating indices.

These restrictions tend to off-set the cost advantages of a single L_LMax concept for noise monitoring applications and a graphic level recorder was investigated as an alternative. This approach proved to be infinitely more flexible as the results are presented in the form of a complete time history of the acoustic climate from which not only is it possible to read the L_LMax along with its time of occurrence but the curve may be inspected and statistical (L_n) and integral (L_{eq}) noise parameters can be calculated for any period of particular interest. With

ENVIRONMENTAL NOISE MONITORING

experience it is possible to interpret a lot of information about the character of an acoustic climate from the level against time trace and to correlate the significant events with particular occurrences such as steam vents, railway movements, *etc.*

From a cost and data presentation viewpoint a graphic level recorder seems to be ideally suited; the problem, however, is in selecting a suitable instrument. As we require readout answers in dB it is necessary to select an instrument having a logarithmic amplitude response in order that a linear scale in dB may be obtained over a sufficiently wide range. This should cover a minimum dynamic range of 40 dB in order that the wide range of noise levels occurring at different times of day can be measured without the need to change ranges. This means that it is necessary to select one of the recorders that have been specifically developed for use in acoustic work as these instruments have the necessary dynamic range. A complication associated with these devices is that they invariably have a constant velocity writing response; hence, the trace will differ from the indication given by a sound level meter which, of course, has an exponential response. The magnitude of the error that this would introduce is dependent upon the rate of change of the signal and this will be an important consideration where the results provided by the monitoring installations have to be in conformity with a national or international standard that calls for measurements made with sound level meters. Suitable recorders are, in fact, now available but can be relatively costly.

Microphone System

The most important consideration in respect of the microphone system is that it must be suitable for long-term outdoor installation in the prevailing climate. The introduction of precision electrecet measurement microphones has greatly simplified matters in this respect as, for a slight reduction in performance in other areas, they are able to operate within the International Precision Specification in relative humidities up to around 95%. This will obviously result in considerable savings in power consumption, as heaters are not then required; making this type of microphone the first choice in battery powered systems preferably with a type of preamplifier which can be fully sealed against moisture.

Logically, too, the microphone should be of outdoor type contained within a custom-designed enclosure for complete protection against the environment, and particularly rain and moisture.

Locating the microphone it is also important to ensure that the signal received reflects the noise in question and there is no interference from wind or other environmental conditions. The microphone should be located away from the corners of buildings or any other sharp objects likely to generate wind noise and also such that it is not the target for vandalism. To prevent localized noise such as voices of people standing very close to the microphone from affecting the result, the usual procedure is to mount the complete microphone assembly high on a pole.

An example of a complete low-cost monitoring system is shown in Fig 1.

Fig 1 Low cost permanent environmental noise monitoring installation. (Computer Engineering Ltd).

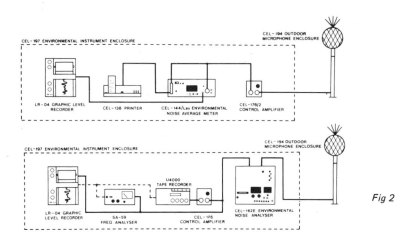

Fig 2

Hybrid Systems

The diagram given in Fig 2 outlines some typical systems that can be constructed from standard units and are designed to provide data for specific applications. The first system was constructed as a low cost package designed to provide data on noise events in the vicinity of an airport. It has facilities that provide the L_{AX} of each individual aircraft over-flight whilst the level recorder provides the time sequence of each event. Based on the results obtained by this configuration it was possible to provide a detailed analysis of the effects of aircraft noise on the community.

The second system is based on an Environmental Noise Analyzer designed to obtain data for a noise map in a busy urban area. The full range of a L_n percentiles was produced at regular intervals along with both period and cumulative L_{eq} values. The level recorder provided a continuous trace against time whilst the meter on the frequency analyzer functioned as a wide dynamic range sound level meter to monitor levels. The control amplifier was used in its time sequence mode to provide discrete samples of the noise level at regular intervals for more detailed analysis. These were subsequently fed into the Frequency Analyzer in order to provide information on the signals in their frequency domain.

Custom Designed Systems

For a permanent installation it is better to take a different approach and provide a fully engineered system. In such cases requirements vary from situation to situation and it becomes practical to design and install a specific system. Many airports and large industrial plants have already such equipment operating and these are used to control and sometimes penalize individual operators. The considerations involved in specifying such a system revolve around the size and usage of the airport or plant, the current noise control regulations and, of course, the size of the budget allocated for the installation. Typical systems therefore vary between one and twenty monitoring points and the data logged could be simply maximum levels through continuous monitoring on graphic level recorders to a fully computer controlled system.

Most installations are of the latter type and the main elements of a system are the microphone system, the remote monitoring terminal, data transmission link, control and computing system and readout. Each monitoring point obviously needs its own microphone and noise monitoring terminal whilst there are several possibilities in respect of the data transmission. However, once the distances involved are in excess of 1 or 2 km, digital data transmission is by far the most effective.

A single speech quality telephone line will be capable of handling in the order of 20 individual monitoring terminals.

The considerations relating to the microphone are as outlined previously; whilst in the Noise Monitoring terminal there will be circuits arranged to amplify and frequency weight the signal. It is then necessary to condition the signal such that it is suitable for transmission over the data link. The usual form is to feed it to a wide dynamic range RMS detector to give a sound level which is then sampled, digitized and fed over the link *via* modems. There are problems associated with this approach mainly related to the crest factor limitation of wide range RMS detectors and sampling errors introduced prior to the analogue to digital conversion.

In order to overcome these problems true energy integrating systems have been developed that cover ranges as wide as 90 dB with no crest factor limitation. These can be transmitted over the data link as short period equivalent continuous noise levels thereby eliminating the chance of sampling errors. These short period results are then processed by the mainframe computer to provide the required noise rating parameters. Also contained within the Noise Monitoring Terminal are all the necessary circuits to enable the system to check its performance and accuracy.

Quick checks on sensitivity can be made by means of an insert voltage level injected into the microphone; however, it is always possible that a noise meter can give the correct answer at one point on its dynamic range but be seriously in error at other points. In order to check the electronics thoroughly, from the a.c. noise signal back to the computer, a test signal generated at 1 kHz and varying in 150 steps, 2 seconds at a time, over a range of 85 dB, should be applied to the input. This signal is applied by a control command from the computer and the computer can check the varying levels against time to prove that the Noise Monitoring Terminal and transmission line are thoroughly correct.

The Noise Monitoring Terminal electronics and transmission modem should be housed in a weatherproof box of the type used for housing traffic light electronics for example. The internal temperature needs to be controlled by a thermostat and heater such that it never falls below +10°C. The design of the electronics should be such that they will work satisfactorily up to 60°C; a level which is unlikely to be exceeded; however, in certain extreme cases cooling fans may be found necessary. Manual controls should provide selection of frequency weighting, test or measurement modes, *etc*. A digital display giving a continuous monitoring of noise levels will be required during setting up and any subsequent servicing.

The computer and up to 20 Noise Monitoring Terminals can be connected to a single telephone line *via* modems allowing data transfers in both directions between Noise Monitoring Terminals and the computer. Only one Noise Monitoring Terminal transfers data at a time and then only at the command of the computer. This command addresses the particular Noise Monitoring Terminal and also sends a central command such as measure, test or calibrate. The addressed Noise Monitoring Terminal then replies with a two word data transfer. The first word consists of the two most significant digits of sound level and also echoes back the control command from the computer. The second word returns to the two least significant digits down to 0.1 dB accuracy.

Should the central command be corrupted by the interference on the telephone lines, the error must be detected in the Noise Monitoring Terminals in order that none will reply. The computer will then repeat the command until it is received properly and acknowledged by a reply. Similarly, if the return data is corrupted, the computer will request a second transfer. Using this error checking system it is possible to transmit sound levels to an accuracy of 0.1 dB over many kilometres of telephone cable without faulty data being accepted. Data can be requested from each terminal as fast as the computer takes to service the total number of terminals.

Typical transmission rates are 1 200 bits per second, so the three words needed to service each unit take 30 msecs (12 bits per word). Consequently, 16 terminals could be serviced in 0.5 seconds, for example.

The mainframe computer used in these systems is usually a standard unit manufactured by a major computer company. Systems usually consist of a fast 16 bit microprocessor with 8 K upwards and 16 bit random access words backed up, if required, by a twin tape cassette system or a twin floppy disc memory. Input and output is *via* a 30 character per second silent printer with keyboard. It is also usual to have a facility for a digital and an analogue data output for driving ancillary units such as graphic level recorders, *etc.*

A computer system is necessary and economically advantageous when several points are to be monitored at an installation. The cost of the computer can then be shared between the total monitoring points. Although a separate Noise Monitoring Terminal is required at each monitoring point, the distance from many of the points to the central station is usually large enough to make digital transmissions a necessity anyway. The computer system has the great advantage of complete flexibility in dealing with the sampled sound levels. The programme can be easily modified to deal with almost any requirement of the customer or of the local regulations. Examples of the type of parameters that may be produced are as follows:-

— Maximum sound level reached and the time of day
— Duration and extent of the sound level above a pre-determined level
— Co-relation between the amount levels are exceeded and the flight number
— L_{eq} (Noise Average) levels for set periods during the 24 hours
— Statistical parameters such as the percentage of the time for which various noise levels are exceeded
— L_{AX} (Single Event Noise Exposure Level) for each flyover
— Histograms of individual significant events
— Automatic calibration and Self-Test checking

The output details that are required immediately are typed on to a printed record whilst further details can be stored in the computer's memory banks for later processing to obtain long term trends. A digital to analogue converter followed by a suitable graphic level recorder will reconvert the digital data to an analogue voltage and give a graphic recording of a particular flyover should this be necessary. Any noise monitoring point can be selected and the data from past flyovers can be brought out from memory and re-recorded on the graph paper for comparison whenever required.

Single event noise exposure level analyzer. (Computer Engineering Ltd).

Audiometry

AUDIOMETRY, OR the measurement of hearing, is based on subjective response to noise which, to be quantitative, needs to be related to objective data. The standard method of determining hearing efficiency or *acuity*, for example, is by measurement of the sound pressure level at which the subject can first hear a sound. This is called the *auditory threshold*. Normally this is 0 dB, but hearing is frequency-sensitive and so normal hearing will show a departure from a straight line or norm. Instead of measuring absolute levels at different frequencies, normal practice in audiometry is to determine the difference between the threshold level at each frequency and 0 dB, this difference being called the threshold shift. Such a record is known as an audiogram. A typical subject with normal hearing will yield an audiogram with a threshold shift of not more than ± 5 dB, the most likely departures from the 0 dB norm occurring at 200 Hz, 4 000 Hz and 8 000 Hz.

An audiogram (*eg* see Figs 1–3) will show hearing loss as a function of frequency. Presbyacusis (advancing age) will produce a permanent threshold shift, as will exposure to noise. The latter may be reversible (temporary threshold shift) or irreversible (permanent threshold shift).

In industry there are definite advantages in carrying out a monitoring audiometry programme, particularly as current recommended noise limits protect only a percentage of a group of people, not all individuals. This is because the susceptibility of a group of persons to receiving heavy damage from noise varies quite considerably. Thus an arbitrary 90 dB(A) noise limit does not protect 100% of a group. There will always be a number of undetectable persons who will acquire

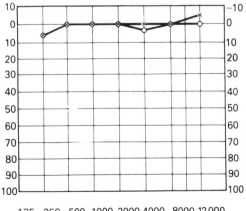

O *Right ear*
X *Left ear*

Fig 1 Typical audiogram of person with normal hearing.

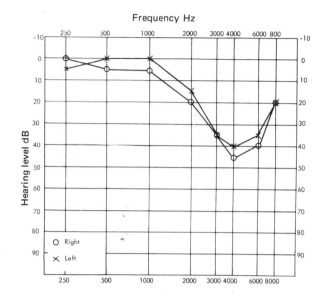

Fig 2 An audiogram chart illustrating the threshold of hearing of a person with a mild noise-induced hearing loss, measured with a manual audiometer.

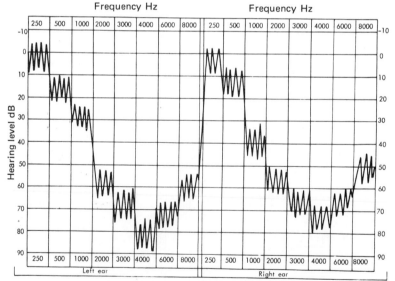

Fig 3 An audiogram chart illustrating the threshold of hearing of a person with a severe noise-induced hearing loss, measured with a self-recording audiometer.

auditory damage if exposed. Regular monitoring audiometry would allow such persons to be discovered before permanent hearing damage is done.

Measurements to produce an audiogram are made with an audiometer. Basically this is an instrument capable of generating pure tone signals of specific frequencies fed to earphones worn by the subject under test. The subject adjust the sound pressure level of each signal received until that tone is just perceptible. The instrument is calibrated to a normal threshold level, where the difference or threshold shift at each frequency can be read, indicated or recorded. Whilst this is an objective measurement of subjective response the technique does admit differences, notably in the ambient noise conditions during the test, and whether the samples employed to determine the calibration norm are representative. Standard calibration for threshold levels is specified in ISO R 389 – 1974.

Audiometers

An important feature in audiometric measurement is the difference between ascending and descending thresholds, the former being obtained as the test signal is increased in level from below audibility, while the latter is obtained by reducing the signal intensity from a clearly audible level. The threshold of hearing is usually defined as a function of these two values. The results are then plotted on an audiogram chart, as illustrated in Figs 2 and 3.

In the case of self-recording audiometers, the read-out is presented in the form of a series of 'peaks' and 'valleys' – see Fig 3. The threshold of hearing is usually taken as the average of the mid points between the peaks and valleys of the most regular tracings at each test frequency.

Simple measurement/recording techniques involve delivering all seven main frequencies, starting with the lowest frequency first to one ear then the other, with the tone presentation lasting about 30 to 35 seconds. The subject adjusts his control to minimum level for perception. The attenuator should switch to 0 dB between each charge, without a 'click'. Alternatively, with automatic audiometers, all seven tones are delivered first to the left ear in sequence,, and then the right ear in sequence. In this case it is particularly important to present the tones in exactly the same manner to the right ear as to the left ear.

Types of Audiometers

Audiometers can be categorized under three headings:

(i) *Monitoring audiometers* which provide simple indication of threshold shift and indicate whether a subject is suffering from hearing loss. These have the widest application and may be manual (indicating, requiring subsequent manual plotting of the audiogram), or automatic (presenting the complete audiogram during the test).

(ii) *Diagnostic audiometers* used by audiologists and medical specialists for more detailed analysis of hearing loss and to assist in determination of the type of corrective action necessary.

(iii) *Research audiometers* used by otologists for basic research.

Test signal frequencies for the signal tones should be 250, 500, 1 000, 2 000, 3 000, 4 000, 6 000 and 8 000 Hz, although the 250 Hz tone is not always specified. It is preferable that the tones be pulsed to avoid confusing a tone with tinnitus, the pulses being 250 ms long with a 250 ms gap between them. Rise and fall times should be of the order of 50 ms measured between the −1 and −20 dB points down from the maximum steady value of the pulse signal.

Simple Audiometers

A simple monitoring audiometer incorporates an oscillator as a signal source, an amplifier and attenuator to vary the intensity of the signal, and pulse shaping network to modify the rise and fall times of the signal to remove switching transients. The test signals are presented through cushioned earphones (preferably high grade moving coil type). The subject is also presented with a separate

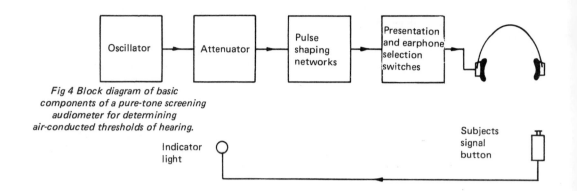

Fig 4 Block diagram of basic components of a pure-tone screening audiometer for determining air-conducted thresholds of hearing.

TABLE I — AUDIOMETER TONE FREQUENCIES

FOR AIR CONDUCTION		FOR BONE CONDUCTION	
Frequency Hz	Maximum intensity dB	Frequency Hz	Maximum intensity* dB
250†	80–90	250	45
500	80–100	500	60
1 000	80–100	1 000	60
1 500†	80–100	1 500	60
2 000	80–100	2 000	60
3 000	80–100	3 000	60
4 000	80–100	4 000	60
6 000	80–90		
8 000	80		

* Typical values for monitoring audiometers
† Not necessary for monitoring audiograms

signalling method to indicate when he or she is hearing a signal. A block diagram of a basic audiometer of this type is shown in Fig 4.

For industrial purposes a self-recording audiometer may be preferred, with the advantage that operator errors are reduced to a minimum. Monitoring audiometers are used for routine air conduction measurements, and particularly for rapid screening. They may also include basic diagnostic features — *eg* 'white noise' masking and additional frequencies for bone conduction — see Table I. Facilities may also be provided to supply pulsed tones, useful in detecting nerve tumours and CNS lesions, and in detecting malingerers (in the latter case a continuous tone threshold which is appreciably better than a pulsed tone threshold indicates a malingerer).

Calibration

Where an audiometer is in constant use, a daily monitoring check on calibration is recommended. This can usually be done with a relatively inexpensive calibration set comprising a sound level meter earphone coupler to measure the output level and frequency response of the audiometer, and a sound level calibrator to check the readings of the sound level meter.

AUDIOMETRY

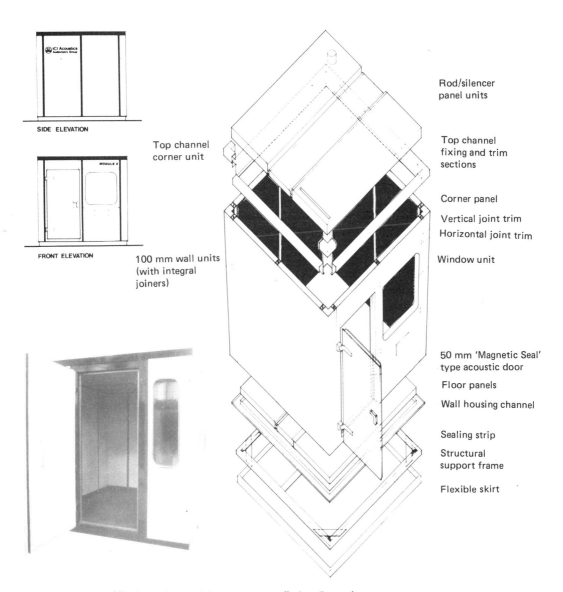

ICI Acoustics modular system medical audiometric room.

In any case both audiometer and earphones should be accurately re-calibrated every year at least, or preferably at 3 monthly intervals (depending on usage), using a standard calibration conforming to IEC303 and artificial ear to IEC318 respectively. Such a calibration service is usually provided by the manufacturer of the audiometer.

As a rough check for large calibration errors the operator can use his or her known threshold of hearing as a standard. This will not show up slight or progressive errors in calibration however.

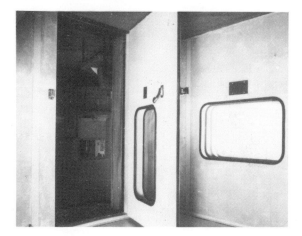

An IAC Audiometric Facility installed at Harrogate Hospital in Yorkshire in the late 60's.

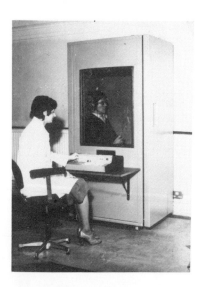

IAC 250 Sound Shelter being used for industrial audiology screening programmes.

The same installation upgraded with increased height, high quality wall finishes, aluminium framed acoustic windows and architecturally pleasing ceiling finishes. Throughout the original panels were totally incorporated into the updated design.

Audiometer Booths

To be valid audiometric examination must be conducted in a room where the background noise level is minimal. Satisfactory quiet areas may exist in a plant or office for monitoring tests, but as a general rule tests should only be conducted in a special acoustically tested room or enclosure (audiometric booth). This requires higher attenuation than that normally provided by a sound shelter. Recommended maximum ambient octave-band sound levels within audiometric booths

TABLE II
LIMITING AMBIENT OCTAVE BAND SOUND LEVELS FOR AUDIOMETRIC ENCLOSURES

31.5	76
63	62
125	48
250	36
500	14
1 000	16
2 000	29
4 000	37
8 000	32

TABLE III — LIMITED AMBIENT OCTAVE BAND SOUND LEVELS FOR EARPHONES FITTED WITH NOISE REDUCING MUFFS

Centre Frequency Hz	Sound Level dB
31.5	76
63	62
125	95
250	44
500	31
1 000	31
2 000	43
4 000	50
8 000	44

TABLE IV — NOISE LEVEL LIMITS FOR AUDIOLOGY ROOMS
(Department of Health and Social Security, 1974)

1/3 octave centre frequency Hz	dB SPL re 2×10^{-5} N/m^2	Octave bands
50	50	
63	46	51.5
80	42	
100	39	
125	36	41.4
160	33	
200	30	
250	28	33.0
315	26	
400	25	
500	23	28.3
630	22	
800	21	
1 000	20	25.1
1 250	20	
1 600	19	
2 000	19	23.7
2 500	19	
3 150	19	
4 000	19	24.1
5 000 and above	20	

are given in Table II. Any masking of the threshold of hearing by excessive ambient noise will show as an apparent hearing loss at the lower test frequencies (*ie* usually below 1 000 Hz). If satisfactory ambient levels cannot be achieved, then the use of earphones fitted with noise-reducing ear muffs may enable satisfactory audiometric tests to be conducted in a somewhat 'noisy' environment — see Table III. It should be noted, however, that muffs may also be required in an audiometric booth in order to achieve the required ambient noise levels where the noise is particularly loud in the vicinity of the booth.

Mobile IAC Audiometric facility for the Hong Kong Government in course of installation.

IAC Mobile Audiometric facility for the Hong Kong Government. Shown right in the course of construction.

Audiometric Programmes

A typical industrial/commercial hearing conservation audiometric programme could comprise the following:

Pre-employment audiogram, preferably measured over a wide range of audiometric frequencies including 250, 500, 1 000, 2 000, 3 000, 4 000, 6 000 and 8 000 Hz. A detailed pre-employment noise history and description of present work and noise environment. Individual instruction by the audiometrician or nurse about the hazards of noise and the need to wear hearing protection all the time, followed by the individual fitting of the appropriate protectors.

Monitoring audiometry after six months, one year, and then every year, together with further education and individual discussions about the noise environment and the need for hearing protection. If the first six-months audiogram indicates that the worker is acquiring a hearing loss, a further audiogram should be obtained after another six months, and so on until no further apparent loss occurs, after which audiograms may be obtained annually.

Immediately a deterioration in hearing is detected, an individual should be vigorously re-instructed in the need for protection of hearing. This aspect of monitoring audiometry is a most valuable contribution to the overall effectiveness of the educational aspects of a hearing conservation programme. In cases where individuals continue to lose their hearing, for whatever reason (apparently susceptible hearing, not wearing protection properly, refusing to wear protectors or medical cause) consideration should be given to the possibility of moving the individual to quieter work.

See also chapter on *Sound Level Meters*.

THE BEST PERIODICAL DEVOTED TO NOISE & VIBRATION CONTROL

Packed with practical, up-to-date, worldwide news, information and data, NOISE & VIBRATION CONTROL Worldwide covers all aspects of this important subject. This journal deals with its causes, effects, measurement, and methods of control. As legislation in all countries becomes more stringent, so too does the demand for this detailed specialist journal.

TRADE AND TECHNICAL PRESS LTD.,
CROWN HOUSE, MORDEN, SURREY, SM4 5EW, ENGLAND.

FOS.3B

TEL : (0635) 40240   TELEX : 847868 ENVAIR G

ENVIRONMENTAL EQUIPMENTS LTD.

TEST LABORATORY ENVIRONMENTAL FACILITIES USED TO ESTABLISH AND MAINTAIN OUR QUALITY ASSURANCE STANDARDS

- ✱ *3.4 KN Peak Thrust Vibration System with Digital Sine Control*
- ✱ *10,000 'g' Shock Testing Machine with Programmes for Half-Sine, Sawtooth and Square-Wave Pulses*
- ✱ *100,000 'g' Dual Mass Shock Amplifier*
- ✱ *20 - 10,000 'g' Half-Sine Shock Calibrator*
- ✱ *Sine-Wave Accelerometer Calibration System with BCS & NPL Certified Reference Standards*
- ✱ *Cross-Axis Accelerometer Sensitivity Machine*
- ✱ *50 lbs (22.5Kg) & 250 lbs (112.5Kg) Bump Testing Machines*
- ✱ *50 to 350°C Temperature Ovens* ✱ *50 to 600°C Fluidised Baths*
- ✱ *-70 to 150°C and -80 to 200°C Hi-Lo Temperature Chambers*
- ✱ *-70 to 150°C Thermal Shock Chamber*
- ✱ *10 to 90°C Humidity Chamber* ✱ *Thermal Vacuum Facility*

OUR RANGE OF ENVIRONMENTAL TESTING PRODUCTS.
Note: VIBRATION SYSTEMS USE PERMANENT MAGNET VIBRATORS. MAX. 450N.

- • NON-LINEAR ANTI-VIBRATION MOUNTS & PADS FOR INDUSTRIAL MACHINERY

- ■ PIEZO-ELECTRIC ACCELEROMETERS
- ■ CHARGE AMPLIFIERS
- ■ BUMP TESTING MACHINES
- ■ PERMANENT MAGNET VIBRATORS
- ■ AMPLIFIERS FOR VIBRATORS
- ■ ACCELEROMETER CALIBRATION SYSTEMS [BCS & NPL References]

- • STROBOSCOPES
- • STROBE' MOTION ANALYSER
- • RECORDING SPECTRUM ANALYSER
- • HELICOPTER ROTORS and AIRCRAFT PROPELLERS
  [Dynamic Tracking and Balancing Systems]

Note: • Indicates Agency Products with restricted sales areas.

FLEMING ROAD   LONDON ROAD INDUSTRIAL ESTATE
NEWBURY   BERKS   RG13 2DE   ENGLAND

SECTION 3b

Vibration Measurement

THE THREE components of vibration that are significant are amplitude, frequency and phase. Amplitude, in simple terms, is the amount of vibration; frequency is its repetition rate; and phase the relationship of one vibration to another.

Amplitude and Frequency

Amplitude parameters are defined in Fig 1, the measurable quantities being displacement, velocity and acceleration. The most common measurement is displacement, although this is not necessarily the best way of measuring vibration. Vibration velocity or acceleration can be more significant in certain cases. A particular point to note here is that equal displacements at two different *frequencies* do not result in equal velocities or accelerations. However, displacement, velocity and acceleration are all inter-related, as shown by the following formulas. (See also chapter on *Principles of Vibration*).

$$D = 0.318 \frac{V}{Hz} \qquad d = (1.910)(10^4) \frac{V}{CPM}$$

$$D = 19.607 \frac{A}{(Hz)^2} \qquad d = (7.059)(10^7) \frac{A}{(CPM)^2}$$

$$V = \pi(Hz)(D) \qquad V = (5.236)(10^{-5})(CPM)(d)$$

$$V = 61.440 \frac{A}{Hz} \qquad V = (3.696)(10^3) \frac{A}{CPM}$$

$$A = 0.051 (Hz)^2 (D) \qquad A = (1.417)(10^{-8})(CPM)^2 (d)$$

$$A = 0.016 (V)(Hz) \qquad A = (2.704)(10^{-4})(CPM)(V)$$

Simple Measurement

For measurement of overall vibration levels, simple instrumentation can be used comprising vibration pick-up, an amplifier to boost the signal strength, with readout through a meter calibrated in suitable vibration units of displacement (*eg* μmm or μin), velocity (*eg* mm/sec or in/sec) or acceleration (g). The main requirement for satisfactory working is that the frequency response of the pick-up and amplifier be flat over the range of vibration frequencies likely to be involved.

Many sound level meters can be adapted for this type of work when fitted with an accelerometer, particularly impulse sound level meters. If associated with octave filter set, such meters can also be used for vibration frequency analysis.

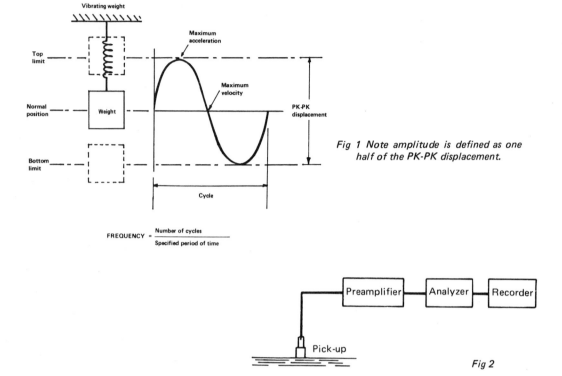

Fig 1 Note amplitude is defined as one half of the PK-PK displacement.

Fig 2

One distinct limitation of a sound level meter which uses a vibration pick-up system as a vibration meter, is that it is limited to the audio frequency range, unless the meter was originally designed as a dual purpose type with frequency ranges extended below 20 Hz and above 20 000 Hz. Extension of frequency above 20 000 Hz is seldom necessary, however, because vibration pick-ups themselves are limited to lower maximum frequency ratings by their resonance.

For vibration analysis the same basic set-up applies, except that instead of being directly indicated, the amplified signal from the pick-up is analyzed and then read out or recorded — Fig 2.

For types of pick-ups used, see chapter on *Vibration Transducers.*

Determination of Direction of Vibration

A mechanical system can be expected to vibrate in a direction corresponding to the least stiffness of the system, although oscillations in other modes may also be present. It is often difficult to decide which is the predominant plane or mode of vibration, so that the pick-up can be oriented for maximum sensitivity and reading range.

Where it is necessary to determine the component vibrations, or establish the orientation and value of the resultant total vibration, three pick-ups (or three separate readings) can be used, with readings obtained along three mutually perpendicular axes. The result can then be analyzed vectorially, but to do this the relative phase of the components must also be determined. This can be done by repeating the three sets of measurements with the axes rotated from the first set and determining the phase sums and differences by comparison.

The advantages of being tiny.

At Entran® we build a line of accelerometers that can best be described as tiny.

Being tiny means you don't interfere with the integrity of the system being measured.

Being tiny gives you greater handling ease in hard-to-get-at places, which adds up to more applications flexibility. And, being tiny means less mass to distort your test scale, more room for other instruments. All at no sacrifice in test data.

Entran accelerometers give you high output, low impedance readings, both static and dynamic. As a result, they can usually be interfaced directly with scopes, recorders and milliammeters.

Entran accelerometers come complete and ready to use. All you supply is the excitation voltage on two leads and read output in millivolts on the other two leads.

All units give you direct linear output in millivolts/g and read both static and dynamic acceleration. There's no need for charge or voltage amplifiers. Calibrations are traceable to The National Bureau of Standards.

Entran Ltd. Exclusive agent: THORN EMI Datatech Ltd., North Feltham Trading Estate, Feltham, Middlesex TW14 0TD. Phone 01-890 1477. Telex 23995.

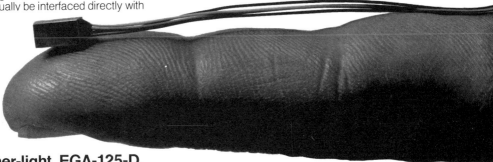

Tiny and feather-light, EGA-125-D.

This accelerometer weighs only one half a gram. Is less than 0.006 cu. inches in volume. Yet gives you static & dynamic readings of acceleration, vibration and shock. And comes with .7 critical damping to eliminate resonance. ("D" is for "damped.")

Typical ranges	± 5g	± 10g	± 50g	± 100g	± 500g
Sensitivity nom.	15 mV/g	12 mV/g	4 mV/g	2.5 mV/g	0.5 mV/g
Useful frequency nom.	150 Hz	200 Hz	500 Hz	600 Hz	1200 Hz

Excitation: 15V, Weight: ½ gram without leads, Linearity: ± 1%,
5 other mounting styles available at no extra charge.

Tiny but tough, EGC-500DS.

Guaranteed overrange of 1000%. And only a half inch cube to boot.

Static & dynamic measurements of acceleration, vibration and shock in ranges from 5 to 1,000 g. Fully damped to eliminate resonance and fitted with mechanical stops to impede overload response.

Typical ranges	± 5g	± 10g	± 50g	± 100g	± 500g
Sensitivity nom.	25 mV/g	20 mV/g	4 mV/g	2 mV/g	0.5 mV/g
Useful frequency	150 Hz	200 Hz	600 Hz	1000 Hz	1500 Hz

Excitation: 15V, Damping: 0.7 cr. nom., Linearity: ± 1%
Output impedance: 450 Ω nom., Input impedance: 1000 nom.
Size: 0.5" x 0.5" 0.6" width, 10-32UNF tapped hole

Entran Ltd.
(Today, miniature makes sense.)

VIBRATION TRANSDUCERS (B)

BBN is number one in accelerometers. Twice.

OUR 500 SERIES
smart accelerometers operate over a temperature range of −65°F (−54°C) to 250°F (121°C)
- Internal electronics with lowest noise floor.
- Completely eliminates costly charge amplifiers.
- Drives several hundred feet of cable.
- Mains or battery operation for laboratory or field environments.
- Monitors aircraft compressor bearings, gas turbines and motors.
- Smallest available sub-miniature circuited accelerometers.
- New inverted shear design, thermal shock insensitive.
- High performance underwater and shock accelerometers available.

Not only did an independent survey find us to be the number one manufacturer of accelerometers (we offer the broadest piezoelectric accelerometer line available for quality, value, and performance), but also to be number one in preference as a supplier. We won't rest on our laurels. We'll tackle any problem you can throw at us: off-the-shelf or custom, for OEM or end-user. We also offer signal conditioners, power supplies, amplifiers, systems and accessories. Do business with number one. For two good reasons and more.

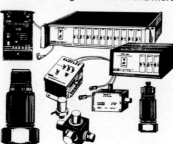

OUR 400 SERIES
hot accelerometers with cables and charge amplifiers operate to 750°F/400°C
- High sensitivity and frequency response.
- Case electrically isolated.
- All welded hermetic construction.
- Nuclear-hardened models.
- Ideal for monitoring turbines, compressors, aircraft engines and nuclear power plants.
- New inverted shear design, thermal shock insensitive.

We respond.

BBN Instruments

50 Moulton Street,
Cambridge, MA 02238 U.S.A.
Phone: (617) 491-0091/Telex: 92-1470
Cable: BBNCO
A Subsidiary of Bolt Beranek and Newman Inc.

THE PORTABLE CF300 FFT ANALYSER

Bigger inside than out.

The CF300 is small. And light. Weighing only 12.5kg it's the worlds smallest portable FFT Analyser with CRT display. And it's small in cost.
However, when it comes to performance, the CF300 assumes new proportions. An impressive list of **standard** features includes GP-1B interface bus, time axis averaging and histograms, power spectrum three dimensional display, phase spectrum, amplitude histograms — the list goes on and on. And when you take into account the largest memory for its type (32k words) and the add on option facilities its a veritable giant.
Put the CF300 into perspective for yourself.
Also available, the dual channel CF500

Just circle the enquiry number or contact:-

Hakuto International (UK) Ltd.,
Opto Electronics Division,
159a, Chase Side,
Enfield, Middx, EN2 OPW.
Tel: 01 367 4633 Telex: 299288

Agents Addresses
Telonic/Berkeley U.K.,
2 Castle Hill Terrace,
Maidenhead,
BERKS.

J.T.Sinclair & Co.,
8 Dixon Place,
College Milton North,
East Kilbride
GLASGOW.

VIBRATION MEASUREMENT

Hand-Held Pick-Ups

A pick-up held in position by hand is seldom capable of yielding accurate vibration measurements, although it may be the obvious approach for exploratory analysis. It can also give reasonably satisfactory results where the vibrating system is massive and the vibration frequencies are below 1 000 Hz.

For hand-held measurement, the pick-up is normally fitted with a probe. This will inevitably modify the frequency response of the pick-up — see Fig 3. The reading obtained is also likely to be affected by 'hand tremor' imparting additional vibration to the probe. Largely for this reason, repeated hand-held measurements are often inconsistent.

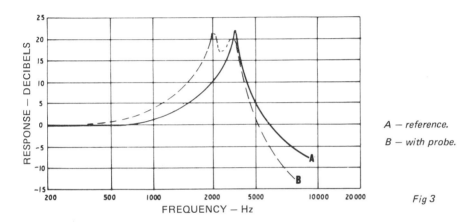

A — reference.
B — with probe.

Fig 3

Readout

Readout from displacement- or velocity-measuring pick-ups is given directly in engineering units. In the case of accelerometers, readout may be displayed in a number of ways, *eg:*

Peak value, $V_p = V_{max}(t)$

Average (absolute) value $V_A = \dfrac{1}{T} \int (VT) \, dt$

RMS value $V_{RMS} = \dfrac{1}{T} \int_{t_1}^{t_2} (VT)^2 \, dt$

where

T = time interval from t_1 to t_2

Crest factor $= \dfrac{V_p}{V_{RMS}}$

Form factor $= \dfrac{V_{RMS}}{V_A}$

Modern circuits are capable of detecting the RMS, average absolute and half peak-to-peak values of the input signal on the same linear scale. The typical frequency range of such instruments is 2 to 200 000 Hz. The response scale is linear, but a logarithmic scale is also usually provided to give equivalent readings in dB; RMS indication provides valid values where the waveform is periodic (sinusoidal). For non-sinusoidal waveforms the crest factor and form factor will vary, calling for modification of the circuit design to indicate true RMS values.

Note that velocity values are obtained by integrating the acceleration signal once. An integrator may be employed with an accelerometer and sound level meter to enable vibration velocity readings to be obtained from the accelerometer. This principle is used with certain portable sound level meters where the addition of an integrator and accelerometer enables the instrument to be used either for direct measurement of sound pressure levels or vibration frequency (with suitable conversion from the dB scale).

Level Recorders

Level recorders are designed to record vibration levels, frequency response characteristics, spectrograms, *etc,* in a convenient and permanent manner, normally by writing on calibrated recording paper. A single recorder is usually designed to accommodate a variety of analyzers, the combination chosen depending on the specific programme undertaken. Recorders are precision instruments of some complexity. A block diagram of a level recorder is shown in Fig 4.

Other recorders include tape recorders and oscilloscopes. Tape recorders are normally of two-channel type, and specially designed. Phase interlocking of the two channels is usually provided to permit analysis of two time-interdependent phenomena.

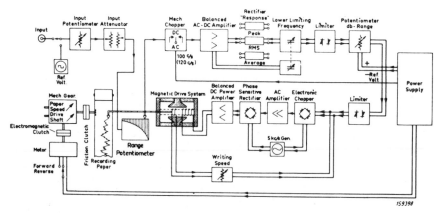

*Fig 4 Block diagram of level recorder.
(Bruel and Kjaer).*

Microphone Amplifiers

A microphone amplifier is a multi-purpose instrument for use in vibration and sound measurement and comprises an input amplifier, weighting networks, an output amplifier or a meter circuit. In the vibration field it is used for linear and selective vibration measurement in conjunction with a suitable accelerometer, with a capability of reading both velocity and displacement of vibrations, as well as RMS, average or half peak-to-peak values of the input signal. Frequency ranges of typical readout devices are shown in Table I.

VIBRATION MEASUREMENT

TABLE I – FREQUENCY RANGE OF READ-OUT DEVICES

Device	Frequency Range Hz
Electronic voltmeters	2–200 000
Frequency analyzers	2–40 000
Frequency spectrometers	2–45 000
Microphone amplifiers	2–200 000
Sound level meters	10–20 000
Level recorders	2–200 000

Calibration

Vibration meters commonly have a built-in calibrator, but this will serve only to check the electrical system. A vibration calibrator should be used regularly to check the complete system. Manufacturers of vibration measuring instruments invariably supply calibrators 'matched' to the requirements of their products, with detailed instructions as to their use. It can be mentioned, however, that the accuracy of calibration achieved will depend partly on the accuracy of the calibration data and so is not covered by calibration check.

Pick-ups also normally need to be returned to manufacturers periodically for frequency checks. Long term stability can be estimated as 5% per year or better. This figure represents the change in calibration data and so is not covered by calibration checks.

The transience sensitivity of piezoelectric pick-ups may also suffer a temporary shift if the pick-up is subject to shock. Unless the shock has been severe enough to cause permanent damage to the pick-up, transience sensitivity should revert to its original figure within 24 hours.

Vibration level meter. (Rion Co Ltd).

Vibration Frequency Analysis

Frequency analysis provides spectrum analysis of vibration frequencies in the same manner as noise frequency analysis. In fact identical instruments can be used to analyze both noise and vibration signals, provided they have the necessary frequency range (vibration analysis may extend to

GenRad's Vibration Control Systems.

lower and higher frequencies than the audio range). Fourier analysis is now the most common method employed — see chapter on *Dynamic Analysis of Vibration*.

Mention can be made here of the difference between time domain and frequency domain analysis, as an introduction to dynamic analysis. Virtually all practical vibrations are of complex waveform nature, but can be represented by a composite of separate sine waves (this is the basis of Fourier analysis). Effectively, therefore, the spectrum can be represented as a three-dimensional 'picture' with co-ordinates in amplitude, time and frequency — Fig 5.

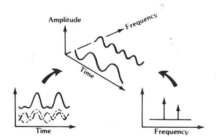

Fig 5 Relationship between time and frequency domains.

The conventional two-dimensional diagram shows the complex waveform related to amplitude and time co-ordinates only, and as such, referred to specifically as the *time domain* view. If, however, the three-dimensional 'picture' is viewed looking along the time axis (rather than the frequency axis as previously), the resulting two-dimensional plot now shows 'end-on' views of the individual frequencies present, which now appear as straight (vertical) lines. This is called the *frequency domain view*.

It is obvious that whilst signal analysis in the time domain shows the overall nature of the input waveform (the conventional spectrum), presentation in the frequency domain provides a unique picture of the individual components making up the original complex waveform. In particular, it can show the presence of individual signals of small signal strength which may not be apparent in the time domain presentation.

See also chapter on *Modal Analysis*.

Vibration Transducers

THE THREE types of transducers normally used for vibration measurement are:
  (i)   Non-contact pick-ups, normally called *proximity probes,* which work on the eddy-current principle (and can also measure rotational speed).
  (ii) The *velocity pick-up* which is a seismic device employing either a moving coil or a moving magnet.
  (iii) *Accelerometers,* which produce an output signal proportional to acceleration.
        Advantages and disadvantages of each type are summarized in Table I.

Proximity Probes

A proximity probe comprises, basically, a probe, a driver and the probe-to-driver connecting cable. The driver produces a high frequency signal which is fed into a coil in the probe which generates a magnetic energy field around the probe tip. When a conductive material (shaft, *etc*) is approached by the probe tip, eddy-currents are produced on the surface of the material, and power is absorbed which affects the field strength (Fig 1). The amount that the field strength is affected is proportional to the distance from the object to the probe tip. The closer the object, the more power is

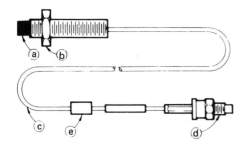

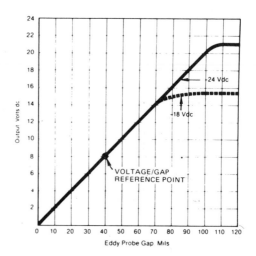

Fig 1 Eddy probe — *functional diagram.*
a) *probe tip*
b) *lock nut*
c) *coaxial cable (flexible armour optional)*
d) *coaxial connector*
e) *floating sleeve*

Typical response.

VIBRATION MEASUREMENT (A)

Castle-Microair

Sound Level Instrumentation

Castle-Microair serves the noise testing, monitoring, and analysis sciences. Our products give a total capability in noise and vibration testing, control and recording.

SOUND LEVEL METERS Hand-held sound pressure and also integrating/averaging/impulse types.

MODEL GA301 Simple survey meter to IEC651 Type 3, for colleges, universities, hospital training and monitoring applications. Used extensively by machine installation and commissioning engineers.

MODEL GA201 General use/Industrial Grade meter for factory noise monitoring applications for the assessment of hearing damage risk.

MODEL GA101 Precision Grade meter for the precise measurement of noise from machinery under controlled conditions for noise control assessments.

MODEL CS192(B) Professional model, for acoustic engineers and research applications, assessment of noise from office machinery, vehicle survey work, and building acoustics.

Full supporting accessories are available including calibrators, tripods, boom microphones, audio headsets, octave-band filters, vibration adaptors, level recorders, tape recorders, chart recorders and XY plotters.

Contact Geoff Faragher for further details at Castle-Microair Ltd., Unit 1, Hogwood Industrial Estate, Finchampstead, Wokingham, Berks. RG11 4QW. Tel: (0734) 730050

dynamic instrumentation from Endevco

Shock and vibration measurement accelerometers and instrumentation

Pressure and acoustic measurement transducers and instrumentation

Engine vibration monitoring systems

Acoustic emission sensors and instrumentation for laboratory and industrial applications

Loose particle detection for semi-conductor quality assurance

Short range wireless data couplers

Packaging testing equipment

Mechanical shock testing machines

Vibration generators for modal and seismic testing

Accelerometer and signal conditioner calibration service

ENDEVCO U.K. LTD.

Melbourn, Royston, Herts. SG8 6AQ.
Tel. Royston (0763) 61311   Telex 81522

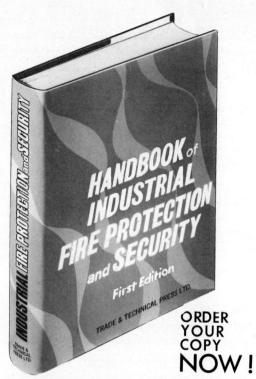

ORDER YOUR COPY NOW!

A 600 page Handbook, specially produced for those concerned with the safety and security of factories, works, warehouses and offices. A complete reference work, of paramount importance in these days of high fire risk, breaking and entering, vandalism, arson and political anarchy. Contents include:- 1. Fire: Fire Prevention; Fire Training; Fire Engineering; Fire Alarms; Fire Fighting Equipment Appliances; Fire Extinguishers — all types; CO_2 Installations; Fire Protection Equipment, Blankets, Safety Clothing, Eye Protection, etc. etc. 2. Hardware: Locks; Grilles; Safes, Strongrooms; Safety Containers, etc. 3. Detection: Fire Alarms; Fire Detection Systems; Heat detectors; Smoke Detectors; Gas and Fume Detectors; Burglar Alarms. 4. Security: Industrial Security; Alarm Systems; Guard Systems; Private Police Forces; Security Vehicles. 5. Buildings: Materials; Fire Resistant Structures; Fire Ventilation; Fire Doors; Escape Systems. 6. Materials: Fire Resistance; Fire-resistant and Non-Combustible Materials; Fireproof Materials; Fireproofing of Materials; Plastic in Fires. 7. Environmental: Electrical Fires; Hazardous Environments; Hazardous liquids; Hazardous Stores. 8. Emergency: First Aid; Ambulances; Escape Equipment; Rescue Equipment; Emergency Lighting and Power; Emergency Services; Communications Equipment; Public Address Equipment. 9. Reference: Industrial Fire Brigades; Fire Prevention Acts and Regulations; National Authorities; Publicity Material; Rescue Corps; Salvage and Wrecker Services; Security Services. 10. Trade Names Index, Classified list of equipment and materials, manufacturers. etc.

**TRADE & TECHNICAL PRESS LTD.
CROWN HOUSE, MORDEN, SURREY**

TABLE I — VIBRATION TRANSDUCERS

Type	Advantages	Disadvantages
Proximity probe	(i) non-contacting (ii) measures motion directly (iii) measures in engineering terms (iv) solid state device with no moving parts (v) measures dynamic motion and average position simultaneously (vi) excellent frequency response (vii) small size (viii) well suited to machine environments (ix) easily calibrated (x) accurate low frequency amplitude and phase angle information (xi) high level low impedance output.	(i) tends to be excessively sensitive to shaft run-out. (ii) can be sensitive to shaft material. (iii) requires external power source. (iv) can be difficult to install. (v) limited maximum service temperature (typically $175°C$ ($350°F$) for the probe and $100°C$ ($212°F$) for driver).
Velocity pick-up	(i) ease of installation (ii) strong signal in mid-frequency range (iii) no external power required (iv) may be suitable for high temperature environment	(i) relatively large and heavy. (ii) sensitive to input frequency. (iii) relatively narrow frequency response. (iv) moving part device (subject to wear). (v) difficult to calibrate. (vi) measures dynamic motion only.
Accelerometer	(i) good frequency response, especially to high frequencies (ii) small and light weight (iii) strong signal in high frequency range (iv) may be suitable for high temperature environment	(i) sensitive to input frequency. (ii) relatively expensive. (iii) difficult to calibrate. (iv) requires external power source. (v) sensitive to spurious vibrations. (vi) impedance matching necessary; also some filtering for monitoring applications.

absorbed. The driver then measures the change in field strength and converts it into a standard calibrated output. This output is normally 200 mV per mil (0.001 m inch) of air gap between probe and object.

An eddy-current proximity probe is used to measure displacement — static and/or dynamic — and can be used to measure shaft vibration (usually with respect to the journal bearing housing), as well as the axial position of a rotor to establish, for example, the axial thrust bearing condition. Although there are other types of pick-up that can measure static and dynamic displacement, they are neither as accurate nor as reliable as the eddy-current probe.

The output of the eddy probe is linearly proportional to the shaft displacement, as shown in Fig 1. The probe is driven by an oscillator and the resultant probe signal is demodulated and amplified. The unit containing the oscillator/demodulator/amplifier is called the 'driver'. A typical eddy probe is linear over a range of 2.5 mm (0.1 in) with a sensitivity of approximately 200 mV/m inch and a frequency range of 0–10 000 Hz. The eddy probe having no moving parts, is very robust and therefore very reliable.

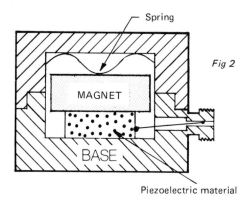

Fig 2

Piezoelectric material

1510 Industrial accelerometer — totally sealed to operate submerged if necessary. (Condition Monitoring Ltd).

Velocity Pick-Ups

A typical velocity pick-up is shown in Fig 2. A cylindrical coil is mounted on the case of the pick-up and a permanent magnet is spring-suspended within the coil. If the case of the pick-up is firmly placed against the vibrating body, the case vibrates at the same magnitude as the body. The spring-suspended magnet tends to stay stationary. The relative motion between the coil and magnet causes the magnetic lines of force to cut through the coil, inducing in it a voltage proportional to the velocity of vibration. Damping is required in the pick-up to lower the natural frequency of the spring mass (magnet) system, and to keep the magnet from vibrating more than the actual vibration.

The velocity pick-up is often used to measure bearing cap vibration. This type of pick-up is a seismic device, either moving coil or moving magnet, and generates an output voltage proportional to velocity.

Sometimes it is required to monitor the bearing cap vibration in terms of displacement. In that case the velocity signal is electronically integrated so that the pick-up will read out in micrometers or μin of displacement. The frequency range of this type of pick-up is typically 20–2 000 Hz.

A velocity pick-up that is intended for use in an electro-magnetic field, *eg* of an electric motor or generator, must be shielded to avoid the generation of a false signal.

An important consideration when using a velocity pick-up is the mounting angle; some probes are designed to be mounted at one particular angle only, *ie* vertically or horizontally or at some other specific angle, while some designs are adjustable.

Excessive temperature is a common cause of failure on pumps and gas turbines with this type of transducer. The maximum ambient temperature of the machine should not exceed the maximum operation temperature of the pick-up which is typically +250°C (+480°F)

Accelerometers

There are four major types of accelerometer — piezoelectric, strain gauge, capacitance and servo. Of these the *piezoelectric* is normally used for vibration measurement.

The major advantage of this type of transducer is its superior frequency range, which can cover up to 25 kHz or higher, although 5 kHz is usually sufficient for rotating machinery monitoring. Since there are no moving parts in this transducer, it is robust and therefore very reliable.

The piezoelectric accelerometer is a high impedance device and requires an impedance matching amplifier to be connected in close proximity, so that the output signal cable connects to a low impedance. This ensures that the signal cable length between accelerometer and monitor is not

VIBRATION TRANSDUCERS

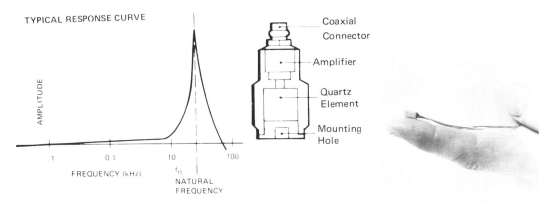

Fig 3 Typical accelerometer.

Entran miniature strain gauge accelerometer. (Vibromfilter UK)

critical and that the system is not prone to extraneous noise pick-up problems. Typically, an accelerometer can be up to one hundred metres away from the read-out instruments. The ideal monitoring accelerometer contains the matching amplifier (emitter-follower) within the same body (as in Fig 3).

Measurement of Shaft Speed

A simple way to measure speed is to generate pulses using an eddy probe in proximity to a discontinuity on the shaft — a bolt head or dimple. Another very effective way to measure shaft speed is to use a fibre optic tachometer. This consists of a flexible light guide about 1.2 metres long connected to an electronic unit. The flexible guide contains mixed light transmitting and light receiving fibres terminating in a focussing optical head. Light is carried from its source through the guide and lens to the target. The target is a reflective strip (or strips) on the shaft. Light reflected from the target is picked up by the lens and carried back to the detection circuit in the electronics unit, which converts the light pulse to a noise-free electric pulse whose amplitude is constant, regardless of speed. Since the transmitting and receiving optical fibres are mounted in the same head assembly, alignment problems are eliminated. In this way a train of constant-amplitude pulses is generated, the frequency of which is directly related to shaft speed. This type of fibre optic probe works at a distance of 6.35–483 mm from the target and generates pulses over a range of 0.02 pulse/sec to 6 000 pulse/sec. A typical fibre optic tachometer is shown in Fig 4.

Thus, the versatile eddy-current probe which can be used for the measurement of static and dynamic displacement as well as rev/min, the velocity pick-up, the accelerometer and the fibre optic tachometer are typical basic transducers for the machine health monitoring engineer.

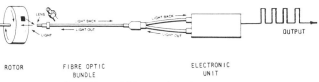

Fig 4 Typical fibre optic tachometer

TABLE II – SELECTION PARAMETERS FOR TRANSDUCERS

Transducer Selection Parameters		Preferred Pick-up		
		Non-contact pick-up	Velocity pick-up	Accelerometer
Mode of measurement	(i) acceleration (ii) velocity (iii) displacement (iv) axial position	 X X	 X X 	X X X
Type of bearings	(i) sleeve (ii) anti-friction (–ball) (iii) anti-friction (–roller)	X 	 X X	 X X
Speed of machine, or dominant frequencies to be measured	(i) 1–15 Hz (ii) 15–1 500 Hz (iii) 1 500 Hz up	X X X	 X 	X X X
Temperature at mounting point	Note temperature limitations of each type of probe	175°C (350°F) Max	200–250°C (400–500°F) Max	70–290°C (160–550°F) Max
Cable length, strength of signal	Note typical pick-up to monitor cable length limitations	Up to 450 m (1 500 ft)	Up to 300 m (1 000 ft)	30–300 m (100–1 000 ft)
Installation requirements		Easy	Hard	Hard
Relative mass, rotor to case	Rotor mass $>$ case mass Rotor mass $<$ case mass	 X	X 	X
Unusual installation problems	Environment, chemicals, *etc*	Depends on application		

Transducer Selection

A basic guide to transducer selection is given in Table II. The foremost governing factor is the mode of measurement – *eg* only an accelerometer will measure acceleration value of a vibration. However, velocity can be measured by a velocity pick-up, or an accelerometer (by integrating the signal in the latter case). Displacement can also be measured by a velocity pick-up when its signal is integrated once, or by an accelerometer whose signal is integrated twice. Axial piston, (a non-vibrating air gap), can be measured only by a non-contact probe.

Other selection parameters are:

(i) *Speed of Machine, or Dominant Frequencies to be Measured.* This parameter basically serves to indicate the limited range of the velocity pick-up, and its insensitivity to low frequency vibrations. Various problems within a machine can generate vibration frequencies from ½–50 times rotating speed. This must also be taken into account when determining the frequency to be measured.

(ii) *Temperature at Mounting Point.* The typical maximum operating temperature limitation is listed for each type of transducer. In general, if there is a range listed for the maximum operating temperature, the higher temperature units are more expensive. The 20°C (160°F) listed under accelerometers reflects an accelerometer with an integral charge amplifier.

(iii) *Cable Length, Strength of Signal.* Table II lists the general limitations of lengths of cable which can be used between the pick-up and the monitor. All three types of pick-ups require a good grade of twisted, shielded, transducer cable. The accelerometer is listed as 30–300 mm (100–1 000 ft) and the 30 m (100 ft) limit is for a transducer with a charge output, the 300 m (1 000 ft) limit refers to an accelerometer with an integral charge amplifier. It should be noted here that all three types of transducers generate a relatively low level a.c. signal. Proper transducer cable installation is critical to the overall operation of the system. Transducer cable runs near or parallel to high voltage or high current cable can induce false signals into the system.

(iv) *Installation Requirements.* Accelerometers and velocity pick-ups are normally installed *via* a ¼–28 stud on the machine and thereby rate 'easy' for ease of installation. The non-contact probe with its necessary probe tip clearances, *etc,* rates 'hard' for hard to install. The non-contact probe is difficult to install on a retrofit programme or a machine that is in service. In addition to complexity of installation, physical space limitations for mounting the transducer must also be considered.

(v) *Relative Mass.* When the mass of the case of the machine is much greater than the mass of the rotor, such as in a boiler feed pump, the shaft forces may not be sufficient to cause significant case vibration, and a non-contact probe would be preferable. With light case machines, the case tends to follow the shaft vibration and a velocity pick-up or accelerometer is adequate.

(vi) *Machine Limitations/Past Machine Problems.* Past machine problems or specific machine problems that are being protected against should also be kept in mind. If a particular machine has been destroying motor mounts or connecting ductwork, a pick-up which measures overall case motion (velocity pick-up or accelerometer) would be preferred. If you are trying to monitor a rotor's position within a housing very accurately to protect against possible mechanical interference, a non-contact probe would be preferred.

(vii) *Unusual Installation Problems.* Various environmental factors can affect transducer selections. Is the machine located in an unusual place such as on a ship or very flexible platform? Is the machine subject to many starts and stops? Will the transducer be subjected to salt air, corrosive chemicals or other unusual substances? Do the pick-ups need to be protected against physical damage? These are all important questions when applied to transducer selection.

(viii) *User Experience.* An often ignored factor is past experience with a certain type of pick-up. If a plant and its crew are experienced in the installation and characteristics of a velocity type pick-up, there will be fewer problems with the installation than if a different transducer were selected.

See also chapter on *Machinery Health Monitoring.*

Dynamic Analysis of Vibration

ESSENTIALLY TWO forms of dynamic analysis can be used to obtain the data from which inferences can be made concerning the prescribed properties of the system or structure under investigation. The generic terms for these two forms of dynamic analysis are 'time domain analysis' and 'frequency domain analysis' respectively.

Time Domain Analysis

Tools available to facilitate the time domain analysis of a system or structure include transient capture and display instruments, correlators, and sophisticated parameter estimation algorithms for implementation on substantial real-time computer installations. Generally these tools provide rapid estimates of the time-domain parameters or response characteristics of the object under investigation. Where the object under investigation is operating in a noisy environment, or exhibits significantly non-linear behaviour, or where both conditions pertain, the rapid estimates thus obtained are at their best only very approximate; in the worst case they may be totally incomprehensible or ambiguous. The effects of noise on estimates of the parameters or response characteristics of the system can generally be reduced at the expense of a much increased measurement time. Other effects, for example non-linearity in the system, can grossly distort estimates of the response characteristics making it impossible to infer any relationship between the measured characteristic and the prescribed property being used to define the behaviour of the system.

Time domain analysis techniques should not be dismissed lightly for there are many instances where these techniques can be used with considerable effectiveness. It must be said however, that as a general rule measured time domain response data require considerably more skilful interpretation in terms of the parameters of any prescribed mathematical model used to quantify the behaviour of the system or structure to a given class of stimuli.

Solartron 1200 signal processor.

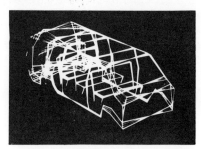
Modes of vibration of a van body. (GenRad Ltd).

DYNAMIC ANALYSIS OF VIBRATION (A)

Laser Vibrometer

The DISA Laser Vibrometer is an integral unit for the remote, non-contact point measurement of vibration.

The Vibrometer has a linear transfer function, high sensitivity and a very wide dynamic range. It features rapid response and operates with a remote range of up to 20 m.

DISA is a leading manufacturer of Hot Wire/Film and Laser Anemometry equipment for the measurement of flow, turbulence and temperature.
If you would like to know how we can solve your measuring problem, please contact us.

DISA ELEKTRONIK A/S, Mileparken 22, 2740 Skovlunde, DENMARK
DISA, Techno House, Redcliffe Way, Bristol BS1 6NU, UNITED KINGDOM
DISA ELECTRONIQUE S.A.R.L., 2 bis, rue Leon Blum, F - 91120 Palaiseau, FRANCE
DISA ELEKTRONIK G.M.B.H., 7500 Karlsruhe 41, Postfach 410267, FEDERAL REPUBLIC OF GERMANY
DISA ELETTRONICA, (Italiana) SLR, Viale Farmagosta 75, I - 20142 Milano, ITALY
DISA ELECTRONICS, 779 Susquehanna Avenue, Franklin Lakes, New Jersey, U.S.A.
DISA ELECTRONICS LTD., 140 Shortings Road, Scarborough, Ontario M1S 3S6, CANADA

DYNAMIC ANALYSIS OF VIBRATION (B)

The Ultimate Signal Processor
"in a world of its own"

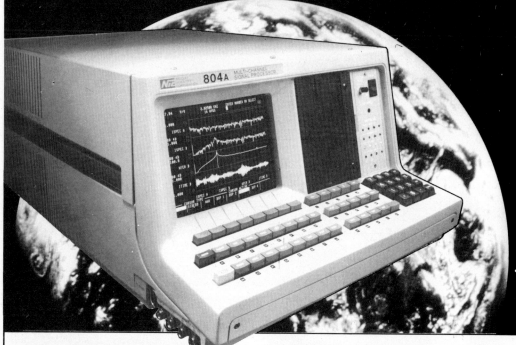

The new era in signal processing has begun with the Nicolet 804A multi-channel signal processor.

The power of a computer with the simplicity of an instrument.
- 4 Channels processed simultaneously.
- Over 500K bytes of built-in memory plus mass storage on disc.
- User friendly through soft key control.
- Programmable for solving specific problems.
- Large built-in display with video output.
- Plots in 8 colours directly to most digital plotters.
- Optional Nicolet Zeta 8 pen digital plotter.
- Optional full computer key board.
- Standard Modal Analysis.

Nicolet continue to lead the world in Spectrum Analysis.

 Nicolet Instruments Limited

A Nicolet Instrument Subsidiary   Budbrooke Road, Warwick, England, CV34 5XH. Tel: (0926) 494111 Telex: 311135

Frequency Domain Analysis Techniques

Essentially frequency domain analysis techniques are based on 'Fourier Transform', which relates an arbitrary time function x(t) to a continuous complex frequency function X(ω) as follows:

$$X(\omega) = \int_{-\infty}^{+\infty} x(t) e^{-j\omega t} dt \tag{1}$$

If x(t) is a complex periodic time function, with period T, then the frequency domain function X(ω) reduces to a discrete complex frequency spectrum, termed the Fourier Series, the k-th component of which is expressed as follows:-

$$X(\omega_k) = \frac{1}{T} \int_{-T/2}^{T/2} x(t) e^{-j\omega_k t} dt \tag{2}$$

When the time signal takes the form of a sampled signal $x(t_n)$, the transform becomes the Discrete Fourier Transform resulting in a spectrum which is both discrete and periodic in the frequency domain, as follows:

$$X(k) = \frac{1}{N} \sum_{n=0}^{n-1} x(t_n) e^{-j \frac{2\pi nk}{N}} \tag{3}$$

where N is the number of time samples taken

Thus any arbitrary time function can be described either in the time domain, or alternatively, as a continuous or discrete spectrum in the frequency domain.

A further function that has some considerable significance in frequency domain analysis relates the input stimulus of a linear dynamic system to the resulting output. This relationship, known as 'Convolution Integral', is expressed as follows:

$$y(t) = \int_{-\infty}^{+\infty} h(\tau) \times (t - \tau) d\tau \tag{4}$$

where x(t) is the input function (stimulus)
y(t) is the response function

and h(τ) is termed the impulse response function of the system connecting the stimulus x(t) to the response y(t).

It can be shown from equations (1) and (4) that the Fourier Transform Y(ω) of the response function y(t) is related to the Fourier Transform X(ω) of the input function x(t) as follows:

$$Y(\omega) = H(\omega) \cdot X(\omega) \tag{5}$$

where H(ω) is the Fourier Transform of the impulse response function h(τ) of the interconnecting system. The complex frequency response function H(ω) is commonly referred to as the 'Transfer Function'.

Examination of equations (4) and (5) shows that, by virtue of the Fourier Transform relationship, the complicated mathematical operation of convolution in the time domain is transformed to the relatively simple mathematical operation of multiplication in the frequency domain. This is significant in that it simplifies considerably the interpretation of the frequency response data in terms of prescribed parameters quantifying the behaviour of the system to a given stimuli.

Basically, the practical implementation of the Fourier relationships has resulted in the evolution of two very distinct measurement techniques. One approach is based upon the Fourier Series relationship, equation (2), whereby the response of the system or structure is measured relative to a single frequency sinusoidal function synchronous with the source of excitation. The alternative approach uses the Fourier Transform, equation (1), to calculate the frequency components of the response of the system or structure to a broad-band signal which may be either deliberately imposed, if the system is passive, or self-generated. Generally these two techniques are termed single sine analysis and FFT analysis respectively.

Superficially it would seem that irrespective of the analysis technique chosen, the end product will be the same, *ie* the frequency domain characteristics of the system or structure under investigation. In practice however, the two distinctly different approaches to the determination of the frequency response data each have their own particular merits and limitations. It is the understanding of these particular advantages and disadvantages which is the essential ingredient in the judicious selection of the right technique for a given application.

Single-Sine Analysis Techniques and Tools

It was shown previously, equation (2), that the Fourier Transform of a complex periodic time function, x(t), with period T, reduces to a discrete complex frequency spectrum $X(\omega_k)$. Using Demoivres Theorem, equation (2) can be expressed in the alternative form:

$$X(\omega_k) = \frac{1}{T} \int_{-T/2}^{T/2} x(t) [\cos \omega_k t + j \sin \omega_k t] \, dt \tag{6}$$

where j is the complex operator

It can be shown, by calculating the Inverse Fourier Transform of equation (2), that the complex periodic function x(t) may be expressed as the General Fourier Series:

$$x(t) = \frac{b_0}{2} + \sum_{n=1}^{\infty} (b_n \cos(nt) + a_n \sin(nt)) \tag{7}$$

where

$$b_n = \frac{1}{T} \int_{-T/2}^{+T/2} x(t) \cos(nt) \, dt \tag{8}$$

and

$$a_n = \frac{1}{T} \int_{-T/2}^{+T/2} x(t) \sin(nt) \, dt \tag{9}$$

Substituting the RHS of equation (7) in place of x(t) in the RHS of equation (6) and evaluating the integrals, yields the result:

$$X(\omega_k) = a_k + jb_k \tag{10}$$

where

$$a_k = \frac{1}{T}\int_{-T/2}^{+T/2} x(t) \sin(\omega_k t) dt \tag{11}$$

and

$$b_k = \frac{1}{T}\int_{-T/2}^{+T/2} x(t) \cos(\omega_k t) dt \tag{12}$$

Consequently it is seen that by evaluating the integrals expressed by equations (11) and (12) the real and imaginary components respectively of the complex Fourier Series at the single frequency ω_k are determined.

Essentially all single-sine Frequency Response Analyzers (FRA's) are practical implementations of the equations (11) and (12) and are used to resolve the real and imaginary components of the complex Fourier Series at the selected single measurement frequency ω_k. By incorporating the facility to vary the measurement frequency over a number of decade ranges, the FRA is made a useful tool for evaluating the Fourier Series of a time signal x(t) over a wide frequency spectrum.

Further examination of equations (11) and (12) reveals some additional features of FRA operation of significant interest to the vibration engineer. Firstly, it can be seen that if the signal x(t) contains, in addition to a fundamental component, any harmonic components of the measurement frequency ω_k, the effects of these harmonic components will be averaged to zero by virtue of the computation of the integrals over a complete number of periods of the measurement frequency. Consequently, the single-sine FRA technique is seen to exhibit excellent harmonic rejection characteristics. Clearly, if the FRA is set to measure at an harmonic of the fundamental frequency, then the instrument will act as a very high Q resonant filter and reject all other harmonic components, including the fundamental.

A second interesting feature of the implementation of the equations (11) and (12) in the single-sine FRA concerns measurement noise. Most signals derived from systems or structures contain, in addition to the desired frequency response information, some random noise component. Again, the integrating operation performed in the implementation of the equations (11) and (12) serves to 'average out' this random noise component such that, if the integration period is chosen sufficiently long (ie a large multiple of integer periods of the measurement frequency ω_k), the effects of the noise on the measured component values can be reduced to an insignificant level. The diagram (Fig 1) illustrates the improvement in frequency selectivity to be gained by increasing the integer number of periods of the measurement frequency ω_k over which the integration operation is performed.

To summarize, the single-sine FRA has two particular merits, good harmonic rejection and, provided a relatively long measurement period can be tolerated, good noise rejection. On the minus side, the single-sine FRA only measures the components of the Fourier Series at one frequency at any one time. Consequently, the evaluation of the Fourier Series components over a

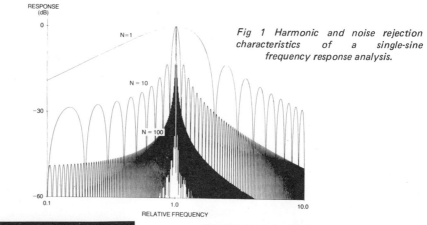

Fig 1 Harmonic and noise rejection characteristics of a single-sine frequency response analysis.

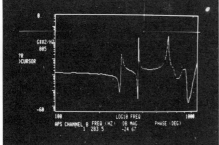

An interactive display of an auto power spectra using GenRad's Time Series Language (TSL).
(GenRad Ltd).

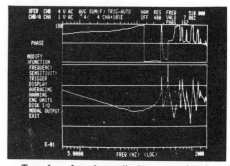

Transfer function display on signal analysis system.
(GenRad Ltd).

sizeable spectrum may require many individual measurements to be taken, involving a considerable amount of time. The vibration engineer must therefore assess the trade-offs between achieving a suitably high level of measurement accuracy and the time available in which to reasonably complete the measurement programme. Such trade-offs become all the more pertinent in situations where the parameters of the system or structure under investigation are time-varying, *eg* the vibration characteristics of a turbine casing during run-up or run-down.

The twin (or multi-channel) single-sine FRA can be of inestimable value to the vibration engineer in many different ways. One very obvious application is the determination of the resonant modes of a structure using an experimental set-up such as that illustrated in Fig 2. The structure is being excited sinusoidally by the generator of the FRA and monitored at several points of interest simultaneously using the multi-channel analyzer facility. Using the processing capability of the 1250 FRA, it is possible to compute the transfer characteristic between the generator input and any monitor point on the structure, or alternatively, the relative transfer characteristic between any two prescribed monitor points on the structure. Thus, by sweeping the measurement frequency through the spectrum of interest, it is possible to determine the relative frequency response characteristics of the nodes of interest, (typically as shown in Fig 3), from which the damping and stiffness of the elements coupling the node points can be determined by conventional modal analysis techniques.

DYNAMIC ANALYSIS OF VIBRATION

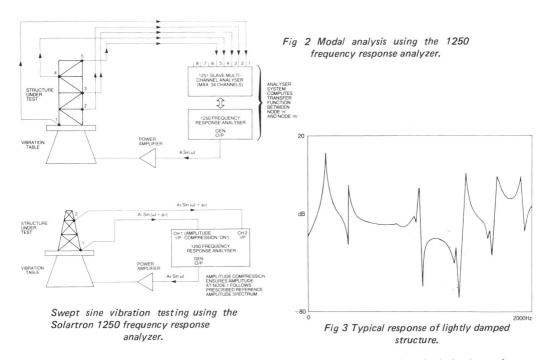

Fig 2 Modal analysis using the 1250 frequency response analyzer.

Swept sine vibration testing using the Solartron 1250 frequency response analyzer.

Fig 3 Typical response of lightly damped structure.

Swept-sine vibration testing has proved a useful method for determining the behaviour of components of structures subjected to prescribed power spectra. Often there is a requirement with this type of testing to excite a certain point on the structure under investigation with a controlled spectrum and to monitor the response resulting from the excitation at a second point located elsewhere on the structure. A problem arises however, either because the vibrator used to excite the structure has an inherent frequency response characteristic which modifies the spectrum input at the driving point, or it is not physically convenient to excite the structure with the prescribed spectrum at the defined driving point. Hence the spectrum input at the actual driving point is modified by the inherent transfer function between the actual driving point and the desired driving point. Either situation results in failure to achieve the required power spectrum at the desired driving point, thereby invalidating the measurement.

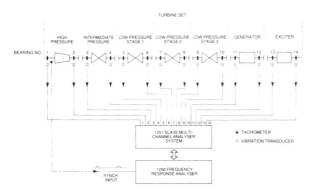

Synchronous vibration analysis of a turbine using the Solartron 1250 frequency response analyzer.

DYNAMIC ANALYSIS OF VIBRATION

In applications concerning the vibration analysis of rotating machinery, there is often a requirement to analyze the vibration response of a component of the machine under investigation at a frequency synchronous with the fundamental frequency of rotation (or a prescribed harmonic thereof) of a given shaft. A typical example, the determination of the vibration characteristics of a turbine during run-up/run-down, is depicted in Fig 4.

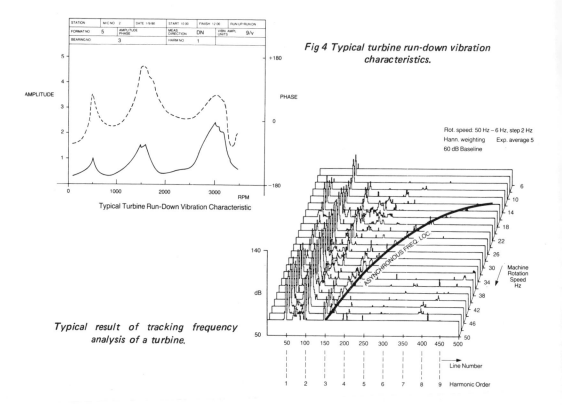

Fig 4 Typical turbine run-down vibration characteristics.

Typical result of tracking frequency analysis of a turbine.

The applications cited here are but a few of the many guises in which single-sine frequency domain analysis tools can assist the vibration engineer in determining the behavioural characteristics of vibrating systems or structures.

Fast Fourier Transform Analysis Techniques and Tools

Fast Fourier Transform (FFT) analysis techniques aim to measure a large number of spectral components of a broad-band signal simultaneously. The broad-band time signal derived from the system or structure under investigation is sampled and the sampled-time data is processed using an FFT algorithm to compute the Fourier coefficients.

Where the system or structure under investigation is passive, it is necessary to provide some form of external broad-band excitation to obtain a measurable response. The form of broad-band excitation used for a particular test can significantly affect the results that are obtained; therefore, it is necessary for the vibration engineer to be aware of the implications of selecting a certain type of excitation for a given application.

Continuous broad-band excitation may be either a truly random non-periodic signal or a deterministic periodic signal. The truly random signal exhibits a Gaussian amplitude probability function and, theoretically at least, an infinite uniform power spectrum. The latter poses a problem in that a necessary condition for the error-free computation of the Fourier Transform is that the time waveform be exactly periodic within the measurement window, otherwise leakage will occur giving rise to side lobes in the frequency spectrum and hence noise on the spectral display.

Pre-processing the random time record with a weighting function, (eg a Hanning Window), such that the time record approximates more closely to a periodic signal considerably reduces the leakage effects (Fig 5). Since, however, the pre-processed signal still exhibits a small aperiodic component the leakage effects can never be completely eliminated from the measured spectrum.

Periodic noise signals can be created using multi-level pseudo-random generation techniques, or alternatively, by synthesizing a signal in the frequency domain exhibiting a uniform power spectrum, but with a random phase spectrum, which can then be Fourier transformed into a time-domain signal to provide a source of excitation. These generation processes ensure the test signal is periodic within the sample window and hence leakage effects are eliminated. Such signals do introduce other problems however, particularly where the system or structure under investigation exhibits non-linear behaviour.

Non-linearities cause inter-modulation of the individual frequency components of the test signal such that the contributions of these individual components to a given spectral line of the measured response signal are inseparable. This makes it impossible to obtain and measure the transfer characteristic of the system or structure under investigation.

The problem of leakage can also be resolved using multiple sine waves as an excitation signal, generated simultaneously and coincident with the spectral lines of analysis. Again this signal is periodic and can be used with a rectangular sampling window with no corruption of the signal.

In addition to continuous signals, transient signals are often used for excitation. With this form of excitation, a broad-band response is generated from the system by inputting a transient, such as a rectangular pulse, or commonly for vibration analysis, a hammer blow. A rapid swept sine, or 'chirp', is also sometimes used. Since all of these signals start and finish at zero amplitude and are, in effect, periodic over the sample window, leakage problems do not arise. A major problem of 'transient' signals is that the power distribution in the signal is not even over the signal spectrum.

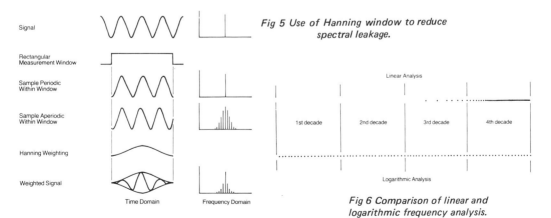

Fig 5 Use of Hanning window to reduce spectral leakage.

Fig 6 Comparison of linear and logarithmic frequency analysis.

Consequently certain modes of the system or structure under investigation are excited more than others and this can lead to problems in obtaining measurable data concerning secondary modes of vibration in the presence of dominant resonant modes which are more heavily excited due to the characteristics of the test signal used.

Frequently, because of dominant resonant modes in the system or structure under investigation, the overall amplitude level of the broad-band excitation signal must be restricted to a very low limit. In such cases it is not uncommon to be operating under conditions where the signal-to-noise ratio is less than unity. If pure random noise is used as the source of broad-band excitation, then any sample record taken of the response of the system or structure under investigation will be unique. Thus, if several records are ensemble averaged, noise, non-linear effects, and distortion will be reduced to a mean level. The consistency of the results is therefore dependent upon the number of averages taken and a compromise is necessary between accuracy and the time taken to complete the measurement.

An important consideration in performing broad-band frequency response testing is the accuracy and resolution required. This must be considered in relation to the time available to complete the measurement. With a conventional spectrum analyzer the frequency resolution (for a given number of analysis lines) is proportional to the bandwidth of analysis, usually with a linear distribution of spectral lines (constant bandwidth). Where the bandwith of the analysis is greater than two decades the low frequency resolution becomes very poor. This is clearly demonstrated by the diagram (Fig 6) which shows a four decade bandwith analyzed on linear (constant bandwidth) and logarithmic (constant percentage bandwith) bases. It can be seen that low frequency resolution is improved by logarithmic analysis, but this is not generally available on spectrum analyzers.

Summary

It is evident that several factors must be taken into account in the selection of the most appropriate analysis tool for a particular application. Trade-offs must be made concerning the quality and quantity of data required; the test time either available or acceptable; whether an excitation signal is required, if so, which type to use; whether the effects of environmental noise or system non-linearity are likely to affect the measured result.

Undoubtedly the greatest understanding of the advantages and shortcomings of any particular analysis tool is gained by experiences of using the tool in real applications. It has frequently been the experience of those who manufacture such tools that their true value is only recognized when an unusual application arises and it is found that with a little ingenuity the basic tools can be readily adapted to meet the measurement requirement. The fact that the Fourier Transform approach has been applied to so many widely different disciplines, ranging from the study of the behaviour of delicate physiological systems to the analysis of the vibration characteristics of multi-megawatt turbines, is perhaps an indication of the flexibility of these techniques. Similarly, the frequency domain tools available to the vibration engineer have proved easily adaptable and flexible in their application. The art is in understanding thoroughly the capabilities of the tools available and being able to exploit this capability skilfully in the context of the measurement problem at hand.

Modal Analysis

ANY REAL waveform can be represented by the sum of much simpler waveforms. In a similar manner, any real vibration in a structure or body can be represented by the sum of much simpler vibration modes. *Modal analysis* is the tool for determining the shape and magnitude of the structural deformations in each vibration mode. From this, it is usually apparent how the overall vibration can be changed.

There are two basic techniques for determining the modes of vibration in complicated structures:

(i) Exciting only one mode at a time
(ii) Computing the modes of vibration from the total vibration.

Taking a tuning fork as a simple example, to excite just the first mode two shakers are needed, driven by a sine wave and attached to the ends of the tines as in Fig 1. Varying the frequency of the generator near the first mode resonance frequency would then give its frequency, damping and mode shape.

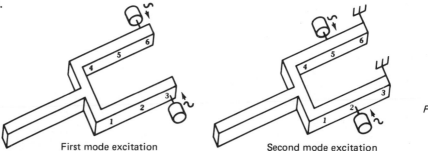

First mode excitation   Second mode excitation   *Fig 1*

In the second mode, the ends of the tines do not move, so to excite the second mode the shakers must be moved to the centre of the tines. If the ends of the tines are anchored, vibrations will then be constrained to the second mode alone.

In more realistic, three dimensional problems, it is necessary to add many more shakers to ensure that only one mode is excited. The difficulties and expense of testing with many shakers has limited the application of this traditional modal analysis technique.

To determine the modes of vibration from the total vibration of the structure, it is necessary to determine the frequency response of the structure at several points and compute at each resonance the frequency, damping and what is called the residue (which represents the height of the resonance). This is done by a curve-fitting routine to smooth out any noise or small experimental

MODAL ANALYSIS (A)

THE COMPUTER-AIDED TEST SYSTEM

* MODAL ANALYSIS
* STRUCTURAL MODELLING
* OPTIMISATION OF STRUCTURAL MODIFICATIONS
* SIGNAL ANALYSIS
* VIBRATION CONTROL
* PROGRAMMABLE
* ACOUSTIC INTENSITY

MULTI-CHANNEL — DISC STORAGE — RAPID REAL TIME GRAPHICS — MANY SOFT-WARE PACKAGES COMMUNICATIONS LINKED TO COMPUTER AIDED ENGINEERING SYSTEMS

GenRad

GenRad Ltd
Norreys Drive
Maidenhead
Berks SL6 4BP
0628-39181

GenRad Inc
2855 Bowers Avenue
Santa Clara
Cal. 95051 USA
408-727-4400

GenRad (France) 01-797-07-39
GenRad (Netherlands) 020-4998-74240
GenRad (Switzerland) 01-55-24-20
GenRad (Italy) 02-84-66-541
GenRad (Germany) 089—41690

MODAL ANALYSIS (B)

Linear motor car developed by JNR

RION TECHNOLOGY ADVANCES AT HIGH SPEED

Rapidly advancing transportation technology is guided by the need for vehicles that save energy, have higher speeds but less pollution, noise and vibration. From its high level of electronic and transportation technology, Japan has developed the linear motor car and parallel with this, the necessary advanced measurement technology. Because its instruments are well-known for high reliability, precision and accuracy, Rion instruments have been selected to evaluate the sound and vibration phenomena of the magnetic levitation, linear motor car.

To meet the severe criteria of advanced technology, Rion engineers are constantly developing more sophisticated measuring instruments and systems.
Accordingly, Rion know-how is advancing at the leading edge of the sound and vibration measurement frontier into new and unexplored regions of scientific research.

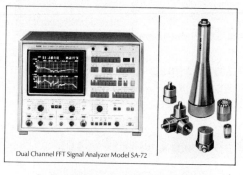

Dual Channel FFT Signal Analyzer Model SA-72

For more information write for our catalog.
RION CO., LTD.
TOKYO JAPAN
Ikeda Bldg., 7-7, Yoyogi 2-chome,
Shibuya-ku, Tokyo 151, Japan Telex: J28437

errors. From these measurements and the geometry of the structure, the mode shapes are computed and drawn on a CRT display or a plotter. If drawn on a CRT, these displays may be animated to help the user understand the vibration mode.

From the aforementioned description, it is apparent that a modal analyzer requires some type of network analyzer to measure the frequency response of the structure and a computer to convert the frequency response to mode shapes. This can be accomplished by connecting a Dynamic Signal Analyzer through a digital interface to a computer furnished with the appropriate software. This capability is also available in single instruments called Structural Dynamics Analyzers. In general, computer systems offer more versatile performance since they can be programmed to solve other problems. However, Structural Dynamics Analyzers generally are much easier to use than computer systems.

Frequency Response Function

The frequency response function of a structure *ie* the transfer function as measured by the Fourier transform is merely the Laplace transform evaluated along the frequency axis. The poles of the Laplace transform occur in pairs and each pair corresponds to a mode of vibration of the structure and they are located at:

$$p_K = \sigma_K \pm j\omega_K$$

where σ_K is the modal damping coefficient
ω_K is the natural frequency

In general, a pole location, p_K, will be the same for all transfer functions of the structure because a mode of vibration and its associated natural frequency and modal damping are global properties of an elastic structure. It is possible to derive from any one frequency response function the pole locations and hence the global natural frequencies and damping of the modes of vibration together with the amplitude and phase of the response at the measurement point in each mode. Therefore, the modal deformation together with its associated natural frequency and damping can be determined by measuring the frequency response function at a sufficient number of points over the structure.

The validity of the final mode shape will depend upon the accuracy of the measurement of the frequency response functions and it is therefore worthwhile considering the various test techniques available:

(i) Random Excitation

Random excitation can be further subdivided into three types, (a) pure random, (b) pseudo random and (c) periodic random.

(a) *Pure random* has a Gaussian distribution and is non-periodic and this excitation is typical of many naturally occurring phenomena. Usually the excitation signal spectrum will be flat across the frequency range of the measurement. However the spectrum of force applied to the structure may be distorted by an impedance mismatch between it and the shaker which in severe cases may lead to noise problems. These can be overcome by using a closed loop control system employing the analyzer computer.

The other drawback of using pure random noise is the necessity to use a Hanning window to reduce 'leakage' caused by the non-periodicity of the signal within the sampling window of the analyzer. The Hanning window reduces the frequency resolution of the analyzer and thereby reduces its capability to reject noise.

After the time records have been transformed the frequency domain data is normally ensemble averaged in order to reduce non-linear effects, noise and distortion from the measurement. This averaging produces a far better measure of the linear least squares estimate of the response of the structure. This is particularly important as the subsequent digital parameter estimation technique is based on the assumption that the structure behaves linearly.

Bruel & Kjaer dual channel signal analyzer with 801 lines resolution and six different course functions can measure and display 34 different time domain, frequency domain and statistical functions.

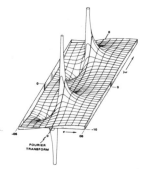

Fig 2 Real part of the Laplace transform of a single degree of freedom model.

This technique is fairly fast and its performance on a simple single degree of freedom model with and without distortion is shown in Fig 2.

(b) *Pseudo random* noise is usually generated by the analyzer and has the same record length as the analyzer's measurement window. The same signal is repeatedly output to the shaker and is therefore periodic in the measurement window and hence there is no error due to leakage and no need to Hann the data, thereby providing better frequency resolution.

As the signal is usually generated by specifying a single amplitude spectrum with a random phase and then transformed into the time domain, it is relatively simple to compensate for impedance mismatching of the shaker by adjustment of the amplitude spectrum.

Pseudo random noise provides not only a leakage free measurement but it may also be performed very quickly as, if there is no extraneous noise, there is no benefit in averaging more than

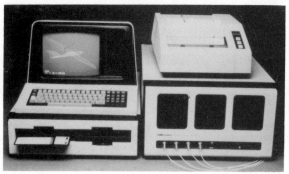

Portable Signal and Modal Analysis system.
(GenRad Ltd).

one result as each will be identical. However, this can be a disadvantage in some cases for as it always provides the same stimulus for the structure it does not exercize it over a wide amplitude range and therefore averaging will not reduce errors due to non-linearities or rattling components.

(c) *Periodic random*. In this case the same record of random noise is repeatedly output 2 or 3 times by the computer to excite the structure and a measurement of the frequency response is made during the last repetition. A new block of random noise, uncorrelated with the previous sample is then output and the experiment repeated and the frequency results averaged. As the measurements are only taken after the structure's transients have decayed, the signal can be considered to be periodic within the measurement window, thereby eliminating the leakage error and as each new block of output data is uncorrelated, ensemble averaging produces a good estimate of the linear response. Its superior performance is shown in Fig 3.

Its only disadvantage is that it is two to three times slower than the other random methods.

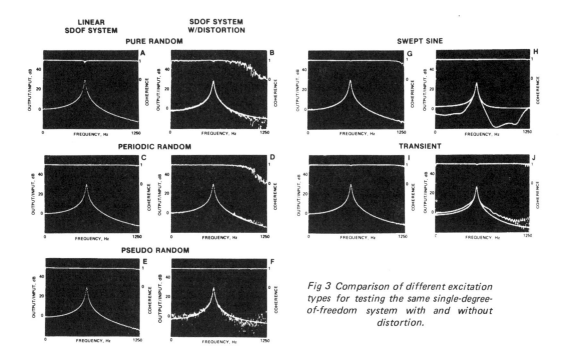

Fig 3 Comparison of different excitation types for testing the same single-degree-of-freedom system with and without distortion.

(ii) Sinusoidal Testing

The main advantages of a sinusoidal testing are that it enables high input forces to be fed into the structure and that the force level may be easily controlled. However, a normal swept sine test can be time consuming if good resolution is required and also non-linearities and distortion are not coped with satisfactorily.

A method of shortening the test time is to use a 'chirp' signal which is a logarithmically swept sinewave that is periodic in the analyzer measurement window. The important advantage of this type of signal is that it is sinusoidal and it has a good peak-to-rms ratio which assists in obtaining maximum accuracy and dynamic range from the signal conditioning electronics of the test set up.

There are many other alternative schemes for using sinusoidal excitation with a Fourier analyzer. However, none are as fast as the 'chirp' signal, neither do they overcome the problems due to non-linearities.

(iii) Transient Testing

There are two types of transient testing, (a) impact testing and (b) step relaxation.

(a) *Impact testing.* The method normally employed for transient testing is to use a hammer with a load cell mounted close to the striking face to measure the force applied to the structure, with an accelerometer mounted on the structure to measure its response. This technique has some important advantages:

(1) The structure requires no elaborate mountings.
(2) It is extremely fast, up to 100 times faster than some sinusoidal tests.
(3) No vibrator is required.

The major drawback with this method is that the input force power spectrum is not as easily controllable as when a vibrator is used and there can be significant variations between successive blows which can cause non-linearities to be excited. To some extent the bandwidth of the input force spectrum can be selected by changing the material of the hammer head; a softer head will give a longer impulse and hence more energy at the lower frequencies. If a hard head is used the total energy will be spread over a wide frequency range and the excitation energy density will be low, leading to measurements with poor signal to noise ratios particularly for massive, heavily damped structures.

Another problem occurs on lightly damped specimens, for if the transient response of the system does not decay within the length of the measurement window, it is essential to apply a window to reduce the leakage error. However the Hanning window would destroy most of the signal as it occurs at the start of the record and so an exponential window has to be utilized. This artificially increases the damping in the record thereby ensuring that the response decays within the measurement window, but this known increase in the damping must be taken into account later if the correct modal damping is to be ascertained, and it also tends to smear modes together making it more difficult to separate close modes.

(b) *Step relaxation.* In this method the structure is loaded to an acceptable strain level prior to the test and then the loading is suddenly released triggering the measurement. This method has the advantage that the loading condition is repeatable and can be used on fragile specimens but, of course, requires more setting up.

Typical Modal Analysis system by Scientific Atlanta based on real time two-channel FFT processor and a processor system incorporating a DEC-based CPU, high speed display terminal and powerful modal analysis software package.

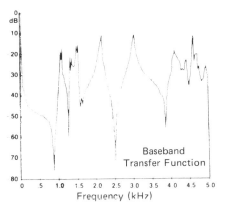 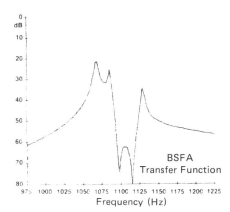

Fig 4 Base band vs Zoom transform

Zoom Transforms

Probably the most important factor affecting the accuracy of the modal parameters is the frequency resolution of the frequency response measurement as all the algorithms used to identify the modes are very dependent upon the frequency resolution. The Fourier transform is limited to baseband spectral analyses, that is, those extending from d.c. to a maximum frequency. Therefore, in order to improve the frequency resolution it is essential to increase the number of spectral lines in the baseband which in turn means increasing the block size ie lengthening the measurement window. This approach soon proves unsatisfactory as a very large computer memory is required just to store the data and its transformed products and also the processing time for the data becomes long for the larger blocks. However, since the development of the 'zoom' transform or band selectable Fourier analysis it has been possible to improve the frequency resolution by only looking at the frequency band of interest as the lower limit no longer needs to be d.c. All the spectral lines can therefore be inserted into a narrow frequency band giving frequency enhancements of up to 256 times, typically, and with the hardwired preprocessors now available BSFA can be performed on line on signals up to 100 kHz. Fig 4 shows the typical enhancement feasible. Another benefit of BSFA is that it can increase the dynamic range of the measurement to 90 dB's by reducing the effect of the quantization noise of the ADC's.

The Fourier analyzer can therefore produce high quality estimates of the frequency response function and provided a little care is taken in defining the experiment, satisfactory data can be passed to the modal analysis programme.

Vibrator Attachment Points

If the engineer has some prior knowledge of the structure he should be able to select an attachment point for the vibrator which will excite all the modes of vibration or at least all those of interest. If the specimen is unknown it may be necessary to use two excitation points to ensure that one is not at a node of a significant mode, and the analyzer is capable of utilizing the information generated from both excitations in order to obtain the best estimates of the modes. In the modal identification process it is beneficial, although not essential, to have all the modes of interest contributing to each transfer function and therefore, it is sometimes better not to place the response accelerometers along axes of symmetry as these are normally nodal for some modes. For example, on a uniform beam having distinct vertical and horizontal modes, if the accelerometers were placed at $\pm 45°$ to the vertical, both the vertical and horizontal modes would appear in both transfer functions and the fact that they are inclined to the geometric axis may readily be fed into the programme. It is also essential that sufficient measurement points are taken to define the structure adequately, although extra points necessary to improve the mode shape around sharp discontinuities discovered by the analyzers may be added later. The number of measurement locations which can be handled by modal analysis systems is typically about 500 with storage for parameters of over 65 modes.

The selection of the technique utilized for mode identification is dependent upon the structure's measured frequency response functions, for if there is light modal overlap, then a simple routine such as 'circle fitting' would be perfectly adequate, while with heavy modal overlap one of the more sophisticated multidegree of freedom fitting routines would be required.

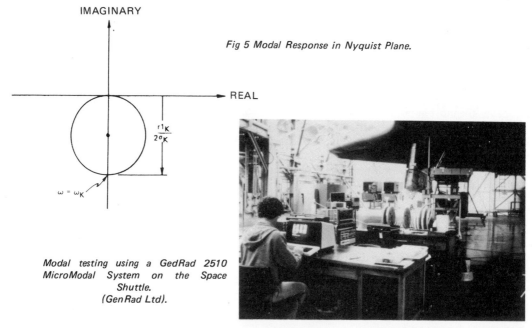

Fig 5 Modal Response in Nyquist Plane.

Modal testing using a GedRad 2510 MicroModal System on the Space Shuttle. (GenRad Ltd).

First consider the circle fit method, which is based on the method of Kennedy and Pancu and which depends on the fact that a modal resonance represented in the Nyquist plane is a circle, see Fig 5. As the response around the resonant frequency is almost entirely due to the single mode, if a circle is fitted through the response point corresponding to the resonant frequency and a few points either side of it, then a good estimate of the response in that mode may be obtained. From the fitted circle it is possible to obtain estimates of the modal damping and the amplitude and phase of response of that measurement point in that mode.

The advantages of this technique are that it is very quick and due to its relative simplicity may be implemented on a comparatively small system without requiring any mass storage devices.

For a structure with more heavily coupled modes a multidegree-of-freedom curve fitting routine is essential and at present there are two being used. The first operates upon the inverse Fourier transform of the transform function, the impulse response, and is a fully automatic system. The second, which may also be automated, does permit the engineer to specify the location of the resonant frequencies and then the damping and the amplitude and phase of the response at each of the measurement points.

Vibration Testing

VIBRATION TESTING is now an established technique for assessing the performance of products that are subjected to vibration (and as a consequence fatigue) in the normal working environment. Testing enables the product designer to isolate troublesome resonance and identify potential vibration problems and simulate realistic working conditions to see how the product stands up and thus assess performance and reliability. Mechanical vibrators can provide a relatively simple method of generating impulses over a wide range of frequencies but have a tendency to produce 'chatter' waveforms and have a distinct lack of facilities for programming and control. They have, therefore, been almost entirely replaced by electronically controlled electro-dynamic or hydraulic vibration testers.

Basic components of a vibration tester are shown diagrammatically in Fig 1. Heart of the system is a *vibration generator* capable of producing the form of vibration(s) required. This normally operates in a closed loop where a control system monitors the mechanical motions of the

Environmental test chamber for combined climatic testing in conjunction with wide band vibration generator. (Ling Dynamic Systems Ltd).

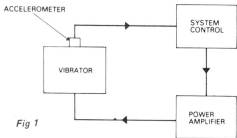

Fig 1

Ling Dynamics model 860 vibration generator for production testing to MIL STD 781C test specifications.

Model 1506 Electrodynamic shaker with maximum acceleration of 110 g maximum velocity 1.2 m/s and maximum displacement 25 mm (Instron Environmental Equipment)

Vibration generator with associated linear power amplifiers (left) and programme controller/monitors (right). (Ling Dynamic Systems Ltd).

vibrator (as determined by an accelerator) and corrects the power supplied to adjust for amplitude variations as the frequency is changed. The loop can also include automatic frequency sweep facilities, spectrum shaping and any other complex functions required for the test programme. In fact the control system may comprise a complete microprocessor or computer.

Electro-Magnetic Vibrators

The electro-magnetic vibrator works on the principle that the interaction between a steady magnetic force across a conductor and the circular magnetic field produced by a current flowing in this conductor, results in a proportional force. By suspending the conductor (in the form of a coil) within an annular gap, the force can be transmitted to a lightweight frame, the moving assembly being constructed to give the maximum possible strength compatible with lowest possible weight. Such a system can provide a practical means of testing many different pieces of equipment over a wide frequency range at high force or acceleration levels. A further advantage is that electro-dynamic vibrators can readily be completely controlled and programmed *via* electronic circuitry. Sizes can range from small magnetic transducers up to shake tables with a vector force of the order of 18 000 kg (40 000 lbs) or more.

VIBRATION TESTING (A)

The World of Vibration

* Electro-dynamic Vibrators, from hand size to man size. 9N to 160,000 Newtons Force. Long Stroke version introduced in 1980 for computer controlled Shock and Bump Operation.
* Solid State Amplifiers 25 to 192,000 Watts
* Combined Environmental and Vibration Test Enclosures.
* Specialist magnesium support Jigs and Fixtures.

Find out more today, write for the new Short Form Catalogue.

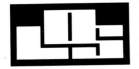

LING DYNAMIC SYSTEMS LTD.,
Baldock Road, Royston, Herts. SG8 5BQ England
Tel: Royston (0763) 42424, Telex: 81174

VIBRATION TESTING (B)

LOCATE RELATE QUANTIFY
your torsional vibration levels

If you build or operate:
- DIESEL ENGINES ● PETROL ENGINES
- COMPRESSORS ● TURBINES ● GEARBOXES
- FLEXIBLE COUPLINGS ● DRIVE SHAFTS
- CLUTCHES, BELT OR CHAIN DRIVES

You need to take your torsional vibration problems seriously.

If you check through the list below and return it to us, we will advise how our instruments can help solve your Torsional Vibration problems.
A brief description of the potential application would be helpful.

- Indefinite life compared to short life Seismic Transducers.
- Response from 10 Hz to 500 Hz. (Range can be extended or compressed to order).
- High rotational speed capability — 10,000 or 20,000 R.P.M. (Top limit to choice).
- Low parasitic inertia. (Where accurately pitched gear teeth can be observed, parasitic inertia is nil).
- In-line as well as free-end measurements are possible in hostile environments (i.e., at high temperatures, with oil and dirt contamination — no loss of accuracy).
- Remote operation is possible (Up to 45 metres between instrument and transducer).
- The instrument is fully portable and battery powered for use in vehicles, etc.
- Instantly compatible with X-Y plotter, Wave Analyser, Tracking Filter, Tape Recorder, Oscilloscope, etc. High speed data acquisition under dangerous conditions is possible with high speed tape recording techniques. Linear output means quick and easy calibration of charts, etc.
- Protection of vibrating system is possible utilising variable set level trip facility.

Manufactured by:

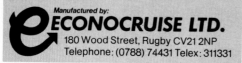

ECONOCRUISE LTD.
180 Wood Street, Rugby CV21 2NP
Telephone: (0788) 74431 Telex: 311331

DON'T LET THE NOISE PROBLEM FALL ON DEAF EARS

CONTINUOUS LOUD NOISE CAN CAUSE PERMANENT DAMAGE TO THE INNER EAR!

SPECIFY AND INSTALL

BARAFOAM
THE SOUND ABSORBER

A RANGE OF ACOUSTIC PRODUCTS, PERFORMANCE PROVED OVER MANY YEARS IN APPLICATIONS AS DIVERSE AS FETTLING BOOTHS, COMPRESSORS, AIR HANDLING UNITS, BOATS, LORRY CABS, AIRCRAFT AND LUXURY COACHES.
FULL TECHNICAL ADVISORY SERVICE AVAILABLE.
SEND FOR FULL DESCRIPTIVE LITERATURE NOW.
Kay-Metzeler Ltd., New Mill, Park Road, Dukinfield, Cheshire SK15 5LL. Tel: 061-330 7311

Experience And Research

KAY-METZELER

INSTRUMENTATION
HIRE & SALES

For sound, vibration & signal analysis

GRACEY & ASSOCIATES

Threeways, Chelveston,
Northants, NN9 6AJ.
Telephone 0933 624212
Telex 312517

VIBRATION TESTING (C)

TICO MANUFACTURING COMPANY LIMITED

TICO WORKS, HIPLEY STREET, OLD WOKING, WOKING SURREY GU22 9LL
☎ (04862) 62635/6

TICO STRUCTURAL BEARINGS AND RESILIENT SEATINGS

Ground borne noise and vibration emitted by underground and main line railways can be eliminated by constructing the building on TICO anti-vibration bearings.

They are maintenance free and designed to isolate office blocks, hotels, theatres, studios, anechoic chambers, floating floors, oil rig accommodation modules and any other structure where quiet is of paramount importance.

The TICO range of elastomeric products also includes resilient seatings, sliding bearings, adhesives, sealants etc., providing a full service to the civil engineering industry.

Our technical advisory service is available to make recommendations employing TICO structural bearing products.

◁ *(Irish Centre, Camden)*

VIBRATION CONTROL SYSTEMS

- ★ RANDOM
- ★ SINE
- ★ SHOCK
- ★ SINE ON RANDOM
- ★ RANDOM ON RANDOM
- ★ ANALYSIS
- ★ EXPANDABLE TO A COMPUTER-AIDED TESTING SYSTEM WITH MODAL AND STRUCTURAL ANALYSIS

GenRad

GenRad Ltd
Norreys Drive
Maidenhead
Berks SL6 4BP
0628-39181

GenRad Inc
2855 Bowers Avenue
Santa Clara
Cal. 95051 USA
408-727-4400

GenRad (France) 01-797-07-39
GenRad (Netherlands) 020-4998-74240
GenRad (Switzerland) 01-55-24-20
GenRad (Italy) 02-84-66-541
GenRad (Germany) 089—41690

VIBRATION TESTING (D)

Britain's quality range of electro-dynamic shakers and amplifiers

Caption Left: Electro-dynamic shaker undergoes final performance tests before shipping.
Caption Above: A special purpose Instron horizontal shaker system with slip table.
Caption Below: Instron's 1508 electro-dynamic shaker and matching amplifier system.

Instron, one of the world's leading test equipment specialists, introduces a comprehensive range of electro-dynamic shakers and amplifiers.

High performance, low distortion systems with built-in operational reliability.
- High Acceleration
- Low Distortion
- Low Vertical Stiffness
- Low Transverse Motion
- Load Compensation
- Low Frequency Isolation Trunnion

Write, telephone or telex now for fully detailed specification literature.

Designed and built in Britain by
Instron Environmental Ltd
Coronation Road, High Wycombe,
Bucks HP12 3SY, England
Tel: (0494) 33333 Telex: 83222

Combined Test Cabinets for Vibration, Temperature, Humidity and Pressure.

Typical applications are:
- SPACE AND DEFENCE
- VEHICLE TESTING
- ELECTRONIC SYSTEMS
- PACKAGING
- BUILDING AND CONSTRUCTION
- TRANSPORTATION

Ever growing quality consciousness forces industrial manufacturers to recognize and react at an early stage to potential failure sources for the function of their products, presented by environmental conditions.

The following conditions can be simulated by dynamic tests, leading to the recognition of areas of weakness and subsequent remedies:
— Dangerous vibration of construction elements
— Duration of joints
— Adhesion of surface treatments
— Influence of increased friction
— Effectiveness of damping measures

as well as all problems arising from mobile application of apparatus and equipment as well as with — generally speaking — all power machines.

According to the range of utilization of the units to be tested, some test standards, such as DIN 40 046, DEF 133, IEC 68-2-6, MIL STD 810B, VG 95332 or MIL STD 781B have been established in Europe and USA.

Amongst other things, these standards demand for a temperature range of −65 up to + 125 °C. Most tests are made for cyclic operation in the temperature range of −54 up to + 55 °C and −54 up to +71 °C with temperature variation velocities up to 10 °C/min., including two periods of vibration test of 10 minutes each to find out whether in the range of 20 to 60 Hz a resonance occurs of the specimen.

Although, no additional climatic conditions are expressly prescribed in the above standards, WEISS TECHNIK — in cooperation with renowned domestic and foreign companies — have designed test equipment suitable for tests beyond the above standards so that parameters such as temperature, humidity, pressure and vacuum can positively be set.

handbook of INDUSTRIAL INSTRUMENTS MICROPROCESSORS AND COMPUTERS

1ST EDITION

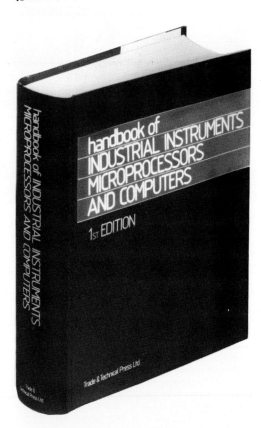

Divided into six sections for easy reference, this is the first Handbook specifically written for the engineer to introduce and explain existing and future use of instrumentation, microprocessors and computers in industrial design, control, monitoring, measurement, protection, and replacement in plant and machinery.

Chapters include:-

Section 1 — Techniques: Linear measurement, Measurement of angles, Surface measurement, Measurement of mass, Measurement of temperature, Measurement of pressure, Polarimetry, Calorimetry, Colorimetry, Microscopy, Dynamic analysis.

Section 2A — Mechanical Engineering: Measurement of physical properties, Form measurement, Displacement measurement, Precision workshop instruments, Gauges and gauging, Force, torque and power, Radiation gauging, Noise and vibration, Tachometers, Non-destructive testing.

Section 2B — Fluid Engineering: Measurement of air and gas flow, Measurement of liquid flow, Measurement of viscosity, Humidity measurement and control, Moisture meters, pH measurement, Flue gas and smoke density, Gas detectors, Gas meters, Turbidity meters, Analyzers.

Section 3 — Process Control and Automation: Automation and process control, Level indicators and controllers, Counters and timers, Calibrators, Control instrumentation, Recorders, Accelerometers, Balancers, Transducers, Machinery health monitoring.

Section 4A — Electrical/Electronic Metering: Voltmeters, ammeters, ohmmeters, Digital instruments, D-to-A and A-to-D converters, Potentiometry, Rate measurement, Telemeters, Data recorders, Regulators, Data loggers.

Section 4B — Systems and Instrumentation: Signal analysis, Spectrum analysis, Dynamic analysis, Modal analysis, Real time analysis, FFT analyzers, Structural analysis, Multi-channel systems.

Section 4C — Microprocessors and Computers: Microprocessors, types and classification, Microprocessors as control programmers, Microprocessors as design tools, Computers, Computer-aided measurement, Computer-aided testing, Computer-aided design, Databases.

Section 4D — Software: Software structures, Software codes, Processing, Management, Displays.

Section 5 — Data: Tables, Charts, Standards, Formulae, etc.

Section 6 — Buyers' Guide.

TRADE & TECHNICAL PRESS LTD
CROWN HOUSE, MORDEN, SURREY SM4 5EW, ENGLAND

ORDER NOW!

OVER 550 PAGES,
1,500 DIAGRAMS,
CHARTS,
TABLES
AND ILLUSTRATIONS.

VIBRATION TESTING

Small systems suitable for testing industrial components, *etc*, may be based on permanent magnetic vibrations. With robust construction and solid state amplifiers they can be extremely reliable. Cooling can be by convection or by forced air if necessary. For increased thrust levels an electromagnetic system is used, the required steady magnetic flux being provided by operating the vibrator as an electro-magnet energized from an external field power supply. Stray magnetic fields can be cancelled by incorporating a degaussing coil and, if the magnetic circuit is operated in a saturated condition, normal supply variations will not degrade the vibrator performance.

Again, smaller vibrators of this type may be aircooled. For increased payload capacities, pneumatic internal load supports or similar devices may be fitted. Medium thrust and large thrust electro-magnetic vibrators are normally watercooled, cooling embracing the armature and field coils and amplifier.

Solid state amplifiers are normally employed, permitting direct coupling of the amplifier to the vibrator for d.c. operation up to frequencies of 10 kHz. However, d.c. amplifiers are susceptible to drift and so, unless operation below 1 Hz is an essential requirement, a.c. coupling is preferred, when drift is virtually eliminated. For operation below 1 Hz, d.c. coupling becomes essential. Modern designs of electro-magnetic vibrators are thus capable of working down to the lowest frequencies which may be required for test purposes. A complete system comprises, basically, a signal source or generator, power amplifier and electro-magnetic vibrator, together with any required read-out, analyzer, programming or control circuits, selected according to the duty required. Larger systems also need a cooling unit.

The performance of an electro-magnetic vibration is commonly given in the form shown in Fig 2, the working regions for displacement, velocity and acceleration.

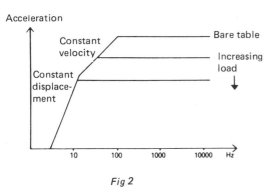

Fig 2

Air cooled permanent vibrators.
(Ling Dynamic Systems Ltd).

Electro-Hydraulic Vibrators

Hydraulic shaker systems are particularly effective where high thrusts are required. They can provide both static and dynamic force levels which allow complex structures to be tested under the same conditions of loading as are experienced during actual operation. Correct values of dynamic stiffness and accurate pin-pointing of resonant frequencies and damping factors are assured when exciting a structure under these conditions. Additionally, structural non-linearities are automatically accounted for when force output can be maintained regardless of structural loading.

Fig 3 shows a typical hydraulic system incorporating analysis of a controlled input force and an output response, say acceleration.

VIBRATION TESTING

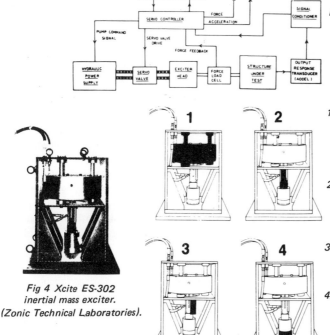

Fig 3 Electro-hydraulic exciter system.

Fig 4 Xcite ES-302 inertial mass exciter. (Zonic Technical Laboratories).

1. The cylindrical mass is guided by three rods using linear ball bearings. This support scheme minimizes side loads on the actuator bearings and seals and permits use in any orientation.
2. A double-ended hydraulic cylinder is used to drive the inertial mass through a large displacement. Control of the hydraulic oil flow is with a two-stage servo-valve.
3. The static centre position of the mass is maintained by the feedback control loop from a displacement signal by a LVDT displacement transducer.
4. The dynamic force of the exciter is monitored by a strain gauge load cell. The feedback from this load cell is used by the controller in maintaining the desired level.

Developments in electro-hydraulic servo-valves have enabled high frequency responses to be achieved whilst still coping with the increased oil flows necessary in high thrust systems. The incorportion of the latest concepts in servo-valves and transducers has given rise to compact exciter heads ideally suited to simulate the level and direction of input forces encountered in complex machinery. Force load cells permit continuous read-out of static and dynamic force levels, and the exciter heads can be position-controlled for load centring using a displacement transducer.

Torsional exciter units with the provision of foot mounting permit easy installation to a wide range of test structures, such as drive trains and torsionally loaded frames. Effective testing of extremely weak structures where large displacements are required to generate desired torque levels is possible with torsion heads having large angular displacement (typically 100°). Torsional units are capable of providing up to 33 500 Kg m (300 000 lb in) static preload and 33 500 Kg m (300 000 lb in) peak dynamic torque.

Inertial mass excitation systems have provided a new dimension to structural testing. High inertial forces can be generated by exciting a mass through a large displacement hydraulic actuator and therefore provide ample input to structures for performing low frequency seismic studies and high frequency rotation equipment analysis. Fig 4 shows an inertial mass exciter. The ability of this system to excite massive structures without the use of back-up fixtures enhances the capability of laboratories to test structures of a more unwieldly nature such as buildings and ship hulls.

Machinery Health Monitoring

MACHINERY HEALTH Monitoring applies to the broad system of specific measurement of significant parameters of a machine in operation whereby its condition can be diagnosed. Its particular significance is that short of instantaneous catastrophic failure occurring (*eg* a turbine shedding a blade with no forewarning), any deterioration of machine condition or 'health' can be detected at an early stage without the machine having to be taken out of service or stripped down for inspection. Thus, impending failure can be diagnosed and located at the onset when remedial action is relatively inexpensive and shutdown time for this work can be planned in advance.

Significant monitoring parameters relative to all types of machines are summarized in Table I. Vibration monitoring is normally the most significant parameter, on the basis that characteristic vibrations are associated with particular phenomena in rotating machines and any change in the vibration signature signifies a change in condition. Acoustic signature analysis can be equally valuable, since noise and vibration are normally generated by the same sources; and in some cases two-channel measurement may be employed to quantify the mutual properties between a power of acoustic and vibration signals and established possible cause-and-effect relationships between these signals. Other systems are also available for condition monitoring in applications where noise or vibration monitoring may not be sensitive enough — *ie* by the time a sufficient change has occurred to be quantified, the component involved is nearing breakdown. This can apply particularly in the case of rolling bearings where mechanical signature analysis techniques can be employed to advantage. Temperature monitoring on its own provides the simplest, most basic technique, but the amount of information it can provide is limited.

The extent of monitory coverage, and whether it be continuous or applied only periodically, depends on the significance of a particular machine. In general, machines can be sub-divided into three groups.

(i) *Critical machines* (*eg* compressors, turbines, motors or generators) that cause a plant to shut down in case of failure.

Here permanent, continuous monitoring systems are required for full benefit, monitoring all the significant parameters. The cost of a protection system is more than paid for if the system protects the machine against major damage only once during its lifetime. In this critical category it is therefore vital that the optimum choice of transducer/monitor for a specific machine has been made, (see Fig 1 for typical system).

(ii) *Semi-critical machines* which, if they fail, may cause part of a plant to be shut down but still allow production to proceed (if only on a reduced scale).

TABLE I — MONITORING PARAMETERS

Parameter	Measuring instruments/transducers	Remarks
Temperature	Thermocouples, resistance type temperature detectors (RTD's)	Monitoring points should include: (i) bearings (as distinct from lube oil temperatures). (ii) lubricant. (iii) stator and rotor windings (in electrical machines). (iv) machine casing. (v) any other significant parts.
Vibration (rotor)	(i) proximity probes. (ii) velocity pick-ups (iii) accelerometers	Measurement should cover all three phases of: (i) shaft motion relative to bearings. (ii) shaft motion relative to free space.
Vibration (non-rotating parts)	Seismic transducers	Monitoring points should include: (i) bearings. (ii) bearing housings. (iii) casings. (iv) foundations or mount. (v) connected ancillaries.
Rotational speed		Shaft acceleration should be considered as an auxiliary measurement.
Shaft phase angle	Keyphasor probe	Can be used directly for balancing
Position	Various sensors	Monitoring points should include: (i) shaft axial position. (ii) shaft radial position (eccentricity ratio) (iii) casing expansion. (iv) shaft and casing alignment.
Process variables	As appropriate	*eg* temperature, pressures and flows of fluids handled by a machine.

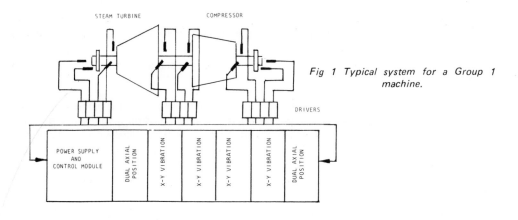

Fig 1 Typical system for a Group 1 machine.

MACHINERY HEALTH MONITORING (A)

MACHINE HEALTH MONITORS

CONDITION MONITORING LTD.
UNIT 2. TAVISTOCK INDUSTRIAL ESTATE TWYFORD
BERKS UK. RG10 9NJ. Tel: (0734) 342636 Telex 847151

We're not making a big noise about coming clean

Just a quiet word to let you know about the new Norgren range of coalescing exhaust silencers for large and small pneumatic systems.

Series CS silencers effectively reduce noise emission from multi valve and cylinder systems to a low level. They also remove oil mist and sub-micron particles to maintain a clean and healthy local working environment.

Olympian 'plug-in' design; one or two main inlet ports, G¼ - G1½; large cartridge minimises 'back pressure'; simple manual drain for collected liquids. Write for details today.

**IMI Norgren Enots Ltd.**
Norgren Works, Shipston-on-Stour,
Warwickshire CV36 4PX. England.
Telephone: 0608 61676. Telex: 83208.

Acoustic and Vibration Technology

Dedicated to a Safer/Quieter Environment.

- Noise Surveys/Control
- Acoustic Design
- Machine Health Monitoring
- Acoustic Emission Monitoring
- Stress/Strain Measurement

AVT can provide practicable cost-effective solutions to your noise and vibration problems.

27 Bramhall Lane South,
Bramhall,
Stockport,
Cheshire SK7 2DN
Tel: 061 440 9392
Telex: 669028

50 Carden Place,
Aberdeen,
AB1 1UP.
Tel: 0224 641666
Telex: 669028

Heavy-Duty Intercoms for Industrial Applicatio

Atkinson intercoms are designed to meet the stringent communication needs of industry—measured by their ability to operate efficiently despite high ambient noise levels, rough usage, or severe conditions of weather, temperature, etc.

Thousands are in use worldwide, many units in locations where conventional equipment previously had failed to perform. Typical locations are areas adjacent to noisy machinery in factory production, in warehouses, and on loading docks, storage yards and garages. They also are used extensively by the petroleum industry for *non-explosive areas* of offshore platforms, drill rig sites, and refineries. Many chemical plants favor their ability to function indoors or out under a wide range of temperature and atmosphere conditions.

In addition to a well proven reputation for delivering clear, dependable voice communication, Atkinson intercoms offer an unmatched simplicity of installation, operation, maintenance and repair.

Heavy-Duty Construction

Each station's solid state amplifier, controls and speaker/microphone are enclosed in a sealed, submergence-proof, cast aluminum case. Units are not intended for installation in explosive areas.

Self-Contained Stations

Since each intercom amplifier is powered by its nearest AC or DC power source — and each individual intercom is a self-contained station both receiving and transmitting the amplified signal — a system can include almost any desired number of stations over very long distances.

Simple Installation

Plug in each intercom to a nearby power source, then connect audio inputs in parallel with ordinary two-wire low voltage cable.

Questions and Orders

Atkinson intercoms may be ordered direct from the manufacturer. Questions regarding special systems or needs are invited; address Engineering Department, Atkinson Dynamics.

Basic Model Specifications

LOW IMPEDANCE MODELS	Model AD-26	Model AD-27
HIGH IMPEDANCE (SOUND-POWER) MODELS	Model AD-42	Model AD-43
Supply Voltage	Nom. 12.6v DC	117v, 50/60 H
Audio Power Output at 10% Distortion	6.0 watts	8.0 watts
Input Sensitivity	1.0V RMS max.	1.0V RMS max
Nominal Power Consumption	.1 A standby 1.0 A full power output	9.0 VA standby 25.0 VA full po output
Operating Temperature	0°F to 150°F	0°F to 150°F
Shipping Weight	11 lbs.	12 lbs.
DIMENSIONS Identical for all models	A — width 6½ inches B — height 10⅜₁₆ inches C — depth 4¹⁵⁄₁₆ inches	

To Order: Additional Information

1 — Select basic model, see above (example: AD
2 — Select letter group designation, see below ample: A)
3 — Select number group designation, see below ample: -9)   **Sample: AD-2**

Letter Group

"A" models — have a call button for sending an internally g ated tone signal out on the audio line to all intercoms

"B" models — are primarily for monitor or paging use. They no provision for talk-back. This unit can also be used extra amplifier and speaker on existing paging systems

"C" models — monitor "D" model sub-stations and have p sions for actuating the sub-station talk/listen relay.

"D" models — are in hands-free talk position all the time, ex when a model "C" actuates the talk/listen relay to l position.

Number Group

"-1" models — receive power from the other end of system. power and audio will be in the same cable.

"-2" models — have provisions for external talk/listen switch

"-3" models — have a call button with open contacts for actua customer's external signal device.

"-4" models — are made for extreme temperature use (—50° +150°F).

"-5" models — have the call signal on a separate circuit. Recei call signal will be loud at all times regardless of volume c trol setting. Recommended for *extreme* noise areas only.

"-6" models — have separate talk and listen lines, for selec switching.

"-7" models — have provisions for a hand held noise cancel microphone.

"-8" models — have a pre-amp for use with a remote sl speaker.

"-9" models — have lightning protection.

ATKINSON DYNAMICS

Building 2, 10 West Orange Avenue
South San Francisco, California 94080, U.S.A.
International Desk Telephone Number: (415) 583-98
Telex Number: 34297 ATKINSON SSF

MACHINERY HEALTH MONITORING

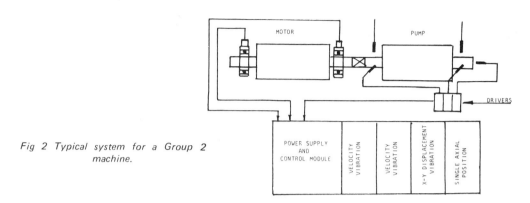

Fig 2 Typical system for a Group 2 machine.

Comprehensive permanent vibration monitoring systems are not essential for machines in this group. The best approach for such machines is often to install a comprehensive set of transducers but use a reduced number of permanent monitors to cut the overall system cost. (See Fig 2 for typical system).

A very good approach is to use periodic preventive monitoring, as a complementary back-up of the limited permanent monitoring system. A well organized and meticulously executed preventive maintenance programme has proved to be very successful, achieving a very high degree of plant reliability.

(iii) *Non-critical* machines whose failure only represents material cost and increased work load for maintenance personnel, but does not affect production significantly.

These are usually checked for vibration on a periodic basis (or when the operator feels that a machine is vibrating more than normal). Since the machines in this category are not critical, a simple portable vibration monitor based on overall level is a good cost-effective approach.

The other major consideration is scale, since there comes a point at which the burden of manual logging — no matter how simple a method is used — becomes too big to handle. At this stage the merits of a computer managed system become self-evident.

The recent advent of computerized monitoring systems provides a very economical means of continuously monitoring many non-critical machines which have previously gone unmonitored.

Basic Systems

The two basic forms of vibration monitoring systems are:

(i) Built-in instrumentation providing a measure of vibration in terms of overall level, coupled to alarm and shutdown devices in the event of the vibration level rising to a certain level. Such a system of monitoring provides continuous protection but no specific information other than that vibration has increased, indicating wear or a potential fault.

(ii) Vibration measurement by analytical instruments to provide a vibration signature. This would normally be done periodically but, in the case of extremely critical machines, could be continuous, although continuous monitoring with built-in instrumentation plus periodic check with analysis machines would be more realistic. Signature analysis provides a much more sophisticated preventative maintenance capability since the deterioration of specific machine components can be isolated whilst the machine is running.

Continuous or Periodic Measurement?

Parameters also fall into two distinct categories; those which should be continually monitored, and those where periodic measurement only is necessary. A parameter not considered important enough for continuous monitoring may be considered important enough to require a very reliable means of measurement on a periodic basis. An example could be machine housing vibration. A machine could be continuously monitored by shaft-observing proximity probes but may require an accurate analysis of shaft versus housing vibrations during certain running conditions, start-up for example. Since a permanently installed transducer usually provides a more reliable measurement than any handheld transducer, the housing measurement transducers can be permanently installed, without continuous monitors, for machine analysis.

The selection of parameters to be monitored depends on the level of sophistication desired for the monitoring system and the various mechanical considerations particular to a specific machine design. Equally, it is important to note that a transducer chosen for monitoring one parameter can sometimes be employed to provide the measurement for monitoring a second parameter. Examples of this are an axial position sensor which can be used to measure axial vibration as well, and a shaft observing radial vibration proximity probe which can also be used to measure shaft radial position, an indicator of alignment conditions.

Specific Machine Requirements

In general, it is important to recognize that in order to determine the optimum protection system for machinery, each piece of machinery must be evaluated individually. Often insufficient data is

TABLE II — RECOMMENDED MONITORING FOR MACHINES

Machine	Monitors	Parameters monitored
Electric motors	X-Y proximity probes Keyphasor probe Temperature indicators	(i) axial vibration. (ii) position measurements (periodically). (iii) casing vibration. (iv) speed, phase angle and timing. (v) bearing and oil temperatures. (vi) rotor and stator winding temperatures.
Pumps	X-Y proximity probes Keyphasor probe	(i) axial vibration. (ii) shaft motion relative to bearings. (iii) shaft phase angle (unless directly coupled). (iv) bearing and oil temperatures. (v) casing vibration. (vi) casing temperature.
Fans	X-Y proximity probes	(i) shaft vibration. (ii) bearing housing vibration. (iii) casing vibration.
Gears	X-Y proximity probe at each bearing	(i) axial vibration. (ii) input shaft. (iii) output shaft. (iv) thrust loads (axial probes). (v) gear teeth interaction. (vi) casing vibration. (vii) bearing and oil temperatures.

available for a detailed analysis of a particular machine's expected behaviour under normal and malfunction conditions. It then becomes necessary to use best engineering judgment and experience in determining what should be monitored. Often the user company has a machinery specialist group to provide the function of monitoring system specification. However, the user can also rely on the machinery manufacturer, the engineering consultant/contractor, and/or the machinery protection system manufacturer to accomplish this function.

As a general guide, some specific recommendations for common machines are given in Table II.

Continuous Monitoring Equipment

This type of equipment is usually modular so that protection systems can be matched to specific machine requirements. Typical modules with meter indication of levels are as follows:

Power Supply/Control

Displacement Monitor — say, for thrust or quasi-static displacement monitoring — used with eddy current proximity probes.

Vibration Displacement Monitor — Single or Dual channel — used with eddy current proximity probe.

Vibration Velocity Monitor — used with moving coil pick-up.

Vibration Acceleration Monitor — used with piezoelectric accelerometers.

Rev/min Monitor — used with either eddy probe or fibre optic tachometer. Indicates rev/min, and trips at pre-set overspeed.

Continuous vibration monitors are usually based on the measurement of overall vibration level, *ie* the vibration signals are rectified and smoothed, resulting in a d.c. level. Typically both quasi-static and vibration monitors contain facilities for:

Adjustment of pre-set levels for alarm and shutdown and meter indication of these trip levels when required.

Continuous comparison between set levels and monitored level and consequent contact closure and indication of alarm trip status by lamps.

Voting logic where necessary for auto shutdown.

Check and Control Circuits for bypass (for calibration checks *etc*), re-set, scale multipliers (for run through criticals), indicator lamp tests *etc*.

First failure alarm.

Modules of this type can be powered by a common power module or can be individually powered. Signal Conditioning such as low/high/band pass filtering is often included to improve the signal-to-noise ratio, or pinpoint particular frequency bands. Sometimes tuned band-pass tracking filters are incorporated to allow monitoring of vibrations related to particular shaft frequencies.

Periodic Monitoring Equipment

Periodic monitoring consists of logging measurements at pre-determined intervals from transducers identical in type and location to those used for permanent monitoring systems.

Typical monitoring equipment ranges from relatively simple overall reading meters to relatively complex vibration analyzers — usually frequency analyzers. The data is collected at periods appropriate to the machine and its previous history, *ie* monthly, weekly, daily, hourly or even continuously in critical situations. The measurements can be logged manually, plotted in analogue form or processed digitally in a computer system.

Most equipments in this category are portable and can be transported from machine to machine and site to site. Some typical methods, in order of sophistication are:

(a) Hand-held overall vibration level meter with low or high or fixed band pass filtering to suit the signal-to-noise requirements and/or to pinpoint particular frequencies.

(b) Manually tunable band pass filter with level meter to pinpoint specific frequencies.

(c) Tracking filter based equipment which can provide both frequency analysis at a given machine running speed (rev/min) and the vibration level of a given machine order versus speed. This type of equipment is based on the principle of automatically tuning a narrow band pass filter (fixed bandwidth usually) in such a way that the centre frequency of the filter is locked to an external tuning signal originating either from an oscillator (for frequency analysis) or from a tachometer signal generated at a given multiple of the rotational speed of a shaft (for order analysis).

(d) Time compression Real Time Analyzer (RTA). Typically, this type of analyzer can produce one amplitude versus frequency spectrum in 50 ms which can be plotted in analogue form on a standard X-Y plotter in a few seconds or fed to a computer for further processing, *ie* scaling, comparison, storage, readout, *etc*. This method of measurement is therefore very fast and over the last few years has become widely used as a preventive maintenance tool. An extensive methodology has developed around the real time analyzer for this purpose.

Accessories which can extend the analysis capability of the basic analyzer, include:

Ensemble Averager — for signal enhancement and averaging time varying data.

Signal Ratio Adapter — for order analysis and order tracking.

Frequency Translator — for high resolution 'zoom-in'.

The RTA provides a relatively simple means of gathering a sequence of standardized plots for a given machine/transducer location so that the trend of levels of specific vibration frequencies and/or orders can be simply determined.

It plays another important role in the care of rotating machines. It is often a major tool in the R and D and early installation stages of a machine, leading ultimately to the establishment of criteria of acceptable machine vibration levels for subsequent monitoring purposes.

(e) Digital Signal Processor (DSP). A 'hard-wired' FFT based dual input analyzer can be used in the same way as a time compression real time analyzer, its analysis capability is more comprehensive including, for instance, a time averaging capability which is a powerful form of machine vibration analysis for such things as gear boxes etc.

(f) Computor Aided Vibration Monitoring System. Both the time compression RTA and DSP can be used in conjunction with digital computers so that automatic scanning of large numbers of transducers on a machine complex can be handled effectively in this way. The computer system can be programmed to manipulate and store data from the RTA or DSP. The high speed and economy of dedicated analyzers are thus combined with the flexibility of a digital computer and related storage peripherals to handle very efficiently large scale machine monitoring installations.

(g) Lastly, the analysis manipulation and storage of machine monitoring data can be performed on a purely software based computer system. The analyzing functions are slower on this type of system as compared with dedicated analyzers (such as the RTA or DSP) on a cost comparative basis and, of course, the software system requires the back-up of experienced computer personnel.

It should be noted that, although methods (a) to (e) lend themselves to portable operation, methods (f) and (g) are essentially static systems and hence more applicable to large machinery installations.

It will be seen that for periodic monitoring, there is an emphasis on frequency analysis in most of the methods listed previously. The ability to monitor vibration levels at particular frequencies for a given machine condition, *ie* speed and/or load, is very powerful because often one transducer, for example a velocity pick-up or accelerometer on a casing, can be used to indicate the status of several parts of the machine, *eg* rotor unbalance, bearing and gearbox.

This same information can be used for malfunction diagnosis since it offers a very positive method of identifying a potential failure mechanism within the frequency range of the transducer used in the measurement.

Thus, a spectrum analyzer may be used to relate the frequency components of the noise spectrum to some specific mechanical event or pattern in the machine as it operates. Single-channel real time spectrum analyzers are used to obtain this amplitude *vs* frequency, or amplitude *vs* order, information.

If several noise sources exist within common surroundings, the analysis problem becomes more difficult. To identify which noise source is contributing the most to the overall noise measurement requires two-channel analysis capability for mutual-property investigation of the data signals. Cross-correlation techniques have been used to separate noise sources in a composite noise signal. Recently, frequency-domain mutual-property characteristics, coherent output power and the coherence function have provided the capability to identify noise sources and their respective contribution to a total power measurement.

Multiple Noise Sources

To determine which specific machine, when more than one machine is operating in an area, is the major contributor to the total measured noise level, may require other techniques. Due to multiple transmission paths, reverberation characteristics and nearly identical frequencies produced by individual machines, it may not be possible to ascertain the relative noise contributions based on magnitude measurements obtained with a single-channel analyzer. Stand-alone, two-channel FFT signal processors provide the capability to assess mutual properties between two signals by utilizing time delay and phase information.

The most direct approach to determine if two signals are related is to measure the cross-correlation coefficient, which is a direct indication of any mutual properties that two signals may share in the time domain. A normalized correlation coefficient varies between ± representing maximum correlation. For two completely unrelated signals, the average cross-correlation coefficient will tend toward zero.

If the noise signals are predominantly random, cross-correlation gives a direct indication of time delay between the signals. Fig 3, where a minimum time delay has occurred between transducer locations, shows this clearly.

If periodic functions are mixed with random noise in two signals, care must be exercised in interpolating the correlation coefficient. For example, combined in Fig 4, there are several

Nicolet FFT spectrum analysis system designed specifically for solving machinery vibration problems.

Kurtosis bearing damage detector testing motor bearings on a cut-off saw. (Condition Monitoring Ltd).

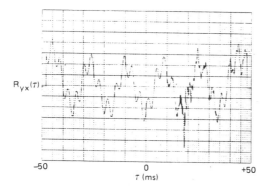

Fig 3 Cross-correlation between two predominantly random signals.

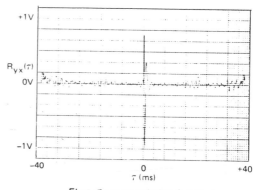

Fig 4 Cross-correlation function, periodic plus noise.

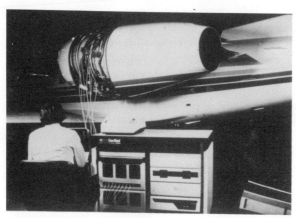

Vibration and Health Monitoring of a modern jet engine using powerful signal analyzer. (GenRad Ltd).

periodics present with random noise. The combined effects of the broadband time delay and the periodic components are illustrated.

If more than one noise source exists and the sources are not independent — that is, they are somewhat correlated — the direct interpretation of the correlation coefficient for relative contributions can be in error because the cross-correlation function does not necessarily give a causal relationship. In this case, it is desirable to investigate the mutual-properties of signals in the frequency-domain utilizing the concept of coherent output power spectrum.

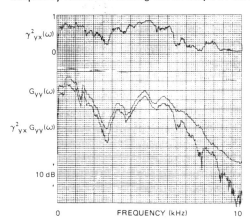

Fig 5 Total power, coherent output power and coherence function spectra with independent noise sources.

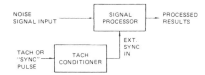

Fig 6 Concept for measuring synchronous power spectra.

Synchronous Power Spectra

Total-power spectrum measurement gives a direct indication of the composite signal existing at a specific measurement location. Coherent output power measurements, on the other hand, show the contribution of an independent noise source in a composite-power measurement (Fig 5).

However, when analyzing vibration and noise signatures from rotating machinery, there is another type of signal analysis that is very useful when broadband characteristics, such as flow-induced vibration, are present in addition to rotational vibration components in a machine signature. In this case, it is desirable to remove signal components that are not directly related to the machine's running speed or any of its harmonics. An analysis technique, signal enhancement or signal averaging, has been developed which discards data signals, either periodic or random, that are not synchronous with a specific reference or tach signal. The concept for this technique is presented in Fig 6.

The use of signal enhancement is extremely valuable in situations where a basic fundamental rotational speed signal or sync pulse is available and other periodic phenomena that are non-synchronous with the primary reference are also present. Analysis of a relatively complex mechanism can be simplified if each major assembly operates such that one or more sync reference signals are available. An averaged time-domain representation of actual motion with reference to a known spatial orientation can give a direct indication of vibration levels caused by specific machine operations within a total cycle, such as one revolution of a flywheel.

After a signal has been time-averaged, its Fourier transform can be taken, creating a 'synchronous' spectrum that is free from any non-speed-related components. If the signal is time-domain averaged, it is not necessary to spectrum average also. The result of averaging in the time-domain with the aid of a sync reference signal will be either a periodic function or a single that averages to zero.

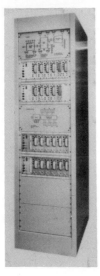

Rack assembly of Scientific Atlanta M700 series monitors.

M6000 machinery health monitoring sytem by Scientific Atlanta features automatic continuous surveillance of the condition of bearings, rotor positions, blade and gear meshing, lube oil and bearing metal temperatures; together with monitoring and visual presentation of machinery data.

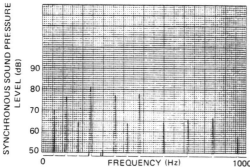

Fig 7 Synchronous sound pressure level spectrum.

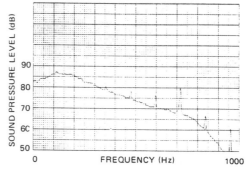

Fig 8 Total sound pressure level spectrum.

The averaged time-domain representation is important in many applications, but it is sometimes more revealing to observe the Fourier transform of the time-averaged waveform. Consider the set of machine-operating conditions where the total sound pressure level as measured by a nearfield microphone, was analyzed and presented (Fig 7).

The synchronized portion of this spectrum using a tach reference sync signal is presented in Fig 8. Observe the complete absence of broadband noise components in the synchronous spectrum. Although the random-noise portion of the total spectrum represents real power and possible destructive forces, it is power that is not synchronous with the fundamental operational speed of the machine. Other rotational components related to some other fundamental speed, such as a gear-reducer output, will also be rejected in the computation of the synchronous power spectrum.

An important distinction to be made concerning the signal processing that takes place for the coherent power spectrum and the synchronous power spectrum is that coherent power spectrum, which can be thought of as the product between a single-signal power spectrum and the coherence function between two signals, is a two-channel measurement. However, the synchronous power spectrum obtained by performing a Fourier transform on the averaged time-domain waveform is a single-channel measurement in which special sampling characteristics are used to process a single data channel. Even though a synchronous spectrum analysis will provide more signal-to-noise enhancement than coherence spectrum analysis, each processing function has its application in obtaining the maximum amount of information from data signals.

SECTION 4a

Machines

THE NUMBER of potential noise and vibration sources present in machines depends on the type of machine and its construction. The actual noise source may or may not radiate airborne sound. In the latter case it can still generate noise by transmitting vibration to a mechanically or acoustically resonant system within the machine, or coupled to it, which acts as an efficient radiator of airborne sound. The operating speed of a machine can also materially affect noise levels. High-strength, lightweight materials used in machines operating at high speeds have little inherent damping and are easily excited, creating noise due to vibration. In addition, the higher speeds create greater impact forces on bearings, joints and gear teeth, inducing more noise and vibration.

Basic 'generators' in machines are:

		vibration	noise
(i)	rotating imbalance	√	√
(ii)	reciprocating imbalance	√	√
(iii)	friction	√	√
(iv)	bearings	√	√
(v)	gears	√	√
(vi)	fans	√	√
(vii)	windage		√
(viii)	air turbulence		√
(ix)	impacts	√	√

Note that the majority of these sources initially generate *vibration* as the primary response. The strength of the related noise developed is not necessarily in proportion to the strength of the vibration; also the resulting spectra can be quite different — *eg* see Fig 1.

Gear Noise

Noise caused by gears is likely to be obtrusive since tooth-meshing frequency in most gear drives is normally in the audio frequency range. Tooth-meshing frequency for simple countershaft gearing is:

$$f_g = nN/60 \text{ Hz}$$

where   N is the number of teeth on the gear
    n is the relative speed in rev/min

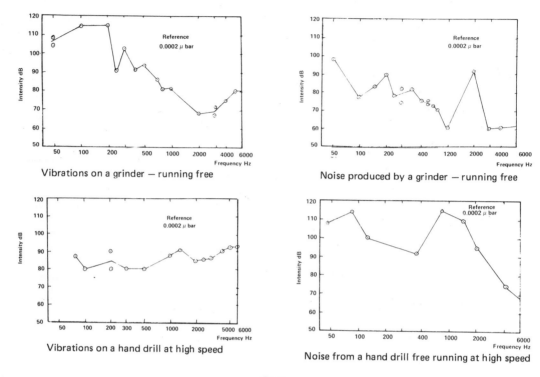

Fig 1

Gear noise is also caused by variations in compressive stress loading on the individual tooth, or a combination of this and tooth-meshing. Cyclic stress loading is characterized by gear 'whine' at gear tooth contact frequency and/or its second or third harmonics. In many instances this can be aggravated by a degree of 'play' present, allowing a pair of gears to develop very small oscillations, or promoting forced vibrations of gear shafts and even the casing. In the former case the 'whine' may be sharply peaked, whilst in the latter broad band noise may be generated.

The elimination of gear whine may be particularly desirable in specific applications — but the general problem of gear noise is usually far less important. Gears invariably form part of mechanical transmission systems and are thus associated with machines or drivers which may themselves be inherently noisy. It is thus generally quite unnecessary to reduce gear noise below the noise of the associated machine. However, gear noise may vary widely under different operating conditions or loads. It may, therefore, be necessary to establish an acceptable level for gear noise consistent with minimum machine noise rather than maximum load or normal operating conditions. If that is not done, gear noise may predominate under such conditions.

Any inaccuracies in gear manufacture will necessarily tend to promote vibration and noise, the most common fault being a periodic error in tooth spacing which will normally be aggravated if it mates with a gear with a similar flaw. The problem is complicated by the fact that the load capacity requirements of most gears are such as to call for the use of very stiff, hard materials which have low damping and are thus sensitive to excitation by impact. Nor is it possible to produce gears entirely free from geometric errors.

If a periodic error exists on a tooth in both the pinion and gear, maximum noise will occur when the bad teeth meet, the resulting tooth repeat frequency being given by

$$f_{tr} = n_g/60N_p$$

where suffix g refers to the gear
suffix p refers to the pinion

'Ghost' noise components which may be present in the noise spectrum stem from inaccurate transmission in the gear cutting machine itself, (see Fig 2).

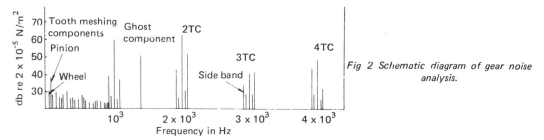

Fig 2 Schematic diagram of gear noise analysis.

The significance of manufacturing accuracy also varies with the type of gears involved. Thus it is very important in the case of spur gears, herringbone and double herringbone gears and hypoid gears, but less important in the case of helical gears. Bevel and worm gears fall between the two in this classification.

In the case of planetary gears with a fixed ring, tooth meshing frequency is given by

$$f_p = N_r (n_r \pm n_c)/60 \text{ Hz}$$

where N = number of teeth in any reference gear
n_r = rotational speed/rev/min of that gear
n_c = rotational speed (rev/min) of cage

The sign in the bracketed term is negative for the same direction of rotation and positive for opposite direction of rotation.

Effect of Damping

For the nearest possible approach to quiet operation, the alternative solution of employing non-metallic gears, or breaking a metallic gear chain with an intervening non-metallic gear is often worth considering — eg a laminated reinforced thermoset plastic, or an engineering plastic (such as nylon or polyacetal). These materials have high internal damping, and thermoplastics in particular are effective vibration dampers. They do have certain mechanical limitations, however, and so cannot begin to become competitive with hardened steel gears for high load capacities. For intermediate capacities, cast iron or non-ferrous alloys may be considered because of their damping properties.

In particular cases where a high strength, low damping material has to be employed for smaller, high speed gears, considerable reduction in noise can be achieved by 'detuning'. This is an adjustment of the mass of the gear, eg by varying the thickness, so that its natural frequency is outside the range of exciting frequencies. This can often eliminate 'resonant noise' experienced at particular speeds, although the casing should first be checked to see that it is not this, rather than the gears, which is resonating.

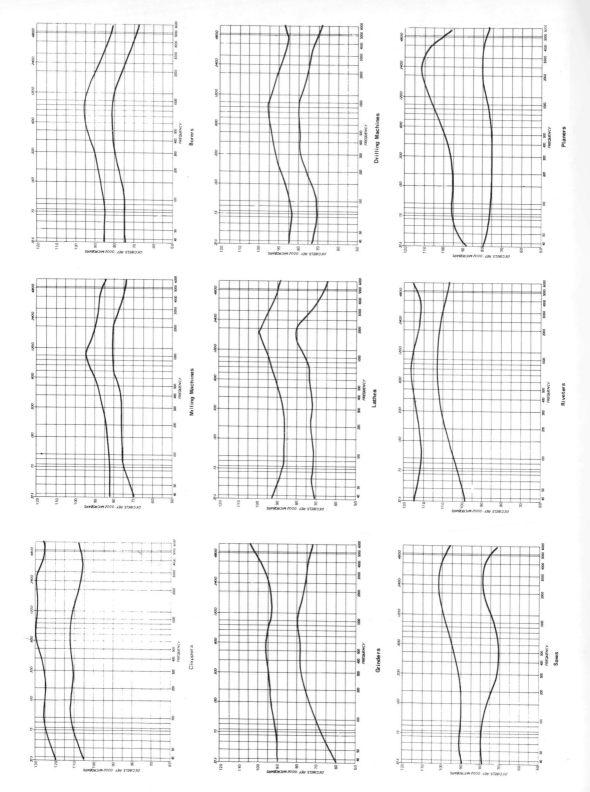

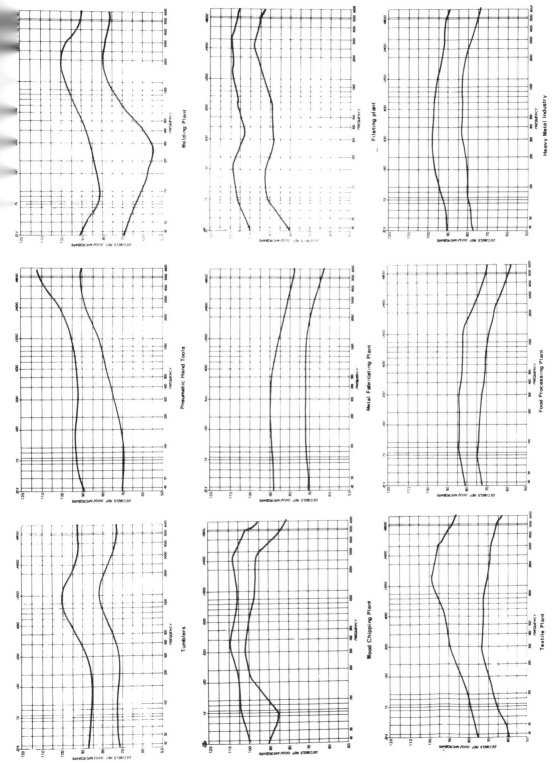

Typical machine noise spectra.

General Requirements

In general, the primary requirements for high capacity and/or high speed gearing in a rigid material are —

(i) a suitable tooth form, and
(ii) high accuracy of manufacture

With increasing accuracy of manufacture there is a sort of 'levelling off' of relative noise and so there are logical limits to the degree of accuracy when, with suitable lubrication, noise level will be a practical minimum and virtually incapable of being improved upon. Poorly made gears, however, will always tend to be noisy.

Stress loading, and the generation of compressive stress waves, become increasingly important with higher load capacities and higher tooth loads and thus may become an unavoidable source of noise in such types. Modifications of tooth forms providing stress relief — and particularly avoiding localized stress concentrations — can reduce noise and are also desirable to improve material resistance to fatigue.

The type of gear largely governs the noise level likely to be generated in gear drives. For general references, sources of gear noise and gear noise parameters are summarized in Tables I and II. Other forms of power transmission —*eg* chain drives and belt drives — are generally quieter than gear drives, but may develop noise with increasing wear on the elements (*eg* chain wear and 'stretch' and bearing wear).

TABLE I — SOURCES OF GEAR NOISE

Cause	Form of Excitation	Remarks
Geometry — (i) inaccuracies	Impact	(a) Due to surface irregularities (b) Due to unbalanced radial forces generated
(ii) eccentricity	Impact	(a) Unbalanced radial forces
(iii) elasticity	Deformation	(a) Elastic deformation of teeth under cyclic stressing
(iv) friction	Tooth contact	(a) Independent of accuracy
Stress loading	Compressive stress waves	(a) Magnitude dependent on gear geometry and tooth load
Oil pocketing	Hydrodynamic shock	(a) Excess of oil trapped between pressure surfaces
Oil splash	Turbulence	(a) More noticeable at higher speeds with splash lubrication
Air pocketing	Compression expansion of trapped air	(a) Only likely to be experienced with high speed gearing
Bearing noise	Resonance	(a) Amplification of basic gear-generated noise
Casing	Resonance	(a) Amplification of basic gear-generated noise

Fan Noise and Windage

Fans are a common means of generating cooling airflows on many machines, and are inherent noise generators. Every time the blade of a fan passes a stationary part an impulsive noise is generated behind it, thus yielding a fundamental fan noise with a frequency of N x rev/min Hz where N

TABLE II – GEAR NOISE PARAMETERS

Parameter	Requirements or Treatment	Remarks
Gear type	Suitable mechanical performance, but see 'Remarks' for inherent noise level	Inherent noise level: Spur – moderate to high Helical – low Herringbone – low Hypoid – low to moderate Bevel – moderate to high Worm – low
Gear geometry	(i) large number of small teeth preferred (ii) adequate gear stiffness (iii) accurate manufacture	Slim gears may be subject to disc vibration, very important parameter
Gear material	(i) optimum material selection (but may have to be based on mechanical requirements) see also Damping	Non-metallic gears may provide quiet operation in gear trains
Bearing	Good bearing design	Must give a rigid system
Lubrication	EP lubricants may be desirable, or necessary	Excess lubricant can lead to noisy operation
Damping	(i) gear material with high internal damping preferred, if practical (ii) damping rings or damping plates on gears (iii) isolation of gear rim	Resilient materials (*eg* nylon) provide excellent damping Most effective when applied to rim of gear Not necessarily effective and has practical limitations
Isolation	(i) resilient mounting can be used to reduce structural-borne sound (ii) enclosure can be used to reduce airborne sound	Treatment applicable in specialized or very bad cases of gear noise
Housing	(i) rigid housing essential (ii) good hydrodynamic internal shape (iii) large mass preferred	Housing material should have high damping if possible To provide efficient circulation and retention of lubricant Natural frequency should be well above rotational frequency range

is the number of fan blades. This is usually accompanied by strong harmonics. Increasing the number of blades will tend to decrease the number of audible harmonics, but will not reduce the overall fan noise.

See chapter on *Fan Noise*.

Windage

Windage noise is a fairly common feature on high speed machines but is not always noticeable in the presence of other noise components. It is caused by stationary parts close to rotating elements producing local acceleration of air and turbulence. The intensity of windage noise varies approximately as the square root of air acceleration and has a fundamental frequency which is a function of the rotational speed and the clearance of rotor and stationary member(s). With simple obstructions and fixed clearance, a 'siren' note can be produced. Variations on the size of the obstructing stationary member and clearances tend to generate broad band noise.

Vortex Noise

The trailing edges of rotating fan blades cause vortices, the size and intensity of which are a function of blade speed, blade geometry and enclosure geometry. These result in a fluctuating pressure pattern in the airflow creating random broad band noise. Vortices may also be shed from the trailing edge of downstream obstructions in the airflow although at the lower linear velocities involved their effect may be negligible compared with fan-shed vortices.

The subject of windage and vortex flow is described in more detail in the chapter on *Electric Motors*.

Impact Noise

Impact noise may be generated as the consequence of the operating performance of a machine (ie one having mechanically impacting output movements); or by rapidly changing output load characteristics. The latter can be a characteristic of link systems where there is shock movement at pivot points. In this case the noise frequency is the product of the rotational speed and the number of pivot points producing acceleration 'jerks'.

It is a general characteristic of impulse noise of this type, that airborne noise radiated from the actual source (pivot points) is generally small. Vibration generated by acceleration jerks may, however, travel through the system and supporting structure to be radiated from some source of resonance in the system.

Noise Radiators

Basically a noise radiator is any semi-rigid object of a size equal to or greater than the wavelength of the vibration frequency transmitted to it. Its efficiency as a noise radiator is then inversely proportional to the self-damping properties of the material from which it is made.

The relationship between wavelength and frequency is:-

$$\text{wavelength (metres)} = 340/\text{frequency (Hz)}$$
$$\text{wavelength (feet)} = 1117/\text{frequency (Hz)}$$

It follows that to be an efficient radiator at, say, 100 Hz, a panel would need to be at least 3.4 metres long; at 1 000 Hz, at least 340 mm long; and at 10 000 Hz, only 34 mm long. The practical probability of panel resonance thus increases characteristically with increasing vibrational frequency.

Airborne noise is emitted from any resonant point, whether an efficient radiator or not, when the excitation frequency coincides with the natural frequency of that point or with one of its modes defined by its resonant frequencies.

The paths of vibratory mechanical-energy transmission within a structure may be infinite, depending on the source and the complexity of the machine. Structure-borne energy has a tendency to spread among the elements at a joint. It also has the characteristics of changing direction and reflecting at joints, surfaces, and interfaces as well as being absorbed by the material it is travelling through. Hence, the energy of vibration decreases as the distance from the source increases, and for two panels that have the same resonance, the closer to the source will emit the greater airborne noise.

Machine Noise Reduction

Whilst noise reduction at source is an obvious (and the best) method of treatment at the design stage, this can involve complex procedures and high costs. Possibilities to be considered include:-

(i) mechanical re-design
(ii) improved balancing
(iii) reduced tolerances
(iv) reduced bearing clearances
(v) higher bearing finishes
(vi) reduction in mass of rotating parts
(vii) improved gearing (where applicable)

A major difficulty can be determining which are the greatest noise sources of a machine. Techniques applicable here are:-

(i) probing or 'listening in' at various points on a prototype machine using a stethoscope for subjective analysis; or preferably, a microphone for objective analysis.
(ii) Disassembling the machine a step at a time and recording and comparing the noise spectrum at each step.
(iii) Applying dynamic analysis techniques to establish and relate the frequency of the peaks in the noise spectrum to operating parameters. This demands specialized equipment and a specialized knowledge of the subject.

Failing noise reduction methods applied at source, (ie in the construction of a low-noise machine), or if further noise reduction is necessary, treatment can embrace one or more of the following:-

(i) isolation of the machine from all potential sound radiators (resonators)
(ii) damping of resonators
(iii) reduction of airborne noise from radiating source(s).

As a general rule, isolation is the best treatment for low frequency noise, provided it is applied as close to the source as possible. Damping is most effective when applied to panels. The wide range of characteristics offered by various damping materials make it possible to cover a wide range of frequencies.

Reduction of airborne noise from a radiating source is specified separately for there is often failure to realize that there are two sources of airborne noise — those generated by a vibrating surface (which can be treated by isolation and/or damping); and those which are aerodynamic in origin. Unless eliminated, or reduced to acceptable levels, at the design stage, aerodynamic noise can only be treated by ensuring reduction of that noise from the radiating source.

See also chapters on *Bearings, Internal Combustion Engines, Fan Noise, Factory Noise, Bearings* and *Machine Balance.*

Bearings

THE MAIN cause of friction noise generated by sliding or rubbing surfaces is lack of lubrication. It is essentially a form of high frequency vibration resulting from rapid intermittent contacting of the surfaces. Where this is transferred to a resonant structure this is heard as a 'screech'. In the case of plane sliding surfaces the frequency of the noise is related to the surface finish of the surfaces, and its intensity to both surface finish and surface loading.

In addition to plane sliding surfaces, all mechanical bearings are subject to friction. Vibration can also be a source of bearing noise — either separately or by virtue of increased friction.

Some types of bearing will generate noise if excited by vibration. Noise levels in such cases will increase with wear, accelerated by vibration. Machine balance is therefore of considerable importance, both for reducing wear and controlling noise levels which may be inherent.

Plain Bearings

Parameters affecting plain bearing noise are summarized in Table I. Lubrication plays a significant part in controlling friction and thus noise. Under complete hydrodynamic lubrication conditions friction is low and the noise generated negligible. In the case of high speed bearings, however, there is the possibility of oil film whirl developing which can lead to vibration and an increase in noise level, particularly if any resonant effects are present. The speed of whirl, and thus the frequency of vibration produced, is generally low and never more than half the rotational speed of the shaft. Such noise is thus seldom objectionable unless amplified by resonance. It can, in any case, be eliminated if necessary by employing a discontinuous bearing surface — *eg* a segmental or pad type of bearing — or even a modification of the clearance in the original bearing. Other 'direct' cures may be produced by a change in lubricant viscosity or a change in bearing pressure (although this will usually necessitate replacement of the bearing by one of a different surface area).

In the absence of complete hydrodynamic lubrication, intermittent local seizure may occur through metal-to-metal contact.

The general case is where smooth rotation is inhibited by a series of relatively small drag forces due to localized friction or localized welding. Whilst this may be damaging to the bearing surfaces there may be no apparent effect on the running of the shaft in the majority of cases, although on occasions marked vibration may occur. This is because any form of stick-slip motion tends to excite the moving (in this case the rotating) system at its natural frequency. The extent to which this is noticeable as vibration and increased noise is dependent on the elasticity of the mass involved. Thus an increase in noise and vibration, as distinct from 'chatter', from a plain bearing is a likely indication that stick-slip motion is present, *ie* lubrication is inadequate. The fact that the bearing is tight will eliminate 'chatter' due to excessive clearance (*eg* in a badly worn bearing). Thus a distinction can be drawn between these two types of vibration.

BEARINGS (A)

BURGESS...THE SPECIALISTS WHO OFFER A COMPLETE NOISE CONTROL SERVICE

Burgess's experience in over 50 years specialisation in the design, manufacture and installation of noise control equipment for all types of power generation plant is only a telephone call away. If you need to silence diesel engines, gas turbines, high pressure steam or process gas discharge valves, our expertise in this field could solve your problem.

Burgess field engineers are available to carry out surveys on site, and with back-up from Burgess's design and engineering team, specialist solutions can be evolved.

A wide range of noise control equipment is available:
A standard range of absorptive and reactive silencers for use with all types of internal combustion engines, blowers and compressors; special designs where applicable.
A range of standard intake and exhaust silencers for gas turbines.
Special combined packages for gas turbine and diesel engine generator and compressor sets.
Silencers for venting process gases and steam.
Acoustic enclosures and ventilation silencers for all types of application, including a range of standard modular panels, doors and windows.
For further details please consult Burgess sales staff, who are located at the addresses below.
Alternatively for overseas enquiries our appointed agents will be pleased to assist.

Agents:
Akustikprodukter AB
Atterbomsvagen 50
112 57 Stockholm
Sweden.

Azeta S A R L
20124 Milano
Via Locatelli 6
Italy.

De Boer BV
PO Box 4105
Cruquusweg 118
Amsterdam.

Claus Kettel
Laurentsvej 28
2880 Copenhagen
Denmark.

P T Srijaya Pasaka
J L Kartini Raya
41C Jakarta
Indonesia.

Burgess Industrial Silencing Ltd.

Company Address: Burgess Industrial Silencing Ltd. Shaftesbury Avenue, Simonside Industrial Estate, South Shields, Tyne & Wear NE34 9PH. Telephone: South Shields 0632 566721/5. Telex: 537404
Midlands Sales Office: Burgess Industrial Silencing Ltd. Brookfield Road, Hinckley, Leicestershire LE10 2LN. Telephone: Hinckley 0455 637701. Telex: 34549.

This is part of a purpose designed and manufactured silencing system installed on the roof of the new Leicester Building Society computer headquarters.

WHY NOT LET TANTALIC PUT YOUR PROBLEM UNDER LOCK AND KEY

* FULL RANGE OF NOISE CONTROL EQUIPMENT
* FULL DESIGN MANUFACTURING AND INSTALLATION SERVICE

Contact:-
TANTALIC ACOUSTICAL ENGINEERING LTD., DEE-CEE HOUSE, PRINCES ROAD, DARTFORD, KENT
A MEMBER OF THE DEE CEE GROUP
TEL: DARTFORD 77700   TELEX: 895 3600

BEARINGS (B)

VIBRATION MONITORING SPECIALISTS

CML

CONDITION MONITORING LTD.
UNIT 2. TAVISTOCK INDUSTRIAL ESTATE TWYFORD
BERKS UK. RG10 9NJ. Tel: (0734) 342636 Telex 847151

Britain loses more each year, in financial terms, from noise than from fire.

A shocking and surprising fact. Silencing factory pneumatic control equipment can improve health, safety, productivity and working conditions. It helps save money too.

- The noise from a single, unsilenced air exhaust port is reduced from about 90 dB(A) (The danger level for prolonged exposure) to between 60 and 70 dB(A) when fitted with a Vyon silencer.

- Vyon silencers screw directly into the exhaust ports of control valves, using existing threads, so fitting is simple.

- Vyon silencers are inexpensive and made in six BSP sizes up to 1".

Porvair Limited, Industrial Division,
King's Lynn, Norfolk,
England. Tel: (0553) 61111

Fit Vyon silencers now

Vyon porous plastics are available for a wide range of industrial applications including filtration and powder fluidisation.

TABLE I – PLAIN BEARING NOISE CONTROL

Parameter	Frictional excitation	Oil film whirl	Use of parameter for quiet design
Shaft speed	Likelihood increases with increasing speed.† Also increases at low speeds with oscillating loads.	Frequency less than half shaft speed. Generally not significant at high speeds.	Not usually an adjustable parameter.
Bearing pressure	Possibility increases with increasing pressure.	Tends to increase with bearing pressure.	Reduce – *ie* use larger bearing area and lower bearing load.
Clearance	Small clearances increase possibility.†	Small clearances reduce oil film whirl.	0.001 in per 1 in shaft diameter recommended to eliminate oil film whirl.
Bearing diameter	Increasing diameter increases rubbing speed for same shaft speed.†		Larger diameter decreases bearing unit load.
Lubricant viscosity	Generated if viscosity too low.	Varies with viscosity.	Change lubricant if necessary to maintain full hydrodynamic lubrication.
Lubricant film strength	Positive if film strength inadequate.		EP lubricants may be required with high bearing loads.
Lack of lubricant	Minimized with non-metallic bearings.		Select low friction materials.

†In absence of full hydrodynamic lubrication

Bearing 'squeal' is the same form of vibration generated by friction, although a distinction can be drawn between seizure and localized overload conditions. In the case of metal shafts running in metal bearings, stick-slip motion will tend to generate high frequency vibrations in the skin of the shaft, the frequency increasing with the degree of seizure. Thus a bearing which is tending to seize will commonly generate a high frequency 'squeal'. On the other hand, in the case of a non-metallic bearing, squeal may be generated by localized overloading of the bearing without necessarily being an indication of potential seizure, particularly in the case of elastic materials. The bearing may be perfectly capable of adjusting itself to such conditions because of the relatively large bearing area available (*ie* because of the lower design unit load with such materials).

In the absence of non-metallic bearings, a number of materials are suitable for running dry. The addition of lubricant may have little or no effect on friction developed, although it can be effective as a coolant. This can be important in controlling the expansion of the bearing and the running clearances involved.

The load-carrying capacity of plastic bearings is relatively low, calling for large bearing surface areas; the material itself is usually resilient. These two factors tend to provide inherent damping of vibrations. Such bearings, however, are not necessarily completely free from 'squeal', particularly if excessively tight or subject to a degree of misalignment relative to the shaft. In the former case, the coefficient of expansion of plastic materials is always considerably higher than that of metals, so that tightness can readily arise from differential expansion. It is usual to allow far more generous clearances to take account of such possibilities. Also the dimensional stability of many plastics is relatively poor, so that changes in clearance can occur with differences in humidity.

Lubrication can help where excessive tightness has developed, although this is usually a frictional loss rather than a noise problem (for tightness in a plastic bearing is not necessarily accompanied by noise). A 'solid' type lubricant such as molybdenum disulphide is usually the most effective in such cases. For general lubrication, water is usually quite effective, the requirement being more for a coolant than an actual lubricant.

Certain non-metallic bearings must be operated only in a fully hydrodynamically lubricated condition. The segmental rubber bearing is a typical example. It is normally lubricated by having water continuously flushed through cooling channels in the bearing surface itself. If run dry the bearing will rapidly become very noisy (and also be irreparably damaged in a very short time). Rubber bearings are usually of fairly long length relative to their diameter and are prone to develop 'squeal' if badly misaligned. They will normally accept a small amount of misalignment without becoming locally overloaded and have excellent damping characteristics against shaft vibration. They are, therefore, basically quiet bearings.

In general, noise or vibration should never be a problem with any suitably designed plain bearing provided the bearing operates under conditions of full hydrodynamic lubrication. Plastic bearings can operate dry, but also act in the manner of isolating elements as regards vibration and noise.

Rolling Bearings

Although having much lower friction, rolling bearings tend to be noisier than plain bearings. Full hydrodynamic lubrication is seldom present because of the higher unit loadings and rolling and skidding metal-to-metal contact is usually present to some extent. The amount of vibration generated is largely dependent on the geometry of the bearing and geometric inaccuracies. Thus, for example, the running surfaces of a rolling bearing always have a certain roughness. When the bearing rotates this roughness will cause mechanical impacts between the rolling elements, the magnitude of these impacts being dependent on the surface condition and the peripheral speed of the bearing. Ball bearings are more critical than roller bearings in this respect.

In a good bearing the level of vibration due to normal surface roughness will be very low, but will be aggravated by improper installation, overloading, lack of lubrication, contamination of the lubricant, cavitation, progressive wear or actual damage. Initially, however, the vibration level, and thus bearing noise, will be mainly dependent on race geometry, surface finish, waviness, eccentricity and parallelism, ball geometry (spheriatry) and groove wobble.

The amplitudes of all excited frequencies will, in general, tend to be very small, unless serious irregularities are present or there is resonance. Resonance is the primary cause of noisy rolling bearing operation, and can be reduced or eliminated by employing a housing material with good damping characteristics. It can also be advantageous if the outer ring material has good damping properties.

The fundamental vibration frequency which will appear when any unbalance or eccentricity is present is:

$$f = N/60 \text{ Hz}$$

where N = vibrational speed in rev/min

If the outer ring is stationary and the inner ring is rotatory, then a second fundamental frequency will be developed due to the rotation of the train of rolling elements:

$$f_i = \frac{f}{2}\left(1 - \frac{d}{E}\cos\beta\right) Hz$$

where d = diameter of rolling elements
       E = pitch diameter   } in same units
       β = contact angle, degrees

The suffix i denotes the inner ring rotation

If the inner ring is stationary and the outer ring is rotating:

$$f_o = \frac{f}{2}\left(1 + \frac{d}{E}\cos\beta\right) Hz$$

The suffix i denotes the inner ring rotating.

Excessive noise at either of these frequencies would indicate an irregularity of a rolling element or a cage.

There would also be a spin frequency (f_s) for the rolling elements.

$$f_s = \frac{E}{2d} f\left(1 - \left(\frac{d}{E}\right)^2 \cos^2\beta\right) Hz$$

An irregularity (indentation or a rough spot) on a rolling element would generate a frequency of $2f_s$ because the spot contacts the inner and outer rings alternately, once per revolution.

If there is any irregularity on the stationary rollway, the resulting fundamental vibration frequencies generated are:

irregularity on rotating way : $f = N(f - f_r)$ Hz
irregularity on stationary way : $f = N f_r$ Hz
where $f_r = f_i$ or f_o as appropriate

All fundamental frequencies generated by irregularities can give rise to harmonics. If there are are several such irregularities then the harmonic frequencies will be more pronounced.

General recommendations applicable to noise reduction in rolling bearings are:-

Eccentricity — which should be less than 0.001 inches.

Tolerances — which should be within 0.001 to 0.0075 inches.

Surface Finish — with particular emphasis on the elimination of irregularities.

Fit — an optimum value, as recommended by the manufacturers, so that inherent damping characteristics are not reduced or eliminated.

Elastomeric Bearings

The elastomeric bearing is a structure comprising alternating layers of rubber and metal laminates, with a configuration designed to accommodate various modes of load and motion. The elastomer is vulcanized and bonded to the metal laminates as well as to the attachment metal components (inner and outer 'races'). Attachment metal component configuration is optional and designed to attach the 'bearing' to the structure. The elastomer may be bonded to virtually any material including steel, aluminium, stainless steel, titanium, and many non-ferrous materials. Vulcanization of the elastomer to the metal laminates and to the attachment metal components eliminates potential brinelling problems, assures structural stability due to the integral construction, and provides a permanent, indestructible seal to keep out contaminants such as dirt.

The particular advantage of elastomeric bearings is that they can be made insensitive to vibration, shock or impact loading and can provide a noise barrier by proper selection of spring rates. On the mechanical side, they are self-lubricating, cannot seize and are free from pitting, galling or brinelling as well as offering very low friction and long life.

Configurations

Basic configurations for elastomeric bearings are:

(i) *Sandwich*: The sandwich type bearing as illustrated in Fig 1 is designed to react to high axial loads while accommodating torsional or lateral motions through compression and shear of the elastomer respectively. This type of bearing is very stiff in the axial and soft in shear mode.

(ii) *Spherical sandwich*: The spherical bearing shown in Fig 2 is designed to permit motions around three axes through shear of the elastomer while carrying high axial loads in compression. The stiffness of the bearing is high in compression and low in the shear mode. This type of bearing can replace three or more separate conventional bearings.

(iii) *Cylindrical*: The cylindrical type bearing reacts to high loads radially through elastomer compression while accommodating torsional or axial motions in shear. High stiffness in the radial compression direction while maintaining low stiffness in the axial and torsional modes is characteristic of this bearing, shown in Fig 3.

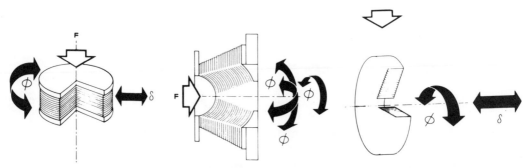

Fig 1 Typical axial bearing   Fig 2 Spherical sandwich bearing   Fig 3 Typical radial bearing

(iv) *Spherical:* Fig 4 illustrates the spherical (ball-joint) type bearing which accommodates torsional or cocking motions about three axes, in addition to some axial motion, while reacting to high radial loads. The high compressive stiffness and low shear are effectively utilized to accommodate the radial loads and torsional, cocking and axial motions respectively.

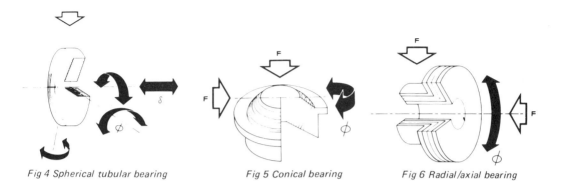

Fig 4 Spherical tubular bearing   Fig 5 Conical bearing   Fig 6 Radial/axial bearing

(v) *Conical*: High axial and radial loading capability in compression-shear loading of the elastomer and a low torsional shear stiffness are characteristic of the conical type bearing. Some cocking load or motion is possible with this configuration, as illustrated in Fig 5.

The types illustrated provide the basic foundation for elastomeric bearings. Various combinations of the basic types can be integrated for virtually unlimited usage. For example, the sandwich and cylindrical types can be combined in order to react to axial and radial loads while permitting torsional motion as illustrated in Fig 6.

Bearing Mounts

The chief source of noise from metallic bearings is resonant vibration. The noise level developed by the secondary generator may be further amplified by virtue of being directly coupled to a resonant panel or vibrating system, or the unit containing the bearing may simply act as a rigid coupling to another resonator. Logically, therefore, damping should be provided at the source as far as possible, *ie* in the bearing support. This calls for a satisfactory design of housing or mounting, particularly where there is a degree of flexibility in how the bearing may be mounted, and in considerations of material specification.

Where the mounting is integral with a casing or end cover, sufficient rigidity is usually ensured by a generous thickness of metal around the bearing. Consideration can also be given to the selection of a material with good damping qualities, although cost and strength may be equally important. Thus manganese-copper alloys with a manganese content of about 80% have high inherent internal damping, but are appreciably more costly than steel or cast iron and lower in strength (about 80% of the strength of cast iron). Cast iron is a low cost material with good damping qualities, but is by no means universally acceptable.

In a design for quiet operation where a material with high internal damping cannot be utilized for the bearing support because of other considerations, the same effect can be realized by restricting the use of such a material to damping rings or damping plates in contact with the bearing or, if necessary, even going to the extreme of isolating the main body of the machine or unit with a resilient material. Some bearings, in fact, are produced on this principle — an example being the elastomeric coating of a plain journal sleeve or the outer face of a rolling bearing. When fitted, this elastomeric layer provides isolation of the metallic bearing elements from the rest of the unit. Thermoplastic bearings provide similar isolation but are necessarily limited to lighter duty applications. However, the use of non-metallic rolling elements in rolling bearings cannot be overlooked as a method of providing isolation and resilience in the bearing itself.

Flexural rigidity of the bearing housing is less important in the case of rolling bearings than with plain bearings. A common design fault in the latter case is failure to provide adequate support for the length of bearing concerned (particularly in the case of bearings intended to support long shafts). As a consequence the bearing has a certain amount of freedom to flex under the shaft, when it will readily develop a barrel shaped longitudinal section allowing the amplitude of the shaft vibration to increase. This can only be avoided by stiffening the bearing support. If the initial vibration is high, bearing life and performance may both be improved by resilient mounting, but this can again destroy the original requirement of flexural rigidity if it permits the bearing to rock, as it will again wear to a barrel shape. A better solution may be a resilient type bearing (surface) such as a segmental rubber bearing, provided it can be lubricated satisfactorily, but much depends on the particular application involved.

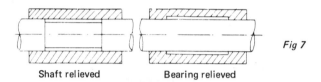

Shaft relieved   Bearing relieved   *Fig 7*

Wineglassing of plain bearings can result from excessive clearances, especially where the shaft may be subject to a radial load (*eg* in a single cylinder i/c engine crankshaft). The freedom to rock provided by excessive clearance can rapidly lead to bellmouthing of the ends of the bearings with an acccompanying increase in vibration and noise, although in a practical application this will be masked by the noise of the prime mover. This is likely to be further aggravated by typical production geometry, such as light initial 'waisting' of the bearing as finished, and slight initial 'barrelling' of the shaft. A better design for a long plain bearing subject to radial loading is shown in Fig 7, where the shaft diameter is deliberately relieved, the effective bearing areas then being restricted to two relatively short lengths at each end of the bearing. Such a bearing is less likely to develop vibration noise due to wear.

See also chapters on *Machine Balance* and *Machinery Health Monitoring.*

Internal Combustion Engines

THE GREATEST source of noise in conventional internal combustion engines is exhaust noise. This can be reduced to virtually any required level by suitable silencers, although practical silencing is almost inevitably a compromise between attenuation and loss of engine performance and efficiency, except in the case of tuned pipe systems.

Once exhaust noise has been silenced, noise levels generated by other sources within the engine may become objectionable and need further treatment. Direct sources of vibration may also need isolation or further forms of control. All these are usually more significant in diesel than petrol engines. It is generally possible to achieve some reduction of engine noise by minor changes in detail design, but this is not likely to be greater than 5–6 dB and may well be less. In most cases it is more cost-effective to apply noise treatment to the vehicle rather than the engine, to meet current legislation or customer requirements. The one exception here is that the extensive research undertaken by various major diesel engine manufacturers is now beginning to yield a new generation of compression-ignition engines which are much quieter than all previous diesels.

Exhaust Noise

Exhaust noise is the major parameter in determining how objectionable the noise is to an observer. When the engine is installed in a vehicle, exhaust noise can readily be reduced to levels fully acceptable by the occupants by engine silencing, although it may still remain objectionably high to the outside observer, particularly by virtue of the noise spectrum. This, in turn, is related to the type, size, speed and power of the engine and the circumstances in which it is used. The latter largely governs the performance of the typical silencer fitted. This may range from virtually unsilenced (*eg* in a 2-stroke or 4-stroke lawnmower engine) to virtually complete silencing (*eg* in a luxury limousine).

Some generalized examples are shown in Fig 1, but these must be regarded as diagrammatic only, serving to indicate the main differences in frequency content between different vehicle engine noises.

Exhaust noise is produced by the periodic expulsion of hot combustion gases through the exhaust. The lowest frequency of the noise spectrum is given by the number of exhaust charges per cylinder per second, which normally corresponds to the firing frequency. In theory, at least, this suggests that combining the exhausts of several cylinders into one exit can provide some cancellation of exhaust pressure energy, but this can be only partially realized in practice. Further, the noise spectrum would be expected to show a long series of harmonics peaking at integral multiples of the firing frequency, which is in fact the case. However, other intermediate peaks may be

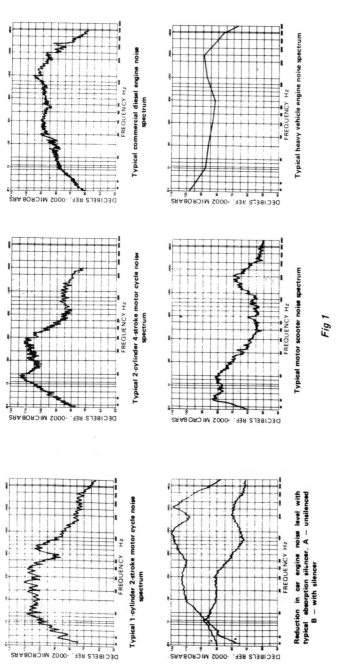

Fig 1

Air intake silencer for a large diesel engine. (Burgess Industrial Silencing Ltd).

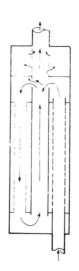

Fig 2 Automotive type silencer.

apparent in the full spectrum, and these can considerably modify the 'quality' of the sound. It is also unfortunate that the strongest peaks tend to occur at frequencies most disturbing to the human ear.

The only practical method of controlling exhaust noise is by fitting a silencer. A reactive type is usually chosen, although this may be associated with resonator or dissipative elements. Basically a silencer can serve two purposes:

(i)   reduce the overall sound pressure level or overall noise, and

(ii) apply selective filtering or tuning to improve the 'quality' of the final noise.

As a consequence the final design can differ appreciably from that of a simple automotive type silencer configuration shown in Fig 2. Size alone is no criterion for silencer efficiency, although silencer volumes have tended to grow with the evolution of more sophisticated geometry. Automobile engine silencer volumes may range from less than the engine displacement up to five times the engine displacement or more (or twice this figure with twin exhaust systems). The interpretation of 'quality' too, is extremely variable and essentially subjective.

The degree of tuning and/or silencing achieved will vary with gas temperature and with engine speed (*ie* governing the actual firing rate). This means that a silencer can only be 'tuned' to a hot engine (or, if tuned to a cold engine, would not be tuned when the engine reaches its normal running temperature), and that the tuning and level of silencing achieved will be speed dependent — *ie* the noise spectrum will 'shift' with speed. From these comments it will be obvious that no silencer will provide consistent silencing over the potential operating range of these engines. The easiest case is that of a stationary engine or similar application, where the engine is run at constant speed. The bulk of silencer design is largely a compromise, and generally tackled semi-empirically, bearing in mind the conflicting requirement that excessive back pressure must be avoided. Thus 'over-silencing' must be avoided as far as possible, if engine efficiency is to be maintained, although internal 'bleeding' can help reduce back pressure without seriously reducing silencer efficiency.

Other Parameters

The total noise content also includes contributions from:

(i)   Noise generated by combustion pressure transients and mechanical sources generated within the engine and radiated by the engine surfaces.

(ii) Air intake noise.

(iii) Cooling fan noise.

When fitted into a vehicle, other calibrators of noise are tyre noise, gear and bearing noise generated in transmissions and aerodynamic noise generated by the passage of the vehicle.

Diesel Engine Noise

Diesel engines are by far the greatest offenders as regards noise generated under category (i) above. Combustion and mechanical noise sources generate vibrations which can effectively turn various structural components into noise radiators — through resonant vibration (see Fig 3).

Sources of noise are, basically, combustion noise, timing gear rattle, piston slap and fuel injection noise. Related analysis of such noise sources is well documented and need not be elaborated on here. Fig 4 is interesting, however, in showing a comparison between cylinder spectra and external noise spectra for three direct injection engines of approximately 1 litre/cylinder and one

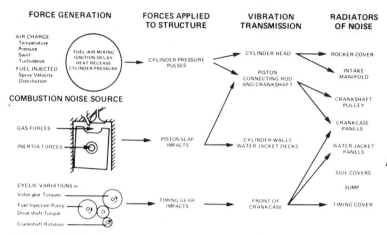

Fig 3 Noise generated in diesel engines.

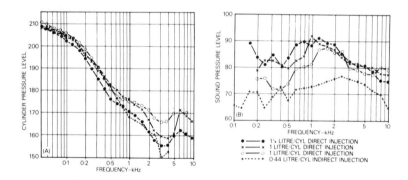

Fig 4 Comparison between spectra of four naturally-aspirated automotive engines: (a) cylinder pressure spectra; (b) external noise spectra.
(Lucas Industries Ltd).

Silencer for British Rail Class 58 diesel locomotive.
(Burgess Industrial Silencing Ltd).

INTERNAL COMBUSTION ENGINES

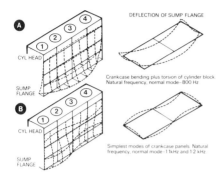
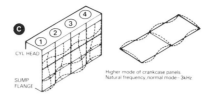

Fig 5 Typical modes of vibration of a small diesel engine structure. (Lucas Industries Ltd).

indirect injection engine of approximately 0.5 litres/cylinder. All four combustion pressure spectra have a trough between 1 and 4 kHz, which is surprising in view of the fact that most engines exhibit external noise spectra with some sort of spectral hump or peak in this region. Either the structure response has a peak which more than compensates for the decrease of combustion noise between 1 and 4 kHz, or another source of noise is responsible. The peak in the cylinder pressure spectrum of direct injection engines (near 6 kHz in Fig 4) which is due to gas oscillations in the cylinder, rarely appears in the final noise spectra.

As regards transmission of vibration through the spectra, Fig 5 shows typical modes of vibration of a small four cylinder diesel. Specifically, the crankshaft vibrates as a beam simply supported by its main bearings, the crossbeam panels responding accordingly.

In the region of the combustion chamber, the engine structure is normally designed to contain the gas pressure and is therefore sufficiently stiff to respond minimally to the gas forces developed within the chamber. This results in a considerable attenuation of the low frequency components of combustion noise. This basic attenuation is increased at low frequencies by the inefficient radiation of noise by the external surfaces of the engine structure.

In the medium and high audio-frequency ranges, resonances of the connecting rod and crankshaft as well as resonances of the crankcase and water jacket panels, sump, timing gear covers, front pulley *etc*, are all excited by vibration transmitted through the structure. This increases the response to forces generated within the structure and detracts from the attenuation provided by the engine structure adjacent to the combustion chamber.

Noise Control Techniques

Three basic approaches are possible for noise control:
(i) Treatment at source of all major sources of noise;
(ii) Development of quieter, less responsive engine structures;
(iii) Acoustic enclosure.

Treatment at source involves combustion, piston shaft and timing gears as major parameters. The aim of treating combustion noise is to produce first a smoother cylinder pressure development and secondly a cylinder pressure development with a low peak cylinder pressure. There are several techniques already in use on diesel engines to produce a smoother cylinder pressure development, such as turbocharging and indirect injection; and pilot injection has been used from time to time. With conventional turbochargers, the boost pressure, and therefore the temperature of the air charge into which the fuel is injected, varies with speed and load. This makes the turbocharger an effective means of reducing noise under full load conditions at high speeds but the noise levels tend to increase as load is reduced.

When modifications to the combustion process to reduce noise are being considered, account must be taken also of the compatibility of such modifications with future requirements for exhaust emissions, smoke, economy and other dimensions in which legislation is being proposed. Figure 6 shows the relationship between noise and smoke (Fig 6a) and the relationship between noise and nitric oxide (Fig 6b) for various fuel injection equipments on a conventional direct injection engine with constant air-fuel ratios. The noise levels quoted are for combustion noise only. The different points in the figures represent various dynamic injection timings in 5° (engine) steps, covering a range of approximately 30°.

The noise versus smoke curves exhibit an 'elbow' at the timing for maximum brake mean effective pressure. This elbow may be emphasized by different timings and there seems little likelihood of an all-round optimum injection system for both noise and smoke, utilizing only conventional components. However, there are points on the high-rate curve with less noise for the same smoke level as points on the low-rate curve (current conventional fuel injection equipment).

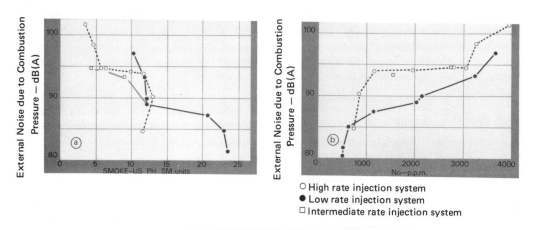

○ High rate injection system
● Low rate injection system
□ Intermediate rate injection system

Fig 6 Relationship between noise and emissions of (a) smoke and (b) nitric oxide from a direct-injection engine running at 1100 rev/min. Readings at 5° intervals over 30° range of dynamic timing.

Piston Slap

The reduction of impacts between piston and cylinder bore during the reciprocating engine cycle has become a major study in its own right. It is rarely possible to reduce the clearance between the piston and the bore under full-load running conditions and, since the piston and cylinder block are made of materials with different temperature coefficients of expansion, the clearance will be expected to increase at lower loads. However, devices which distort the piston skirt as it cools to follow more nearly the dimensional changes of the bore offer a means to prevent the increase of piston slap under cold start and cool running conditions (such as in urban streets or traffic jams, *etc*). Many other devices have been postulated and are being developed to control piston slap, such as articulated pistons, offset gudgeon pins, angled piston rings, oil-cushion pistons, pressure pads incorporated in the skirt, *etc.*

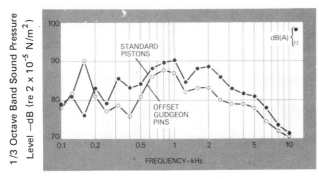

Fig 7 Reduction of external noise by fitting offset gudgeon pins to diminish piston slap. Engine running at 1500 rev/min on light load. (Lucas Industries Ltd).

Figure 7 shows the reduction in noise spectrum external to a truck engine, running on a test bed at 1 500 rev/min under residual brake load, as a result of fitting pistons with gudgeon pins offset by 1/16th inch towards the thrust side.

Timing Gears

Noise from the timing gear drive fitted to most truck and industrial diesel engines may be reduced if the backlash is kept to a minimum or is eliminated. If the timing drive is made to drive a constant (non-fluctuating) load which is greater than the torque fluctuations from the fuel pump, valve gear *etc*, then the noise from the timing gears may be reduced. Alternatively, backlash eliminators may be incorporated into the timing drive. For smaller engines, suitable for passenger cars and light vans, a toothed rubber belt (with non-rectangular teeth) offers an inexpensive alternative which generates far less noise. With such belts it is important that the teeth are fed into their sprocket wheels at the correct angle, and of course that the toothed-rubber belt has sufficient 'wrap-around' on each sprocket.

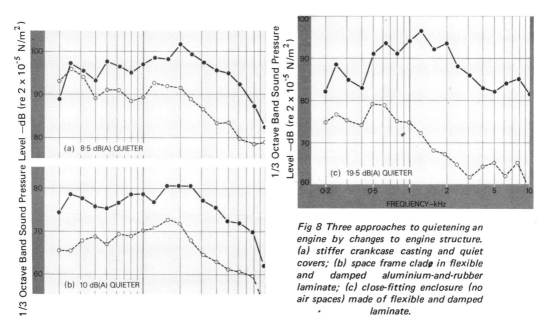

Fig 8 Three approaches to quietening an engine by changes to engine structure. (a) stiffer crankcase casting and quiet covers; (b) space frame clade in flexible and damped aluminium-and-rubber laminate; (c) close-fitting enclosure (no air spaces) made of flexible and damped laminate.

Development of Quieter Structures

The vibration velocity of the thin external surfaces of conventional engines may be reduced by a combination of the following treatments:

(a) Stiffening and dividing cast crankcase and water jacket panels to move the fundamental normal modes into a frequency range where there is less excitation from the noise sources (if possible move to a frequency range well above the resonance of the piston and connecting rod;)

(b) Isolating noise-radiating areas of stiff covers from the vibration of the crankcase and cylinder block casting.

(c) Increasing the flexibility of sheet steel covers together with damping their flexural resonances by a sandwich (constrained shear) technique.

Three approaches to quietening an engine by changes to engine structure are shown in Fig 8 — (a) stiffer crankcase casting and quiet covers, (b) space frame in flexible and damped aluminium-and-rubber laminate, (c) close-fitting enclosure (no air spaces) made of flexible and damped laminate.

Acoustic Enclosure

There are three separate types of soundproof enclosure which can be applied to diesel engines:

(1) 'Close shields' placed over the noisy areas of the engine surface supported on anti-vibration mountings to isolate them from the engine.

(2) Complete 'airtight' enclosures. The enclosures are made from thin flexible sheet material to maximize the inefficiency of the acoustic radiation and they are built to be literally airtight, that is, no cooling air is allowed in the space between the engine and the enclosure. These enclosures do not use acoustic absorbent to prevent reverberant buildup inside the enclosure, rather they are tailored to fit snugly around the engine to prevent low-frequency and medium-frequency standing wave effects. Noise reduction of up to 20 dB(A) has been proved possible by this means.

(3) A tunnel may be fitted around the engine and transmission, this having a silencer in front of the radiator cooling fan and terminating with a silenced air outlet at the rear of the vehicle. Such tunnels must be lined with acoustic absorbent to minimize the reverberant noise inside and to prevent excessive noise escaping through the silencers at either end. The noise reduction available from such designs is limited by the openings required for air flow.

Control of Noise from Diesel Generators

In considering the control of noise from a diesel generator it is necessary to evaluate all the various routes by which sound energy can be radiated and transmitted. Usually this amounts to three main routes: general airborne noise, structure-borne noise or vibration and exhaust noise. Due to the very wide range of audible sound energy (illustrated by the fact that an attenuation of 40 decibels represents a reduction by a factor of 10,000:1 in terms of energy) it is essential that each of these routes is examined in detail to ensure that the overall project is successful.

Structural borne noise can be eliminated by mounting on heavy foundations at ground level, and/or on anti-vibration mounts.

IAC modular acoustic housing with silencing for hospital diesel generator.

Exhaust noise can also be dealt with relatively simply by the fitting of a suitable exhaust system, although diesel silencers have quite specific design requirements. Generally they will comprise a series of tubes, chambers and baffles, tuned to resonate and absorb sound energy at the low frequencies which predominate in diesel exhaust noise. These components can also be designed to act as spark traps to remove particles from the gas stream and collect them in a 'cinder box' at the bottom of the silencer. Frequently the final stage of the silencer will be packed with glass or mineral fibre to absorb the middle and higher frequencies. Mild steel is the most common material for construction, the thickness being sufficient to ensure that noise does not 'break out' excessively through the silencer walls, and that the silencer has an acceptable life despite the corrosive nature and high temperature of the exhaust gases. Stainless steel can be used to give longer life but generally the extra cost is not considered justifiable. The performance of exhaust silencers as described previously is difficult to predict or specify accurately as it depends on many factors including temperature and flow rate of the exhaust gases, and also on the rest of the exhaust system, for instance the length of the tail pipe. Theories and mathematical calculations are available to predict the performance of simple reactive or resonant silencers but practical exhaust silencers are generally more complex, their design being highly dependent on experience and empirical data. Consequently silencers are generally selected from established and proven ranges rather than being designed individually.

Airborne Noise

Airborne noise is generally the most expensive area to treat, not necessarily because it has the greatest intensity or most sophisticated treatment, but because it radiates from a large area rather than being transmitted through a narrow route such as the engine mountings or the exhaust pipe. The whole of the engine casing, fittings and ancillary equipment, including the radiator fan (if fitted) all radiate noise to a greater or lesser extent and generally the only solution is to enclose the entire unit in a sound reducing structure. Close fitting acoustic cladding has been produced for some engines to control the noise from the casing but this is not generally available, has limited effect and has disadvantages with respect to access for inspection or maintenance.

The requirements of an enclosure are numerous. Obviously it must attenuate the noise of the diesel, but also it must allow a flow of air to the engine for combustion and over the engine for cooling, and allow access for operation, inspection and maintenance. Optional benefits that the enclosure may offer include weather protection, security and pleasing appearance, and mobility may be achieved by mounting the generator and enclosure on a road trailer.

A further requirement for insulation is that the enclosure is virtually complete. Even small openings can have disastrous effects locally with significant loss of overall insulation. This immediately raises the problem of access both for personnel and for cooling and combustion air. The personnel access can be provided by special acoustic doors and hatches having similar insulating properties to the general construction. This implies considerable mass and good seals at all edges, so it is important to have well engineered doors which will operate easily despite their mass and will always close firmly and accurately on their seals.

Air ingress and egress is not quite so simple because in the case of radiator cooled units it is necessary to pass large volumes of air, typically 17 000 cubic feet per minute for a 250 kVA generator set and appropriately more for larger units. The air has to pass into and out of the enclosure without allowing noise to pass along the same route. This is generally achieved by using absorptive 'splitter' type silencers where the air is passed through a duct fitted with absorptive panels, or splitters, parallel to the airflow. This ensures that the air passes closely over a large acoustically absorptive area such that the sound energy is progressively absorbed as it passes along the duct. The design of these silencers, involving the selection of airway widths, splitter thickness and lengths, *etc*, is a matter of compromise and optimization to achieve economic silencers which will give the required attenuations without restricting the flow excessively. Auxiliary fans can be used to induce the flow of air but it is possible to use the radiator fan, thus ensuring a safe method of cooling, as the fan is always turning when the engine is running. To ensure that the fan can still provide the required airflow despite the extra load of the silencers it is best to use a 'tropical' radiator and fan, but, whether this is used or not, it is still necessary to pay close attention to the silencers, including the aerodynamic shaping of the splitters and the layout of the system as a whole, to ensure that the pressure loss is not excessive. Figures around 12–15 mm water gauge are typical for the whole silencing system.

Water cooled engines using cooling towers or other external cooling methods require less cooling air in the enclosure, but combustion air is still required as is some ventilation air to remove the heat radiating direct from the engine. In this case auxiliary fans are necessary but it is quite satisfactory to use electric fans which will start up once the generator is running.

Free Standing Diesel Units in the Open Air

The most basic situation is when a diesel generator is required at a particular site, and no facilities other than a suitable open area are provided. This might apply to a temporary work site, or a permanent site where there is a temporary requirement. It also frequently applies to established factories where it is decided that standby power is required and the best site for the generator is in a corner of the works car park. It seems that, almost be definition, the chosen site will be close to houses of the most sensitive members of the local community!

Another favourite site, especially in urban location, is on a flat roof. This means that in addition to other general problems there are often restrictions with regard to access and weight. An untreated diesel generator will give around 105 dB(A) at one metre distance, or about 70 dB(A) at the bedroom windows 50 metres away. Typically levels of around 45 dB(A) would be regarded as acceptable, and lower levels may be necessary for frequent or night-time running, especially in sensitive areas, so an attenuation of the order of 25 dB(A) is required, *ie* a reduction in radiated sound energy by a factor of over 300.

INTERNAL COMBUSTION ENGINES

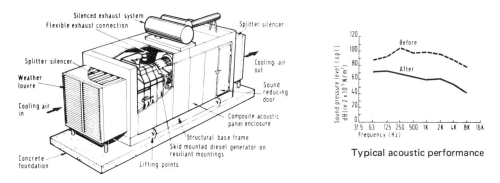

Fig 9 Typical noise control package for a diesel generator.
(ICI Acoustics).

The solution may be a 'lift-off' type of enclosure which is simply lowered by crane over the generator. All necessary features and ancillaries are built in so installation work is reduced to minimum (see Fig 9). The walls and roof of this type of enclosure would probably be constructed from sheet steel panels (to give mass for insulation) stiffened and fitted to some form of heavier steel frame to give strength for lifting. The steel panels would be lined with damping materials and 50–100 mm of mineral or glass fibre (to give absorption inside the enclosure). The inner surface would be of perforated steel to afford a durable surface while still permitting the noise to reach the absorptive layers.

Doors of similar construction would be fitted with double resilient seals and heavy duty hinges and fasteners. To ensure personnel safety in the event of fire it is usual to fit two doors each with 'panic' releases on the inside.

Air silencers would be constructed from similar materials and generally positioned such that the discharge from the radiator is directed straight into the discharge silencer and the inlet air enters either at the opposite end or through the roof. With face velocities often as low as 33.3 m/s (650 ft/min) and lengths up to 2.4 m (8 ft) required for high attenuations these silencers are often large in relation to the rest of the installation and optimum design is essential to ensure that the most economic solution is achieved. Fixed, manual or automatic weather-louvres are generally fitted to silencer ends to ensure weather protection.

The silenced exhaust system would generally be roof mounted externally, but occasionally all or part of the system would be inside the enclosure. This may improve the appearance of the unit and in very high performance applications may be necessary for acoustic reasons but it does increase the heat load on the ventilation system.

A wide range of additional features can be incorporated in this type of installation to suit individual requirements, one of the most notable being external cladding to match adjacent buildings or to suit the surroundings. Double skinned enclosures can be used to achieve attenuations of 55 dB(A) or more, or close-fitting enclosures with acoustic louvres in place of air silencers can be used to give modest attenuations within confined space limitations.

Permanently Sited Diesel Power Units

When a diesel generator or compressor is expected to be sited permanently in one place it may be that the durability of bricks and mortar has a certain appeal. Acoustically the same principles apply. Mass for insulation would be provided by the brickwork of the walls but it is important to remember that the requirements for the roof are just as stringent, so that a concrete or similar roof construction is advisable. Internal absorption can be ignored if the high internal noise levels and effective loss of insulation can be accepted, or absorptive wall and/or ceiling treatments can be applied. If wood wool slabs can be used as permanent shuttering on the underside of a concrete roof and the open structure of the wood wool is left exposed this will give a very useful degree of absorption.

The ventilation and access requirements, the exhaust system and vibration isolation requirements all remain as before so air silencers, sound reducing doors, exhaust silencers and possibly anti-vibration mounts will be necessary.

Diesel Power Units Inside Buildings

If the chosen site is inside a building all the problems and requirements mentioned previously still apply, plus a few more. Noise transmission from room to room within a building is more complex than noise transmission outside and if acceptable levels are to be maintained inside the building extra care must be taken. Firstly the reverberant build-up mentioned earlier in relation to the diesel enclosure can also take place at the receiving end, *ie* noise levels can build up in the supposedly quiet areas due to lack of absorption there. Secondly there are many more routes along which noise and vibration can be transmitted. The structure-borne vibration aspect is obviously much more important when the source is inside the same structure, (*ie* the building), as potentially sensitive areas. As before isolating mountings are important, but other vibration routes must also be treated. Equipment has been installed in a basement with pipework connected directly both to the machinery and the ceiling. In the offices immediately above, the effects were obvious both as vibration in the floor and as raised airborne sound levels. Even distance is no certain barrier as in another case vibrations were transmitted from a basement plant room to a first floor laboratory at the far end of the building some 150 ft away.

Similarly with airborne noise the building can provide numerous transmission paths, *via* ventilation ducts, service ducts, and of course passageways, stairwells, *etc*. Each of these will need attention using doors, acoustic panels or silencers.

Construction Site Equipment

PNEUMATIC TOOLS and appliances are inherently noisy devices if unsilenced, since compressed air at a pressure of the order of 7 bar is normally exhausted directly to atmosphere with consequent high exhaust velocities. In many cases the overall noise level is further raised by vibration or the impact noise generated by the mechanical parts of the tool or device. This can limit the degree of successful silencing which can be applied, particularly with percussive tools, even though proper exhaust silencing can eliminate any excessive noise being generated by the compressed air itself. Where this is still objectionably high, the erection of noise barriers to screen the noise is a simple, practical solution.

Compressors

Compressors for construction site working are commonly designed as mobile units, truck or trolley mounted, together with a diesel engine driver and a simple canopy for weather protection.

Noise reduction of the order of 15–20 dB is readily possible by improved driver silencing (*eg* to motor car standards) and by fitting a complete enclosure, including trays under the driver and compressor, especially if the enclosure is suitably damped to eliminate panel resonance and lined with a sound absorbent material. Further improvement is possible by proper attention to the cooling air ducts, isolation or flexible mounting of the canopy so that machine vibrations are not transmitted to it, and similar attention to detail design. Reduction in noise levels achieved on modern portable compressors incorporating noise treatment in their design normally meet the general recommendation that the noise levels on site working should not exceed 70 dB(A) in rural, suburban and urban areas away from main traffic and industrial noise; or 75 dB(A) in urban areas near main roads and in heavy industrial areas.

In practice actual noise affecting people is related to the siting of the compressor. In a built-up area 7 metres is about the maximum distance it is possible to get from the nearest dwelling, when a reduction in noise level to 70 dB(A) can demand extensive acoustic treatment.

Percussive Tools

Basic treatment to reduce the noise emitted by percussive tools is to silence the exhaust. The first silencers were in the form of bags made of canvas and plastic, hung from the handles, surrounding the body of the machine and tied to the front end. These were very effective in reducing noise and did not cause any problems of choked exhausts due to ice formation, but were easily damaged or mislaid and generally did not last long in service. They also tended to impede moil point changes due to the front end fixing, which added to the problems of getting them used consistently.

To a large extent this type (which has the advantage of being readily fitted to a variety of different machines) has been superseded by specially designed mufflers fitted closely around the machine cylinder, including the exhaust ports. These are usually made from a tough, resilient plastic, such as polyurethane, and are held in place by metal clips. They have proved much more durable and, far from being a nuisance to operators, are easier on the knees than is the bare machine.

The noise reduction obtained is about the same as with the bag types overall, though the low frequency attenuation is better. In very adverse conditions there is some risk of ice formation in the exhaust passages — this occurs rarely and can usually be dealt with by the use of an anti-freeze oil additive, or in extreme cases by simply removing the muffler *pro tem*.

When the exhaust noise has been greatly reduced by these means, the remaining noise, largely emanating from the struck tool, has proved very irritating and penetrating. Though the overall noise level is reduced, this residual component often appears more intrusive and great efforts have been devoted to its reduction. This has proved difficult, not so much because it is difficult to damp the 'ring' of a struck steel, but because of the problems of doing this without weakening the steel (as by putting in damping inserts) and of keeping the damping element in place under severe service conditions if it is attached to the outside of the steel. Various damped steels are now on the market and most of them rely on the use of a rubber sleeve held on to an additional collar on the steel by means of a swaged-on steel ferrule. Naturally these are more expensive than the plain steels, but together with the exhaust mufflers they bring the noise down to an acceptable level for most purposes.

A further recent development stems from the problem of sharpening moil points, which entails heating and re-forging the working end. Not only is this a nuisance, but it is more difficult to achieve satisfactorily on a silenced steel, without damaging the rubber collar. Partly as a result of this problem, partly because of its inherent advantages, a throw-away detachable tip has recently been put on to the market, which makes claim to considerable overall cost savings.

Another aspect which has received attention is the vibration produced by the machine and transmitted to the operator. It is well known that machines of similar weight and output can have quite different handling characteristics and that the detailed internal design of the machine can affect them considerably. Apart from this, machines have begun to appear on the market which have special external vibration damping systems incorporated to prevent the transmission of impacts and vibration to the operator. Combined with effective silencing these are no doubt the most comfortable for operator and public yet available, but they tend to be more cumbersome and considerably more expensive than the simpler types.

Hydraulic Tools

Another approach to a hand-held machine for concrete breaking is that of the hydraulic powered type. Over many years, a number of attempts have been made to use hydraulic fluid instead of air as the working medium. The main motivation has usually been to reduce noise, due to the absence of any exhaust from the machine. Though this was a reasonable approach, success in producing a viable hydraulic machine roughly coincided with the introduction of the muffled pneumatic machine. Generally speaking, for machines with similar output characteristics, there is now little difference in the noise emissions from the two types. The ring of the struck steel is present in both and can be dealt with by the same means.

A particular advantage obtained by the hydraulic machine is the fact that it uses significantly less power from the prime mover. This is due fundamentally to the fact that it is difficult to use a

Vibration, noise and shock control

Structure-borne vibration, noise and shock control by

  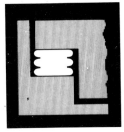 

Unit isolator systems   Area isolator systems   Special purpose isolator systems   Viscous dampers

Airborne noise control by

Silencing units   Acoustic tiles, wall panels and ceilings   Acoustic enclosures, screens and booths   Sound resisting/fire doors

The disturbing and damaging effects of vibration, noise and shock on human beings, buildings and machinery have become increasingly evident and costly to us all. Yet almost invariably such problems can be overcome by expert treatment.

The solution may lie in the use of isolators to control structure-borne vibration, noise and shock, or in the use of acoustic equipment to control airborne noise, or in both. In any case correct diagnosis and application of the most effective control media are essential to success, and these demand specialised knowledge and great experience backed by a very wide range of isolators and acoustic equipment.

At Christie & Grey we have been providing a unique service all over the world for more than sixty years and we use all types of control media. Our complete service includes investigations and surveys, design, detailed drawings and site supervision.

W Christie & Grey Limited

Engineers in vibration, noise and shock control
Associated with
Christie & Grey Acoustics Limited
Sovereign Way Tonbridge Kent TN9 1RJ
Telephone Tonbridge 366444 (STD 0732)

- **Design & Development Services for CAB and ENGINE enclosures**
- **Sound Suppression/Trim Kits**
- **Urethane Mouldings & Floor Mats**
- **Polyether Flame Retardant Foams**
- **'High Tec' Low Weight Damping Pads**
- **Technical Advisory Services**

Manufacturers:
Supra Chemicals & Paints Ltd.
Acoustic Division, Globe Street
Wednesbury, West Midlands WS11 0NN
Tel: 021 502 2857

COBRA ACOUSTICS LTD

1. We Specialise in All Types of Marine, Industrial and Automotive Noise Control.

2. We Supply All Types of Sheet and Pre-cut Acoustic Materials.

3. We Provide a Complete On-site Service Including Design and Development.

RING: COBRA ACOUSTICS LTD TELFORD (0952) 616109

compressible fluid, such as air, as efficiently as a non-compressible one over the pressure range which is needed to give sufficient output within the size and weight limitations necessary to keep the machine to a manageable size.

Reduced power usage tends to help in noise reduction overall as a smaller, or slower running, engine can be used to drive the oil pump as opposed to the compressor.

Rotary Tools

Straightforward exhaust silencing using a simple expansion chamber with an outlet diffuser is effective on rotary machines. The expansion chamber provides a volume in which the pressure ratio can be reduced to a suitable figure (preferably below 1 atmosphere), whilst the diffuser spreads the outlet flow over a large area and at the same time acts as a filter for high frequency noise. In the latter respect the diffuser material should be chosen to provide effective attenuation at the predominant frequencies of the exhaust. Materials commonly used include porous plastics, ceramics, metal gauze, fibrous materials, felts and metallic felts.

Since the noise energy is proportional to AV^8, by diffusing the same mass flow over a larger area so that the quantity VA is a constant, it follows that:-

Sound power is reduced by (area ratio)^7

Sound pressure level is reduced by (area ratio) $A^{3.5}$

A primary requirement is that the expansion chamber volume be adequate to reduce the strength of the shock waves formed by sudden (supersonic) expansion and also to eliminate excessive back pressure which would affect the working of the machine. At the same time the use of a large volume will reduce the risk of ice formation and accumulation which could clog the diffuser.

Depending on the type of rotary machine involved, various sources of mechanical noise may be present, *eg* gears. These can easily be dealt with by normal machine silencing treatment.

Construction Machines

A number of manufacturers already produce low-noise-level construction machines embodying acoustic treatment. A study of conventional machines in the 60–2 000 horsepower class with water-cooled four-stroke diesel engines shows that, in general, the major contributor to the external noise levels of these machines was the engine cooling fan. Noise from the fan was also found to make a significant contribution to the noise at the operator's position on most machines, whether a cab was fitted or not.

A typical example of the relative contributions of noise sources on the machines investigated is shown diagrammatically in Fig 1.

The exhaust system ranked as the next dominant source of noise after the cooling system on many machines. However, it was found that on most machines the muffler volumes were small compared with those of trucks powered by similar engines, and that substantial exhaust noise reductions could be achieved simply by fitting alternative proprietary mufflers.

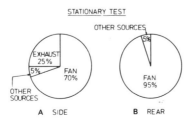

Fig 1 *Relative contributions of external noise sources on typical wheeled loader. (Perkins Engines).*

TABLE I – TYPICAL SOUND POWER LEVELS FROM SITE EQUIPMENT
(BS5228 : 1975)

Plant	Measured range of A-weighted sound power level dB ref 10^{-12} W
Batching plant	107 to 121
Bulldozers	
hp	
55	101 to 108
90	— to 106
95	109 to 118
100	98 to 110
120	— to 106
140	105 to 120
180	107 to 116
270	105 to 128
385	118 to 128
Concrete mixers	
5/3¼ Petrol 1.6 hp	89 to 93
10/7 Diesel 11 hp	107 to 112
10/7 Electric	— to 93
18/12 Diesel	— to 119
Compactors	
hp	
300	117 to 124
Compressors	
m³/min	
3½ to 4½	
Standard	91 to 106
Partially silenced	89 to 100
Super silenced	— to 95
7	
Standard	100 to 111
Partially silenced	98 to 100
Super silenced	— to 95
8½ to 13	
Standard	— to 113
Partially silenced	— to 105
Super silenced	— to 100
17	
Standard	108 to 119
Partially silenced	105 to 110
Super silenced	— to 100
Above 17	
Standard	114 to 120
Partially silenced	— to 100

Plant	Measured range of A-weighted sound power level dB ref 10^{-12} W
Compressors, stationary	
¼ hp to 1 hp	76 to 83
1½ hp to 7½ hp	83 to 92
10 hp to 25 hp	88 to 100
50 hp to 75 hp	93 to 107
100 hp to 300 hp	83 to 103
Double acting class C	83 to 92
Double acting over 200 hp	99 to 106
Rotary positive compressor	88 to 102
Cranes, crawler	
50 hp	112 to 123
55 hp	98 to 115
60 hp	— to 115
80 hp	100 to 115
133 hp	102 to 115
160 hp	102 to 110
75 t steam	— to 106
Crane, tower	
24.5 hp	104 to 108
Crane, lorry mounted	
10 t	118 to 120
Dump trucks	
20 t	102 to 107
25 t 150 hp	— to 114
Dumpers	
1.1 t 8 hp	93 to 97
1.5 t	94 to 114
1.75 t 18 hp	— to 109
2.5 t 45 hp	105 to 107
3 t	— to 113
Excavators	
hp	
Wheeled 44	114 to 117
Tracked 60	— to 103
Wheeled 60	103 to 116
Tracked 62	— to 107
Tracked 70	102 to 118
Tracked 80	105 to 116
Tracked 90	106 to 116
Tracked 95	100 to 112
Tracked 100	103 to 112
Tracked 110	— to 106
Tracked 125	— to 108
Tracked 140	112 to 113
Tracked 150	106 to 111
Tracked 170	111 to 117
Tracked 220	— to 109

cont...

TABLE I – TYPICAL SOUND POWER LEVELS FROM SITE EQUIPMENT
BS5228 : 1975   (contd.)

Plant	Measured range of A-weighted sound power level dB ref 10^{-12} W	Plant	Measured range of A-weighted sound power level dB ref 10^{-12} W
Fork lift trucks		**Hoists**	
1 t	98 to 103	Pneumatic, silenced	— to 108
Generators		Electric	— to 92
For power vibrator		Petrol scaffold hoist	— to 100
2.8 hp	100 to 103	**Loaders**	
For power		hp	
1.25 kVA	— to 112	Tracked 50	107 to 118
2.8 kVA	102 to 109	Tracked 70	— to 117
3.5 to 4 kVA	99 to 111	Wheeled 70	— to 108
6 kVA	106 to 108	Tracked 85	107 to 110
110 kVA	— to 108	Wheeled 90	— to 110
250 kVA	112 to 119	Tracked 100	108 to 116
For welding		Wheeled 150	102 to 112
300 A	107 to 113	Wheeled 200	— to 116
		Wheeled 235	114 to 117
Graders		**Lorries**	
hp		t	
126	— to 109	10	103 to 120
166	— to 115	20	103 to 108
225	112 to 120	Concrete mixer 24	107 to 116
		Miscellaneous	
Hand tools		Oxy-acetylene welder	— to 96
Electric		Thermic lance	— to 94
Drill	— to 94		
Hammer drill	101 to 112	**Piling**	
Grinder	101 to 106	Diesel hammer (sheet piles)	126 to 147
Breaker 36 kg	105 to 110	Single acting air hammer	
Petrol		(precast concrete piles)	118 to 139
Breaker 36 kg	103 to 100	Double acting air hammer	
Power float 5 hp	— to 100	(sheet piles)	130 to 145
Pneumatic		2 t drop hammer (precast	
Auger 35 mm to 70 mm dia.	100 to 111	concrete piles)	114 to 128
Breaker 14 kg standard	112 to 116	2½ t to 5 t drop hammer	
silenced	100 to 108	sectioned precast/(cased piles)	116 to 131
27 kg standard	120 to 123	Resonant (sheet/H section)	— to 126
silenced	— to 118	Impact boring (driving case)	
27 kg standard	116 to 125	tripod mounted	— to 120
silenced	108 to 113	Vibratory system (sheet piles)	119 to 130
silenced +		6 t internal drop hammer	
muffled steel	105 to 110	(cased pile)	103 to 113
Chipping hammer/chisel	101 to 119	Trenching hammer (sheet piles)	114 to 120
Impact wrench 12 mm cap	94 to 97	Rotary bored	112 to 124
19 mm to 25 mm cap	105 to 109	Screened drop hammer (sheet	
Plug drill	101 to 116	piles)	— to 94
Rivet buster	— to 118		
Rotary sinker drill	— to 106	Vibratory system (H section)	101 to 130
Screwdriver (¼ hp to ½ hp)	76 to 95	Screened drop hammer	
Hydraulic		(H section)	— to 91
Breaker 36 kg	108 to 112	Jacked system (sheet pile)	78 to 82

TABLE I – TYPICAL SOUND POWER LEVELS FROM SITE EQUIPMENT
BS5228 : 1975 (contd.)

Plant	Measured range of A-weighted sound power level dB ref 10^{-12} W	Plant	Measured range of A-weighted sound power level dB ref 10^{-12} W
Pumps		**Saws**	
Water (petrol 1.4 hp)	99 to 105	Stone	— to 118
Water (diesel 3.5 hp)	— to 108	Circular bench/hand	112 to 114
Diesel 5 hp	98 to 108	Circular bench/hand (no load)	96 to 114
Petrol 7 hp	89 to 103	Chain saw	105 to 126
Concrete pump	102 to 107	Diamond (30 hp electric motor)	— to 109
Rock breakers, excavator mounted		**Scrapers**	
Pneumatic up to and including		hp	
75 mm diameter	114 to 124	450	118 to 128
over 75 mm diameter	124 to 140	560	119 to 124
Hydraulic	110 to 119	600	125 to 128
		950	— to 120
Rock drills, crawler mounted			
Pneumatic 63 mm diameter	118 to 129	**Tractors**	
127 mm diameter	115 to 117	hp	
over 127 mm diameter	94 to 109	25	— to 108
Rotary and percussive		40 to 60	103 to 110
over 127 mm diameter	96 to 109	85	110 to 111
Rollers		**Vibratory**	
Vibratory roller	100 to 104	Compactor	— to 100
Single drum 2.75 hp		Poker 2.8 hp	99 to 112
Double drum 7.7 hp	108 to 111	Screen	105 to 106
Diesel road 10 t	98 to 106		
Diesel 25 t	— to 110	**Winches**	
100 t	110 to 119	Pneumatic	102 to 108
		Electric	— to 94

NOTE. 1 hp = 745.70 W.

TABLE II – CONSTRUCTION SITE NOISE SOURCES AND POSSIBLE REMEDIES
BS5228 : 1975

Machine	Source of noise	Possible remedies (to be discussed with machine manufacturer)		Possible alternatives
Piling equipment	Pneumatic/diesel hammer or steam winch vibrator driver	Enclose hammer head and top of pile in acoustic screen, acoustically dampen sheet steel piles to reduce vibration and resonance		(1) Alternative method of piling. (2) Alternative methods of soil retention and ground improvement e.g. diaphragm walls, ground anchors, shafts formed of pre-cast concrete segments sunk into the ground under Kentledge, use of pre-treatment prior to excavation such as dewatering freezing soil injection etc.
	Impact on pile	Use resilient pad (dolly) between pile and hammer head e.g. 2 layers of asbestos cloth stuffed with glass fibre or mineral wool and protected by plywood. Packing should be kept in good condition		
	Crane cables, pile guides and attachments	Careful alignment of the pile and rig.		
	Power units or base machine	Fit more efficient silencer or exhaust. Acoustically dampen panels and covers. When intended by the manufacturers engine panels should be kept closed. Use acoustic screens where possible		
Bulldozer Compactor Crane Dumptruck Dumper Excavator Grader Loader Scraper Shovel	Engine	Fit more efficient silencer or exhaust Enclosure panels, when fitted, should be kept closed		
Compressor Generator	Engine	Fit more efficient exhaust silencer	Screen the compressor or generator	Use electric motor in preference to diesel or petrol engine for compressors. If there is no mains supply, a sound reduced compressor or generator can be used to supply several pieces of plant. Use centralized generator system
	Compressor or generator	Acoustically dampen metal casing. Enclosure panels should be kept closed		
Pneumatic concrete breaker and tools	Tool	Fit a muffler or silencer, this will reduce the noise without impairing efficiency	Use the breaker inside a portable acoustic enclosure	Use rotary drill and burster. Hydraulic and electric tools are also available. A thermic lance can be used to burn holes in concrete and to cut through large sections of concrete; any reinforcement helps the burning process. For breaking large areas of concrete, equipment which breaks concrete in bending could be used
	Bit	Use dampened bit to eliminate 'ringing'. Little noise once surface is broken		
	Air line	Leaks in air line should be sealed		
	Motor	Fit muffler to pneumatic saws		
Power saws	Vibration of blade or material being cut	Keep saw sharp. Use a damped blade. Clamp material while cutting with packing if necessary		
Rotary drills diamond drilling and boring	Drive motor and bit	Use machine inside an acoustic cover		Thermic lance

cont...

TABLE II – CONSTRUCTION SITE NOISE SOURCES AND POSSIBLE REMEDIES
BS5228 : 1975 (contd.)

Machine	Source of noise	Possible remedies (to be discussed with machine manufacturer)		Possible alternatives
Riveters	Impact on rivet	Enclose working area in acoustic screen		Design for high tensile steel bolts instead of rivets
Cartridge gun	Explosion of cartridge	Use a sound reduced gun		Drilled fixings
Pump	Engine pulsing	Enclosure in acoustic screen (allow for engine cooling and exhaust)		
Batching plant. Concrete mixer	Engine	Fit more efficient silencer on diesel or petrol engine. Enclose engine	Locate static mixing plant as far as possible from those likely to be inconvenienced by the noise	Use electric motor in preference to diesel or petrol engine
	Filling	Do not let aggregates fall from an excessive height		
	Cleaning	Do not hammer the drum. See Department of the Environment Advisory Leaflet 'Making concrete' Appendix J		
Hammer	Impact on nail			Screws
Electric impact chisel	Impact			Rotary hand milling machine
Materials handling	Impact of material	Do not drop materials from a height. Screen dropping zones especially on conveyor systems		Cover surface with resilient material or unload elsewhere
Steam cleaning	Escaping jet of steam	Pass escaping steam through silencer or screen the outlet zone		

Care should be taken to ensure that when selecting a quiet process the ancillary equipment such as cranes and compressors do not become the major source of noise.

CONSTRUCTION SITE EQUIPMENT

TABLE III – PILING ALTERNATIVES (BS5228:1975)

Bearing piles

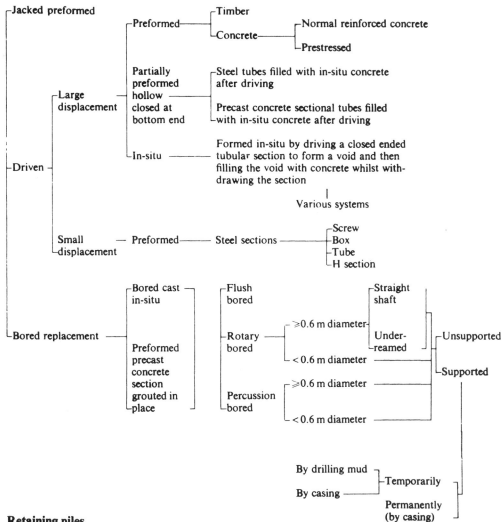

Retaining piles

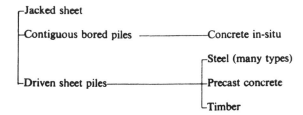

The remaining noise sources including engine structure, transmissions, hydraulics, and induction system were variable in magnitude, but in general, their combined contribution was considerably less than the individual contribution of either the cooling system or the exhaust.

Some typical data relating to construction site noise are summarized in Tables I, II and III.

TABLE IV – NOISE OF TYPICAL EARTHWORK OPERATIONS

Operation	L_{ax} dB(A)	R m	t(R) s	Noise level at 20 m
Scrape by 25 m³ capacity scraper with dozer assistance.	102	20	60	CUT-AND-FILL AREAS
Scrape by 15 m³ capacity scraper with dozer assistance	100	20	60	97–102 dB(A)
Excavation by 5 m³ tractor shovel	97	20	40	
Compaction by 225 kW motor compactor.	100	20	60	
Drive-by of 15 m³ capacity scraper	90	20	10	HAULING AREAS
Drive-by of 16 m³ capacity dump truck	88	20	10	88–90 dB(A)

Site Analysis

Noise levels at short-term work sites are commonly measured with a simple sound level meter, recording maximum values of dB(A) at specified positions. Tolerable maxima are 70–75 dB(A), as previously noted, but such data are relatively useless unless noted to specific positions near the site.

For long-term work sites it may be necessary to determine noise levels at the site boundary based on continuous noise measurement, following standard procedures. Further useful information can then be gained by plotting a histogram of one hour L_{eq} values at the site boundary. Equally, L_{eq} values determined for individual work cycles can be recalculated as an event energy noise level L_{ax} which is a measure of the total sound energy radiated to the reference position during one complete cycle of the process and is thus dependent on the duration of the process. It is the noise level which, if maintained constant for a period of one second, would cause the same sound energy to be received as is actually received from the event. It is defined as:

$$L_{ax} = L_{eq}(R) + 10 \log t(R)$$

where

$L_{eq}(R)$ is the energy-equivalent continuous noise level of the cycle
R is the reference distance
t(R) is the duration of the cycle

Table IV shows a summary of L_{ax} values for typical earthwork operations, together with related noise level values at a distance of 20 metres.

Air Distribution Systems

THE PRIMARY source of acoustic energy in an air distribution system is the fan, the energy it generates moving along the system with the air through the main transmission path. At the end of the system this is radiated to produce a sound pressure level in the receiver area (see Fig 1). Additionally, acoustic energy is generated directly as the result of airflow past the solid surfaces of the ducting system. This is classified as secondary noise. Such an input of secondary noise energy is likely to occur at any point along the air distribution system depending upon its design. It can be represented on the energy flow diagram by a number of separate inputs which have to be added to the fan noise energy at that point.

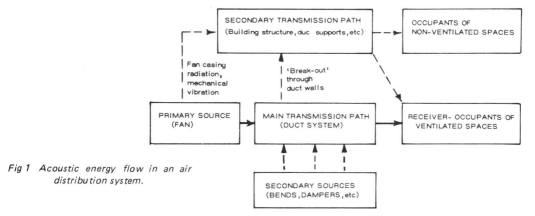

Fig 1 Acoustic energy flow in an air distribution system.

Finally energy can reach occupants of the building by paths other than the distribution system, the most obvious source being mechanical energy transmitted from the fan casing into the plant room and subsequent transmission through plant room walls to adjacent areas, and break-out of duct borne energy through the walls at various points where the system passes through noise sensitive areas. These are labelled collectively 'flanking transmission'.

The control of noise in air distribution systems thus first means identification of all the possible noise sources and transmission paths, followed by analysis of the significance of and treatment required for each:

(i) Determine sound power level of the fan;
(ii) Calculate attenuation that will be provided by system as designed, both for system side and atmospheric side, and hence determine sound pressure level at receiver's position;

(iii) Compare this sound pressure level with appropriate criteria and select suitable silencing treatment for main fan;

(iv) Estimate secondary sound pressure level generated at various critical points along the system, compare this with the *silenced* fan noise at that point in the system, and prescribe suitable attenuation measures if the fan noise is significantly increased.

(v) Ensure all flanking transmission paths are adequately treated to prevent excess energy reaching the occupied spaces by indirect means.

Transmission Path

Determination of fan sound power level is covered in detail in the chapter on *Fan Noise*. The figure used for any direct calculation must include all possible contributions arising from the installation design. That is the sound power level must be representative not only of the basic vortex noise produced by the fan, but also of any additions due to intake turbulence or interaction noise.

Starting with the fan sound power level, the attenuation that can be expected from the distribution ducting system as designed is then subtracted to give the sound power actually radiated into the room. Given this sound power, the sound pressure at any point in the room can be calculated from the standard expression:

$$SPL = SWL + 10 \log \left(\frac{Q}{4\pi r^2} + \frac{4}{R} \right) \text{ dB}$$

with r in m, and R in m² units

First, system attenuation needs to be calculated and then subtracted from the input fan sound power level to provide 'SWL' in the above equation.

It may not be immediately obvious to the system designer, but as soon as he has laid out the system of ducting to distribute the required air throughout the building, he has in fact designed a form of fan silencer. Although acoustic energy produced by the fan is, as was stated earlier, all constrained to pass along the duct system, the transmission of energy between the fan and the ventilated space is not entirely without loss. Most elements of conventional ducting systems do provide some attenuation of acoustic energy through them, as the following sections indicate.

Plain Duct Runs

Real duct walls are of course flexible to some extent. As sound energy passes through the duct in the form of a succession of pressure waves, its action is to physically move the duct wall. The work required to overcome resistance of the walls to movement is provided by extraction of energy from the sound field itself, resulting in a progressive loss of sound power along the duct run. The order of magnitude that these losses may be expected to amount to is shown in Fig 2.

Bends

Energy is lost at bends by a process of diffraction and reflection rather than by pure absorption. The effect is most pronounced for plain 90° mitre bends, when the attenuation to be expected is as indicated in Fig 3. Note here the resonance effect which gives maximum attenuation at the frequency whose wavelength is twice the duct width. For long radius bends, or bends fitted with long chord turning vanes, the loss of energy will not be so pronounced, and it is normally the practice to ignore these completely in so far as providing any attenuation is concerned.

AIR DISTRIBUTION SYSTEMS

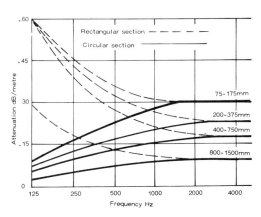

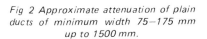

Fig 2 Approximate attenuation of plain ducts of minimum width 75–175 mm up to 1500 mm.

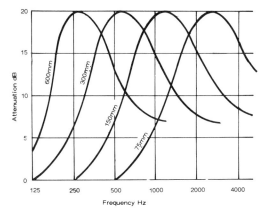

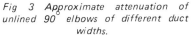

Fig 3 Approximate attenuation of unlined 90° elbows of different duct widths.

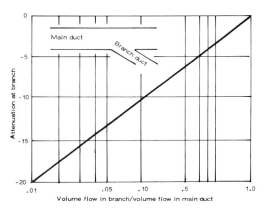

Fig 4 Attenuation at a branch

Branches

Strictly speaking, there is no nett loss of energy at a branch; rather it simply divides between the main and branch duct in much the same proportion as the airflow. Although the total energy in all branches immediately after a junction is the same as that in the main duct approaching it, the energy actually arriving at the grille at the end of a single branch is less than was present before the branch and must therefore be counted as attenuation. If, eventually, all the branches were to lead to the same room then all the individual branch energies are simply added back together in the course of the reverberant part of the room calculation. If not, we are concerned only with the energy reaching a particular grille. Attenuation to be expected by taking one single branch is determined only by the airflow in that branch as a proportion of the total flow approaching the junction. This is shown graphically in Fig 4.

Plenum Chambers

A large cavity in the system such as builders' ducting or even the fan chamber itself, with the fan either pressurizing or extracting from the room, and the duct system starting at the walls, can be used to form an acoustic plenum. Energy fed into a plenum tends to regard the chamber as a small room, and as in the case of a room, the sound pressure level at the entrance to the ducting on the other side which determines how much sound power is transmitted, has two components.

Direct sound energy is radiated straight from the source and reverberant sound energy is all the other energy fed into the chamber but reaching the outlet duct after reflection of the chamber's internal surfaces. Clearly the first component will then be more dependent upon the geometry than anything else, while reverberant energy will be determined largely by the absorption characteristics of the plenum surfaces. Fig 5 defines the geometric parameters in the calculation, and the expression for attenuation across the plenum is:

$$SWL_1 - SWL_2 = -10 \log S_2 \left[\frac{\cos \theta}{2\pi d^2} + \frac{1}{R} \right] \, dB$$

$$\text{where } R = \frac{S_T \cdot \alpha}{1 - \alpha}$$

In fact, the above expression underestimates the attenuation in the lower frequency bands as no account is taken of reflection at the inlet to the plenum. A value for this to be added to the attenuation calculated above, can be taken from Fig 6.

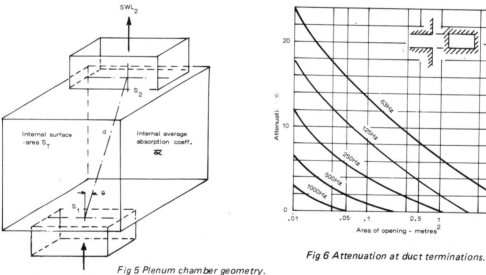

Fig 5 Plenum chamber geometry.

Fig 6 Attenuation at duct terminations.

Proprietary Units

Not many in-duct elements supplied as a package serve to give any appreciable attenuation of sound energy. Such items as heaters, cooling coils or even filters are normally neglected, except with regard to their ability to *produce* sound as mentioned later. There are one or two however, where attenuation of sound may be significant, and notable among these are the constant volume or constant pressure controllers of the type employed in high velocity systems. It is not possible to give a general expression for the loss as this varies with individual manufacturers' designs, but a statement of any attenuation should be obtained from the manufacturer and included in the duct calculation.

AIR DISTRIBUTION SYSTEMS

Terminal Units

When the attenuation provided by all the various duct elements have been added to the fan input sound power level, the result is the sound power which reaches the grille. Not all of this energy is radiated into the ventilated space however because of the phenomenon of reflection which always occurs when waves travelling in a medium of given acoustic impedance, in this case the confines of the duct section, meet a sudden change in impedance, here represented by the emergency into the large volume of the ventilated space. The amount of energy reflected and hence "lost" to the occupants of the room is dependent upon the magnitude of the change. In the case of a ductwork termination into a room, the smaller the duct the greater the change of impedance. More precisely the parameter is the duct size compared with the wavelength (frequency) being considered. A summary of the attenuation to be expected is shown graphically in Fig 6.

Noise Control

Referring back to Fig 1, it can be seen that the noise control engineer has a number of options available to ensure that the correct amount of acoustic energy reaches the occupants of the ventilated spaces. They may be summarized as follows:

1. Reduce strength of the primary source (the fan).
2. Attenuate fan noise in the main transmission path (the duct system), between fan and area affected.
3. Ensure all sound power generated by secondary sources is kept down to a level which is insignificant compared with the level of fan noise, silenced or unsilenced, at that part in the system.
4. Eliminate the transmission of all flanking and indirectly produced energy.

(1) Reducing Fan Noise

The objective of the system designer is of course to control to an acceptable level the amount of acoustic energy reaching an occupant of the served space. As far as the energy produced at the fan is concerned, if less can be fed into the system, obviously less will arrive at the end.

Summary of Design Rules for Fans

(a) For a given duty, a fan producing its energy at higher frequencies will be more amenable to absorptive silencing than a low frequency fan.

(b) Size and speed of fan should always be chosen to bring the design duty point as close to the fan's maximum aerodynamic efficiency as possible.

(c) Intake design should be such as to provide the smoothest turbulence-free flow possible into the fan.

(d) Where possible, multi-staging of axial fans, or the addition of close coupled guide vanes should be avoided.

(e) Where multi-staging, guide vanes, standby, or close coupled intake flow control dampers are inevitable in axial flow systems, the aim should be to provide the maximum possible spacing between fixed and moving vanes, or between impellers in series, and in any case not less than two chord lengths of the upstream obstacle.

(2.) Reducing Noise in Transmission Along the Duct System

Once the minimum possible fan sound power has been achieved by careful selection, the next point of attack is the transmission path by which it reaches the ventilated space. The obvious way to reduce the amount of fan noise passing along the system is to employ a fan silencer.

It is often the case however that some savings in size and hence cost of a proprietary silencer, can be effected by careful design of the air distribution itself. Most duct elements, for example, offer a certain amount of attenuation. It should always be the initial aim to note by what means it may be maximized. For example:

(a) Plain duct runs provide more low frequency attenuation when rectangular in section (Fig 2). A disadvantage here is, of course, that much of the energy extracted from inside the duct often appears as airborne energy outside ('break-out') but providing this can be accepted, or dealt with as described later, the benefits to be gained by using rectangular section can be appreciable.

(b) Plain mitre bends can give quite significant lower frequency attenuation if correctly sized (see Fig 3). The additional pressure loss is of course an important consideration, but the addition of short chord turning vanes can alleviate this to some extent.

(c) Any plenum that can be incorporated in the system will be of considerable benefit, as a trial calculation using Fig 5 and the associated equation will reveal. Again, however, pressure loss at inlet and outlet, together with the volume required to form the chamber, must be balanced against the benefits.

(d) Reflection of energy at the duct termination can be seen from Fig 6 to vary inversely with duct area, at least for the important low frequency energy. A number of small terminations is better than a few large ones.

When all has been done along these lines to ensure maximum 'natural' attenuation in the system, it may well be found that additional attenuation is required. For the main fan silencer, a packaged splitter type will probably be required, but for more moderate attenuation, particularly for terminal silencing of secondary noise (see later), a simple lined duct or bend may suffice. Performance of lined ducts and attenuation for some representative bend sizes and linings is given in Fig 7.

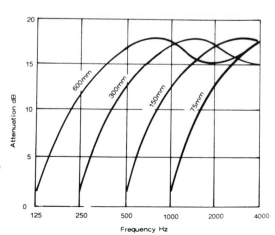

Fig 7 Typical attenuation of lined ducts — duct widths as shown. (Duct lining thickness approximately 10% of duct width).

AIR DISTRIBUTION SYSTEMS

(3) Secondary Noise

It is important to compare the secondary sound levels generated in a system with the main fan noise in the same part of the system. If significantly lower — say more than 6 dB difference — then the secondary noise may be ignored. If on the other hand secondary noise is comparable with — or greater than — the fan noise at that point in the system, then the uplift due to it is bound to create an excessive sound pressure level in the room. In that case, some secondary silencing will be required.

System components which generate secondary noise are conveniently examined in two groups. The first comprises the in-duct elements which, whilst being exposed to higher airflow, have at least one advantage in that the energy they produce is confined within the duct where, if excessive, it is reasonably amenable to conventional attenuation techniques. The second comprises the various types of terminal units, which usually cause the greater problem because the energy they produce actually starts in the ventilated space. The possibilities of attenuation, therefore, before it reaches occupants of the same space, are limited.

In-duct Elements

The elements of the system which have to be examined for secondary noise are plain duct runs; bends; branches; contractions; dampers and proprietary fittings (heater coils, mixing boxes, constant volume or pressure controllers *etc*).

With such elements the following formula can be used for design estimates:

$$SWL = K + 10 \log S + 55 \log V - 45 \text{ dB}$$

where K is a constant for the particular duct element considered, and is given in Table I.

S is the cross sectional area in m^2 of the duct, across which velocity is V.

V is the flow velocity in section S, m/sec

TABLE I — AIRFLOW GENERATED NOISE IN DUCT FITTINGS
(after Stewart)

Fitting	Overall sound power level dB (octave bands 250–8000 Hz) K
Straight duct	38
3:1 contraction (to 12 in x 12 in)	547
90° bends radiused	48 (est)
mitred with turning vanes	56
mitred without turning vanes	57
Duct with 90° tee (5% draw-off)	55
Damper : Open	44
15° closed	53
45° closed	65

Corrections to give octave band SWL

Octave Band Centre Frequency	Hz	250	500	1000	2000	4000	8000
Spectrum correction to overall SWL	dB	−7	−7	−8	−10	−17	−29

Distribution of the overall sound power across the frequency spectrum is also indicated in Table I.

Consideration must also be given to the noise generating propensities of a turbulent flow on to the duct element in question although relative velocities for duct elements are rather less than those for rotating blades, so the effect may not be so drastic. If, however, a turbulent inflow is likely, for example in the case of a damper located immediately after a bend, it would be prudent to add 5 dB to the estimated sound power level across the spectrum.

With some proprietary units, such as dampers, the estimates given previously should be sufficiently accurate for a design estimate, although it must be said that there is no substitute for laboratory based manufacturers' test figures. With others though, such as individual manufacturers' designs of mixing boxes or constant volume units, their test figures are at present about the only data available.

If the test entry conditions were substantially smoother than those which will occur in practice, a 5 dB addition should be made for safety.

Terminal Units

Secondary noise generation at terminal units, grilles, diffusers, induction units and the like, is of critical importance. Unlike the noise of in-duct elements, there is no opportunity to attenuate energy produced by the terminal unit before it reaches the occupant of the same room.

In plain grilles and diffusers, the mechanism of producing noise is essentially the same as the vortex noise mechanism set up by moving fan blades.

First an estimation has to be made of how much power is likely to be produced. As with so many other components of the system which are of proprietary supply, there is no substitute for the manufacturer's figures based upon test data, providing the background to the figures is examined to establish that test conditions were not too dissimilar to those of the installed product. In the absence of such information, one can for some units obtain a design estimate from one of the empirical expressions which are available. One estimate expresses the overall sound power level in terms of the minimum neck area in the case of diffusers, or free area in the case of grilles, $A(m^2)$ and the maximum flow velocity in neck or between vanes, $V(m/sec)$, as follows:

$$SWL = 13 \log A + 60 \log V + 33 \text{ dB}$$

For design purposes, the spectra corrections given in Table I may be assumed to apply.

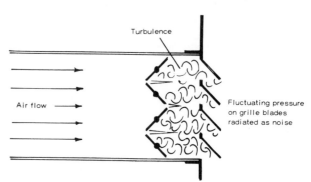

Fig 8 Noise generated by dampers.

AIR DISTRIBUTION SYSTEMS

Fitting any sort of control damper to a grille or diffuser will add two more noise sources. First there will be the vortex shedding component from the damper blades themselves, and second the action of the damper blade wake turbulence upon the grille or diffuser vanes as illustrated in Fig 8. The second is by far the more important.

In current grille design, it is the practice to manufacture the damper as a multi-leaf opposed-blade pack which clips on to the rear of the grille. The system has a number of advantages of course, including the ability to balance flow through a number of grilles easily, adjusting the damper by turning a screw reached through the front face of the grille. Unfortunately, the system also pays a penalty, sometimes a heavy one, in additional noise generation.

Nearly all manufacturers of grilles and diffusers can supply quite detailed figures for the sound power level produced for their grille/damper assemblies, as a function of either face velocity or volume flow through, or, more usually, as a function of pressure drop across the unit.

In the absence of such information at the design stage, one can assume an increase in the aforementioned estimated sound power level, of 5 dB in each octave band for the addition of an internal damper vane fully open.

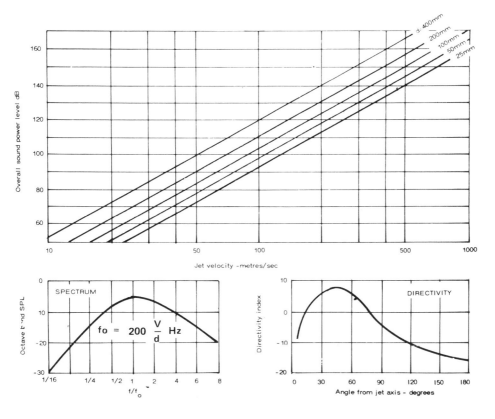

Fig 9 Design chart for estimating induction unit jet noise.

For subsequent flow adjustment, *ie* increased drop across the damper, it can be assumed that the overall sound power level will increase approximately by the following:

$$SWL = 30 \log \frac{p}{p_0}$$

where

p is the new static pressure drop across the unit

p_0 is the pressure drop with damper pack added but vane fully open, with the same volume flow.

Induction units are rather less amenable to a general estimate of the noise they produce. The generating mechanism itself is basically one of free jet noise, but with the possibility of additional output from turbulent flow impingement on solid surfaces.

If free jet noise is the only source, its sound power output may be estimated from a design chart such as Fig 9. Note here that the power level estimated applies to one jet. For the total jet noise from one unit, an addition of 10 log N dB should be made, where N is the number of nozzles. Note also that the directivity of the free jet will not apply in general, being affected by the casing of the unit to a very marked degree.

In addition to the noise of the free jets, additional sources will be present on most induction units. Any damper into the unit plenum will of course produce noise as discussed earlier in this chapter, and secondary generation at the air entry is to be expected. Sound power from both of these sources, however, may be expected to be reduced by some 15 dB across the spectrum in emerging from the unit to the room. Additional noise of the turbulence interaction type may be expected if the primary nozzle jets are allowed to impinge upon any part of the casing of the unit — or if there is a protective grille placed directly in line. Under these conditions about 5 dB should be added across the spectrum.

It should always be the aim to obtain from the manufacturer induction unit noise levels which appertain to the particular design to be used, and which are based upon test figures for that design.

Control of Secondary Noise

The most important factors which determine the output of secondary noise are velocity and turbulent flow. The importance of reducing both at the design stage cannot be over-emphasized.

Where noise is likely to be at all critical, the following general guide lines should be followed:
(a) Keep velocities down to 5 m/sec in main ducts and 3 m/sec in branches.
(b) Bends should be long radius or be fitted with turning vanes (but note conflicting advantage of high half-wavelength attenuation in mitre bends). See also Fig 3.
(c) Branches should be radiused or at least chamfered.
(d) System resistance should be calculated very carefully to obviate the need for high pressure drop across balancing flow control dampers when commissioning.

It may well be found however, as with fan selection and system design, that further attenuation is required after the above guidelines have been followed. With most of the duct elements likely to produce significant secondary noise, the source is located inside the ducting system. Excess energy is then very amenable to attenuation by the simple expedient of lining the terminal branch duct, or replacing it with a small packaged splitter silencer.

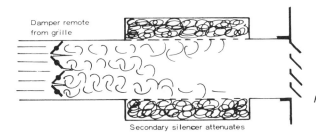

Fig 10 Remedy for damper noise.

The need for all significant secondary energy to be located far enough back into the system for a terminal silencer to be effective, is particularly important for flow control dampers on grilles and diffusers in the room. If it is essential to provide a damper for each grille, then it must be located on the system side of the terminal silencer, as indicated in Fig 10.

With regard to secondary noise generation at the terminal unit itself there is, almost by definition, virtually nothing that can be done to prevent the energy from being radiated into the room.

Induction units, in particular, will almost certainly be limited in their application to noise sensitive areas. A room criteria of some NR 35 is about the minimum that can be achieved with these devices.

With grilles, the only recourse is to oversize to reduce the discharge velocity to the absolute minimum consistent with the throw and distribution required. In extreme cases, air is sometimes fed to the room through a large hole covered with a wide mesh screen. Obviously though, the level of secondary noise generation there has to be balanced against a smaller low frequency reflection loss (see Fig 6).

(4) Indirect and Flanking Transmissions

Figure 11 shows pictorially the more important indirect transmission paths which have to be examined carefully if all the work for ensuring acceptable fan noise and secondary noise levels in the ventilated space, is not to be negated. They are:

- (i) Breakout of duct-borne fan noise energy through duct walls into rooms carrying the ducting.
- (ii) Radiation of fan casing and drive motor noise into the plant room and hence to adjacent area.
- (iii) Mechanical excitation of the duct system and the building structure.

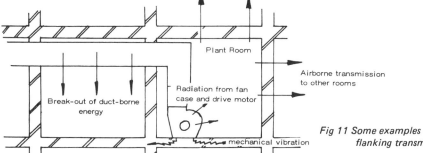

Fig 11 Some examples of indirect and flanking transmission.

(i) Noise Breakout from Ducts

It was pointed out earlier in the chapter, when discussing energy losses during transmission of fan noise along plain duct runs, that the sound waves passing a fixed point in the duct vibrate the duct wall. Of the energy expended in overcoming the wall's resistance to movement, some is transformed into heat by whatever internal damping is in the material of the duct wall, and the rest is re-radiated as airborne noise outside the duct.

The amount of energy radiated through a duct wall, or, as it is commonly referred to, 'breakout' sound power level, the material of which has a Sound Reduction Index R, can be estimated from:

$$SWL_{(break\ out)} = SWL_D - R + 10 \log \frac{S_w}{S_d} \quad dB$$

where SWL_D is the sound power level inside the duct
S_w is the total surface area of the run of ducting concerned
S_d is the cross sectional area of the duct

It will be found in many cases of typical air distribution system ducting, that the light gauge sheet metal used will, at low frequencies at least, provide a value of R which is numerically less than $10 \log S_w/S_d$. This suggests that either all the sound power in the duct breaks out, or that more breaks out than was originally present in the duct.

This is clearly not possible, and in fact the original assumptions which lead to the above equation, cease to be valid under such conditions. Instead, it should not lead to great error to assume that half the sound power in the duct breaks out and the rest carries on along the system. In other words, the maximum value that SWL_B is allowed to assume is $(SWL_D -3)$dB. It should be noted that the expression applies only to transmission of airborne noise from inside to out. Noise outside due to internal 'buffeting' from the turbulence is not calculable but is likely to dominate at duct velocities in excess of 10 m/sec.

(ii) Fan Casing and Drive Motor Noise

Noise radiated from the fan casing has two components. The first is from airborne excitation of the casing by the acoustic energy generated at the fan impeller, in other words a similar 'breakout' to that discussed previously for duct borne energy.

A very approximate figure for the amount of energy to be expected from this source can be obtained using the same formulas as for duct breakout, if SWL_D is taken as the sound power level of the fan, R is the sound reduction index of the fan case, S_w the total area of fan casing and S_d the cross sectional area based upon the impeller diameter for an axial fan, or the casing diameter for a centrifugal fan.

Purely airborne breakout of this kind is likely to be less than the sound power level due to the other component, mechanical excitation of the casing. Belt drives, bearings, direct drive motor support arms, all produce a degree of mechanical vibration energy at frequencies virtually across the spectrum, which transforms into airborne acoustic energy at the relatively large radiating surfaces of the fan casing. Unfortunately, the amount of mechanical vibration at the radiating surface is very dependent upon the detailed mechanical design of the fan and drive motor assembly, rather than being related to overall performance, as is aerodynamic sound power. This means of course that resultant sound power level is almost impossible to predict, being potentially different, for example, for two identical fans with same power but different type and mounting arrangement of the drive motor, or with different residual out-of-balance tolerances.

AIR DISTRIBUTION SYSTEMS

If the fan is driven *via* belts by an external motor, an estimate of drive motor noise needs to be made, but it must be remembered that mechanically induced vibration of the motor support structure may prove to be the dominant source.

If a fan runs open with intake in the plant room, the impeller aerodynamic sound power level radiated from the intake will undoubtedly dominate. If, on the other hand, the intake noise has to be attenuated then it would be prudent to assume — at the design stage — that when more than 15 dB attenuation across the spectrum has been provided by the intake silencer, casing and drive motor noise will start to become significant. Any further attenuation of intake noise will then probably have to be matched by an equal amount of attenuation of the flanking radiation.

(iii) Vibration Isolation

The principles and practice of vibration isolation of machinery have been discussed in considerable detail in other chapters. There are, however, one or two points to be made specifically in the context of air distribution plant.

The first is that where an indirect drive motor is coupled to a slower speed fan, the two must be connected together by a rigid base frame, which is then isolated from the building structure by suitable anti-vibration mounts. Under no circumstances should indirect driven machines have fan and drive motor mounted on separate sets of anti-vibration mounts.

Second, where a slower speed fan is indirectly driven by a higher speed motor, or occasionally *vice versa*, the anti-vibration mounts should always be selected for the speed of the slower unit on the base-frame.

Third, on a centrifugal fan set producing appreciable pressure, allowance must be made for reaction torque due to the pressure reaction at the fan discharge. Fig 12 shows diagrammatically the effect and two methods of alleviating it. One alternative is to choose spring lengths at the rear of the fan which pre-tilt it forwards when stationary. Reactive torque then exerts higher loads at the rear which produce the higher deflection required to bring the fan back to the level position when running.

A rather more elegant variation on that theme is to choose the same spring length, but with increased stiffness at the rear. The higher loads there then produce the same deflection as the lower load on the softer front spring.

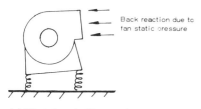

(a) Effect of equal stiffness springs

(b) Mounts of equal stiffness but unequal length

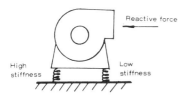

(c) Unequal mount stiffness for equal deflection

Fig 12 Corrections for fan reactive pressure.

Finally, because of the very efficient manner in which mechanical vibration can be transmitted along the walls of light sheet metal ducting, it is essential to provide flexible duct connectors as close to the fan as possible. It is always best to position flexible connectors after fan silencers. In that case, of course, the silencers must also form part of the 'sprung' weight which for silencers of any appreciable length could be difficult. In fact, providing the connector is manufactured in one of the high density materials, there is usually not too much to be lost in practice by locating it directly on the fan flange, before the silencer.

Control of Flanking and Indirect Transmission

Two options are available to achieve a reduction of the amount of sound power expected to break out through duct walls, assuming the ratio Sw/Sd has to remain constant. One is to reduce the sound power level inside the duct, and the other is to increase the sound reduction index of the duct wall. The first has already been discussed in the context of fan noise reduction and fan silencers.

Various methods of increasing the sound reduction index of partitions are available. Air distribution ducting is not, however, normally amenable to a straight increase in weight, even if this gave sufficient increase to provide the required decrease in outside sound power level.

It is of course possible to enclose completely a duct run across a room by, for example, boxing in with plasterboard and studding type construction. Provided sealing is adequate and internal absorption is required, it can be very effective, giving an increase of up to 20 dB average across the spectrum.

A more convenient treatment, however, is to lag the outside of the relevant duct run with 50 mm glass fibre quilt or polyurethane foam, and wrap outside this a mass skin of lead foil, Keene cement, or one of the proprietary loaded PVC sheets, of superficial density not less than 5 kg/m^2. A reduction in breakout sound power level of some 7 to 10 dB across the spectrum can be expected from this treatment.

The same treatment is eminently suitable for acoustically lagging the outside of axial flow fans, and the same order of magnitude of improvement may be expected.

For centrifugal fans, there is no alternative to at least partial, if not complete acoustic enclosure, for any significant degree of improvement in casing noise levels.

A weak point, as far as breakout is concerned, is the flexible duct connector which must be used in conjunction with vibration isolators. The traditional canvas connectors, while providing excellent vibration isolation, provide virtually zero sound reduction index. This problem can be fairly easily overcome however by manufacturing the connectors in one of the proprietary loaded PVC compounds, of superficial density up to 10 kg/m^2, which give a sound reduction index in this application about equal to that of light gauge ducting.

Cross-Talk

With a system where two separate rooms share the same air distribution duct, there is always the danger of noise in one, such as music or speech, being transmitted along the air duct to the other. This effect is commonly referred to as "cross-talk".

The calculation of the degree to which it is likely to occur and its control is exactly the same as for the estimation and control of fan noise described elsewhere in this chapter.

AIR DISTRIBUTION SYSTEMS 245

Noise Transmission to Atmosphere

It must be appreciated that fan noise can be transmitted just as readily upstream as downstream. Fan sound power levels quoted by manufacturers or estimated should be assumed to apply to either side of the fan, and the same figures should be used for calculations in both directions. In other words whether the system is for supply or extract, it is as important to carry out the same exercise for the atmospheric side of the fan, as for the system side. The objective is of course precisely the same in that the sound power level produced at some point likely to be affected by the atmospheric intake or discharge, has to be estimated and compared with the appropriate criterion.

Two cases are considered, the first being where the atmospheric side of the system is ducted in the same way as might be the room side except that the terminal duct leads to an outside louvre; and the second being when the fan runs open in the plant room, and draws in or discharges air through atmospheric louvres in the plant room wall.

(i) Ducted System to Atmosphere

The calculation to be followed here is precisely that outlined previously for the room side. From the fan sound power level must be subtracted the attenuation provided by the various duct elements including reflection at the atmospheric louvre. The sound pressure level then at a distance r metres from the louvres and at angle θ to its axis, resulting from the sound power level SWL which actually leaves the louvres may be estimated from:

$$SPL = SWL - 20 \log r + DI(\theta) - 11 \text{ dB}$$

$DI(\theta)$ is the Directivity Index of the radiation pattern from the louvres. For design purposes one can assume the following values for DI providing the louvre area is in excess of 1 m^2 and the louvre is located roughly centrally in a large wall.

	$DI(\theta)$
0	+ 8
30	+ 6
45	+ 4
60	+ 1
75	− 5
90	− 15

Add 3 dB to the aforementioned figures if the louvre is near the junction of two walls or between wall and ground, or 6 dB if the louvre is near the corner formed by two walls and the ground.

(ii) Fan Running Open in Plant Room

If the fan is drawing air in directly from the plant room, the primary interest will be in the sound pressure level inside the room. From this, sound pressure levels outside or in adjacent rooms can be calculated.

The formula for plant room sound pressure level is the usual one given earlier, *ie*:

$$SPL = SWL + 10 \log \left[\frac{Q}{4\pi r^2} + \frac{4}{R} \right] \text{ dB}$$

Here all the room parameters apply to the plant room, and SWL, the source sound power level, is that of the fan, less a reflection loss appropriate to the size of the fan inlet.

See also chapter on *Fan Noise*.

Fan Noise

NOISE GENERATED by fans comprises broad band noise resulting from intake turbulence and vortex generation, on which is superimposed a number of pure tone components, the frequency of which is directly related to the fan geometry and rotational speed. It is generally considered that the total sound power level generated by fans is proportional to (fan speed)^6. More specifically, sound power level may be estimated from any two of the three parameters rated motor power, static pressure developed and volume flow, using one of the following formulas:

SWL = 77 + 10 log kW + 10 log P dB
SWL = 25 + 20 log Q + 20 log P dB
SWL = 130 + 20 log kW − 10 log Q dB

where

SWL = overall sound power level in the octave frequency bands 31.5 Hz to 8 000 Hz
kW = rated motor power, kW
P = static pressure developed by fan, mm wg
Q = volume flow delivered, m^3/h

For conventional fans, these expressions should yield an answer for the overall sound power level within 5 dB. For very high pressure systems, say above 100 mm, or where otherwise the fan may need to be unconventional (such as a multi-stage axial, high pressure centrifugal blower, or radial bladed particulate transporter,) the error may be as high as 10 dB. For a design estimate of how the overall energy may be distributed across the frequency spectrum, Table I gives the correction to be applied to the overall level to obtain the octave band sound power levels.

TABLE I – CORRECTIONS TO OBTAIN OCTAVE BAND SPECTRA FOR VARIOUS FAN TYPES

Octave Band Centre Frequency (Hz)	Add to overall sound power level dB							
	63	125	250	500	1000	2000	4000	8000
Centrifugal:								
Backward curved blades	−4	−6	−9	−11	−13	−16	−19	−22
Forward curved blades	−2	−6	−13	−18	−19	−22	−25	−30
Radial blades	−3	−5	−11	−12	−15	−20	−23	−26
Axial	−7	−9	−7	−7	−8	−11	−16	−18
Mixed flow	0	−3	−6	−6	−10	−15	−21	−27

vibro-meter
"made to measure"

VIBRATION

Vibration measurement and monitoring is the key activity of Vibro-Meter.

In international civil **aviation,** 87 airlines throughout the world have chosen Vibro-Meter systems. For all the wide body jets of today and for the future, such as the B 767 and A 310, Vibro-Meter has been selected as a basic supplier.

Industrially, Vibro-Meter systems are used for monitoring turbines, alternators, electric motors, compressors, pumps etc., particularly in hydraulic, thermal and nuclear power stations and in the petro-chemical industry. More than 10'000 measurement and monitoring channels are in operational service in these sectors.

The wide range of high quality **eddy current transducers and accelerometers together with electronic monitoring and management systems** which are adaptable to specific requirements, permit multiple combinations of **vibration, temperature and pressure measurements.** Micro-processors ensure rapid and accurate parameter calculation and management.

Profit from the accumulated experience of Vibro-Meter!

Switzerland	Austria	Federal Republic of Germany	France	Great Britain	USA
VIBRO-METER SA	VIBRO-METER Ges.m.b.H	VIBRO-METER GmbH	VIBRO-METER FRANCE SA	VIBRO-METER Ltd.	VIBRO-METER Corp.
Route de Moncor 4	Khuningasse 17	Hamburger Allee 55	43, rue de Châteaudun	Newby Road — Hazel Grove	22109 South Vermont Avenue
CH-1701 Fribourg	A-1030 Wien	D-6000 Frankfurt (Main) 90	F-75009 Paris	Stockport/Cheshire SK7 5EF	Torrance, California 90502
Phone: 037/82.11 41	Phone: 0222/78.55.86 Serie	Phone: 0611/77.80.65-66, 77.60.08	Phone: 01/526.39.38 and 526.37.61	Phone: 061/483.08.11/4	Phone: 213/320.84.10
Telex: 36 232	Telex: 13 20 34	Telex: 41 25 64	Telex: 64 05 29	Telex: 66 80 12	Telex: 67 35 28

MACHINERY PROTECTION SYSTEMS

CML

CONDITION MONITORING LTD.
UNIT 2, TAVISTOCK INDUSTRIAL ESTATE TWYFORD
BERKS UK. RG10 9NJ. Tel: (0734) 342636 Telex 847151

TRY US AND KEEP IT QUIET!

As a manufacturer of polyurethane foam we have now developed a competitively priced range of **Acoustic Composites** including:
- Brushed nylon to foam
- Metallised film to foam
- Foam and barrier materials
- Embossed foams
- Perforated and plain PVC to foam
- Fire retardant foams

All products can be supplied with **Self Adhesive Backing** and are available in sheet form or as shaped parts.

Harrison & Jones (Acoustic Division), Chaul End Lane, Luton, Beds. LU4 8HB. Telex: 826613 Tel: Luton (0582) 595151
For further information telephone Mr. N.C. Blythe, Technical Manager

FAN NOISE

The most accurate figures for the sound power level produced by a particular fan for a given duty will be those obtained from the manufacturer. Most of the major manufacturers of fans have access to appropriate laboratories where sound power produced by their products can be measured directly in accordance with the procedures laid down by one or other of the national or international standard test codes, (eg BS848 Part II).

Whatever code has been used, it is almost certain that the actual test rig used will have provided for the fan impeller a reasonably clean airflow; most of the available standards have this in common. This means of course that what has been measured is substantially all vortex noise produced by the fan without any additional components arising from the intake turbulence. It is in fact the *minimum* noise the fan can be expected to produce for that particular combination of volume flow and pressure development. It may be that the fan already incorporates in its design such features as inlet guide vanes or multi-staging, in which case the noise levels operated are a combination of vortex noise and interaction noise.

In the case of design estimates obtained from the aforementioned formulas, these must be regarded as being representative of vortex noise only.

Basically, therefore, it becomes necessary to determine corrections to sound power level relating original or test figures to form installation design — specifically intake design (as affecting intake turbulences); the form of running, the effect of guide vanes; and, if appropriate, the effect of multi-staging.

Vortex Noise

Considering, first, the mechanics of fan noise generation, it is a characteristic of translational airflow over any solid surface that a shear gradient will be established in which the velocity varies from zero infinitesimally close to the surface, to the free stream velocity of the bulk of the air at some distance away from it. The region in which such a shear gradient exists is known as the boundary layer. At some distance along the surface, the velocity shear gradient becomes such that a laminar flow can no longer be maintained and the flow breaks down to form a turbulent boundary layer. If the surface is of finite length, the turbulent boundary layer leaves the top and bottom surfaces to form the wake. Fig 1 illustrates these flow régimes diagrammatically for an aerofoil shape typical of the blade section of a wide variety of fans.

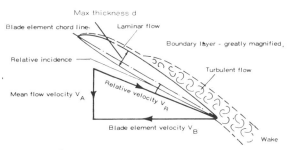

Fig 1 Flow around aerofoil blade section — some definitions.

The passage of this turbulent flow over the after end of the aerofoil and into the wake sets up fluctuating forces on the aerofoil surfaces which in turn act as sources of acoustic energy. Because the turbulent flow is random in the nature of its eddy size and velocity variations, the pressure fluctuations are similarly random, as is the generated sound power. The result is a frequency spectrum which covers a considerable range of the audio spectrum. It is not flat as would be the case if the process were truly random but has a maximum value extending over perhaps as much as two full octaves, and sloping off at about 5 dB per octave on either side. The frequency of maximum acoustic energy generation is governed by the velocity of air relative to the section and the thickness of the section projected in the direction of airflow. This maximum frequency is given by:

$$f_{max} = 200 \frac{V_r}{d}$$

with V_r in m/sec and d in mm as defined in Fig 1.

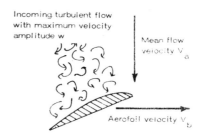

Fig 2 Impeller blade section traversing turbulent flow.

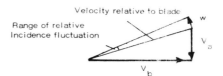

Intake Turbulence

The basic mechanism described previously by which any surface in a flow generates noise will occur to some degree or other in a fan, even if the entry conditions to the impeller are perfect. It is frequently the case however that in industrial installations at least, the intake flow to the impeller is itself turbulent. If we assume the turbulent region to be one in which a particular particle of air has a mean transport velocity Va, and an instantaneous perturbation velocity superimposed on it, which at any instant can be of magnitude up to a maximum of say w in any direction, the effect upon the velocity vector diagram is as shown in Fig 2. The addition of the perturbation vector alters the direction rather than the magnitude of the relative velocity vector V. Remembering that W is continuously varying in magnitude and direction, it is clear that the blade section relative incidence is also varying, as then must be the total lift on the section. Because the initial perturbation in the turbulent flow is random in nature, the sound power resulting from the lift fluctuation must also be random. Hence the frequency spectrum is broad band.

Again the spectrum is not flat but tends to a maximum at a frequency similar to that of vortex noise, as estimated from the previous expression, with V_r equal to the relative velocity of air past the blade, and d the characteristic size of eddy in the turbulent field. Unfortunately d is not easy to determine. In a very simple case, such as turbulence induced by an up-stream duct support or instrumentation probe, it is not too much of an error to take d as being equal to the cross-stream dimension of the obstacle. In many ventilating fan installations, however, of which Fig 3 shows quite typical examples for an axial fan, the characteristic size of eddy passing over the blade tip is much more difficult to estimate.

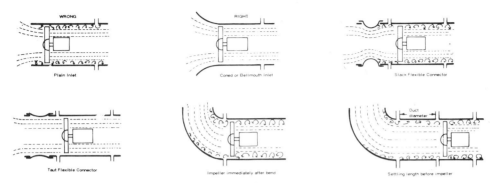

Fig 3 Examples of fan installations producing intake turbulence.

As a practical guide, it is a matter of observation that the additional noise due to this mechanism nearly always occurs at frequencies up to and including the 500 Hz band. Estimates given later for excess in sound power level due to intake turbulence apply to these bands unless otherwise stated.

One reason for the preponderance of low frequency noise is that the lift fluctuation of a blade section is strongest for eddy sizes which are comparable with, or greater than, the blade chord. To take a very simple model, if eddy size is much smaller, then at any instant a number of eddies will be spread across the blade chord. Since the individual velocity and pressure perturbations exerted locally on the aerofoil are varying randomly both in amplitude and in phase relative to each other, the net or effective force fluctuation on the surface which produce the acoustic energy will statistically average out to zero, or at least to a very small residual. This effect can be easily demonstrated by observing the difference in additional noise output when turbulence is introduced first by, say, a very fine wire gauge mesh, then by a flat bar of cross stream dimension equal to the blade chord.

Interaction Noise

The physical mechanism of generating this component of the spectrum is virtually the same as that for intake turbulence. If the flow onto the impeller blading contains non-uniformities which are random in nature such as turbulent flow then, as stated earlier, the resulting additional noise it generates has a frequency spectrum which is broad band. If on the other hand the flow disturbance is periodic, the resulting noise generated will appear in the form of pure tones at the fundamental blade passage frequency and its harmonics where,

$$\text{Blade passage frequency} = \frac{\text{rev/min}}{60} \times \text{number of impeller blades}$$

One way of generating such an intake flow is illustrated in Fig 4 for the specific case of a rotor blade passing behind a single fixed obstacle such as an upstream bearing support, or a silencer pod support. Immediately behind the obstacle there is a decrement in the mean axial velocity of oncoming air in the region of the wake. The effect upon the impeller blade velocity triangle of a change in axial component, can be seen to be very similar to that of the addition of a turbulent velocity as in Fig 2. Again the effect is to vary the direction of the relative velocity onto the moving blade with a consequent change of relative incidence and lift on the section, and hence acoustic energy is radiated.

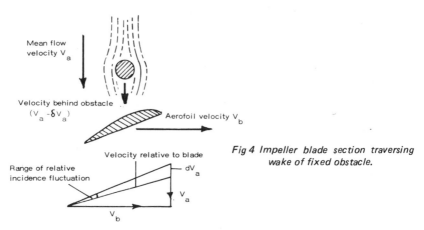

Fig 4 Impeller blade section traversing wake of fixed obstacle.

As mentioned, the strongest component is nearly always produced at the fundamental blade passage frequency, with harmonic content depending upon the pressure wave form of the variation of axial velocity across the region of the disturbance. With industrial fans this mechanism of discrete frequency tone generation is nearly always due to the addition of inlet or outlet guide vanes. It is also prevalent in contra-rotating, two-stage axial fans. Occasionally it can be introduced by setting such upstream duct components as silencers too close to the impeller.

One other interactive effect producing discrete frequency noise, but at levels probably 10—15 dB higher than the wake effect described earlier, is the interaction of potential flow fields. If moving and fixed surfaces are very close together, say within one third of their chord, each will receive an impulsive loading from the static pressure field of the other as they pass. The magnitude then of the force change which produces the acoustic energy output, is usually very much larger than that induced by the relative incidence change described previously.

The effect is marked in multi-stage axial compressors of the gas turbine type, high pressure developing centrifugal blowers, and 'non-streamlined' bladed machines such as radial paddle-bladed fans primarily for dust and other particulate removal. In the last two, the reaction is between the blade tips and the scroll cut-off.

Corrections for Installation Design

(i) Intake Turbulence

Intake designs which characteristically produce turbulence onto the fan impeller have already been shown for the specific case of an axial fan in Fig 3. If any one of these effects, or the installation of a fan silencer immediately upstream of the fan without any "setting" length as described in the chapter *Air Distribution Systems*, are present, 6 dB should be added to each octave band spectrum.

(ii) Form of Running

The form of running for an axial fan is conventionally designated A if the flow passes over the motor before the impeller, and B if it passes over the impeller first.

Clearly in the light of the earlier descriptions of noise generating mechanisms, Form A might be expected to be the noisier because of the turbulent flow induced by the motor and its supports. The extra noise will probably be broad band but in designs where motor support struts are within one strut diameter of the rotor, some discrete frequency interaction noise may also be present.

FAN NOISE

In the absence of specific data to the contrary from the fan manufacturer, it is prudent to assume their published figures are for Form B running, and if Form A is to be installed, add 5 dB for each octave band up to and including 500 Hz.

(iii) Inlet Guide or Damper Vanes

One way of upgrading the performance of a fan is to fit a row of guide vanes to the impeller inlet; these have the effect of pre-swirling the air onto the impeller. For maximum efficiency, the blades need to be as close as possible to the moving blades, a condition which unfortunately maximizes the interaction noise they produce, which appears at the rotor/blade passage frequency and its harmonics.

A similar effect can occur if flow control dampers are placed close to the impeller. Where the added guide or damper vanes are located within a distance from the impeller equal to the length of one stationary vane chord, add 6 dB in the octave bands up to and including 500 Hz.

(iv) Multi-staging or Standby Axial Fan Configuration

Another way of upgrading a system is to add a second fan stage in series with the first. In centrifugal fan systems, there should be no noise penalty, but in axial fan systems where the second stage will usually be contra-rotating, there almost certainly will be added noise due to the downstream impeller operating in the wake of the upstream one.

A similar situation arises in standby systems. Even if the upstream impeller is not powered, its windmilling will cause sufficient disturbance of the airflow onto the downstream one to generate the additional noise components.

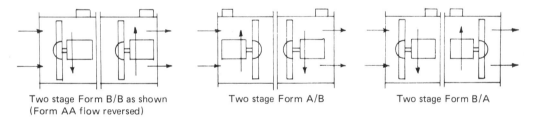

Two stage Form B/B as shown (Form AA flow reversed)   Two stage Form A/B   Two stage Form B/A

Fig 5 Forms of running for multi-stage fans.

The amount to be added to the sound power level of the single fan depends upon the forms of running of the industrial fans, with Form A/B being the worst. Possible configurations are shown in Fig 5. Again, in the absence of specific data from the manufacturer it would be safe to assume the following additions to apply over all octave band levels:

        Form A/B + 10 dB
        Form B/B + 6 dB
        Form A/A + 6 dB
        Form B/A + 4 dB

It must be recognized that Form B/A may not be acceptable to give full multi-stage performance, the impellers being too far apart. On the other hand, it should be a perfectly acceptable arrangement for standby configurations where the fans will in any case be operating as individual units.

Designing for Minimum Noise

In most commercial fan systems, it is true to say that the most common method of achieving a required reduction in fan noise is by the addition to the system of a fan silencer or silencers. The system designer can go some way, however, towards minimizing the size, and hence cost of the silencer that may be required, if not obviating the need for one altogether, by careful design and selection of both fan and system.

It is also true to say that designing the fan to produce minimum noise is more properly the preserve of the machine manufacturer. Some of the following points however may prove useful to the fan designer, and in any case the system designer will have to be aware of possibilities, not so much to design a quiet fan, but at least to ensure that at the end of the day, the required duty is being performed by the system with the least possible noise as a side effect.

(A) Vortex Noise

It has already been pointed out that every fan will produce a certain minimum level of aerodynamic noise, the amount of which is determined primarily by its duty, and by implication by its size and speed. This noise will be at a minimum when:

(a) the only mechanism operating is that of vortex noise.

(b) the impeller blading is operating at its maximum aerodynamic efficiency.

Condition (a) requires of course that there is no induced turbulent flow onto the impeller and that the fan has been designed to avoid any discrete frequency noise. Avoidance of these is discussed later.

Condition (b) requires that the combination of volume flow through the impeller and blade speed is such that each blade element is at its optimum relative angle of incidence. Fig 6(a) shows diagrammatically how the flow will pass over the blade element under such conditions. Here the main flow follows the top and bottom surfaces well, growth of the turbulent boundary layer is small with turbulent intensities being correspondingly small. The consequent lift fluctuations are also lower in magnitude and an appropriately lower level of sound power is generated.

Comparing this with Fig 1 and noting the velocity vector diagram, it can be seen that if for any reason, volume flow and hence the velocity vector V_a is reduced, the relative velocity V_r which the blade "sees" is altered not so much in magnitude, but definitely in direction. The relative angle of incidence is in fact increased for a decreased volume flow. In Fig 6(b) the effect of high incidence upon boundary layer flow is indicated, again diagrammatically. It can be seen that the boundary layer has grown considerably thicker. Turbulent velocity and pressure intensities behind the transition point are correspondingly higher, as is the magnitude of lift fluctuation and hence sound power generated.

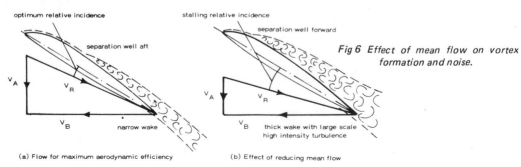

Fig 6 Effect of mean flow on vortex formation and noise.

(a) Flow for maximum aerodynamic efficiency

(b) Effect of reducing mean flow

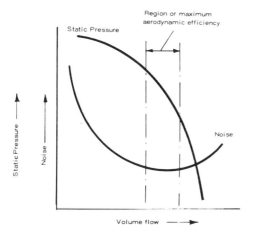

Fig 7 Effect of fan duty.

The effect is summarized diagrammatically in Fig 7 which shows both pressure development and overall sound power produced by a given fan at a given speed, as a function of flow through the fan. The curve marked pressure is the normal pressure — volume characteristic which is produced by all fans. The noise curve is also typical, showing that as volume is reduced, a slight fall-off is evident from free air conditions to a well defined miminum at maximum aerodynamic efficiency; followed by a sharp rise as volume flow decreases further.

The relative importance of the individual sections of the noise curve varies from one type of fan to another but the general shape is common to all fans.

It is also to be noted that the thick boundary layers producing the noise increase generally mean a larger characteristic eddy size and hence lower frequency of noise generation. Such fans as large chord backward curved centrifugal, sheet metal bladed propeller, and mixed flow all produce characteristically low frequency noise increases when called upon to work against pressure.

The lessons to be learned here are:

(i) Always choose the fan type, size and speed which produces the system duty point closest to the fan's maximum aerodynamic efficiency.

(ii) If off-design operation is at all likely, ensure the manufacturer gives the range of sound power produced at each end of the duty range likely to be encountered.

(B) Intake Turbulence Noise

Noise generated by the action of a turbulent intake flow has already been stated to be entirely additional to whatever vortex noise the blade was producing for the same mean velocity smooth flow. The object must therefore be to avoid completely any turbulence at the intake. A check list for the installation designer might be as follows:

(i) Fit coned or bell mouth intakes to all open running fans, or fans drawing direct from plenums.

(ii) Check that all flexible connections are taut, and are unable to form 'bellows', especially on the suction side.

(iii) Axial fans should always be run Form 'B' wherever possible.

(iv) Avoid locating fans — particularly axial flow fans — immediately behind heater batteries, cooling coils, transformation sections or any in-flow components producing turbulent wakes or separated flow. Always aim for at least one fan diameter between turbulence generator and impeller.

(v) Upstream silencers should always incorporate settling lengths.

For the two-stage or standby axial fan installation, the aim should be Form B/A if possible (see Fig 5), or at most Form A/A.

(C) Interaction Noise

Although much has been said about the mechanism and amount to be expected of discrete frequency interaction noise, it is not so much of an installation problem as are the broad band turbulence effects, at least with industrial fans. Its avoidance is a fairly straightforward procedure of ensuring maximum possible spacing in the flow direction between fixed and moving surfaces. In particular, to avoid the acoustically disastrous effects of potential flow interaction between rotor and stationary guide or control vanes in axial fans, and between impeller blades and scroll cut-off in centrifugal blades, separations should never be less than half the blade chord.

Beyond this, the wake effect previously described will predominate, and to ensure its contribution to overall sound power is as low as possible a separation of more than two chords of the wake producer should be the aim.

See also chapter on *Air Distribution Systems.*

SECTION 4b

Factory Noise

FACTORY AND workshop noise falls into three main categories: (i) noise in the vicinity of machines, etc; (ii) overall background noise levels; and (iii) noise generated within the factory which escapes and is radiated to the neighbourhood.

Category (i) directly affects machine operators, as well as contributing to the general overall noise. All machine operators are primarily affected by the noise from their particular machines. Other noisy machines are usually at such a distance that their noise does not add appreciably to the local noise level. Thus if a machine is appreciably noisier than the background noise from other machines (say by 4 or 7 dB), the total noise level is not appreciably increased as far as the operator is concerned. Equally, if one machine in a shop is considerably noisier than other machines running simultaneously, the total noise level will not be increased appreciably by the presence of the other machines. Two equally noisy machines operated simultaneously, on the other hand, will produce a 3 dB rise in total noise level and three equally noisy machines a 5 dB increase in total noise level.

Work area noise levels comprising the cumulative noise of several machines are generally lower than those of individual machines, and with a broader spread — *eg* see Fig 1. Overall or background average noise can also be reduced by acoustic treatment of walls and ceilings, although this will be of no benefit to the individual machine operator since an appreciable decrease in the average sound level will only become noticeable further from the sound source than he normally stands — see Fig 2. Noise reduction at source is thus the only efficient method of treatment as far as the operator is concerned, the most effective treatment being complete enclosure of the machine, with the operator performing his normal work from a position outside the enclosure. If the enclosure provides adequate and complete sealing, and sheet metal panels are adequately damped, noise reduction of the order of 25 dB is possible in the high frequency bands, although the benefits from much lower attenuation may be very real. Much depends on the type and size of machine, and the nature of the working area, as to whether complete enclosure is a practical proposition.

A simpler solution, and one which is more generally applicable, is the use of acoustic barriers and partial enclosures to provide screening in specific directions, to which may be allied acoustic treatment of the building shell. Regions of high noise levels are then 'isolated' with effective reductions of up to 20 dB readily possible in other work areas, although simple acoustic barriers will normally yield about one half this figure as a maximum. The main objection (apart from cost) to barriers and semi-enclosures is that they can interfere with work flow and traffic movement, unless properly designed and positioned. They may also be vulnerable to heavy traffic movements.

In particular cases no method of noise reduction treatment may be effective in reducing the noise level at the operator's point to an acceptable level, and ear protectors may be the only answer.

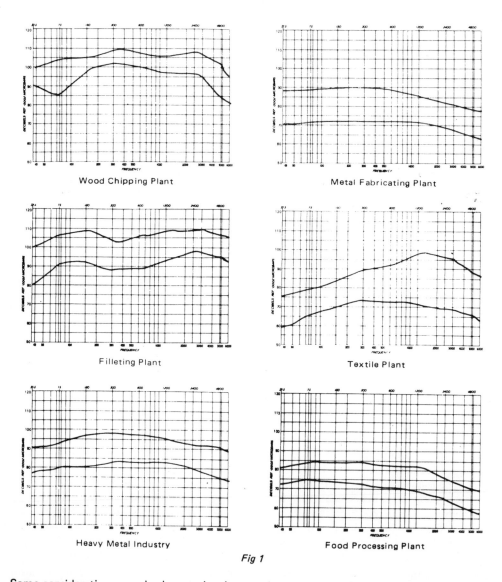

Fig 1

Some consideration may also have to be given to the nature of the noise generated by machines, particularly if they generate impulsive noise, as the true peak noise levels may be appreciably higher than those given by simple meter readings. Thus, Fig 3 shows the noise spectrum of a particular machine measured in three different ways. Curve A shows the spectrum determined by an octave band analyzer on 'slow' setting, and curve B that obtained by the same analyzer on 'fast' setting. There is a difference in excess of 3 dB (equivalent to more than doubling the sound energy) in the most significant frequency range (600–2400 Hz). The true level of (peak) noise is even higher, as shown by curve C measured with an impact meter. For satisfactory analysis it is therefore necessary to take into account the nature of the noise and determine the true noise levels

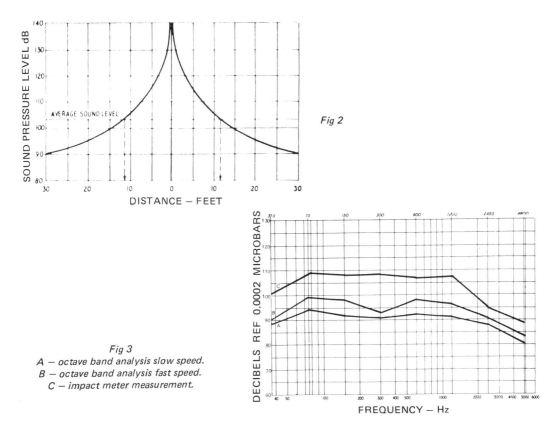

Fig 2

Fig 3
A — octave band analysis slow speed.
B — octave band analysis fast speed.
C — impact meter measurement.

and their significance accordingly. For this purpose noise may be classified under the following headings:

(i) continuous broad band noise.
(ii) intermittent noise, which may be broad band or impulsive noise, or a combination of both, differences modifying the 'exposure time'.
(iii) impact noise with high peak levels of specific duration and frequency. Here there is general agreement that maximum permissible exposure is 140 dB.
(iv) narrow band noise, which can be particularly distressing at the higher frequencies.

In the main, however, L_{eq} measurement should embrace most of the requirements for determining acceptable noise levels within the factory, and the need or otherwise for obligatory noise treatment.

Analysis on a sound spectrum basis may indicate the need for treatment of individual machines, with a view to obtaining a reduction of noise at source. For example, in a large number of cases the major source may be the workpiece rather than the machine itself. This is frequently evident in the case of clipping and riveting machines where the predominant noise is 'surging' of the workpiece. Quite often much can be done to reduce the level of such vibration and noise at source — eg by the use of vibration damping clamps holding the work machines. Thus, in the case of machine tools, 'squeal' is often present. The effect of modifying the operation to eliminate squeal can be quite considerable — eg a reduction in overall sound level of 6 dB (the same reduction which would be given in the original instance if the distance between machine and operator was doubled), with a particular reduction in the significant audio frequency range.

Another point, often overlooked, is that the overall noise level contributed by machines is often appreciably reduced by mounting the machines on vibration-isolating mounts. Whilst this will not necessarily reduce the airborne noise generated by the machine (and thus the noise at the operator's position), it can eliminate (or at least substantially reduce) vibration transmitted through the floor and distributed throughout the whole area as noise.

As regards all noise generated within the factory itself, the Code of Practice for reducing exposure of employed persons to noise lays down a maximum noise exposure of 90 dB(A), determined as L_{eq}. For operating efficiency, reduction of fatigue, and particularly satisfactory speech communication, it is desirable to achieve much lower noise levels at most, if not all, points. Only the former is a specific requirement to be met, however, and if L_{eq} levels of 90 dB(A) or less cannot be readily achieved at particular operators' points, then protective devices are a simple answer, albeit not without their limitations on operating efficiency.

Suspended Noise Absorbers

Ceiling mounted hanging baffles or spatial sound absorbers can greatly improve the noise climate in plant rooms and workshops by absorbing both direct and reflected sound. In particular, they can be used to reduce reverberation time to acceptable levels, which in turn depend on the workshop area — see Fig 4.

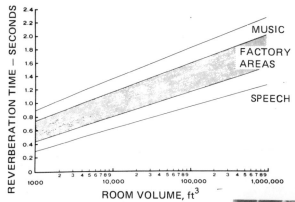

Fig 4 Acceptable reverberation times.

Suspended noise absorbers in a factory. (For acoustic performance see Fig 5) (ICI Acoustics).

Fig 5 shows the acoustic performance of 32 kg/in³ suspended glass fibre noise absorbers spaced at a typical density of one per square metre of ceiling plan area, given both in terms of sound reduction and reverberant noise.

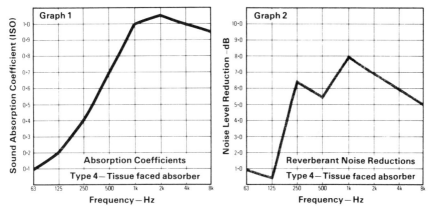

Fig 5 Acoustic performance of suspended noise absorbers. (ICI Acoustics).

Gullfiber System RG panels used for fitting out a complete factory. (Industriakustik Ltd).

Baffles come in various sizes and include flat panels, cylinders, and other shapes. They are usually constructed of a core of 1 in to 2 in porous, acoustically absorbent mineral fibre or glass fibre, polyurethane foam, or wood fibre encased in a decorative or protective wrapper. Some baffles are wrapped and sealed in a thin plastic envelope, making them resistant to moisture and to cleaning by water or steam. The film wrap must be thin and flexible so that it does not inhibit sound absorption by the acoustical core. The baffle may be suspended either by tabs attached to the frame or by integral hanger wires, from overhead bar joists, or beams, or directly from the roof deck using hanger wires, cables, or chains.

The inherent advantage of baffles makes them an attractive means of noise control. Installation is relatively simple, and because of their light weight, 3—15 lb, they can often be put up by plant maintenance workers. They can cause a minimum of interference with lights, air circulation, and sprinklers and do not restrict access to pipes, valves, or switches located in the ceiling. They may be suspended in parallel rows, or with cross rows, or in egg-crate fashion to produce attractive geometric patterns.

The main limitation of ceiling hung baffles is that they are effective in reducing reverberant-field noise levels, but not near-field noise (ie noise received directly or 'line-of-sight'). As a consequence the upper limit of overall noise reduction with suspended acoustic baffles is usually not

IAC 'Quiet Haven' formed from modular panels with acoustic tunnels to provide quiet zone in factory assembly area to check bearing noise.

IAC modular acoustic housings for rolling road test facilities at British Leyland 'Road Train' plant in Leyland, Lancashire. In all, four housings were provided for the rolling road facilities.

above 10 dB and in most cases lower (*ie* 3–8 dB). They are thus only a partial, not complete, noise treatment method.

Escaping Noise

The 'escape' of factory noise and its consequent effect on the neighbourhood is the major problem. There is no specific legislation in this respect although the local authorities responsible may set limits for plant boundary noise which may not be exceeded, to avoid complaints from nearby residents. These are more likely to be stringently applied in the case of new plant. With existing plant in industrial areas the neighbourhood is more likely to be 'conditioned' to accepting a fairly high level of neighbourhood noise during daytime working hours.

Typical industrial noise, as heard at a distance, is a fairly steady low-pitched hum, on which may be superimposed a variety of individual noises, often intermittent in character and high in pitch. The former is the general clatter of the work modified by passing through light building structure, which insulates better against high frequencies than against low.

Examples of individual noise sources are:
a. Doors, windows and ventilators emitting noise from the interior.
b. Fans, compressors, power driven saws, steam hammers and other subsidiary equipment which may be placed outside any building, either in the open or merely under a roof.
c. Loading bays and access roads leading to them.

FACTORY NOISE

Community Reaction

Community reaction must be assessed in terms of:
 (i)   general annoyance
 (ii) effect on people living within earshot
 (iii) long term effects

Tolerance can be expressed in a specific way in terms of an addendum to residential noise levels. The addendum represents the increase in the general background noise level which would normally be regarded as realistic and acceptable over normal or recommended noise levels for typical built-up areas. Thus, in urban areas near industrial sites, an increase in the background noise level of 10 dB could be regarded as reasonable, or even 15 dB near heavy industry.

For single reading measurement of neighbourhood noise, a general level can be recorded, estimating an average value if the general level fluctuates within a range of 10 dB(A). If periods of louder noise are present, the louder level L_2 is also noted as either 'burst' or 'continuous' levels.

The following corrections can then be made to the recorded levels:-

For a continuous tone, deduct 5 dB(A).

For impulsive irregularities in the noise of basically irregular character, add 5 dB(A).

For intermittent noise, the total of the 'on' times is determined as a percentage of the total time, when appropriate corrections can be determined from graphs — Figs 6 and 7.

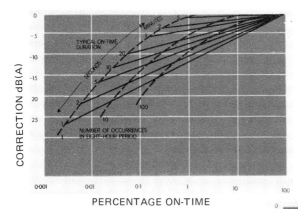

Fig 6 Intermittency and duration correction for night-time.

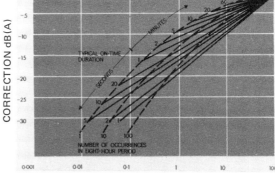

Fig 7 Intermittency and duration correction for other than night-time.

Noise levels corrected for night time are designated $L^1{}_1$, and noise levels corrected for other than night time $L^1{}_2$.

A basic noise level criterion of 50 dB(A) is suggested for factory noise as a reasonable background noise level (although 55 dB(A) or even 60 dB(A) may be acceptable for older, well established factories in industrial areas). This criterion is then subject to the following correction factors:-

Factory Location:	rural area (residential)	–5 dB(A)
	suburban, residential	0
	urban, residential	+5 dB(A)
	predominantly industrial urban areas with some light industry on main roads	+10 dB(A)
	general industrial area	+15 dB(A)
	heavy industrial area with few dwellings	+20 dB(A)
Time of Day:	weekdays only 8 am to 6 pm	+5 dB(A)
	night time 10 pm to 7 am	–5 dB(A)
	6 pm to 10 pm and 7 am to 8 am	0
	Sundays	0
Seasonal:	occurring during winter only	+5 dB(A)

In general, complaints are likely to be common if either the $L^1{}_1$ or $L^1{}_2$ level (or both) exceeds the corrected criterion by 5 dB(A). Subjectively a 5 dB difference is noted as a 'considerable difference', and a 10 dB difference as a 'very noticeable difference'.

Although not considered as establishing the basic noise level criterion, tolerance is also affected by the exposure time. Thus, relatively high noise levels generated for short periods of time may be tolerated, whereas more continuous generation of lower noise levels may not. Specific correction factors based on noise duration during normal working hours are:-

25% of working hours	+5 dB
5% of working hours	+6 dB
1.5% of working hours	+15 dB
0.5% of working hours	+20 dB

This particular method of assessing the probability of complaints is based on BS4142, and may still be employed by some local authorities as a guideline. More generally, however, it can be expected that legislation regarding neighbourhood noise will be developed around L_{eq} or derivatives thereof.

The following is a (simplified) example of calculation of boundary level noise for a proposed plant, the local authority having set a plant boundary noise limit of 40 dB(A). The plant comprises separate machines A1–A6, B1–B4 and C1, positioned as shown in Fig 8. Calculation is simplified by assessing an open site with free field conditions and non-directional noise sources with no impulsive or excessive pure tone noise. Machine noise data are given in Table I.

It is first necessary to calculate the attenuation due to distance for each motor.

FACTORY NOISE

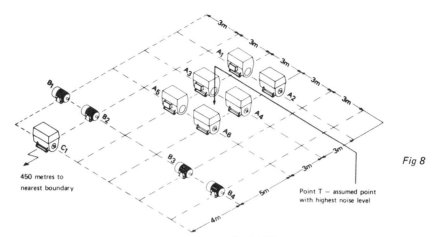

Fig 8

450 metres to nearest boundary

Point T — assumed point with highest noise level

TABLE I — MACHINE NOISE DATA
(Reference Figure 8)
Average sound pressure levels measured at 1 metre

Machine Type	dB(A)	Octave Band Mid Frequencies Hz							
		63	125	250	500	1 000	2 000	4 000	8 000
A	91	74	77	87	112	86	83	78	69
B	88	67	70	76	86	84	81	76	72
C	88	71	79	84	87	84	76	70	65

Using the plant layout dimensions shown in Fig 8 the distance correction is calculated using the basic formula below:

$$\Delta L_p = L_{pd} - L_{pD} = 20 \log_{10} \frac{D}{d}$$

where   d = 1.0 metre

Therefore:-

a) Machines A_1 and A_2 distance D = 465 metres
    ∴ ΔL_p =   53.3490 dB

b) Machines A_3 and A_4 distance D = 462 metres
    ∴ ΔL_p =   53.2928 dB

c) Machines A_5 and A_6 distance D = 459 metres
    ∴ ΔL_p =   53.2362 dB

d) Machines B_1 B_2 B_3 B_4 distance D = 454 metres
    ∴ ΔL_p =   53.1412 dB

e) Machine C_1 distance D = 450 metres
    ∴ ΔL_p =   53.0642 dB

Little error will be introduced if 53 dB is taken as the distance attenuation for all the machines. Grouping all the machines A, B and C and referring to the boundary.

Group A consists of six machines, with a noise level equal to A_1. Total noise level of all group A = noise level of A_1 plus 8 dB.

Similarly, Group B consists of four machines with a noise level equal to B_1. Total noise level of all group B = noise level of B_1 plus 6 dB.

Group C is only one machine.

The dB(A) and octave band levels of the totals of each group are as shown later, together with the grand total of all the machines.

Item Total	dB(A)	Octave Band Mid Frequencies (Hz)							
		63	125	250	500	1 000	2 000	4 000	8 000
Group A	99	82	85	95	99	94	91	86	77
Group B	94	73	76	82	92	90	87	82	78
Group C	88	71	79	84	87	84	76	70	65
Grand Total	100.5	83	86.5	95.5	100	95.5	92.5	87.5	80.5

Subtracting 53 dB from the grand total above refers the total to the boundary, thus:-

Grand Total at Boundary	47.5	30	33.5	42.5	47	42.5	39.5	34.5	27.5

The boundary level is therefore 47.5 dB(A) and fails to meet the specified requirement.

The same example is now extended to determine the highest noise level on the plant site itself — *ie* taken as part T in Fig 8.

Road Traffic Noise

IN THE UK, legislation against noisy vehicles is contained in The Motor Vehicles (Construction and Use) Regulations, with more stringent later requirements still pending. Legislation and strict enforcement of such requirements remains a different matter. Direct attempts to limit the nuisance caused by road traffic noise using quantitative sound levels measured at a specific distance has distinct limitations, particularly as regards the effective enforcement of such legislation and the fact that the standard of measurement for legal limits offers little in the matter of noise *quality* (*eg* see Table I). It also ignores the fact that the major source of annoyance in all denser traffic areas remains the number of noise sources (vehicles) present at a particular time.

TABLE I — VEHICLE NOISE

Type of Vehicle	Order of Significance of Noise Sources				Remarks
	Exhaust Noise	Intake Noise	Engine Noise	Road Noise	
2-stroke motor cycles	1	3	2	V.N.	Differences largely due to differences in intake and exhaust silencing practices.
4-stroke motor cycles	1	3	2	V.N.	Typically 15 dB(A) noisier than cars. Difference in noise largely due to differences in silencing and intake practices.
Passenger cars	1	2	3	4	Excessively noisy automobiles generally have defective exhaust systems.
Sports cars	1	3	2	4	Low frequency 'roar' a common cause of complaint.
Light vans	1	3	2	4	Noise levels often aggravated by loose panels, general clatter and poor attention to maintenance.
Heavy vehicles	1	4	2	3	Major source of traffic noise, often aggravated by lack of maintenance, and/or careless operation. Buses are generally less noisy than commercial goods vehicles.

V.N. = Virtually negligible

GenRad 1982 precision sound level meter and analyzer being used for traffic noise assessment. (GenRad Ltd).

Traffic Noise Level

An original method of assessing the social nuisance caused by road traffic noise was developed by the Building Research Station (UK) in terms of a Traffic Noise Index (TNI). The basis of deriving this index is to separate the noise levels exceeded in a particular proportion of the time covered. Thus, a period of 50% of the time would indicate a mean noise level, and that over a long proportion of the time (say 90%) and average background noise level. The pattern would then be correlated to subjective response.

The mean (50%) level can be expected to show poor correlation over a fairly long period of time and thus generally can be disregarded (except for simple short term analysis, or typical period analysis with fairly constant traffic volume). Equally, the results of detailed analysis would tend to show poor correlation to background noise (90% level) and average maxima (10% level). A much better correlation could be found, however, with a combination of the average 10% and 90% levels — sufficiently high, in fact, to be regarded as a valid and reliable index. The best combination was found to be that of the 10% level less 0.75 of the 90% level, which in a slightly modified form (for easier working) defined the Traffic Noise Index, *viz:*

TNI = 4 × (10% level − 90% level) + 90% level − 30

Basically, the TNI includes the range of noise climate over a period of 24 hours, together with a smaller proportion of the average background noise. The basic combination is multiplied by 4 to eliminate fractional quantities and the quantity 30 is deducted merely to produce a convenient numerical scale.

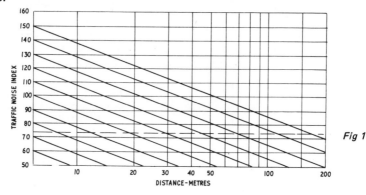

Fig 1

Attenuation of noise with distance follows the relationship shown in Fig 1 from which the following generalization can be drawn:

(i) Over a moderate distance the 10% level falls by about 6 dB(A) each time the distance from the source is doubled.

(ii) Over a moderate distance the 90% level falls by about 3 dB(A) each time the distance from the source is doubled.

Whilst the TNI was designed as a simple design parameter for siting new buildings, *etc*, it has a number of obvious limitations. Thus, the distance required for siting to produce an acceptable TNI value may be impractical, or impossible, in which case either a higher noise level will have to be accepted, or measures taken to insulate the building. Also, the TNI may be subject to change, *ie* is likely to be modified by increase in traffic density, or the area involved may be a project only, in which case no measurement figures can apply, except on a basis of similarity to existing sites. Both these factors can, however, be related to expected TNI values on an approximate basis related to traffic volume (see also Fig 2).

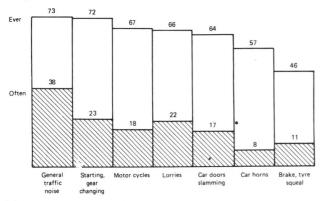

(a) Types of traffic noise heard.

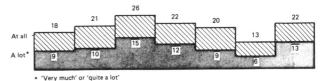

* 'Very much' or 'quite a lot'

(b) Types of traffic noise that bother.

Fig 2 Transport and Road Research Laboratory.

| 25 | 29 | 39 | 33 | 31 | 23 | 48 |

(c) Percentage of those hearing each type of noise who are bothered by it.

L_{10} Measurement

The Noise Insulation Regulations 1973 specified L_{10} (18-hour) as the basis of traffic noise measurement and estimation. 'L_{10}' is the sound level in dB(A) which is exceeded for one-tenth of a period of one hour between 0600 and 2400 hours on a normal working day; 'L_{10} (18-hour)' is the arithmetic average of all the levels of L_{10} during a period from 0600 to 2400 hours on a normal working day, taken to the nearest whole number in decibels.

L_{10} (18-hour) are then adjusted for various factors related to traffic flow and average vehicle speeds.

Table II then shows present reductions for L_{10} levels following the implementation in 1980 of the EEC Directive 77/212/EEC, and future estimates through to 1995.

Worthwhile reductions in L_{10} can be achieved by progressive reduction in vehicle emission standards. However, to achieve this, commercial vehicle noise will have to be reduced considerably from levels presently attainable, calling for extensive further research and development.

Until such standards are reached — and enforced — the most practical means of controlling traffic noise remain:

    (i)   the erection of sound barriers between traffic and dwellings;
    (ii)   severe restrictions on traffic movement in built-up areas;
    (iii) sound-proofing treatment of affected dwellings;
    (iv) re-siting of urban road systems.

Quiet Vehicle Development

The Quiet Heavy Vehicle (QHV) Project is a programme of research and development sponsored by the Transport and Road Research laboratory, the aim being to determine the technical feasibility and costs of producing demonstration heavy diesel-engined articulated vehicles conforming to current regulations, with noise reductions to levels down to 10 dB(A) less than current levels. A lower limit of 80 dB(A) was given as a target to be achieved under all operating conditions provided the cost was not unreasonable. The corresponding noise level target for the inside of the operator's cab is 75 dB(A). British companies co-operating in the scheme include British Leyland, Fodens, Rolls-Royce Motors, the Institute of Sound and Vibration Research and the Motor Industry Research Association (MIRA). Table III shows the target levels specified for major components.

A 'research' engine gave noise levels about 9—10 dB(A) lower than the standard engine — a considerable achievement — but was still about 4—6 dB(A) noisier than the target level. The exhaust system reduced the noise to within 2 dB(A) of the target. The cooling system was substantially developed and met the noise level targets specified. The quietened components are being developed to production prototype standard and the evaluation of manufacturing and user costs is progressing according to programme.

The tyre noise targets were achieved on most road surfaces, but only at speeds below 80 km/h (50 mile/h) and were exceeded by 2—5 dB(A) at speeds up to 100 km/h (52.5 mile/h). The external noise, determined under acceleration test conditions of a research prototype vehicle was 83.5 dB(A). It is expected, however, that further modifications planned for this vehicle should enable a level of 80 dB(A) to be reached under acceleration test conditions and for vehicle speeds below 80 km/h.

Road Surface Noise

The development of safe road surfaces to give good skid resistance at high vehicle speeds has led, in certain cases, to high levels of vehicle noise and, similarly, the use of traction tyres or dual purpose tyres on heavy vehicles is known to increase vehicle noise above that from tyres with circumferentially grooved tread patterns. Tyre and road surface noise is, therefore, a function of both the

TABLE II — REDUCTIONS IN L_{10} dB(A) FOR FUTURE YEARS, FOLLOWING THE IMPLEMENTATION OF STATUTORY VEHICLE NOISE REDUCTIONS IN 1980 AND 1985
(P.M. Nelson)

Forecast Year	Reduction in L_{10} dB(A) with			
	10% Heavy Lorries		20% Heavy Lorries	
	1†	2*	1	2
1980	0 (−0.6)	0 (−0.6)	0 (−0.6)	0 (−0.6)
1985	0.8 (−0.6)	0.8 (−0.2)	0.4 (−0.6)	0.4 (−0.6)
1990	3.9 (2.5)	3.6 (2.2)	4.0 (2.6)	3.4 (2.0)
1995	6.3 (4.6)	5.4 (3.7)	7.2 (5.5)	5.6 (3.9)

† Columns labelled 1 are reductions in L_{10} following the implementation in 1980 of the EEC Directive 77/212/EEC and with vehicle noise limits for all categories reduced to 80 dB(A) from 1985 onwards.

* Columns labelled 2 are the same as for 1 but with the limit for heavy lorries set at 83 dB(A) from 1985 onwards.

N.B. The figures in brackets are adjusted for the anticipated growth in traffic flow.
The other values are for a traffic flow of 2 000 veh/h. Negative values represent increases in noise.

TABLE III — COMPARISON OF UK AND EEC NOISE LIMITS FOR NEW VEHICLES

UK		EEC†	
Vehicle Type	Noise Limits dB(A)	Vehicle Type	Noise Limits dB(A)
Motor cycles ⩽50 cc	77	Motor cycles ⩽ 80 cc	78
50.1−125 cc	82	⩽125 cc	80
>125 cc	86	⩽350 cc	83
		⩽500 cc	85
		>500 cc	86
Cars	84	Cars	80
Light vans	85	Light vans	81
Goods and Buses	89	Buses ⩽200 hp	82
		Buses >200 hp	85
		Goods ⩽200 hp	86
		Goods >200 hp	88

† The EEC limits affected most vehicles entering into service from October 1980.

tyre design and the nature of the road surface. Thus tyre noise is an important component of total vehicle noise. With the advances in vehicle design achieved as part of the QHV project, tyre/surface noise will affect total vehicle noise at most cruising speeds and is practically the dominant source at vehicle speeds above 80 km/h on dry roads. Tyre/road surface noise is even more important for passenger cars and light commercial vehicles, since the contribution from the power train for these vehicles is considerably lower than that for heavy lorries.

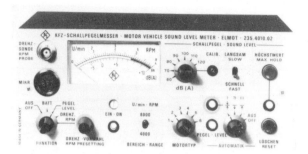

Motor vehicle sound level meter with automatic storage of noise levels at prescribed rpm. (Rohde & Schwarz).

Tyre/road surface noise has been studied at TRRL with the object of achieving an understanding of the mechanism and specifying designs for quieter tyres and road surfaces. The work on the tyres has shown that certain small reductions can be achieved on most road surfaces by using radial tyres rather than cross-ply tyres and by employing circumferentially grooved tread patterns rather than transverse grooves, but the reductions achieved are severely limited by the existing constraints on tyre design arising from the requirements of safety, including cool running durability and low cost. As a result, it seems that there will be no radical reduction in tyre/surface noise due to improvements in tyre design in the near future.

IAC modular acoustic housings for rolling road test facilities at British Leyland 'Road Train' plant in Leyland, Lancashire. In all, four housings were provided for the rolling road facilities.

Aircraft and Airport Noise

THE MAIN noise producer in aircraft is the propulsive system (engine and propeller or rotor, jet engine compressor and turbines and jet flow, *etc*). Noise generated by the motion of the aircraft through the air is largely confined to boundary layer noise which is more truly an aerodynamic phenomenon. This is mainly responsible for secondary noise generated by skin vibration, observed particularly inside the aircraft in flight. Its contribution to an outside observer is largely, if not entirely, masked by 'propulsion noise'. Other pure sources of aerodynamic noise are the static pressure fields moving with very low flying aircraft, and shock waves and sonic boom generated by supersonic flight. These are separate phenomena.

Piston-Engined Aircraft

Piston-engined aircraft probably still form the majority in numbers, although they are now almost exclusively confined to small aircraft engines with relatively low engine powers. With piston-engined aircraft the main sources of noise are the engine(s) and propeller(s). Propeller noise is often predominant except at low tip speeds and/or low power levels.

Propeller noise consists of discrete tones at the blade speed frequency and its harmonics, together with vortex noise or broad band noise which predominates above about 1 000 Hz. At very low tip speeds, it can exceed the noise due to rotation at all frequencies. Mathematical analysis is possible but tends to become complex, especially if forward speed and blade load distribution are taken into account.

The sound field for vortex field generation is directive and characterized by a four-lobe pattern — Fig 1. The sound pressure level under the angle of maximum radiation is usually of the order of 4 dB higher than the space average, this angle being about 25 to 30° inclined backwards from the disc.

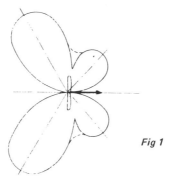

Fig 1

The vortex noise level can be calculated with a reasonable degree of accuracy (plus or minus 10 dB) from the formula:

$$L_p = 10 \log_{10} \frac{k A_b (V_{0.7})^6}{10^{-16}} \text{ dB}$$

where A_b is the total area of the propeller blades in square feet
$V_{0.7}$ = velocity of blade element at 0.7 radius, in feet per second
k is an empirical constant = 3.8×10^{-27}

Peak frequency of vortex noise can be calculated from the following formula:

$$f_{max} = 0.13 \frac{V_h}{0.7R}$$

where V_h is the helical tip speed
R is the propeller radius

Typically vortex noise level contributes about two-thirds of the overall noise level determined as sound power level in dB.

Engine noise is derived almost entirely from the pulsating exhaust flow, with a fundamental frequency equal to the average cylinder firing frequency of the engine. Direct analysis of sound power output treats the engine as a simple source, although this will not hold true for higher harmonics.

The overall sound power level can also be estimated directly from the following empirical formula:

$$L_p = 125 + 10 \log_{10} W_t \text{ dB (re } 10^{-13} \text{ watts)}$$

where W_t is the total horsepower delivered by the engine

By treating the engine as a simple source, the sound pressure level at the distance L then follows as:

$$L_s = L_p - 10 \log_{10} \left(\frac{4\pi}{L^2} \right) \text{ dB re 0.0002 microbars}$$

In general, for low powers, there is a typical decrease in the level for each successive octave of 2.5 to 3 dB, with a levelling off at the higher frequencies. For engines operating at maximum power, and particularly large engines, the power level is substantially the same in all octave bands at a level of the order of 7–8 dB below the overall.

Silencing is the standard method employed to reduce exhaust noise, but the application of silencers is limited to straight-through types to minimize backpressure. The use of silencers may also modify the sound spectrum, suppressing some frequencies and amplifying others, depending on the relative phases of the individual exhaust pulses as they meet in the exhaust system.

The engine may also generate additional noise by engine-excited vibration which can be transmitted through the engine mounts to the airframe. This is only likely to affect noise levels inside

AIRCRAFT AND AIRPORT NOISE

the aircraft, and is generally negligible in modern designs where adequate attention is given to isolation of the engines at the point of mounting. In older aircraft, engine vibration may be responsible for vibrational waves being set up in the fuselage, with frequencies generally below 600 Hz and decreasing with increasing distance from the source. The usual method of meeting such engine excited vibration was to stiffen the forepart of the fuselage (on single-engine aircraft), or the region of the plane of the propeller (on multi-engine aircraft).

Standard treatment is to suspend aircraft engines on rubber-in-shear mounts so as to decouple the six modes of translational and rotational vibration, with a natural frequency for the suspension of less than 50 Hz. For all vibrations above this natural frequency, attenuation increases by 6 dB per octave.

Jet-Prop Aircraft

Where the propeller is driven by a turbojet, the jet exhaust contributes only about 10% of the total thrust and is thus working under comparatively low power level conditions. The jet noise is, therefore, usually far less significant than propeller noise, although it may appreciably modify the overall noise spectrum by 'filling in' between any discrete frequencies which may be present.

For equal thrust, the jet-prop is, in fact, appreciably quieter than the combination of a propeller and a piston engine, typically by a matter of 10 dB.

Under particular conditions, and particularly at low thrust levels, compressor noise may predominate, identified as a high pitched whine due to fan noise radiated from the engine intake. This would be most noticeable to an observer positioned in front of the engine.

Thus, to an observer on the ground, compressor whine is a noticeable characteristic of a jet-prop aircraft approaching to land.

Ground run-up silencer installation for B.A.C. Concorde.

Jet-Aircraft

The world fleet of large jet airliners started to appear in the late 1950's, since which time it has grown at a remarkable rate. Early jets were some 20 dB noisier than the piston-engined aircraft they replaced and serious attempts at noise reduction were not considered until the introduction in 1969 of (US) standards for noise certification. At the same time, larger and more powerful jet aircraft were in the course of development and entering service. A major problem here is the time between design and first introduction into service of a new jet aircraft. Thus aircraft designed in the 1960's can still dominate the noise climate through to the 1990's, although necessarily subjected to 'hushkit' or noise reduction treatment during their life.

Three classifications or 'stages' exist in noise certification terminology. Stage 1 categorizes aircraft that have no noise certificate (older B707's, DC8's and early JT9 powered types); Stage 2 are those that comply with the original 1969/70 standards (hushkitted JT9 types, B747's, etc); while Stage 3 embraces all aircraft that comply with the latest, toughened standards of 1977/78. These include the L-1011, A300, some DC10's and B747's and, by law, the emerging new generation of B757's and 767's.

Figure 2 uses these classifications in projecting the US fleet, and it is clear why current emphasis in international debate is centred upon removing the middle Stage 2 class of aircraft from the fleet earlier than economic forces demand. Over 20 dB separates the noisiest and quietest aircraft in service, while 5–10 dB notionally separates the broad groupings. Because of their high noise levels and greater numbers, the older types always dominate the environmental picture, as is illustrated by the conversion of Fig 3 into noise exposure terms in Fig 4.

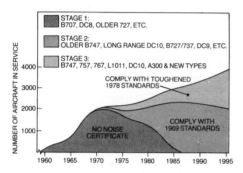

Fig 2 US fleet trends by noise type.

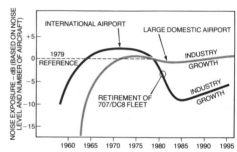

Fig 3 Impact of FAR 36 'Compliance' rule in USA. (Jet powered aircraft: take-off noise only).

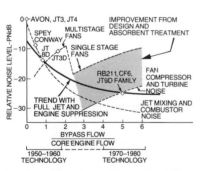

Fig 4 Engine noise sources and variation with bypass ratio.

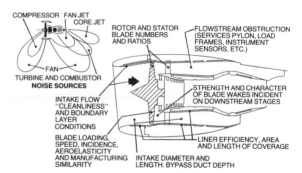

Fig 5 Factors influencing fan noise.

Noise certification, as specified in the Air Navigation (Noise Certification) Order 1979, SI 1979 930, sets down noise limits for aircraft calculated in relationship to their maximum certified weights. Under the Order, aircraft which do not have a Noise Certificate and/or do not comply with the conditions attached to it are prohibited from taking off or landing in the UK. By 1986, it is intended that all non-noise certificated subsonic jets on the UK register will be phased out.

This is quite apart from Noise Abatement Requirements issued under the Civil Aviation Act 1978 which deals specifically with regulation of noise from aircraft. In the main this embraces requirements for minimum noise routes and day and night levels of noise on take-off.

Although the decline of the noisy B707 and DC8 types augurs well for the large international airport in the mid 1980's, the continued high level of noise exposure predicted for the 'domestic' airport explains continuing pressure in the legislative and airport noise control fields. The US 'domestic' airport, in European terms, embraces most major airports on the continent.

One notable step taken within the last decade is the major change from low to high bypass jet engines, with jet noise being replaced by noise from the turbomachinery as the bypass ratio is increased — eg see Fig 5. Turbomachinery noise is internally generated, allowing design treatment to reduce noise hitherto not possible with the external jet mixing noise.

The main source of noise of a turbofan, and within the designer's sphere of influence, is the fan, while the turbine can sometimes run a fairly close second. The factors that influence the noise generated from the fan include, but are not uniquely limited to, those indicated in Fig 5. For the minimum generation, the blade loadings, tip Mach number and operating incidence need to be as low as possible. The relative number of rotor and stator blades will influence the balance of the discrete tone content, while minimizing blade wake intensity at the point of interaction with the next stage will reduce overall levels.

Having an aerodynamically clean inlet flow avoids spurious tones generated by a distortion-interaction mechanism, and a thin boundary layer will minimize noise generation in the high velocity region at the tip of the blade. Absorption of the noise before it radiates from the intake and exhaust nozzle is a function of the efficiency of the acoustic liners, their length, and the depth of the duct in which they are operating. Clearly, the large diameter short intake provides maximum opportunity for fan noise to radiate without being affected by the liner, whereas the comparatively small annulus depth of the bypass duct is a very efficient environment for liners.

These, and other factors affecting the generation of fan noise, mean that many independent variables have to be understood and accounted for in designing the optimum fan/nacelle arrangement without drastically affecting performance.

Jet Engine Noise

The stream pattern developed by a jet engine is shown in simple diagrammatic form in Fig 6. The convergent central core extends for some 4 to 5 diameters downstream and is enclosed by a diverging mixing region which is eventually resolved into a fully turbulent stream. The intervening zone comprises turbulent mixing, generating 'self-noise', and turbulent shear, generating 'shear noise', together with axial velocity fluctuations.

Decay of sound power is very rapid in the developing jet, so that the bulk of the jet sound is developed in the first 8 to 10 diameters downstream.

The complex pattern is further affected by convection, refraction, density and temperature. Thus convection tends to crowd the sound waves in the downstream direction, once the flow is subsonic, producing a Doppler shift of frequency. Refraction has the opposite effect, tending to fan the sound outwards in a downstream direction. Noise level will be affected in direct proportion to the variations in local density whilst temperature changes can generate additional sounds from entropy fluctuation.

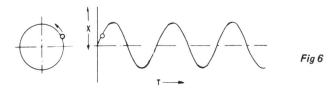

Fig 6

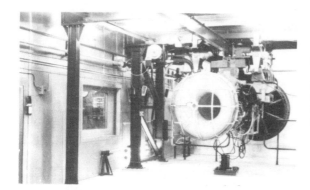

This customised jet engine test facility for the Rolls Royce Turbo-Meca Adour 102 engine, used to power the Jaguar Strike Aircraft, was produced to a total turnkey project and includes Acoustic Enclosure, Control Room, Intake and Exhaust Silencing, Test Bed and full Automatic Data Processing.

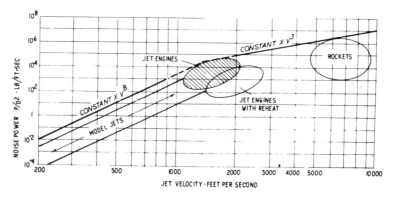

Fig 7

In the case of idealized jet, the total power is proportioned to the eighth power of the nozzle velocity (V_j) and to the square of the nozzle diameter (D). It is also directly proportional to the stream density, which can be assumed constant for simplicity. This proportionality to V_j^8 holds good up to sonic jet speed, after which the noise power is more nearly proportional to V_j^3. This relationship and the transition is shown in Fig 7, verified by practical measurements on model and full size jets. Full size turbojets fall at the upper limit of the V_j^8 range. Jets with after-burners and rockets fall within the V_j^3 range. The latter are working under 'choked' nozzle conditions.

When the pressure ratio approximates to 2.0 or greater, flow through the nozzle is sonic, expanding to a supersonic velocity downstream from the nozzle. The result is the formation of shock waves outside the nozzle and a consequent increase in noise level. Choked jet noise is radiated most strongly at 90° to the axis, which can modify the directivity as well as the sound intensity and spectrum (the choked noise usually being sharply peaked). Many current turbojets operate under choked conditions, but the choked noise is often masked by the normal jet noise.

For simple analysis it is sufficiently accurate to assume that the sound power level is equal to:

$$10 \log_{10} \frac{\pi D}{4V_j} + \text{a constant}$$

Near field analysis is far more complex, since the noise source can no longer be considered as originating from a single point. It can, however, be attempted with the aid of suitable reference contours based on yielding corrections to calculated sound power level with co-ordinates x and y, where $r = x^2 + y^2$. Such reference contours are, however, more or less specific to a particular design and size of turbojet.

Jet Noise Suppression

Various types of suppressors, or so-called mufflers, have been developed to reduce jet noise. These fall into two main categories:

(i) those which can be permanently fitted to the jet engine for inflight operation; and

(ii) those devices developed for ground-running only.

Obviously the latter offer the greater scope, since they are not limited by bulk, weight, potential loss of performance, or even position.

Inflight suppressors have evolved around four basic forms, none of which has yet shown any dramatic results in noise reduction, the best savings being of the order of 5 to 10 dB. These basic types are:

(a) the toothed nozzle or tooth devices around the lip of the nozzle to increase the thickness of the mixing zone;

(b) corrugated nozzles, which increase the perimeter of the nozzle whilst maintaining the same nozzle area;

(c) multiple nozzles of small size replacing a single nozzle; and

(d) ejector devices.

Corrugated nozzles are often a preferred type, for appreciable reduction in sound power level (4 to 7 dB) can be achieved without loss of thrust. Their design needs careful 'optimization' however, both as regards the number of corrugations and corrugation geometry. The number and depth of corrugation have a marked effect on shifting the effective frequency range, and thus to some extent control the 'quality' of the jet noise. Their main disadvantage is the increase in weight (which inevitably implies a performance penalty) and cost involved.

For ground operation, mufflers and screens with a much higher absorption efficiency can be employed, regardless of the high loss of thrust which may result.

Noise Footprints

Direct measurement of sound pressure levels of aircraft flying overhead have only limited significance, although providing a simple source of data for complaints. A more significant parameter is the actual distribution of sound over and on either side of the flight path. This can only be given by sound pressure level contours plotted by measurements taken at a large number of different points. For simplicity this may be rendered in terms of footprint area rather than actual contour shape — some comparisons being shown in Figs 8 and 9.

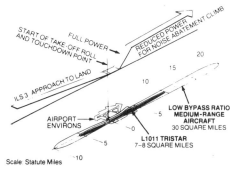

Fig 8 Noise footprint comparison of Tristar and current medium-range aircraft.

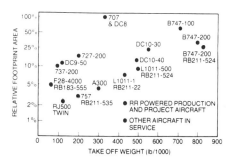

Fig 9 Comparison of footprint areas of major aircraft types (logarithmic scale).

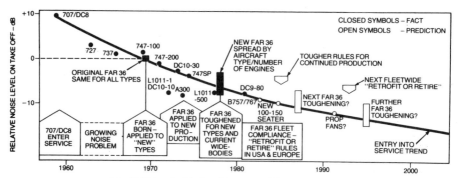
Fig 10 Noise legislation history and future projection.

The Present

All the world's major jet engine manufacturers are actively concerned with programmes with heavy emphasis on both the understanding and suppression of fan noise at minimum penalty. This is also necessary to meet present and projected legislation (shown in Fig 10) although the latter is largely inconsistent with what technology can be expected to provide at realistic performance and cost levels. There is no final solution to such a problem, particularly with the additional growing interest in fuel efficiency. Thus noise control features showing any wanted reduction in noise levels tend only to be achieved at the expense of additional weight and some loss of specific fuel consumption.

At the hot end of the engine, attention is concentrated on low noise turbines, combustor noise and optimum mixing of the hot and cold exhaust flows for minimum jet noise. Nearly all research work accepts the fact that the engine is undoubtedly the major noise source, but natural turbulence generators in the form of flaps and undercarriage systems contribute to the overall noise of an aircraft at low powers on landing. To evaluate the relative importance of these and the main engine sources, together with the amplifications produced by reflections and aerodynamic interference by the airframe structure, a complex source location system based on multi-microphone arrays is necessary.

The Future

Barring unforeseen events, history is likely to repeat itself over the next 15—20 years, taking the form of a downward flight of legislative steps, spaced about eight years apart. The primary steps are standards for 'new' aircraft types, followed by a secondary, yet far-reaching, series of applications of those standards. Such application, firstly to the manufacturer's production line, and then to the operator's fleet seeks to phase out the older, noisier aircraft.

For example, FAR 36 was first applied to new types in 1969, to the production line in 1974 and then to the fleet ten years later. Already ICAO are debating the applicability of the tougher 1977 standards to the production line. It is, therefore, not too difficult to foresee a next round of toughening starting around 1987/88.

On top of this pattern there is growing support for closer control at the airport. Based on a definition of the area contained within a contour of constant noise level, or 'footprint', rather than certified level, airport managers could well set curfews and operational quotas. The UK already does this in respect of night operations; as a secondary form of legislation it could promote a more competitive element between manufacturers to ensure the operators the greatest freedom in route structuring.

Noise in Commercial Buildings

SOURCES OF noise in office blocks and commercial buildings are:
- (i) *Ambient internal background level,* or the irreducible noise level in an unoccupied office normally associated with the air conditioning system. This is usually constant in level and frequency content.
- (ii) *Occupation* noise, being the sound generated in an office during the normal working day. Ideally it should be a steady 'hubbub' arising from conversations, office machinery, personnel movements, *etc*, but it could also include undesirable components such as bangs and thumps from accidental impacts, loud voices, or the monotonous whine of a fan or pump. Occupational noise can be highly variable in level, character and duration, and include talk, typewriters, telex and office machines, footsteps, slamming of doors, lift gates and lift working, *etc*.
- (iii) Noise from *external* sources, such as traffic noise outside the building, and occupation noise in adjacent rooms. The proportion which enters a building structure is moderately variable in level and spectral content.

The consequence of sound on the noise environment in individual offices is exaggerated by the fact that modern buildings of this category are designed to provide the maximum accommodation and occupancy on a minimum area of land, favouring the use of lightweight structures and open-plan offices. This can be offset by sub-dividing the floor area into individual offices, but there may be limits imposed by the structure on the weight of inter-office walls or partitions which can be used.

Basic Requirements

Basic requirements for satisfactory office working are:
- (i) Good communication conditions must be provided in the vicinity of the speaker.
- (ii) There must be privacy outside the group work area.
- (iii) Background noise levels which are comfortable and non-intrusive should be maintained throughout the office.
- (iv) Occupational noise levels should be free from potential irritations such as impulsive noises, *eg* thumps and bangs from slammed doors, loud and distinctive voices outside the group area, or periods of high noise activity from machinery elsewhere in the office.

More specifically there should be a steady background noise in the office; it should be pleasant and easy on the ear; it should be capable of 'masking' distant conversations so that they are

NOISE IN COMMERCIAL BUILDINGS (A)

For the finest Professional Secondary Glazing Insulation .... Consult The Experts

Selectaglaze enjoys an outstanding record of achievement in co-operation with architects, surveyors and building/design consultants. 'Blue-chip' client projects include British Petroleum, Norwich Union, ICI, Unilever and the Royal Institution of Chartered Surveyors.

This private company, established for over 17 years in what is still a young industry, concerns itself exclusively with commercial and industrial projects.

40dB + is the average sound reduction index achieved from Selectaglaze secondary double glazing casement and sliding sash units tested in accordance with BS 2750:1980.ISO.140:1978 — over the 1/3rd Octave Band Frequency Range 100/3150 Hz.*

For those concerned with noise and vibration control, Selectaglaze provides the complete, professional answer to secondary glazing problems — from free consultation service to aesthetic, purpose-designed systems and expert installation.

For an initial discussion, call Meredith Childerstone BA(Oxon) at

*Copy of report No. C/4831/405 by Sound Research Laboratories is available on request.

Sutton Rd., St. Albans, **Telephone ST. ALBANS 37271**(10 LINES)

NOISE IN COMMERCIAL BUILDINGS (B)

slide simply into silence

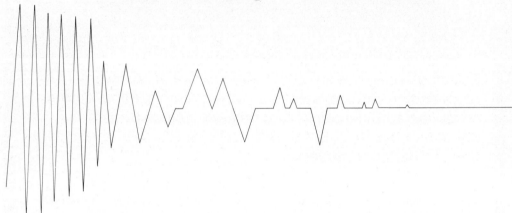

and wave all your noise goodbye

Extensive research and development has resulted in the patented design of the Markus Hermetic Door. It is the solution to the problem caused by noise and lack of space when areas in buildings as diverse as a school, hotel, or a power station have to be separated.
Use of the Markus sliding system guarantees a reduction in noise levels. The complete lack of friction makes opening and closing easy and silent, with an airtight seal when the door is closed.
Valuable space is saved by sliding doors and the Markus Door fits flush to the ground without the inconvenience of a sill.
The simplicity of the design eliminates wear and tear on gaskets, and the doors can work automatically if required.
Two systems, AM2 and AM4 shown below, cover all possible requirements from heavy industrial to light institutional uses.

Light Duty AM4 43dB.
(Max. Size 2600mm x 2500mm high)

Heavy Duty AM2 50dB and 55dB.
(No limit to size)

envirodoor markus

Envirodoor Markus Limited,
Great Gutter Lane,
Hull, HU10 6DT.
Telephone: Hull (STD 0482) 659375
Telex: 527088

relatively unintelligible, and the long distance propagation of speech and noise should be controlled by the room acoustics or the room furnishings.

Acceptable Office Noise Levels

The acceptability of background noise level is essentially a subjective response, but is closely related to the working environment. Thus, in a busy office, including typewriters, a relatively high noise level would be regarded not only as acceptable, but a normal working environment. It would be regarded as intrusive if it interfered with a particular working requirement — *eg* made speech or the use of a telephone difficult. In fact, the overall background noise level may be higher than desirable, but one to which the individual occupants become conditioned. It is desirable to set maximum recommended levels, as conditioning to unduly high levels can be accompanied by loss of working efficiency. Such recommended levels are, therefore, set with regard to maintaining good working efficiency, but at the same time, must be realistic in the matter of what can be achieved economically.

TABLE I – TYPICAL AVERAGE NOISE LEVELS FOR OFFICES

Type of Office	Typical Background Noise Level dB(A)	Recommended Maximum Acceptable Noise Level dB(A)
Board room	30–35	30–35
Conference rooms — small	35–45	35
Conference rooms — large	35–40	30
Managing Director		35
Managing Director's secretary	40–45	40
Other Directors' offices	35–45	40
Personnel Manager		45–50
Plant Manager	50–75	50
Office Manager	50–70	50
General Office		
over 10 people	60–70	50
3–10 people	52–64	50
less than 3 people	50–60	45
Clerical Office	55–65	60
Drawing Office	40–55	50
Large busy office	up to 70	70
Very large, very busy office	up to 80	60
Office machine room	75	
Typing office*	60	60
Typing pool	60–70	60

*A single typewriter has a typical noise level of 73 dB(A) for a manual machine and 66 dB(A) for an electric machine.

Background Noise Levels

Various authorities over the years have operated background noise criteria on broad agreement with the values summarized in Table I. The degree of background noise normally present is significant in determining the attenuation required from dividing walls, ceilings, floor, *etc*, since an intruding noise will not be noticeable (and thus not objectionable) until it equals, or exceeds, the background noise. Equally the background noise should be as characterless as possible so as not to

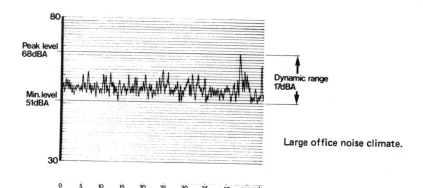

Large office noise climate.

Fig 2
(Sound Research Laboratories).

Small office noise climate.

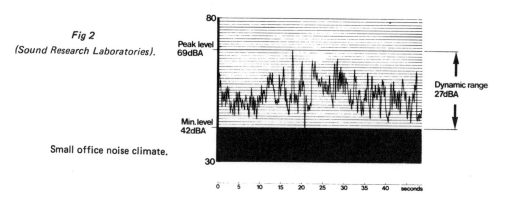

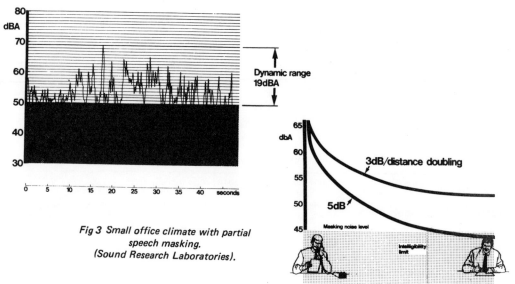

Fig 3 *Small office climate with partial speech masking.*
(Sound Research Laboratories).

Speech masking

be intensive in itself and at the same time mask other intensive noise. Thus the spectral content of the background noise is as impostant as its level. Fig 1, for example, indicates a range of character-less spectra centred on 1 000 Hz which would offer good speech-masking characteristics.

One of the most frequently used measures of speech intelligibility (or conversely speech privacy) is the Articulation Index, which is strongly related to the level of the speech above the background level. AI measures the percentage of syllables intelligible and is related to intelligibility and privacy as shown in Table II.

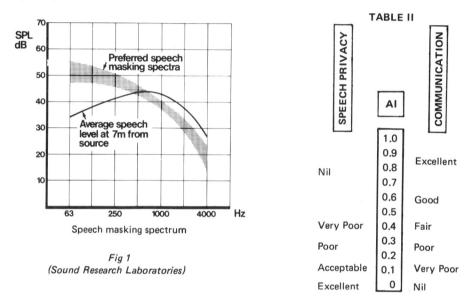

Fig 1
(Sound Research Laboratories)

TABLE II

SPEECH PRIVACY	AI	COMMUNICATION
	1.0	
	0.9	Excellent
Nil	0.8	
	0.7	
	0.6	Good
	0.5	
Very Poor	0.4	Fair
	0.3	
Poor	0.2	Poor
Acceptable	0.1	Very Poor
Excellent	0	Nil

In a semi-reverberant office where sound levels only fall off at 3 dB per doubling of distance, speech sound levels originating at one work station are partially attenuated with distance but are still effectively higher than the background noise level and hence no acoustic privacy is obtained.

In a 'dead' environment where levels fall off by 5 dB per doubling of distance, the speech levels are masked by the pre-existing background noise levels, thereby ensuring a degree of privacy for the speaker.

Speech Masking

Background sound levels must not therefore be allowed to drop too low since sounds are masked most efficiently by sound of the same frequency. This is particularly important over the frequencies critical to speech intelligibility (400 Hz to 4 kHz) — this range also applies to many of the more annoying characteristics of office machinery noise. Unfortunately conditions often work against this aim. External noise breaking in is often of low frequency content, as is residual fan noise and duct breakout in ventilation systems. Air noise in the ventilation system is often of the correct frequency content but is not easily predicted and may be adversely affected when the system is balanced. It is the lack of assured control on ventilation noise which has restricted its development as a source. The fact that this is an identifiable and familiar source is an important factor in favour of using this method. Many rooms are pleasantly sound conditioned by mechanical systems but generally more by fortune than by design. This leaves only occupational noise.

In a smaller open area with lower occupancy or in an area of low occupational density (or even in a large office with many staff absent) the background sound level becomes sporadic and thereby intrusive (see Fig 2). In such a situation where there is an absence of background sound from normal sources, substantial improvements have been achieved by the introduction of electronically generated sound (Sound Conditioning), (see Fig 3). The introduction of sound by such means may seem to conflict with the efforts made to reduce noise but, referring back to the definition of 'noise' as 'unwanted sound', the substitution of characterless background masking sound in place of intrusive sounds is a reduction in the noise. But this cannot be a 'carte blanche' for feeding in more and more sound. Above a certain level the introduced sound becomes of itself intrusive. It is vitally important that the sound is acceptable to listen to in its own right.

It is interesting to examine the preferred spectrum shape of background sound. Research work to date shows that the desirable spectrum is very different from that of the NC or NR curves and closer to the human speech spectrum shape. This is useful since it is therefore also relatively good for masking speech. The sound which has been found to be most effective to date is a broad band random sound, shapes as in Fig 1, evenly distributed (± 2 dB(A)) over the floor area. Work is still proceeding on the assessment of alternative sounds including natural sounds, *eg* the wind, the sea, slowly modulated broad band sources, and contributions of ventilation noise and electro-acoustic systems.

The basic principles of simple electro-acoustic systems are indicated in Fig 4. Typically loudspeakers are set on a grid, *eg* in the void above a suspended ceiling, out of sight, and detailed to ensure an even sound field after transmission through ceiling. The sound generator is equipped with filters capable of compensating for spectrum changes on transmission through the ceiling and enabling the desired spectrum shape to be achieved. Zoning of speakers may be included to give different conditions in different parts of the room. The primary advantage is that of having control.

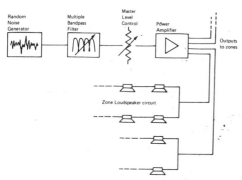

Fig 4 Diagrammatic representation of a typical sound conditioning system. (Sound Research Laboratories)

Printer enclosure (Acoustic Consultants & Engineers).

Reverberation Time

Office rooms can be generally classified as 'quiet' or 'noise producing'. A 'quiet' room is basically one requiring a low level of background noise for normal work to be carried out without distraction, annoyance or interference. It may, however, also be a 'noise-producing' room (*eg* have one or more typewriters in use), when it can require treatment to reduce reflected sound levels within the room (ie sound absorption treatment) to maintain a satisfactory background noise level. Reverberation time is also a significant parameter and the following are specific recommendations (consistent with British Standard Code of Practice CP3):

Small private offices	0.75 seconds or less
General offices Office machine rooms Public offices Banking halls	under 1 second
Canteens	less than 1.25 seconds
Large general offices Large banking halls	1.25 seconds maximum

Sound Field

In a room having reflecting surfaces on the floor, ceiling and walls, a 'reverberant' sound field is created which will have a more or less constant sound level at all points, irrespective of the distance from the source. This is the reason why offices with plastered walls and ceilings and uncarpeted floors sound noisy and acoustically unsatisfactory.

If the acoustically 'hard' surfaces can be replaced by absorbents, a 'semi-reverberant' sound field will be created which will give an improved rate of fall-off in level with distance. In open-plan offices it is obviously desirable that the theoretical maximum rate of fall-off with distance, (6 dB per distance doubling), should be approached, and a value of at least 5 dB per doubling is required. This can only be achieved if all the available surfaces are treated with highly efficient sound absorbents.

The sound absorbing properties of the floor covering should approach the maximum feasible performance. A good quality carpet laid over a felt or foam underlay should ensure that a target absorption value of 0.5 is obtained over the frequency range 500 to 2 000 Hz. The use of a carpet also helps to control surface noise (the noise of footfalls and furniture movements), and impact sound transmission to rooms below.

The ceiling which ordinarily constitutes the principal sound reflection path should be surfaced with a highly efficient sound absorbing material, especially if it is plain unbroken surface. Further improvements in ceiling performance can be achieved if the surface is broken up by coffering or baffles.

The ceiling structure should have an average sound absorption coefficient of at least 0.80 over the range 500 to 2 000 Hz.

In most purpose-designed large open-plan offices the wall area is small in comparison with the plan area and in consequence does not have a measurable effect on the overall sound field in the room.

Nonetheless, where possible, the walls should be treated with acoustically absorbing materials, particularly if a noise source such as a piece of office machinery is located close to it, otherwise there would be a tendency for the acoustically 'hard' surface to reinforce the reflected sound field.

Sound reflections off windows and glazed walls can be controlled by the use of drapes and blinds.

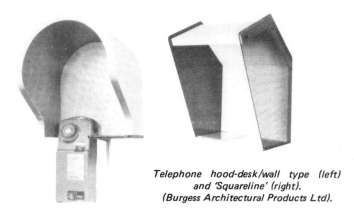

Telephone hood-desk/wall type (left) and 'Squareline' (right). (Burgess Architectural Products Ltd).

In small open offices where the walls are less than 15 m apart and consequently comprise a larger proportion of the total room area, it is essential that a sound absorbent surface treatment is applied.

This may take the form of sprayed acoustic plasters, acoustic tile on battens, or slabs of rockwool or glassfibre faced by slotted wood or plastic covers.

Under no circumstances should two acoustically reflective surfaces face each other, otherwise an objectionable 'flutter' echo will occur between them.

Problems with Common Suspended Ceilings

Modern building technique which conceals mechanical services above a lightweight suspended ceiling can lead to problems of flanking sound transmission through the ceiling void. This can nullify the effects of good sound insulating positioning. In such cases, particularly where demountable partitions are used for flexibility of layout, special attention may have to be given to the ceiling, even to the extent of installation of a plenum barrier above the partitions.

Screens

Screens are a significant feature of most open-plan offices. Their primary function is to break up sightlines and thereby provide privacy. This can sometimes be provided by furnishing such as files and bookshelves or even by large indoor plants. More commonly, however, purpose-built free standing screens are provided which, like other rigid objects in the furnishing scheme, can provide acoustic screening.

All too often in the past, much dependence has been placed on screens as remedial correction for a lack of acoustic privacy, often with disappointing results. To be effective, a screen should be located as close as possible to the source or to the recipient of sound. A screen placed within 50 cm of a noisy typewriter could give up to 15 dB sound reduction for an observer situated on the other side of the screen.

Screens erected in open offices to provide visual privacy seldom ensure acoustic privacy. They are normally used as the demarkation of territorial boundaries within the open plan and, as such, tend to be placed mid-way between the observer and an identified noise source. If the typewriter and observer referred to in the example above were separated by 8 m and a single free standing screen placed on the sightline, the reduction in sound level would amount to approximately 2 dB — a change that is barely detectable to the human ear.

NOISE IN COMMERCIAL BUILDINGS

To be effective, screens should only be used in acoustically treated rooms, otherwise sound transmission via reflections at the room surfaces will flank the unit.

In addition, the screens should always exceed a minimum size. Anything less than 1.25 m high is of little use and, in general, the larger the screen the better. Where possible, screens should be linked together to form a continuous barrier. In this way the sound sources — voices, telephones, typewriters, *etc* — are brought into close proximity with a screening surface and similarly any potential recipient of noise is screened by ensuring that the head position is located close to the continuous vertical surface.

If necessary, further improvements can be made in the screening of known sources such as typewriters, telex machines and similar office machinery. Either small supplementary side screens can be added on to the work stations or a three-sided barrier formed up from adjacent panel units.

The screens can be faced with sound absorbing materials in order to ensure that sound is not reflected away from the work station and, where necessary, arena-orientated layouts and semi-enclosed desks can be created to ensure that open offices in relatively small rooms, or with relatively low densities of population, are entirely feasible.

Individual Offices

Sound reduction treatment for individual offices involves specific attention to sound insulation of the boundary surfaces, the elimination of flanking transmissions and structural-borne sound, and attenuation of the ventilating system.

The typical average noise level in an office will tend to increase with the number of people working in the office. It may be desirable, or even necessary, to apply noise control at source, such as by the erection of partitions or partial barriers, or similar sound absorption treatment.

Some possibilities are:-

Acoustic treatment of ceiling — can be quite effective for simple treatment in large offices.

Partial barriers — locate cabinets, *etc* to mask local sound sources (most effective with acoustically treated ceilings and walls).

Room dividers — effectiveness depends on sound absorption realized. Simple glass partitions are relatively ineffective.

Part-height partitions — limited effectiveness.

Full partitions — more effective but may have local weaknesses.

Ceiling hung blankets — effective, but not usually desirable or practical.

Enclosure of quiet area — most effective with satisfactory acoustic treatment but may have local weaknesses (*eg* at window or door).

Doors — gasket seals around door area most effective. Avoid use of louvred doors.

Acoustic traps — may be applied to louvred doors or wall openings.

Partitions

The sound reduction achieved by any practical barrier is not constant at all significant frequencies in the audio frequency range (100–3 200 Hz), but if the partition is non-permeable to air, the attenuation increases with a rise in frequency at a rate which may be as low as 3 dB or as high as 12 dB for each octave increase in frequency. Partitions of homogeneous construction tend to have a performance similar to that shown in Fig 5, whereas cavity constructions generally have the more rapidly rising sound reduction frequency characteristics of Fig 6. The lightweight demountable types constructed from chipboard, plywood, or plaster panels on a timber or alloy frame, often

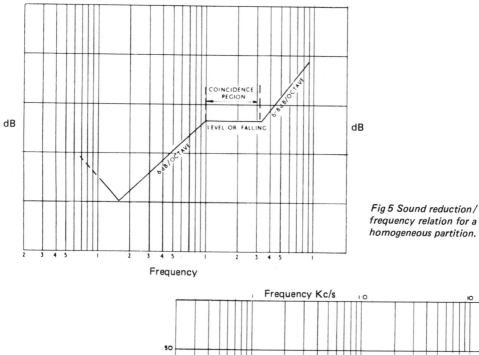

Fig 5 Sound reduction/frequency relation for a homogeneous partition.

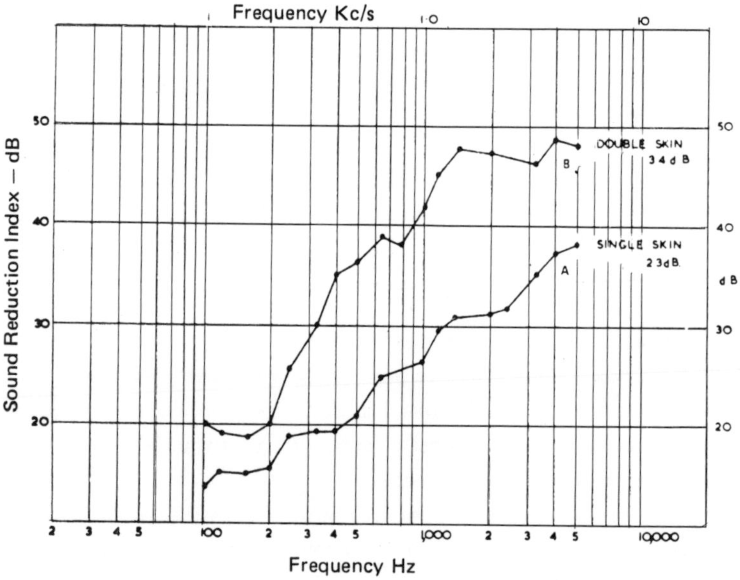

Fig 6 Sound reduction index of single and double leaf partition.

exhibit a phenomenon known as 'coincidence' characterized by the presence of a region in which the sound reduction may remain constant, or even decrease in frequency. The effect is less prominent in panelled partitions of cavity construction, the attenuation provided by the cavity and the sound absorbent material usually included, serving to minimize the attenuation dip so prominent in the homogeneous construction.

It is rarely economic to try to obtain sound reduction much higher than about 40–43 dB from demountable types of office partition. If higher values are suggested as necessary by the design procedures outlined in the earlier paragraphs, then other measures should be considered. An office for the secretary, a store, filing room or waiting room should be interposed between the source of noise and the sensitive area. Attenuation of between 50—55 dB may then be obtained, if a lower limit is not set by the mechanical ventilation system.

For sound reduction between rooms, a minimum attenuation of 45 dB is recommended, particularly between 'noise producing' and 'quiet' rooms. This minimum would normally be provided by a 115 mm (4½ in) thick brick wall, plastered, or a concrete floor weighing 220 kg/m^2 (45 lb/ft^2). Specific recommendations, in the case of adjoining rooms of similar occupancy are:

Rooms requiring quiet, on a quiet site — minimum 45 dB or 40 dB if lower degree of attenuation is tolerable.

Rooms requiring quiet, on a noisy site — minimum 40 dB

Clerical office — 20—30 dB.

Except for the need to achieve uniformity in appearance, there is no point in specifying a partition having a particularly good acoustical performance if doors, sliding hatches, or openable glazing must be installed. Unless of special design, any closeable opening will limit the difference in sound level between adjacent rooms to something in the region of 20 dB.

If high values of sound reduction are required from a partition that must include a door, then special doors, or a short isolation lobby with doors at each end, must be used. An excellent acoustic performance can be secured from a lobby less than 1 m long.

Intrusion of External Noise

However good the sound treatment of internal structures, performance is frequently modified by the presence of opening windows, where, on noisy sites, there is often an unfortunate choice between noise exclusion and ventilation. Here planning can provide the answer by foreseeing the problem and providing sealed noise-excluding windows and air conditioning in offices on the sides of the building exposed to high ambient noise.

Traffic noise, in particular, can be intrusive in city and urban areas, where even a fully sealed facade may not provide a complete answer. This is because although traffic noise is excluded (or at least reduced to insignificant levels) by the noise insulation partition of the facade, traffic indirect vibrations are present in the building.

See also chapter on *Resilient Mounting of Buildings* and *Acoustic Treatment of Floors and Ceilings.*

Noise in Domestic Buildings

BACKGROUND NOISE levels within individual domestic buildings is normally determined by *outside* noise and the degree of attenuation provided by the structure. It is thus very much dependent on the noise climate of the site. Self-generated noise within the individual detached building is unlikely to be objectionable, unless due to an over-noisy appliance, *etc*. Other noise will only be intrusive if it equals or exceeds the background level already present.

Maisonettes, flats and other multi-occupied domestic buildings inherently tend to be noisier due to 'other people's noise' intruding on the normal ambient noise level, and thus demand more stringent noise treatment in design and construction. This must take into account that internally generated noise, intrusive to other occupants, can be both airborne and structural borne. Required standards are set by Building Regulations and Codes of Practice which are derived to give a satisfactory (acceptable) performance.

In the case of multiple-occupancy buildings, concrete floors provide the best performance as regards resistance to the transmission of airborne and impact sound, although variations in design

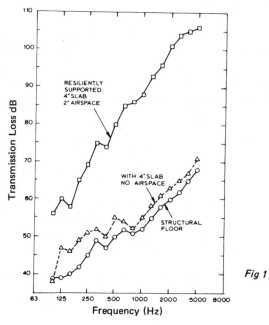

Fig 1

are numerous. Those may range from a basic floor slab through a resiliently supported slab with an air space to a fully floating floor with spring isolators and a resiliently suspended ceiling. Fig 1 shows the relative performance of a basic structural floor slab, a floor slab with an extra 100 mm (4 in) of concrete cast on top and the same structural floor with a 100 mm (4 in) slab resiliently supported to provide a 100 mm (4 in) air space between the two air masses. The enormous increase in performance possible when separating the two masses is clearly illustrated.

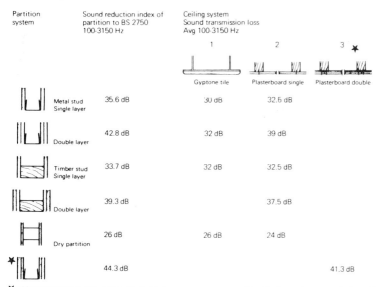

Fig 2 Sound transmission through various combinations of suspended ceiling/partition constructions. (British Gypsum Ltd).

By comparison Fig 2 shows the attenuation provided by various building board constructions for suspended ceilings and partitions. Further data on the performance of wall and partition constructions are summarized in Table I.

Insulation Between Houses

The sound insulation required between two attached dwellings depends on three basic criteria. These are the general background noise level in the area where the dwellings are situated, the noise levels and nature of noise created by the neighbours, and the sensitivity of the occupants to the noise from their neighbours.

The UK Building Regulations attempt to simplify this situation by setting a minimum standard of sound insulation on the basis that if this standard is achieved it will satisfy a good percentage of occupants. Thus, even if this standard was always met there would be a number of dissatisfied people demanding improved sound insulation. The regulations then attempt to simplify the position further by giving a 'Deemed to Satisfy' list of specific constructions which may be used without the need to test whether or not the performance standard is achieved. Fig 3 gives a graphical presentation of the 'Deemed to Satisfy' list. The fact is that apart from the dense concrete block walls, all the other constructions cited will have varying degrees of risk of failing to meet the performance standard.

TABLE I – PERFORMANCE OF WALLS AND PARTITIONS

Construction	Approximate Weight kg/m^3	Average Sound Reduction (100–3200 c/s) dB
Single-Leaf Walls or Partitions		
2 inch compressed straw slab	56	28
2 inch compressed straw slab with skim-coat plaster both sides	88	30
2¼ inch hollow slab consisting of two sheets of ⅜ inch plasterboard joined by cardboard eggcrate core	64	26
2½ inch hollow slab consisting of two sheets of ½ inch plasterboard joined by cardboard eggcrate core	96	29
4 inch hollow slab formed with two layers of ⅝ inch plaster joined by plaster honeycomb core	192	27
2 inch wood-wool slab unplastered	96	8
2 inch wood-wool slab plastered ½ inch both sides	224	35
¾ inch plasterboard with ⅝ inch thick plaster both sides (total thickness 2 inches)	208	34
2 inch thick solid gypsum plaster (reinforced)	320	35
2 inch hollow clay block (unplastered)	144	28
2 inch hollow clay block with ½ inch plaster both sides	272	35
3 inch hollow clay block with ½ inch plaster both sides	352	36
4 inch hollow clay block with ½ inch plaster both sides	400	37
2 inch clinker block with ½ inch plaster both sides	352	37–38
3 inch clinker block (unplastered)	320	23
3 inch clinker block with ½ inch plaster one side	400	39
3 inch clinker block with ½ inch plaster both sides	480	41
4 inch clinker block with ½ inch plaster both sides	600	43
8 inch hollow clinker block with ½ inch plaster both sides	560	42
8 inch hollow dense concrete block with ½ inch plaster both sides	800	45
4½ inch brick (unplastered)	720	35–40
4½ inch brick with ½ inch plaster both sides	880	45
9 inch brick with ½ inch plaster both sides	1600	50
13½ inch brick with ½ inch plaster both sides	2320	53
18 inch brick with ½ inch plaster both sides	3000	55
7 inch dense concrete with ½ inch plaster both sides	1520	50
10 inch dense concrete with ½ inch plaster both sides	2080	52
15 inch dense concrete with ½ inch plaster both sides	3000	55

Construction	Approximate Weight kg/m^3	Average Sound Reduction (100–3200 c/s) dB
Stud Framed Partitions		
½ inch fibre insulation board both sides	24	20–22
⅛ inch hardboard both sides	21	23
¼ inch plywood both sides	24	24
¼ inch asbestos wallboard both sides	48	28–30
¾ inch blockboard both sides	72	30
⅜ inch plasterboard both sides	64	30
⅜ inch plasterboard and plaster skim-coat both sides	96	32
⅜ inch plasterboard and ½ inch plaster both sides	224	35
3-coat plaster on wood or metal lath both sides	256	35–37
½ inch plaster on 1 inch wood-wool slab both sides	256	37
2 inch compressed straw slab both sides	112	33
2 inch compressed straw slab both sides and skim-coat	144–160	35
Double-Leaf Walls or Partitions		
Double 2 inch wood-wool slab with 2 inch cavity. Thin wire ties. ½ inch plaster both sides	320	42
Double 2 inch clinker block with 2 inch cavity. Thin wire ties. ½ inch plaster both sides	600	47
Double 2 inch clinker block with 6 inch cavity. No ties. ½ inch plaster both sides	600	50
Double 3 inch clinker block with 2 inch cavity. Thin wires. ½ inch plaster both sides	800	50
Double 3 inch clinker block with 3 inch cavity. Thin wire ties. ½ inch plaster both sides	800	50
Double 4 inch clinker block with 2 inch cavity Thin wires ties. ½ inch plaster both sides	1000	50
Double 4½ inch brick with 2 inch cavity Thin wire ties. ½ inch plaster both sides	1600	50–53
Double wall with 9 inch brick leaf and 4½ inch brick leaf. 2 inch cavity. Thin wire ties. ½ inch plaster both sides	2300	53
Double 9 inch brick wall with 2 inch cavity. No ties. ½ inch plaster both sides	3000	55

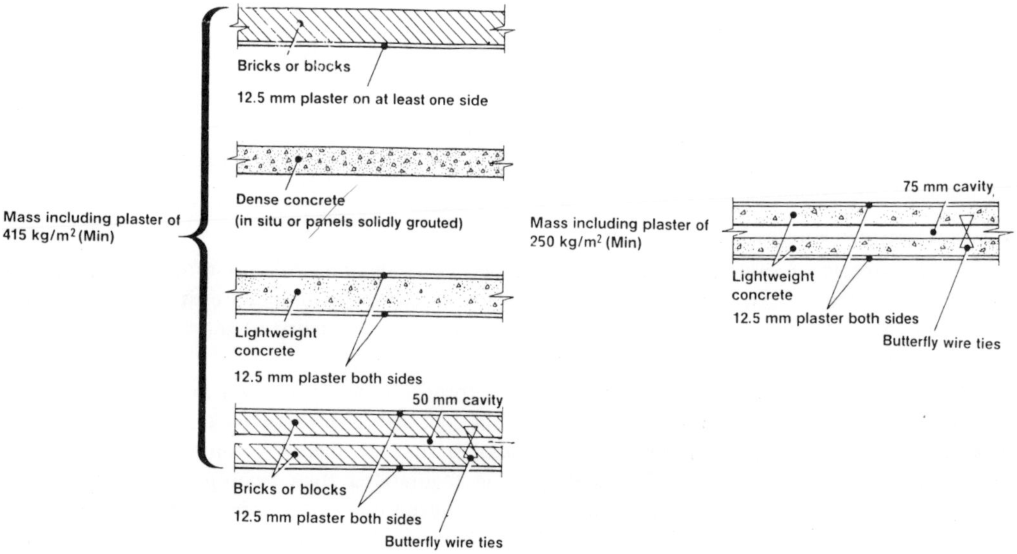

Fig 3 'Deemed to Satisfy' wall constructions. (Building Regulations)

The reason for failure can be generally placed into one of two categories. Either the inherent design of the wall and its surrounding structure is inadequate, even if well constructed, or some aspect of poor workmanship has caused a basically good design to be degraded. In the worst case, both these aspects may be linked — resulting in very poor performance.

It is therefore the case that a high percentage of occupants of attached properties are likely to have some cause for complaint about poor sound insulation.

The method of objectively evaluating the problem involves first of all making airborne sound insulation measurements to BS2750:1980 Part 4 between the two dwellings. This will determine the level of sound insulation between the dwellings which will give some indication of the sort of improvement necessary to solve the problem. An examination of the sound insulation curve may give some idea of the source of the problem. An examination of the sound insulation curve may give some idea of the source of the problem. A plateau at low to middle frequency is common for lightweight masonry walls and indicates a direct transmission problem. In fact, most masonry walls have a reasonably well defined sound insulation curve shape which gives an indication that workmanship is reasonable and that any problem is inherent to the system.

A low performance curve flattening off at high frequencies may be the result of leakage either through windows, ventilation openings or through some weakness in the wall itself, for example, at the base where the wall is unlined behind a skirting board. If the sound insulation curve gives no indication of the source of the problem then the radiation of power into the room by each surface may be investigated with an airborne noise source operating on the other side of the separating wall. This is most simply done using an accelerometer to estimate the sound pressure level L_p at 1/3 octave centre frequency (f) in the room (volume V) due to the sound power radiated by each surface (area S).

$$L_p = 20 \log (a/f) + 10 \log \left(\frac{ST\sigma}{V}\right) + 144$$

where
T is the reverberation time of the room where
σ the radiation coefficient of the surface

The radiation coefficient can be considered as unity above the critical frequency of the radiating panel but it reduces significantly at lower frequencies. Here it depends on the material, the fixing conditions and panel size and is often difficult to predict accurately. If the sound pressure levels of each surface within the room are added together then these should equal the measured sound pressure level in the room. This is a useful check on the accuracy of the prediction. If the prediction is wildly out then this may indicate that the problem is air path leakage.

If one particular surface is shown to be governing the overall sound pressure level of the room then this is where the corrective measures are necessary. If two surfaces are radiating with equal magnitude then it may be necessary to treat both, to improve the situation significantly.

Having isolated the problem it is then necessary either to correct it at source or to provide an additional lining to the surface(s) in question. The actual treatment can be taken from those outlined later which is a description of the course of action to take in a situation where measurements are either not possible or not practical.

The first course of action is to examine a number of basic details which may be the sources of potential weaknesses.

Permanently open window vents, louvred windows or air vents in walls may be the cause of a problem. If these are present they should be temporarily blocked off to see whether any improvement is perceived. If this is the case then action should be taken to reposition or acoustically attenuate the openings if these are essential for ventilation purposes.

Any cracks at ceiling, wall angles and window frame/masonry junctions should be sealed using a flexible acrylic sealant or plaster with a paper reinforcement.

Separating and external wall skirtings should be removed and gaps below plaster or linings filled with plaster or a flexible acrylic sealant.

Where floor joists run into the separating wall or external wall, the section of floor boarding adjacent to the wall should be removed and an inspection made of the joist/wall junction. If gaps are present around the joists then these should be filled with a sand/cement mortar. In order to avoid any possibility of the wall being porous then the whole area should be rendered with a sand/cement mortar or plaster.

Where floor joists run parallel to the separating wall or external wall the floor boarding/wall junction should be sealed with a flexible acrylic sealant.

If ceilings are found to be lightweight boarding, for example, foamed plastic cored board, they should be replaced or underlined with plasterboard or a material of similar surface mass (about 10 kg/m^2). An examination of the separating wall in the roof space will reveal any obvious holes or cracks which should be filled with sand/cement mortar and if necessary bricks or blocks. An amount of glass fibre mat laid on the back of the ceiling for thermal insulation purposes will help to dissipate the sound energy within the roof space.

If on checking and, where appropriate, rectifying the aforementioned details, the situation has not been improved it is highly likely that the separating wall itself, the external wall or both in combination are not capable of providing the level of sound insulation required. If an objective assessment is available then it may be apparent which surface(s) is the problem, but if not, then on the assumption that the houses are of traditional construction, it is generally found that improvement of the separating wall element is beneficial.

There are many ways of increasing the sound insulation of a masonry wall; the most efficient and cost effective employs the principle of erecting a lightweight independent lining adjacent to one face of the wall.

The requirements for the lining are:

It should be capable of providing a reasonable amount of sound insulation in its own right and should therefore have a mass in the region of 20–40 kg/m^2.

It should be structurally stable when free standing, *ie* fixed only at top and bottom such that there is no mechanical connection between it and the wall surface.

It should be well sealed at the perimeter and at joints between panels.

It should be set far enough away from the wall to minimize the effects of mass-airspring-mass resonances. In theory the spacing should be such that the resonance occurs well below 100 Hz but in many cases this is not possible due to practical constraints.

The use of a glass fibre mat hung in the cavity between the lining and existing wall has the effect of damping the airspring at resonance, thus minimizing its influence on the final result. The damping of the airspring also occurs above the resonant frequency resulting in improvements across the whole range.

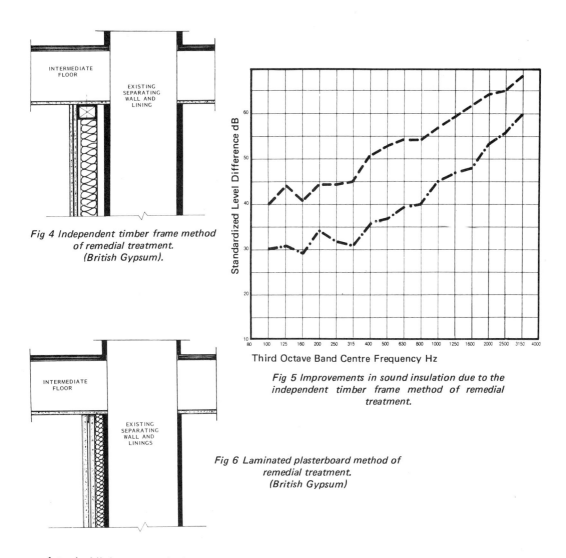

Fig 4 Independent timber frame method of remedial treatment. (British Gypsum).

Fig 5 Improvements in sound insulation due to the independent timber frame method of remedial treatment.

Fig 6 Laminated plasterboard method of remedial treatment. (British Gypsum)

A typical lining system is shown in Fig 4. This involves erecting a timber stud frame adjacent to, but not touching, one side of the separating wall. The frame is fixed to the floor and ceiling only and is set at a distance from the wall that will give the desired cavity width. A glass fibre mat is hung in the stud frame cavities and the frame is then lined with two layers of plasterboard with joints staggered between layers. All external joints and perimeters are sealed by a method which is unlikely to result in subsequent cracking occurring. Fig 5 shows the sound insulation before and after treatment in a case where the checks outlined earlier were carried out prior to the remedial work.

Fig 6 shows another method of remedial treatment. This comprises a double layer of 19 mm plasterboard, the first layer being fixed at the top, bottom and sides in metal channels and the second layer laminated to the first with a plaster adhesive. In an exercise to evaluate this lining,

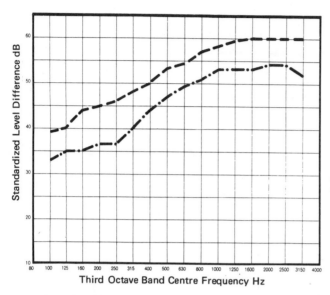

Fig 7 Improvements in sound insulation due to the laminated plasterboard method of remedial treatment.

a number of walls of the same basic construction were treated, the cavity between lining and wall varying from 25–100 mm. In each case a 25 mm glass fibre mat was hung in the cavity. No other examinations or remedial treatment took place prior to erecting the linings. In this instance improvements of similar magnitude were made in all cases, there being no evidence to suggest that the larger cavity was beneficial. Fig 7 shows the typical performance before and after treatment for a number of walls where the lining was set with a cavity width of 25 mm.

If it is found necessary to line the external wall then linings similar to those recommended for the separating wall may be used.

Internal Treatment

The sound climate of any room in a dwelling is appreciably modified by the presence of furniture, curtains, and floor covering; also by any additional treatment applied to walls and ceilings (which is unusual). In particular this modifies the reverberation time of the room; and in the case of floor coverings reduces internally generated impact noise (*ie* footstep noise).

TABLE II – AIRBORNE SOUND COEFFICIENT OF SELECTED FLOOR COVERINGS

Type of Floor Covering	Noise Reduction Coefficient	Absorption Coefficient			
		Low Frequency		High Frequency	
		125 Hz	500 Hz	2000 Hz	4000 Hz
Carpet only	0.35	0.02	0.15	0.58	0.63
Carpet and 40 oz H.D. foam	0.50	0.08	0.58	0.70	0.71
Carpet and 80 oz H.D. foam	0.60	0.13	0.67	0.72	0.74
Carpet and 40 oz hair felt underlay	0.55	0.08	0.55	0.72	0.75
Wood block flooring	0.05	0.05	0.05	0.10	0.10
Linoleum	0.05	0.05	0.05	0.10	0.10
Concrete floor	0.00	0.01	0.02	0.02	0.03

Carpeting, in fact, offers the possibility of achieving a noise reduction coefficient of up to 0.60, or roughly three times that of a wood block floor or hard faced covering — *eg* see Table II.

A further value of carpeting is that its sound absorption is greatest at the higher frequencies, which are usually the most irritating content of the noise spectrum, see Fig 8. Carpet is particularly effective in reducing impact sound transmission through a floor, and surface noise radiation generated in the floor of a room itself (by scraping of furniture over the floor, footsteps, *etc*).

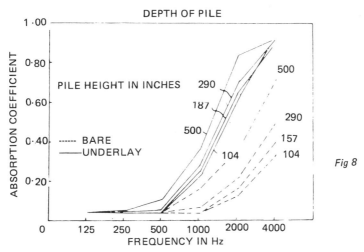

Fig 8

Treatment for External Noise

Recommendations for overall sound insulation treatment of a house against external noise are summarized in Fig 9, showing modifications which could be carried out. These include:-

1. Sound insulation into bedrooms with
    (a) house fully treated (— ·· — ·· —)
    Average SRI = 45 dB

    (b) plasterboard/Cosyfelt treatment replaced by mineral wool pugging (10 kg/m^2) (- - - -)
    Average SRI = 41 dB

    (c) no treatment (· — · — ·)
    Average SRI = 24 dB

2. Sound insulation into bedrooms with
    (a) house fully treated (— ·· — ·· —)
    Average SRI = 45 dB

    (b) plasterboard/Cosyfelt treatment replaced by 50 mm glass wool (········)
    Average SRI = 40 dB

    (c) no roof insulation (———)
    Average SRI = 39 dB

    (d) no treatment (· — — · — — ·)
    Average SRI = 24 dB

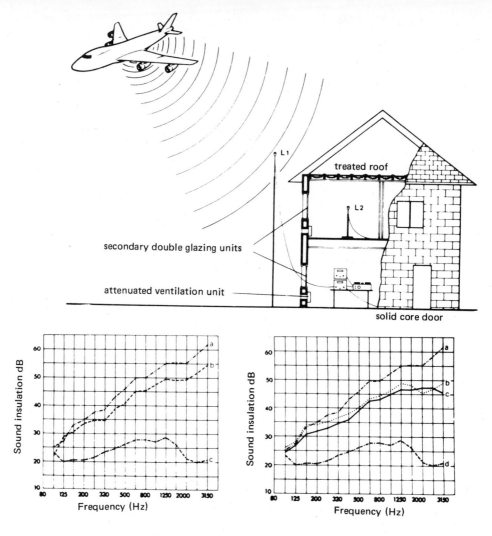

Fig 9 Sound insulation treatment of a house against external noise.

Product Names

Cosyfelt	Standard grade 1 000 mm wide x 50 mm thick (D. Anderson & Son Ltd).
Mineral fibre	Fibreglass Crown 75, 50 mm thick or equivalent.
Plasterboard	Gyproc wallboard or Gyproc lath 12.7 mm thick (British Gypsum Ltd).
Roofing felt	Thermoglass 36 kg/20 m^2 (D. Anderson & Son Ltd).

(Cosyfelt consists of glass wool insulation bonded to Anderson reinforced underslating felt).

See also chapter on *Acoustic Treatment of Floors and Ceilings*.

Auditoria

SOUND ABSORPTION treatment in the case of auditoria, churches, concert halls, *etc* can have a two-fold purpose. First it can be effective in reducing background noise to a suitable level for speech intelligibility. It can also have a marked effect on the quality of the sound generated in the room by controlling reverberation. Only a limited amount of absorption producing a small reduction in loudness is then normally required because excessive sound-deadening within the auditorium would be undesirable, especially for music. Thus the primary application of sound absorbents in such cases is to achieve an optimum reverberation time.

Reverberation time can be calculated by the formula:

$$T \text{ (seconds)} = \frac{KV}{A}$$

where

V = volume of enclosure
A = total absorption in sabines
K is a constant
= 0.05 for V in ft^3
= 1.796 for V in m^3

A range of optimum reverberation times can be broadly defined for rooms of different purposes and volumes. These are shown in Fig 1 for optimum reception of frequencies of 500 Hz and 2 000 Hz (*ie* principally speech). A multiplying factor is also given for improved reception of low frequency sound (125 Hz). An alternative recommendation applying a correction factor over a wider range of speech frequencies is shown in Fig 2.

In simpler terms, noticeable interference with speech intelligibility can be present in larger halls where the reflected sound paths and direct sound paths to the position of the listener differ by more than 17 metres. On this basis there tends to be areas within the hall where speech reception is poor, whilst in other areas it may be satisfactory. To a large extent this is controllable by the interior shape of the hall. Slight out-of-parallelism of walls, for example, can prove extremely beneficial.

In the overall picture, subjection response to reverberation time normally agrees with the following:

Under 0.5 seconds — speech muffled and inaudible; music 'dead'.
0.5–1.0 seconds — speech reception good but music still 'dead'.

1.0–1.5 seconds – speech reception still quite good; music reception better, but only fair.
1.5–2.0 seconds – speech reception only fair but music good (especially choral music).
Over 2.0 seconds – speech reception increasingly poor; music reception also increasingly poor, except for organ music.

See also Tables I and II.

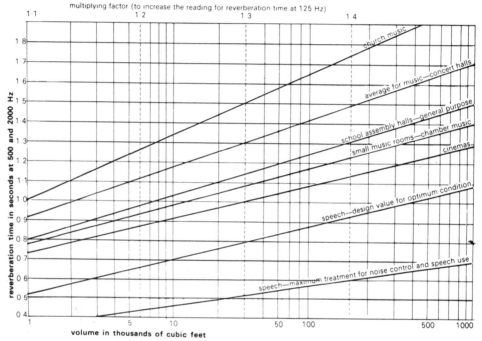

Fig 1 Desirable reverberation times for speech and music related to room volume.

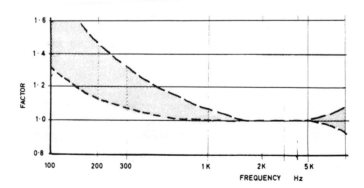

Fig 2 Correction factors for optimum reverberation time dependent on frequency of sound.

TABLE I – MUSICAL AND ACOUSTIC FACTORS RELATED

Musical Factors	Acoustic Parameter(s)	Control	Remarks or Requirements
Presence (intimacy)	initial-time-delay	size of room	less than 20 milliseconds desirable
Loudness (direct sound)	distance	seat arrangement	seats preferably steeply ramped in large halls
Loudness (reflected sound)	(i) intensity of reflections (ii) reverberation time	(i) absorption of surfaces (ii) size and surfaces	optimum balance between volume and reverberation time
Liveness	reverberation time	(a) volume (b) absorption of surfaces	balance achieved particularly affects middle and upper frequencies
Fullness of tone	(i) reverberation time (ii) cross reflections	reverberation time	fullness of base and warmth when reverberation time for lower frequencies is greater than middle or higher frequencies
Timbre (tone quality)	(iii) ratio of loudnesses (iv) speed of playing acoustic environment	direct and/or reflected sound performer(s) all factors affecting acoustic environment	effect related to acoustical conditions ideally should not vary with position
Definition	(i) initial-time-delay (ii) reverberation time (iii) volume (iv) distance	amount of reflections absorption present room size seat placement	should be short should be small should be small should be short
Brilliance	acoustic environment	all factors affecting presence, liveness and definition	
Balance	reflection and diffusion	arrangement of reflective and diffusive surfaces	suitable arrangement of reflective and diffusive surfaces
Diffusion	equal omni-directional reflected sound waves	(a) reverberation time (b) nature of reflective surfaces	
Blend	sound mixture	reflection and diffusion in neighbourhood of source	
Ensemble	acoustic environment in neighbourhood of orchestra	(a) geometric proportions of stage (b) slope of stage floor (c) conductor	musicians ability to hear one another
Attack (immediacy of response)	(i) reflections (ii) reverberation time (iii) diffusion	projection of reflections	
Texture	pattern of sound reflections superimposed on general performance	control or elimination of echoes	later sound reflections should follow short-delayed first reflections uniformly
Echo	degree of echo	reflecting surfaces	echo should be eliminated
Noise	degree of extraneous noise	(i) reduction of interior noise (ii) isolation or insulation from exterior noise	
Dynamic range	spread of audible sounds	control of sound levels	(i) loudness of fortissimo (ii) relation of background noise to loudness of pianissimo
Uniformity	overall acoustic performance		good hearing conditions desirable at all positions

TABLE III – SHAPES FOR MUSIC AUDITORIA

Shape	Acoustic Advantage(s)	Acoustic Disadvantage(s)*	Remarks
Rectangular	good cross reflections	flutter echo likely Room resonance may be apparent	traditional for orchestras
Circular	none	echoes, long time-delayed reflections, hot-spots	often associated with domed roof which can further degrade the acoustical properties
Horseshoe	good definition (but not for orchestral music)	tends to be 'dead' (low reverberation times)*	traditional for operas
Fan	good balance, good presence and loudness	long time-delayed reflections* Echoes and hot-spots	
Irregular	maximum intimacy, definition and brilliance		
Combined Shapes	can include advantages of all types	can include disadvantages of all types*	

*May be minimized by suitable treatment

TABLE II – ACOUSTICAL DEFECTS

Defect	Dependency	Remarks
Echo	(i) size and shape of reflecting surfaces (ii) relative position of sound source and listener (iii) type of sound programme	echo results if time interval between direct and reflected sounds originating from same source is one-tenth to one-twentyfifth second
Short-delay echoes	(i) size and shape of reflecting surfaces (ii) relative position of sound source and listener (iii) type of sound programme	result in blurring or masking of direct sound
Flutter echo	parallelism of reflecting surfaces	sound sources should not be located between parallel reflecting surfaces
Distortion	balance of absorption characteristics of boundary surfaces	sound quality can be damaged by excessive or unbalanced absorption
Shadow	sound shadowing effects by banners or overhangs	reduce sound levels in shadowed areas
Room resonance (colouration)	parallelism of reflecting surfaces at critical distances (ie sub-multiples of the wavelength of sound)	can set standards for room sizes and proportions
Hot spots	sound reflections from concave surfaces	produces local areas of sound concentration. Can also produce 'whispering gallery' effects
Coupled spaces	adjacent rooms with interconnecting air volumes	can modify the reverberation characteristics of the auditorium

TABLE IV – RECOMMENDED ROOM VOLUMES FOR DIFFERENT TYPES OF HALLS
(Room volume per person in m^3)

	Minimum	Optimum	Maximum
Rooms for speech, lectures, etc		2.8	5
Cinemas		3.1	4 4.2
Churches	5.7	7–10	12
Concert halls	6.5	7–7.5	10
Opera houses	4	4–5	6

A fuller or richer quality is given where the reverberation time for lower frequencies is greater than that for middle and higher frequencies, although this is more generally noticeable with music rather than speech. However, variable absorption characteristics, giving reverberation times varying with frequency, can interfere with speech to the extent that there can be excessive absorption of consonant or vowel sounds. For general speech quality, therefore, it is desirable that the absorption coefficients of finishes used in acoustic treatment should be as uniform as possible over the frequency range 250–7 000 Hz.

Subtleties of speech are provided by good musical qualities defined as 'presence' or 'intimacy'. As a general rule this is gained by reinforcement of direct sound waves with reflected waves, with a very short initial time delay — eg of the order of 5 ms, or a difference in sound path of about 10 m (33 ft).

Account must be taken of the directional characteristics of typical speech. Higher frequency sound levels fall off rapidly outside about 140° subtended at the position of the speaker. Hence, ideally, seats should be laid out in a pattern falling within this section (see also Table III).

Reflecting Surfaces

Without amplification, the acoustic power of human speech lies within the range of 10 to 50 microwatts. For adequate reception sound paths to the audience should be as direct as possible to reduce sound losses in the air. Basically this implies a compact room shape with a low volume per seat (eg see Table IV) a raised speaker's platform, and the elimination of all physical obstructions between speaker and audience (eg in an open room, a sloping floor rising to the back). A curved ceiling shape can also help by working as a reflector to provide more uniform distribution of sound energy — Fig 3.

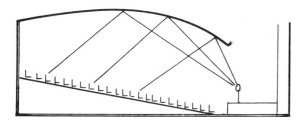

Fig 3 Curved reflective ceiling and inclined seating levels.

In the case of larger halls or auditoria, even with favourable room acoustics the speech level may still be too low for satisfactory listening. In this case a sound amplification system can be installed to achieve a high degree of speech intelligibility. Recommendations in this respect vary considerably and are also influenced by the type of auditorium covered and the normal background noise level.

As a general guide, a sound amplification system is probably desirable for high intelligibility once the room volume exceeds 1 414 m^3 (50 000 ft^3) where the background noise level is likely to exceed 40 dB. If the background noise level is higher, sound amplification may be necessary with smaller volumes, particularly if the background noise content covers frequencies in the normal speech range, the acoustic properties of the room are relatively poor or the walls are treated with sound absorptive materials. On the other hand, for acoustically well designed auditoria with no intrusive noise, considerably larger rooms may prove suitable without resort to sound amplification systems.

Apart from overall increase in sound levels, sound amplification can also be useful for other purposes.

Control of Acoustics

It is possible to provide flexible acoustics by large scale versatility in the building fabric (*eg* swivelling and height adjustable acoustic panels), but the cost of this becomes increasingly prohibitive as the size of the hall increases. It is thus more usual to design auditoria specifically for either speech or music.

The basic requirements for good listening are:

(i) *Absence of background noise* — exclusion or reduction to a suitable level of all extraneous noises.

(ii) *Adequate loudness* — extending to all parts of the auditorium.

(iii) *Uniform distribution of sound* — ideally sound should be equally loud in all hearing positions.

(iv) *Optimum reverberation time* — related to size of room, and the type of performance, (*eg* speech or music).

(v) *Optimum acoustic qualities* — exclusion or reduction to a suitable level of all extraneous noises.

Of these criteria, (i) and (iv) can be directly controlled by sound absorption treatment; and (v) largely controlled by such treatment.

Background noise level is usually rigidly controlled by sound insulation applied to all surfaces of the enclosure, as necessary. The suppression of noise from ventilating ducts and grilles may be particularly important in this respect. Background noise levels may required to be held down to 40 dB, or better, over all frequencies within the audible frequency range.

Lecture Halls

Lecture halls can vary widely in size. Larger lecture halls can be regarded very much as theatres, posing the same acoustic design problems as listed previously. Provided the acoustic design is good, sound amplification systems should not be required until the size exceeds about 1 416 m^3 (50 000 ft^3), or the hall accommodates an audience of more than 500. In all cases the reverberation time is probably best calculated on the basis of a 2/3 audience attendance (*ie* for average conditions which assume that the lecture hall will be two-thirds full).

Modern lecture halls are normally designed with particular attention to the exclusion of external noise, with consequent exclusion of natural light and ventilation. This can place a premium on the design of ceiling units carrying necessary artificial lighting, and on achieving suitable silencing in ventilation and air conditioning services to the room.

Smaller lecture halls forming part of a building may demand special attention to exclude external noise generated in adjacent rooms, with particular study given to flanking transmissions.

Dual-Purpose Halls

The acoustic treatment of halls intended for delivery of both speech and music does not generally present particular problems until the volume exceeds about 3 000 m^3 (10 600 ft^3). In other words, the reverberation time can be a compromise between 'speech' and 'music' requirements — and still give acceptable listening with either — *eg* a reverberation time of between 1 and 2 seconds. If the dual purpose hall is used mainly for speech, the lower reverberation time would be best; if mainly music, a reverberation time approaching or even slightly exceeding 2 seconds, depending on the type of music to be accommodated.

It must be pointed out, however, that a compromise reverberation time is not likely to produce entirely satisfactory results with either speech or music.

Classrooms

Classrooms do not normally present any severe acoustical design problems, except as regards necessary sound insulation. This can be analyzed in terms of occupancy as in Table V and adjacency. The latter is concerned with the minimum noise reduction required between rooms. British Standard recommendations are:-

Between rooms of Class A — 25 dB

Between rooms of Class C or D — 35 dB

Between rooms of Class B or E — 45 dB

Between any two rooms of difference classes — 45 dB

When a room has a dual use, it is classified according to the more severe requirement, *ie* a Class C room, also used at times for Class B duties, would be considered as Class B.

Classroom volumes normally lie between 226 and 340 m^3 (8 000 and 12 000 ft^3), with corresponding floor areas ranging from about 46—93 m^2 (500—1 000 ft^2). (Proportions are usually rectangular). Acoustic treatment is seldom necessary to achieve optimum reverberation times of 0.6 to 0.9 seconds when full.

TABLE V — CLASSIFICATION OF SCHOOL ROOMS

Class	Type	Typical Uses
A	Noise producing	Workshops, kitchens, dining rooms, gymnasia, boiler rooms
B	Noise producing, but needing quiet at times	Assembly halls, lecture rooms, music rooms, commerce and typing rooms
C	Average	General classrooms, laboratories, offices
D	Rooms needing quiet	Libraries and study rooms
E	Rooms needing privacy	Medical and staff rooms

Noise in Ships

IT IS almost traditional that high levels of noise and vibration appear generally acceptable on ships, 'treatment' being to allocate the status of passenger cabins on a broad scale determined by comfort level (or at the lower end of the scale, 'discomfort level'). This attitude has changed in more modern designs which consider the 'health and safety' aspect of crew members as well as passengers, although there are relatively few authoritative guidelines available. Predominant amongst these are Noise Protection in Ships, Swedish Administration of Shipping and Navigation Regulations and Recommendations, originally published in 1973; and Code of Practice for Noise Levels on Ships, Department of Trade 1978.

Whilst hearing damage is obviously a very important factor for some crew members, distraction and interference with task performance can become the prime consideration for others involved in watch-keeping functions requiring long-term concentration, such as the helmsman and navigator. Speech intelligibility will also be very important in some cases, requiring suitably low background noise levels in, for example, the radio room and on the bridge. In messrooms and crew accommodation, ambient conditions should reflect the need for relaxation and freedom from intrusion. However, the margin between practical and desirable criteria in these spaces can be largely due to their use, in many cases, as buffer zones between noisy machinery spaces and farepaying passenger accommodation. Even here, however, there is likely to be a discrepancy between that which is practically attainable and the desirable conditions. Table I sets out a tentative guide to desirable and practicable criteria for maximum noise levels in various spaces on board ship.

It will be seen from this that there can be appreciable discrepancies between desirable and attainable noise levels. The particular problems with a given space will vary from ship to ship and, in some cases, the desirable criteria can be achieved, whilst on many ships the practicable criteria are exceeded.

Noise in accommodation spaces due to the propulsion and auxiliary machinery is probably more readily tolerated than noise from cabin services or from passengers in adjoining cabins. This arises largely from the acceptance of noise as a by-product of the necessary methods of propulsion if the voyage is to be completed in a satisfactory time. Although noise from air conditioning may be a source of irritation, it would be unwise to set too low a noise criterion because of the useful masking properties, which help to ensure privacy. As the sound insulation of inter-cabin partitions and the transmitted machinery noise levels are improved, so then can the cabin services' noise be attenuated towards desirable levels.

The most objectionable form of noise reaching accommodation is that due to conversation in adjoining cabins or in corridors. Bulkhead sound insulation should be such as to ensure that conversation in adjoining space is, at least unintelligible and preferably inaudible. It is known that

TABLE I – GUIDE TO SHIP-BORNE NOISE CRITERIA

Space	Desirable level dB(A)	Practicable level dB(A)
Passenger cabins	35	45–50
Crew accommodation	35	45–55
Day cabins	50	50–55
Sick bays	35	45
Dining saloons	50	55–60
Cinemas/theatres	30	45–50
Lounges	50	50–60
Messrooms	50	65–70
Radio room	50	60–65
Wheelhouse/Bridge	45	60–65
Machinery spaces (continuous occupation)	75	85–90
Workshops	75	85
Machinery spaces (non-continuous occupation)	85	100–105

privacy is determined by the interaction of three complementary factors, namely: the source room speech level, the bulkhead sound insulation, the receiving room background level.

Machinery Noise

For the ships' gear, as for other machines, it is often not practicable to achieve large noise reductions by modification of the machine itself. The control of noise is usually limited to the reduction in the transmission of machine noise to the listener. This may be achieved by the application of sound-absorbing treatments to the surface of the machinery space to reduce the general level of noise at points remote from the particular machine. This procedure is likely to be effective only when the space is initially highly reverberant; reductions in excess of 3 dB are, even then, unlikely. Alternatively, a noise reduction may be achieved by a hood or enclosure surrounding either the machine or the observer. Such a hood or enclosure should carry an effective sound-absorbing treatment internally to reduce the build-up of noise due to enclosure of the machine, which tends to counteract the noise reduction of the enclosure walls. Similarly, an enclosure surrounding a manned station should carry sound-absorbing material to minimize the build-up of sound penetrating the enclosure walls.

Noise may be transmitted from machinery spaces to accommodation in two ways. First, by the transmission of airborne sound in the machinery spaces and, secondly, by the transmission, through the ship's structure, of vibration from the machinery; this is subsequently radiated as noise from the bulkhead and deck surfaces in the accommodation area. That part of the noise which originates as airborne sound in the machinery does not necessarily travel all the way to the accommodation as airborne sound. It may set the casing of the machinery space into vibration and travel in this form. Adjacent to the machinery spaces, the sound insulation of the bulkheads and decks is of paramount importance. If the level of noise in a machinery space is of the order of 100 dB(A), the sound insulation to be provided by a bulkhead, if an adjacent space is to be used as accommodation, must be at least 45 dB. This would be extremely difficult to achieve if the reduction were required at low frequencies — it would be difficult enough on land.

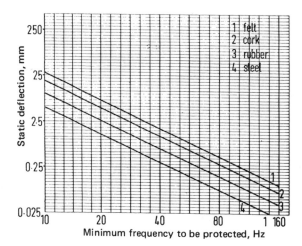

Fig 1 Guide to static deflection versus isolation frequencies for various materials.

Propulsion gas turbine enclosure for LM2500 marine turbines on Spruance class destroyers (Environmental Elements Corporation)

Whenever possible, machinery spaces should be designed so as to be separated from accommodation by quieter machinery spaces or by storage spaces. Hospitals and sick bays should never be adjacent to machinery spaces. With slow speed machinery, such as certain types of diesel or compressor, most of the noise reaches the accommodation as a result of the transmission of machine vibration. At low frequencies, this is particularly difficult to suppress. Fig 1 shows the minimum deflection under static loading which must be provided if vibrations in excess of a certain frequency are to be suppressed. From this, it is clear that to suppress the vibration of machinery rotating at 600 rev/min it is necessary to include mountings with a resilience which would be unacceptable to the marine engineer. At higher speeds, such as 1 500 rev/min and certainly 3 000 rev/min the necessary deflections are much smaller and should be readily acceptable. This point emphasizes the desirability of a trend to higher speed machinery. It should also be pointed out that the control of high frequency noise within the machinery space itself is a much easier proposition than that of low frequency noise.

The suppression of vibration should be extended to all connections between the isolated machine and the ship's structure. Flexible connections or joints should be included as close to the machine as possible. These connections are likely to be least effective where they are included in pipes carrying liquids, since pressure pulsations in the liquid will be transmitted in spite of the flexible coupling. High frequency pulsations will be more quickly damped and this also illustrates the advantage of high speed machinery. The suppression of noise originating as airborne noise in the machinery space may be attempted in two principal ways:

(i) By suppressing the level of noise in the machinery space itself.
(ii) By reducing the excitation of the structure adjacent to the machinery by the application of vibration-damping materials which are effective in the frequency range of interest, or by the use of a lining to the machinery space which is isolated from the main structure, a sound-absorbing material being included in the cavity between the two. For effective noise abatement by this latter method, care must be taken in the details of the fixing of the inner skin.

The transmission of vibration through the structure will also be influenced by the constructional techniques and sections involved in a particular design. This is an area in which continued research is leading to improvements in design details.

Passenger Cabins

From the point of view of reducing cabin noise to a desirable 50 dB(A), it is likely that bulkhead construction will be necessary giving an average sound reduction loss of 35—45 dB. The lower limit can be achieved by single leaf bulkheads of the appropriate weight and mechanical characteristics, but to achieve the upper limit, it is necessary to turn to cavity constructions. Since the interest is essentially in the higher frequency performance, speech being the most important, the cavity may be kept small and the required performance is attainable with an overall thickness of 50 mm and a superficial density of the order of 20 kg/m^2.

Sound insulation of the order discussed cannot be achieved in simple doors, particularly if these include vents or louvres. A slightly lower level of insulation to the corridors may be tolerated but, ideally, a lobby, however small, with a door at each end should be incorporated. The inner door could usefully be a folding door and the lobby have the minimum length to allow the outer door to be opened. The surfaces of the lobby should be treated with a high performance acoustic tile. However, even if a suitable bulkhead partition design is achieved, it is possible that flanking transmission will become an important factor. Some measurements have indicated that in conventional designs, flanking paths may limit the attainable average insulation to about 40 dB. Developments in reducing vibration transmission through the structure may help the partition potential to be attained in practice.

See also chapters on *Acoustic Enclosures, Sound Insulation and Absorption, Acoustic Materials* and *Silencing Gas Turbines.*

FIRST EDITION

Handbook of Industrial Materials
Contents:
Section 1, Ferrous Metals; Cast Iron, Grey Irons, Malleable Irons, Nodular Irons, Alloy Irons, Carbon Steel, Alloy Steels, Chromium Steels, Vanadium Steels, Nickel Chrome Steels, Chrome Vanadlum Steels, Molybdenum Steels, Stainless Steels, High Temperature Steels, High Strength Steels, Selection guides, equivalent specifications, etc. Section 2. Non-Ferrous Metals; Aluminium and Aluminium Alloys, Antimony, Arsenic, Barium, Beryllium, Bismuth, Boron, Brasses, Bronzes, Cadmium, Caesium, Calcium, Cerium, Chromium, Cobalt, Columbium, Copper and Copper Alloys, Gallium, Germanium, Gold, Hafnium, Indium, Iridium, Lanthanum, Lead and Lead Alloys, Lithium, Magnesium and Magnesium Alloys, Manganese, Mercury, Molybdenum, Nickel and Nickel Alloys, Niobium, Osmium, Palladium, Platinum, Plutonium, Potassium, Radium, Rhenium, Rhodoum, Rubidium, Ruthenium, Scandium, Selenium, Silicon, Silver, Sodium, Strontium, Tantalum, Tellurium, Thallium, Thorium, Tin and Tin Alloys, Titanium, Tungsten, Uranium, Vandium, Ytterbium, Yittrium, Zinc and Zinc Alloys, Zirconium. Section 3. Non-Metallic Materials; Acids, Alcohols, Alkalies, Aromatics, Asbestos, Asphalt, Carbon, Carbides, Ceramics, Cermets, Composites, Cork, Elastomers and Plastomers, Felt, Glass, Structural, Glass, Fibre, Cloth and Material, Leather, Mica, Paper, Rare Earths, Resins, Natural Resins, Synthetic, Quartz, Rubber, Natural, Rubber, Synthetic, Silicones, Wood, Natural, Wood, Laminated, Wood, Processed. Section 4. (A) Plastics (Thermoplastic); Acetals, Acrylics, ABS, Celluostic, Fluorocarbons, Ionomers, Polymides (Nylons), EVA, Methyl Pentane (TPX), Polycarbonate, Polyethylene, Polypropylene, Polystyrene, Polysulphones, PPO, Polyurethane, SAN, PPS, Sulphone, PVC and others. (B) Plastics (Thermoset); Alkyds, Aminos, Ephoxides, Phenolics, Melamine, Polyester, PBS, Urea Formaldehyde and others. (C) Plastics (Processed, etc); Laminated Plastics, Filled Plastics, Reinforced Plastics, Cellular Plastics, Plastics Selection guides, etc. Section 5. Adhesives, Bearing Alloys, Casting Alloys, Coatings, Metallic, Coatings, Non-Metallic, Cloth, Composites, Concrete and Cement, Coolants, Detergents, Dielectric Materials, Electrical Contact Materials, Enamels/Paints/Varnishes, Exotic Alloys, Expanded Materials, Fuels, Gasket Materials, Grits and Powders, Greases, GRP and CFRP, Insulating Materials, Electric, Insulating Materials, Acoustic, Insulating Materials, Thermal, Lubricants, Magnetic Materials, Magnetostrictive Materials, Manmade Fibres, Perforated Materials, Petroleum Products, Piezoelectric Materials, Porometric Materials, Powdered Metals, Proprietary Alloys, Refractory Materials, Sealants, Solders and Low Melting Point Alloys, Solvents, Tubes and Tubing, Waxes, Welding Rods, Wire, Electrical Wire, Braid, etc.

OVER 600 PAGES, HARD CASE BOUND, GOLD BLOCKED, WITH DUST COVER

ORDER YOUR COPY
NOW

TRADE & TECHNICAL PRESS LTD., CROWN HOUSE, MORDEN, SURREY SM4 5EW, ENGLAND.

NOISE  CONTROL

As qualified consultants and practical engineers we provide a total noise control service. From comprehensive measurement surveys and analysis to the complete design and installation of sound and vibration reducing systems. Let our experience help you to meet your company's obligations under the new and pending Noise Control Legislation.

- Personnel Enclosures • Machine Enclosures • Control Rooms • Audiological Booths • Acoustic Panels • Suspended Absorbers • Acoustic Doors • Mobile Acoustic Screens • Noise Barriers
- Anti-vibration Systems • Louvres
- Splitter Attenuators • Circular Attenuators

SURVEYS • ADVICE
SOLUTIONS • INSTALLATIONS

St. Anne's House, North Street,
Radcliffe, Manchester M26 9RN.
Telephone: 061-724 9511
Telex: 665449 SEDOL G

ATMOSPHERIC CONTROL ENGINEERS LTD.

SECTION 5a

Sound Insulation and Absorbtion

THERE IS a clear distinction between sound *insulation* and sound *absorption*. A *sound insulating* material (or structure) attenuates sound waves passing through it and thus acts as a 'sound barrier' — *eg* to the passage of sound from one room to another. A *sound absorbent* material, on the other hand, absorbs a proportion of the sound emerging incident on it so that the level of sound reflected from the surface is substantially reduced. Basically, therefore, sound absorption is intended to reduce the loudness of reflected sounds *in* a room or enclosure, and decrease reverberation. At the same time a high proportion of sound energy may be transmitted *through* the absorbent surface, as sound absorbents are, in general, poor sound *insulators*.

This difference is also exemplified in the respective quantitative performance parameter. The performance of a sound *insulating* material is expressed in terms of sound transmission loss (TL) or sound reduction index (R); or as a sound *transmission coefficient* (T). The performance of a sound absorbent material is normally expressed in terms of its *absorption coefficient*.

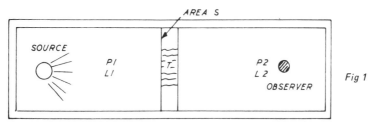

Fig 1

Sound Insulation

The principle of *sound insulation* is shown in Fig 1. A sound source located in one volume generates a sound level P_1 and is separated from another volume by a panel of sound insulating material of area S, with a sound transmission coefficient of T. The observer, in the second volume then receives a lower sound pressure level of P_2, the difference $(P_1 - P_2)$ being the result of sound passing through the insulating material.

The actual sound transmission loss $(P_1 - P_2)$ or sound reduction index (R) is given directly by:

$$R = 10 \log_{10} P_1/P_2 \text{ dB}$$

where

P_1 = sound pressure level on sound source side.

P_2 = sound pressure level on the opposite side.

The ratio P_1/P_2 is thus the ratio of transmitted sound energy to incident sound level, called the *transmission coefficient*.

The aforementioned formula can thus be re-written:

$$(P_2)^2 = (P_1)^2 \times \frac{TS}{\alpha}$$

where
- T = the transmission coefficient of the panel or structure
- S = surface area of panel
- α = total sound absorption of the receiving side volume

Actual performance of a sound insulation panel is subject to modification in a number of ways.

(i) The transmission coefficient, and thus the sound reduction index, is frequency dependent; low frequencies are much more readily transmitted than higher frequencies.

The performance of a sound insulating material is, therefore, only fully described by knowledge of its transmission coefficient at different frequencies. Specific information of this nature may be essential for dealing with critical insulation problems — *eg* when there is a predominance of a particular frequency or frequencies in the incident sound. For normal working *average* values are commonly used.

Average values are based on the determination of the transmission coefficient, or sound reduction index, at a number of specific frequencies. Both the number of individual frequencies chosen, and their values, may vary with different authorities. There may thus be appreciable differences in quoted average values for similar materials, depending on the method of averaging. Averaging over a greater *number* of frequencies, for example, tends to yield a lower numerical average value. The maximum difference involved, however, is usually only of the order of 2 dB.

(ii) The transmission coefficient will depend on the angle of incidence of the sound waves received from the source. Since the great majority of cases of practical application involve random incidence, the sound reduction index is properly defined by sound waves of random incidence equivalent to hemispherical distribution from the source. There may be cases in which this does not apply and hence the transmission coefficient of the insulating panel may be modified.

(iii) At specific frequencies of incident sound, coinciding with the natural frequencies of the panel, resonant vibration will occur, with a marked reduction in sound insulation. This is because the panel itself will now be acting as a generator of sound energy. Loss of performance as a sound insulating material will be most marked at the lowest natural frequency.

(iv) At certain frequencies the phases of incident sound will tend to coincide with the phase vibrations of the panel. This can introduce flexural vibrations in the panel, again substantially reducing the sound insulation. This is known as *coincidence effect,* but can only occur at frequencies greater than a critical frequency (f_e), defined as the frequency at which the flexural wave velocity equals the velocity of sound in air — (see Table I).

SOUND INSULATION AND ABSORPTION (A)

illsonic®

The beautiful way to control noise

illsonic waffle for noise reduction in a compressor house

for any application in automotive ...

illsonic pyramid to control reverberation time in big offices

and computer industry

Sound absorption curve for illsonic waffle

[Hz]	125	250	500	1000	2000	4000
50/75	αs 0,12	0,31	0,78	1,02	1,12	1,20
70/75	αs 0,13	0,56	0,99	1,15	1,15	1,20

Calculation of the degree of sound absorption to DIN 52212 in a large hall.

illbruck international

Germany
illbruck GmbH.
Schaumstofftechnik
Burscheider Straße 454
D-5090 Leverkusen 3
Telex 08515733 illb d
Telekopierer (02171) 391-500
Telefon (02171) 391-0

Australia
illbruck (Australia) Pty. Ltd.
7-11 Antill Street
AUS-Yennora, N.S.W. 2161
Telex 27702
Telephone (02) 6813666

Austria
illbruck Ges.m.b.H.
Treietstraße 10
A-6833 Klaus/Vorarlberg
Telex 52396 illbk-a
Fernkopierer 05523-2119 - 20
Telefon (05523) 2119-0

Belgium
illbruck (Belgium) N.V.
Churchilllaan 69
B-2120 Schoten
Telex 33.926 frimed b
Telephone (031) 58.35.19

France
illbruck S.a.r.l.
35, Avenue d-Italie
F-68110 Illzach
Telex 881792 f
Telephone (89) 54.20.20

Great Britain
illbruck in U.K.
82 The Common
Broughton Gifford, Melksham
GB-Wiltshire SN 12 8 ND
Telephone (02 25) 78 2375

Italy
illbruck (Italiana) SAS
Viale Varese 10-12
I-20020 Lainate/Milano
Telephone (2) 9 37 32 36/7

Switzerland
illbruck AG
Hardstraße 50
CH-4132 Muttenz
Telex 64359 illog ch
Telephone (061) 614566

Spain
illbruck (España) S.A.
Pol. S. Francisco, Calle 9
E-Beniparrell/Valencia
Telex 64429 ille e
Telephone (96) 1201596 7

USA
illbruck (USA) Inc.
3800 Washington Ave. North
USA-Minneapolis, MN 55412
Telex 290 466
Telephone (612) 521-3555

No chance for noise

A worldwide innovation. Worldwide only from DÖRKEN: Delta Sound Absorbers, the plastic film quadrangle that overcomes noise in any room or workshop.

It is often impossible to avoid creating noise. In such cases the Delta Sound Absorber is an essential. Laid over surfaces or secured directly to ceilings – as a suspended ceiling or as a room divider – the three-dimensional cup shaped surfaces absorb the noise. What happens? On impact, the sound waves set the cups in vibration and the flexural vibration of the cup surfaces converts sound energy to heat energy which is absorbed. The reverberation period is reduced and reverberation suppressed.

Delta Sound Absorbers are seamless plastic film quadrangles of unplasticised PVC film. Each 600 x 600 mm section is divided into 42 deep-drawn cups with textured bottoms.

The number of cups, the depth of the cups and the thickness of the plastic film are so matched that a high and constant degree of sound absorption is achieved over a wide range of frequencies, while the low weight per unit area ensures a minimum of stress on the ceiling or supporting structure.

Delta Sound Absorbers are available in the neutral colours, white, light brown and transparant. Special colours on enquiry.

Delta Sound Absorbers as transparant plastic film quadrangles are ideal for rooms and workshops with domed roof lights or wide window areas.

For certain applications the hygienic angle can be of particular importance. The plastic film quadrangles are easy to clean and to disinfect. Use can be made of high-pressure cleaners.

The plastic film quadrangles are easily fixed either directly to the ceiling or as suspended ceilings or room dividers.

Delta Sound Absorbers, the plastic film quadrangle that overcomes noise in any room or workshop.

Detailed information material is obtainable direct from Ewald Dörken AG, Postfach 163, D-5804 Herdecke/Ruhr, West Germany.

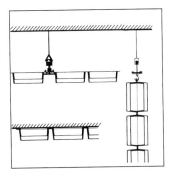

DÖRKEN protects values

Please send us detailed information concerning Delta Sound Absorbers

Name/Company:

Street:

Town/Post code

NEW

Industrial injury and disease is, without doubt, one of the biggest causes of lost working days. If these lost days are to be reduced and industry become more efficient and productive, then a greater understanding of health and safety at work is required. Based on their wealth of experience in the technical publishing field, the Trade and Technical Press Ltd., have published the 1st Edition HANDBOOK OF INDUSTRIAL SAFETY AND HEALTH. More than 650 pages, 1500 illustrations, diagrams and charts, this comprehensive reference work provides essential information vitally important to Top Management, Chief Engineers, Works Managers, Safety Officers, etc. Contents include:-

SECTION 1 – THE CONTROL AND MANAGEMENT OF SAFETY AND HEALTH
Legal Aspects – Standards and the EEC – HASAWA and Enforcement – Product Liability – Safety Policies – Safety and Health Management Techniques – Risk Management – Multi occupancy – Total Loss Control – Compensation Claims, etc, etc.

SECTION 2 – RESPONSIBILITIES AND DUTIES
General duties of Employer – Safety Officials – Representatives and Committees – Authorities and Organisations – Roles of Specialists – Employees' responsibilities – Accident Investigation, etc, etc.

SECTION 3 – THE SAFE PLACE
Design Safety – Safety Inspections – Site Safety – Check list – Electrical Safety – Lighting – Flooring and surfaces – British and European Standard signs and labels – Notices, etc. etc.

SECTION 4 – SAFETY OF PLANT AND SYSTEMS OF WORK
Design of Systems – System safety and analytical techniques – Machine hazards and protection – Safety guards and barriers – Laser safety – Communications equipment and Visual Display Units – Pressure vessels – boilers, etc. etc.

SECTION 5 – OCCUPATIONAL SAFETY AND HEALTH
Safety Training and Instruction at all levels – Hygiene and Welfare requirements – Dust and Fume Monitoring and Control – Noise and Vibration Measurement and Control – Hearing Conservation – Asbestos precautions, handling and removal – Ionising radiation and detection – Gas detection and monitoring – Temperature and Pressure monitoring – Barrier Creams and Cleaners – Janitorial Equipment, etc, etc.

SECTION 6 – THE SAFE PERSON
Personal Protection – Protective Clothing – Eye Protection – Ear Protection – Headgear – Face and Body shields – Footwear – Gloves and Mitts – Safety Harnesses, etc. etc.

SECTION 7 – SAFE USE HANDLING STORAGE AND TRANSPORT
Kinetic Handling – Handling of Materials – Mechanical Handling – Lifting equipment – Ropes, hoists and tackle – Safety tools – The Hazchem Code – Warning devices and perception – Labelling – Securing Containment and transportation – Disposal of toxic waste – Information, advice, abuse and expected misuse, etc, etc.

SECTION 8 – FIRE AND EXPLOSION
General Fire Safety – Fire fighting equipment – Inflammable Substances – Explosion and implosion.

SECTION 9 – EMERGENCY EQUIPMENT
Emergency lighting and power – Breathing and Rescue Apparatus – Resuscitation Equipment – First Aid, etc, etc.

SECTION 10 – BUYERS' GUIDE
Products and Trade Names Index – Alphabetical list of Manufacturers with addresses, telephone numbers, telex and telegram addresses of Head Office, Works and Branches

TABLE I – CRITICAL FREQUENCIES OF COMMON MATERIALS

Material	Density		Critical Frequency x Surface Density	
	kg/m² per mm	lb/ft² per in	Hz x kg/m²	Hz x lb/ft²
Aluminium	2.7	14.1	32 200	6 600
Asbestos cement	1.9	9.9	33 600	6 720
Brick	1.9	9.9	42 000	8 400
Flax board	0.39	2.0	13 200	2 640
Glass	2.5	13.0	39 000	7 800
Hardboard	0.81	4.3	30 600	6 100
Lead	11.2	58.4	600 000	120 000
Partition board	1.6	8.4	124 000	24 800
Plasterboard	0.75	3.8	30 000	6 000
Plywood	0.58	3.0	13 000	2 600
Perspex	1.15	6.1	35 500	7 100
Reinforced concrete	2.3	11.9	44 000	8 800
Steel	8.1	42.2	97 700	19 540

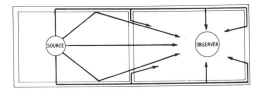

Fig 2

For any given homogeneous panel, the critical frequency is usually proportional to the thickness of the material. The problem of coincident effect, therefore, is most likely to occur in the case of thin panels of homogeneous material with low damping. Increasing thickness or damping can render coincidence effects negligible.

(v) Sound transmission through the insulation panel may be supplemented by secondary transmissions through adjacent structures – Fig 2. These are known as *flanking transmissions,* and can increase the sound level on the receiving side of the panel.

Where the average insulation is of the order of 35 dB or less, flanking transmission is usually negligible. This does not, however, preclude the possibility of sound being transmitted through vibration of the building structure (which may be considered a special case of flanking transmission) or, more particularly, through any gaps or openings which may be present in the receiving room.

With higher levels of insulation, flanking transmissions may have to be considered in detail, as they can become the major factor controlling the actual sound reduction achieved in the receiving room. This applies equally to 'open' receiving areas – *ie* where a relatively small noise source is funnelled into a sound insulating enclosure.

Mass per Unit Area Law

Specifically, the transmission coefficient (and thus the sound reduction index) is dependent on frequency, the value of R increasing with increasing frequency. Individual materials and different constructions may show departures from a steady increase. For practical purposes, therefore, it is necessary to determine values of R at specific test frequencies, which may then be used to determine the sound reduction achieved at specific frequencies, or to arrive at a mean or average value of R, which will give an 'average' performance for the partition from a single calculation.

Preferred test frequencies are those of the one-third octave range 100, 125, 160 . . . . 3 200 Hz, involving sixteen different frequencies. Much early data has, however, been determined at different frequencies. This does not affect the validity of data given for specific frequencies (and near frequencies can be considered similar, or interpolated), but can give appreciably different values for the mean sound reduction index, if based on different frequency ranges and a different number of frequencies in the series. Thus, mean values of R determined on the one-third octave series are usually about 2 dB lower than mean values for the same material determined on a nine frequency series.

The region in which a panel acts as a mass-controlled attenuator is bounded by the resonant frequencies at the lower end and the coincident frequency at the upper end – see Fig 3. The first resonant frequency is usually well below 100 Hz in practical panel sizes and constructions, although this may not always be the case with small, thin panels. As far as behaviour to incident frequencies below the resonant frequency is concerned, the insulation provided is largely stiffness-controlled, with R increasing with decreasing frequency. The actual value of the resonant frequencies will depend on the size, mass, stiffness and method of edge fixing of the panel.

Resonant frequencies of thin homogeneous plate simply supported at the four edges are given by

$$f_f = 0.45 \, V_L \, h \left[\left(\frac{n_x}{L_x} \right)^2 + \left(\frac{n_y}{L_y} \right)^2 \right]$$

where V_L = longitudinal wave velocity
h = plate thickness
L_x, L_y = lateral dimensions of plate
n_x, n_y can have any integral values, starting from n_x = 1, n_y = 1.

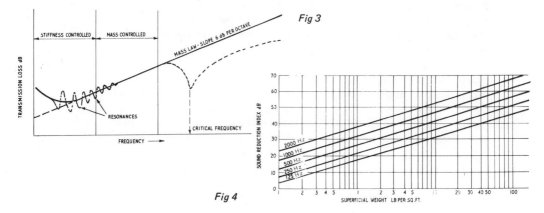

Fig 3

Fig 4

Where the edges are clamped, the first resonant frequency is twice that given by the formula, and all the other resonances are higher in consequence. The frequency range over which these resonances cause marked deviation from the mass-law is governed by the amount of damping on the panel.

Calculation of R on the mass-law basis may be applied to homogeneous materials in the absence of empirical data. A practical formula is

$$R = 14.3 \log_{10} M + 11.4 \text{ dB}$$

where M is the mass per unit weight of the panel or area density in kg/m^2

$$R = 14.3 \log_{10} M + 20 \text{ dB}$$

where M is in lb/ft^2

(See also Fig 4)

The critical frequency of a building material can also be calculated from the Cremier formula

$$f_c = \frac{c^2}{2\pi} \left(\frac{3M(1-\sigma)^2}{Eh} / S \right)$$

where c = velocity of sound in air (343.4 m/s)
M = area density in kg/m^2
E = modulus of elasticity of material, kg/ms^2
h = thickness of material, in metres
σ = Poisson's ratio for the material

Double Surfaced Panels

In the case of double surfaced panels, two types of resonance may occur, a lower resonance where the two surfaces vibrate on the stiffness of the air space, and a higher resonance where the thickness of the air space is approximately equal to a multiple of half the incident wavelength.

At a certain higher frequency, the wavelength of the incident wave may approximate to the wavelength of the free bending waves in the panel, when a substantial change in the sound reduction index may occur. The frequency at which this effect is a maximum is called the coincident frequency, and can be estimated from the formula

$$f = \frac{a^2}{1.8 V_L \, d \sin \theta}$$

where a = velocity of sound
V_L = velocity of longitudinal waves in the partition
= $\dfrac{E}{\rho(1-\sigma^2)}$ for a homogeneous partition

where E = Young's modulus,
σ = Poisson's ratio and
ρ = density

Most practical panels exhibit coincident effects over part of the normal frequency range, and this largely accounts for the departure from the mass-law behaviour.

The extent of this departure again depends primarily on the amount of damping provided by studding and discontinuities in building partitions often reduce coincidence effects to negligible proportions. The significance of coincidence effects in practice also tends to be less than simple theory would predict because, with a random field, only a small proportion of the sound energy strikes at the correct angle to produce coincidence.

The insulation characteristics of partitions vary considerably with the type of construction and so individual characteristics can only be determined empirically. Double panel construction normally provides appreciably higher values of R at higher frequencies and similar values at lower frequencies (although for the same mean value of R, double panels will usually provide a lower value of R at lower frequencies). The type of rendering, or surface finish, can be extremely significant in the case of porous and semi-porous panels, porous materials generally being poor sound insulators, unless sealed.

Composite Panels

In the case of composite panels comprising two or more areas with different values of R, the effective sound reduction index can be determined as:-

$$R_e = 10 \log_{10} \frac{S}{\sum_{1}^{n} t_1 S_1}$$

where S = total area of partition = $S_1 + S_2$ etc
t_1, t_2 etc = transmission coefficients of the respective areas
n = number of areas

In the mass-controlled region the theoretical performance of a partition for any frequency (f) can be determined from

$$R = 10 \log_{10} \left[1 + \frac{\pi f w \cos \theta}{a p}^2 \right]$$

where θ = angle of incidence

For sound normal to the surface ($\theta = 0°$), the level is

$$R_o = 10 \log_{10} \left[1 + \left(\frac{\pi f w}{ap} \right)^2 \right]$$

Where the sound field is reverberant all angles of incidence from $\theta = 0°$ to $\theta = 90°$, are involved when:-

$$R = 10 \log_{10} \frac{\left(\frac{\pi f w^2}{ap} \right)(1 - \cos^2 \theta)}{\log_e \left[1 + \left(\frac{\pi f w}{ap} \right)^2 \right] - \log_e 1 + \left(\frac{\pi f w}{ap} \right)^2 \cos^2 \theta}$$

SOUND INSULATION AND ABSORPTION

In practice, the performance of actual materials may show appreciable departure from this so-called mass law because:-

(i) the relationship is specific to a particular frequency or a mean frequency;
(ii) the panel may be subject to conditions outside the mass-controlled region.

The sound reduction afforded by composite panels can best be expressed in terms of the transmittance value. The effective or total transmittance is the sum of the individual transmittances.

$$\bar{T} S_T = \bar{T}_1 S_1 + \bar{T}_2 S_2 + \bar{T}_3 S_3 \text{ etc}$$

where $\bar{T}_1$ = transmission coefficient for first surface, area S_1, etc

$\bar{T}$ = the average transmission coefficient for the whole panel,

and S_T is the total surface area of the panel

The overall sound reduction index then follows:-

$$R = 10 \log_{10} \frac{1}{\bar{T}}$$

Resonance

Any homogeneous material will have a resonant frequency, the specific value of which can be calculated from the formula

$$f_r = \frac{1}{2\pi} \cdot \frac{gE}{(1-e)Pd} \text{ Hz}$$

where g = acceleration due to gravity (9.81 m/s^2)
E = Young's modulus, kg/ms^2
P = static stress, Pa
d = thickness of material, m
e = compression (%)

In practice, since the lowest sound level frequency likely to be present is about 40 Hz the resonant frequency of any sound insulating material should not exceed 20 Hz, when resonance will not occur. Most practical sizes and thicknesses of sound insulating materials etc, in fact, yield resonant frequencies of below this order, but the problem may arise with more elastic materials.

Effect of Openings

The effect of openings or areas of low insulation can be to decrease the sound reduction achieved by insulation and also limit the maximum degree of sound reduction which can be achieved. As a rough approximation, the performance of areas of low insulation can be estimated on the mass-law basis, when the surface can be analyzed as a composite panel. In the case of definite gaps and openings, the sound reduction will be zero, ie sound will have a direct path through the opening, provided the thickness of the opening is small compared with the maximum dimension of the opening.

The most common cause of loss is a slit or opening around badly fitting doors or windows, particularly a threshold slit at the bottom of a door. The effect of such slits can be estimated from

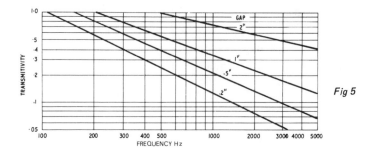

Fig 5

Fig 5, which gives typical transmission coefficients for narrow openings whose length is large compared with the slit width and the wavelength of sound. In the case of threshold slits, the effective slit width may be increased by reflection from the plane (floor) surfaces on either side, and in such cases it is generally recommended that twice the value of the actual slit width be taken in determining the transmission coefficient. For normal working a suitable mean value for the transmission coefficient of a slit derived from Fig 5, is the value at 1 000 Hz, representing a typical average for speech frequencies.

Impact-Sound Insulation

Impact sound represents a special case as far as insulation is concerned. *Impact-sound insulation* values are essentially determined empirically, related to a standard reference, *ie* the use of a standard impact sound generator or tapping machine. The impact-noise level is measured with the tapping machine striking a standard surface and again when striking the insulation surface under test. The difference between the two readings is the impact-sound insulation value of the material on test.

Effectively, to provide insulation against impact sound it is necessary to interface an elastic damping medium between the source of the noise and the main structure. This material must remain resilient (*ie* loading of the material must not exceed its elastic limit) and must not be bridged by possible sound paths. It is also essential that the resonant frequency of the damping material does not coincide with any sound frequency present. In practice this requires that the natural frequency of the damping material should be less than one-third that of the lowest frequency sound present.

Sound Absorption

The basic principle of sound absorption is shown in Fig 6. When a source of sound is in a room or enclosed, sound waves will strike the surfaces of the room at random incidence and be reflected. If the surface is sound absorbent a proportion of the sound emerging will be absorbed so that each reflected wave is weakened.

The overall effect is a little more complicated as far as an observer in the room is concerned. He will receive both direct sound (since he must be between the source and the boundary walls), and reflected sound waves. He has no means of distinguishing between the two and hears only the resultant sound. The intensity of this will vary with his position. If close to the noise source, direct sound will be the predominant component and will be inversely proportional to the square of the distance between sound and observer. At more distant parts of the room the sound level will be substantially constant with almost the whole of the noise emerging consisting of components that have been reflected many times between the room boundaries. Thus the more sound absorbent these boundaries are, the lower will be the sound level at these areas.

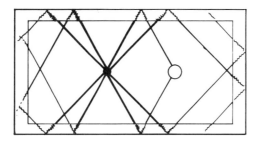

Fig 6

Basically, therefore, the value of sound absorption is in lessening the general noise level within a room, and minimizing excessive background noise. It can be particularly beneficial in large general offices and workshops, *etc*, where noise sources are confined to certain parts of the room. Sound absorption treatment can prevent noise being reflected back into other areas of quieter activity. Here, it should be noted, however, that additional sound *insulating* barriers may be necessary to reduce the noise radiated in a *direct* path from noise sources.

All the generally used surface materials absorb some sound energy from a wave striking the surface. The absorption varies from perhaps 5% for a hard painted surface to 80—90% for some of the specially prepared sound absorbent materials. Surfaces having high sound absorption coefficients will obviously produce a large reduction at each reflection and the final steady uniform level of reflected sound will be lower than would result if the surfaces had low absorption coefficients. In fact, it can be shown that the steady uniform sound level will be inversely proportional to the total amount of sound absorption present in the room. Each time the total amount of sound absorption in the room is doubled there is a reduction of 3 dB in the average sound level.

In principle, the level of the reflected sound can thus be reduced to any desired extent by installing sufficient sound absorbing material. In practice this is not so simple or so effective as it would appear. It has been indicated that a doubling of the total absorption in the room halves the intensity of the reflected sound. When there is only a small amount of sound absorption present in the room it is easy to double the amount. However, when the ceiling has been acoustically tiled and the floor covered with carpet it generally becomes difficult to find surfaces of sufficient area to allow the total sound absorption to be doubled again, even using acoustic treatment of high efficiency. It is this practical situation which limits the amount of noise reduction that can be obtained by the use of sound absorbents.

Several important deductions can be made from the foregoing:

(i) The addition of sound absorbents to the walls or ceiling is particularly effective in reducing noise if the amount of sound absorbent material in the room is small.

(ii) Conversely, if the room is well furnished, the addition of sound absorbents will be of little benefit. For example in rooms in dwellings containing soft furnishings it is unlikely that noise reduction of more than 5 dB will be achieved with acoustic treatment. An improvement of up to 10 dB may well be achieved, however, in acoustically 'loud' rooms such as workshops, school classrooms, *etc*.

(iii) Because the sound absorbent is ineffective in reducing the direct sound, the noise level to which the operator of a noisy machine is exposed cannot be significantly reduced by the addition of any sound absorbent treatment to the walls of a room in which the machine is operated. This is particularly true in large rooms, the average machine shop for example.

(iv) Because the sound absorbent is effective in reducing the indirect sound the noise criticized by staff remote from a noisy machine may be appreciably reduced by the addition of acoustic treatment to the boundary surfaces of the room.

(v) The operator of one machine criticizing the noise from another machine may benefit by the addition of sound absorbent to the walls. There is advantage only during those intervals when his own machine is at rest, but where the machines have a 'time off/time on' ratio greater than unity the advantage may be significant.

Sound Absorption Materials

A sound absorbent is a material (or composite structure) which can provide sound energy absorption by one or more of the following:

(i) friction between the fibres of a porous fibrous material.
(ii) absorption in the voids of a porous, non-fibrous material.
(iii) absorption within narrow entries to an air space.

Absorption Coefficient

The performance of an absorbent (or its effectiveness as a sound absorbent) is usually expressed by its *absorption coefficient* with a minimum value between 0 and 1.0. A coefficient of 0 represents total reflection (no absorption) and a coefficient of 1.0 represents total absorption (no reflection). The coefficient multiplied by 100 thus represents the actual *percentage* of sound energy absorbed.

The absorption coefficient is not an absolute constant quantity for any material or composite structure since it will vary with frequency and also be affected by size and position, method of mounting, *etc.* Thus absorption coefficients determined by different test methods for the same material or composite are not directly comparable. It is also possible with certain methods of measurement to achieve absorption coefficients in excess of unity (*ie* apparently more than 100% absorption), due to diffraction or edge losses.

As a general rule an increase in thickness in fibrous or porous absorbents will improve sound absorption mainly over the low and middle frequency ranges. Thus porous absorbents should preferably be a minimum of 25 mm (1 in) thick unless absorption is only required at high frequencies — *eg* see Fig 7.

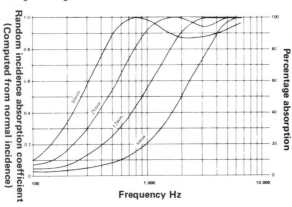
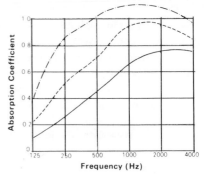

Fig 7 Absorption characteristics of 'Revertex' foam pads.

The beneficial effect of increasing thickness generally applies whether or not the absorbent is mounted over a cavity.

General Classification

A general classification for sound absorbents is based on these absorption coefficients over the middle and high frequency range. This is:

low absorbent — absorption coefficient 0.25 or less

moderately absorbent — absorption coefficient 0.25–0.40

absorbent — absorption coefficient 0.40–0.70

highly absorbent — absorption coefficient over 0.70

Absorbents can also be broadly classified according to their effectiveness over specific frequency ranges, *eg*:

A — general wide band absorption

B — absorption over middle and high frequencies

C — low frequency absorbents

Examples of general classification applied to materials commonly used as absorbents in buildings are given in Table II.

TABLE I — GENERAL CHARACTERISTICS OF SOUND ABSORBING MATERIALS

Material Type	Typical Absorption Coefficient*	Remarks
Porous panels	0.20 to 0.80 generally tends to increase with increasing frequency	Performance may be appreciably modified by surface treatment. Typical surface finishes — muslin or perforated covers, or membranes.
Porous materials (eg mineral, glass wool and felts)	0.25 to 0.90 generally increasing with increasing frequency	Turnover frequency determined by depth of air volume
Porous blocks	0.20 to 0.80 depending on cellular form and particularly thickness	May exhibit optimum performance at mid-frequencies.
Plain panels	Typically 0.40 maximum decreasing with increasing frequency	Low frequency performance improved by addition of absorbent backing. Absorption mainly by vibration causing energy damping.
Perforated panels and tiles	0.20 to 0.80 depending on thickness and size and disposition of perforations. Poor performance at low frequencies	Normally exhibit a maximum value at mid-frequencies. Usually comprise porous material with hard face.
Perforated panels with absorbent backing	0.20 to 0.80 generally increasing with increasing frequency	Thin perforated face panels with absorbent backing behave largely like porous panels. Rigid constructions may act as resonant absorbers.
Carpets	0.05 to 0.65 poor at lower frequencies	Performance generally improved with underfelt or underlay.
Drapes	0.10 to 0.65	Performance varies considerably with thickness and weight of material, also texture and form.

* 125 Hz to 4 000 Hz

Empirical determination of sound reduction coefficients or acoustic absorptiveness is commonly confined to six frequencies in the range 125 Hz to 4 000 Hz, although measurements may be extended to higher frequencies. In general, however, values for frequencies above 4 000 Hz are similar to those for 4 000 Hz. The noise reduction coefficient is commonly quoted as a single number average value, based on the sound reduction coefficients at 250, 500, 1 000 and 2 000 Hz. Values, in all cases, are rounded off to the nearest 0.05.

In the case of porous materials with inter-connecting pores, friction is the predominant factor in absorbing the energy of sound waves by progressive damping. In terms of an electrical analogy, the material offers direct resistance and thus the damping effect is largely independent of frequency. However, there is an optimum value for the resistance. If too high, sound waves will be rejected or reflected rather than penetrating in depth. If too low, there will be insufficient friction to provide enough damping to make the material effective as a sound absorber.

With perforated materials opening into a body of porous material, solid materials and membranes, damping is provided by reaction rather than pure resistance. In consequence, the performance can be markedly dependent on frequency. This in turn is further dependent on the proportion of open area in the case of a perforated surface, or the mass in the case of an impervious membrane.

The total depth of the air volume between the face of the material and the rigid backing can also modify the frequency-dependent characteristics — this air volume includes open pore volumes in the case of porous materials. Basically, this introduces the concept of a 'turnover frequency' at which the low frequency absorption characteristics will deteriorate rapidly. This turnover frequency (f_t) is given approximately by:-

$$f_t = \frac{c}{2d}$$

where c is the velocity of sound in air
d is the total depth of air volume

For d in inches this reduces to:-

$$f_t = \frac{500}{d} \text{ Hz (approximately)}$$

This emphasizes the importance of air volume in the absorption of lower frequencies, ie to produce suitable low values of turnover frequency.

The basic requirements of a sound absorbing material are —

(i) that it should be sufficiently porous to allow sound waves to enter the material, and
(ii) the nature of the material be such that the maximum proportion of sound energy be transformed into heat energy by friction, thus providing dissipation of the sound energy.

Both can be related to the flow resistance, or the ratio of the pressure drop to the velocity of the air passing through the material. The pressure drop must be low enough to provide adequate transparency to sound waves, and high enough to provide sufficient friction.

Flow resistance can be defined specifically in rayls/metre where:-

$$R = \frac{\Delta p}{\delta t v} \text{ rayls/unit length}$$

where R = specific flow resistance
where Δp = sound pressure differential across a thickness δt measured in the direction of particle velocity
δt = incremental thickness
v = particle velocity through sample

For the purpose of practical measurement of porous materials, the flow resistance is generally determined in a steady airflow. The volume velocity U_v is related to the linear velocity U_L by:-

$$U_v = U_L \times S$$

where S is the surface area of the specimen

The pressure drop (Δp) across the thickness of the material is also measured, when the flow resistance (RF) follows as:-

$$RF = \frac{\Delta p}{U_v/S}$$

The specific flow resistance is then equal to:-

$$R = \frac{RF}{St}$$

where t is the thickness

The actual porosity of a porous material is defined as the ratio of the volume of voids present in the material to the total volume. In the case of solid, fibrous materials the porosity can be estimated directly from the density of the fibres and total mass.

$$\text{Porosity} = 1 - \frac{\text{total mass}}{\text{total volume} \times \text{density of fibres}}$$

mass, volume and density being in consistent units.

If binders are present, an allowance for this must be made to estimate the true mass. For materials of mixed or composite structure, porosity can only be determined accurately by direct measurement.

The performance of porous materials as sound absorbers may be analyzed directly in terms of the flow resistance, but the mathematics involved is somewhat complex and lengthy requiring also considerations of the actual structure. Measured values are thus normally preferred, ie empirical values or the random-incidence absorption coefficient for the material, determined under standard test conditions. These conditions are not usually identical in different laboratories and so some differences can be expected when comparing the data from different sources. Agreement is normally good enough for general use, however.

At very high sound levels — *eg* 150 dB or above — sound absorbing media may disintegrate or 'burn out'. This is because high intensity sound waves are non-linear. This non-linearity occurs when the rarefaction of the wave approaches about half atmospheric pressure, causing the wave to become increasingly asymmetrical as the sound pressure level increases above about half 120 dB. At substantially higher levels the wave degrades into a sawtooth form, with the impact on materials in its path becoming similar to that of hammer blows.

The equivalent absorption of a surface is given by the product of the area of the surface and the absorption coefficient. In the case of a room the total absorption is given by the sum of the equivalent absorption of each surface, also including absorption given by furnishings, seats, **occupants**, *etc,* **where applicable**

Thus:

$$\text{Total absorption (A)} = \Sigma \propto S \text{ in Sabines}$$

where $\propto$ is the absorption coefficient (*eg* typically using a mean value)

S = area of surfaces

The average sound pressure level of reflected sound in a room is then given by:-

$$SPL_{av} = 10 \log_{10} w - 10 \log_{10} A + 136.4 \text{ dB}$$

$$= w - 10 \log_{10} A + 16.4 \text{ dB}$$

where w = power level of source, ref 10^{-12} watt and S is in sq ft

$$= w - 10 \log_{10} A + 6.1 \text{ dB}$$

where S is in sq metres

Strictly speaking this formula applies only for diffuse distribution of sound within the room, and for a single sound source.

The total sound level in a room will be the logarithmic sum of the direct and reflected sound levels, *viz*

$$\text{total sound level} = 10 \log_{10} \left(\text{antilog } \frac{Ld}{10} \text{ antilog } \frac{Lr}{10} \right) \text{ dB}$$

This will be only slightly higher than the level of the larger sound, and never more than 3 dB higher. Up to a distance of approximately $0.5\sqrt{A}$ from the source, the direct sound will be the louder. At greater distances, the reflected sound level will be greater than the direct sound and the total level will be substantially the same as that of the reflected level (and constant at that level).

Surface Finishes

Thin glass tissue interleaves are often placed immediately over perforated ceiling panels and a fibrous absorbent directly above. Mineral fibre is often applied in the form of a quilt with a covering of fine scrim or muslin. The glass tissue or scrim cloth prevents any loose fibres from the unfaced absorbent from falling through large perforations.

Alternatively, thin plastic film may be placed over the perforated ceiling trays, to act additionally as a vapour check. It is generally considered impracticable to obtain a complete vapour barrier with suspended ceilings because of the extent of jointing, and requirement for access. The effect of applying a non-porous plastic film to the surface of a fibrous insulant is to

SOUND INSULATION AND ABSORPTION

reduce absorption at high frequencies. Perforated panels similarly influence absorption, and consequently when employed together will minimize the individual effect of the plastic film.

In hospitals and other situations where facility of cleaning is desired, porous absorbent ceiling tiles may be covered with flexible plastic film, without unduly affecting their sound absorption characteristics. It is recommended that the film should not exceed about 0.05 mm (200 gauge) thick and that it be secured only around the edges of the tile, allowing the film to vibrate independently.

The painting of porous ceiling tiles is not normally recommended unless it is assured that decoration would not clog up the pores. Sealing the surface in this way will increase sound reflection and have a deleterious effect on the absorption performance of the ceiling.

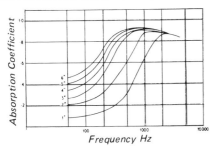

Sound absorption characteristics of different thicknesses of wool felt.

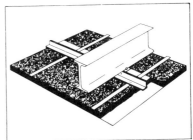

Location of 'Rocksil' rock/wool in suspended ceiling structure.

Effect of Density

Most porous absorbents have an optimum density and flow resistance at which maximum absorption is achieved. Too small a pore structure will restrict the passage of sound waves, while too large a pore structure will offer low frictional resistance. Low absorption will result in both extreme cases. The more porous materials generally have a low density.

The effect of increasing the density of Rocksil (rockwool) is shown in Fig 8 to increase absorption between the *middle* and *high* frequencies. The curves on the two graphs are directly comparable, and are separated only to avoid overcomplication and possible misinterpretation.

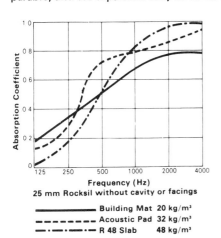

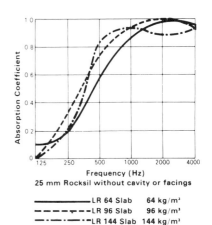

Fig 8

The heavier density absorbents can be employed gainfully in raising the level of sound absorption, when it is not practicable or desirable to include a cavity of appreciable depth within a wall or ceiling structure.

Cavity Depth

An airspace formed between the porous absorbent and solid backing will improve absorption at the *lower* frequencies. Increased cavity depth increases absorption, and it is generally understood that airspaces up to about 400 mm (16 in) have correspondingly higher absorptive power at frequencies below 250 Hz.

By increasing absorption at low frequencies, the airspace has a similar effect to that obtained by increasing the thickness of mineral fibre. However, the use of a cavity alone is not generally sufficient for normal requirements and some absorbent material will usually be included. The absorbent will often increase and extend the absorption coefficient over a broader range of frequencies.

Fig 9 illustrates the influence of cavity depth on sound absorption in a suspended ceiling with 25 mm (1 in) Rocksil Building Mat and perforated Asbestolux panels. The decreasing absorption at frequencies above 500 Hz shown in this example is characteristic of the effect created by the perforated ceiling tiles.

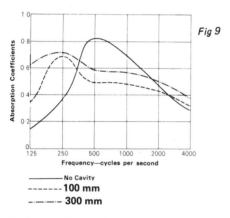

Fig 9

——— No Cavity
- - - - - 100 mm
—·—·— 300 mm

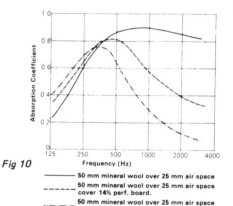

Fig 10

——— 50 mm mineral wool over 25 mm air space
- - - - - 50 mm mineral wool over 25 mm air space cover 14% perf. board
—·—·— 50 mm mineral wool over 25 mm air space cover 4% perf. board

Perforated Panels

Porous absorbents such as mineral fibre are commonly applied behind perforated or slotted hardboard, plywood or plasterboard, mounted upon battens of sufficient thickness to accommodate a minimum 25 mm (1 in) mineral fibre and an airspace. Absorbent materials are also generally used in conjunction with hardwood straps with narrow continuous gaps between them. For maximum efficiency the porous absorbent should be placed close to the vertical or horizontal perforated panel. Low frequency absorption will be improved where a cavity is also constructed.

Acoustic treatment with perforated panels may not be suitable for application in areas of high humidity such as laundries, swimming baths and certain industrial situations, because of the presence of large volumes of moisture vapour which could gain access to the fibrous absorbent.

In the case of tiles the least effective way of mounting a given area of sound absorbent tile is to stick it tightly to the wall or ceiling, somewhere around the middle of the surface being treated. In this position the air flow through the tile is at a minimum, the air particle velocity always being zero at the wall surface. In consequence, the energy dissipation is minimized in those areas of

SOUND INSULATION AND ABSORPTION

material adjacent to the wall. Mounting the material off the wall, leaving a cavity behind the tile, is almost as effective as the use of a sound absorbent with a thickness equal to the total of cavity plus tile. On the other hand if the noise energy is concentrated in the high frequencies a cavity behind the tile is of little advantage.

An absorbent tile stuck in the middle of one of the long walls of a room has an absorption which is only about one quarter that of the same area of tile mounted in the same way in a corner at the junction of three surfaces. The same tile stuck to the wall at the junction of two surfaces is about twice as efficient as the tile stuck to the middle of the long wall. If the tile is used to cover a column well away from the walls, or to face a free standing screen, it may be three or four times as effective as when wall mounted. The same considerations apply when the absorbent is suspended from the ceiling or from the roof trusses.

Perforated Facings

Perforated panels are intended to form a decorative or protective facing for sound absorbent materials, while exerting minimum influence upon their absorptive characteristics. The extent of perforation controls the absorption behaviour of perforated panels. The amount of absorption achievable depends upon the spacings between and diameter of perforations, the relative percentage 'open area' and thickness of panels. With all else equal, the effect of the facing is minimized with open area in the region of 10 to 20%, and the performance of the absorbent structure is influenced mainly by the backing material and cavity.

A reduction in open area alters the overall performance, generally by increasing low frequency efficiency and decreasing absorption efficiency at higher frequencies. Where low frequency noises are the most disturbing and a lowering of efficiency at high frequencies is acceptable, perforated panels with as little as 3% open area may be preferred. Unperforated panels do, in fact, function essentially as low frequency absorbers.

The effect of reducing open area of perforated panels is clearly illustrated in Fig 10, which also reveals progressively diminishing absorption in each case at frequencies above 500 Hz. The absorption performance of perforated ceiling tiles of different designs and open area, employed in conjunction with Rocksil Building Mat are compared in Tables III and IV.

The results indicate that effective control over the pattern of perforations, often for decorative effect, will allow a wide variety of designs to be produced, which can provide equally favourable degrees of absorption. In such cases direct relationship between the 'open area' of different panels is not the critical factor, but the combination of several important features of perforation.

TABLE III — ABSORPTIVE PROPERTIES OF ROCKSIL PRODUCTS MOUNTED OVER 300 mm (12 in) CAVITY AND WITHOUT FACINGS

Rocksil Product (Rock Wool)	Density		Thickness		Absorption Coefficients (ISO)					
	kg/m^3	lb/ft^3	mm	in	125 Hz	250 Hz	500 Hz	1 000 Hz	2 000 Hz	4 000 Hz
Building Mat	20	1.3	25	1	0.45	0.65	0.55	0.65	0.75	0.85
Building Mat	20	1.3	50	2	0.65	0.85	0.75	0.85	0 95	0.95
Acoustic Pads	32	2.0	50	2	0.80	0.95	0.80	0.90	0.95	1.10
R 48 Slab	48	3.0	25	1	0.45	0.70	0.65	0.95	0.95	0.90
LR 64 slab	64	4.0	50	2	0.95	0.85	0.95	1.00	1.05	0.95

TABLE IV – ABSORPTIVE PROPERTIES OF PERFORATED CEILING PANELS WITH BONDED FIBRE FABRIC AND 25 mm (1 in) ROCKSIL BUILDING MAT, AND 300 mm (12 in) CEILING CAVITY

Ceiling Panel	Open Area %	Absorption Coefficients (ISO)					
		125 Hz	250 Hz	500 Hz	1 000 Hz	2 000 Hz	4 000 Hz
10 mm (3/8 in) Asbestolux (A4)	4.83	0.48	0.55	0.49	0.47	0.35	0.27
10 mm (3/8 in) Asbestolux (A5)	7.98	0.52	0.49	0.38	0.31	0.25	0.21
10 mm (3/8 in) Asbestolux (A1)	9.80	0.44	0.49	0.49	0.58	0.46	0.36
10 mm (3/8 in) Asbestolux (A3)	11.00	0.47	0.53	0.44	0.42	0.31	0.18

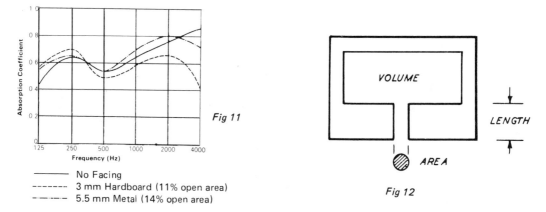

Fig 11

——— No Facing
- - - - - 3 mm Hardboard (11% open area)
—·—·— 5.5 mm Metal (14% open area)

Fig 12

Fig 11 illustrates the relatively high overall level of absorption which may be achieved with perforated panels of adequate open area, used with a 300 mm (12 in) deep cavity and 25 mm (1 in) Rocksil Building Mat.

Resonant Absorbers

A simple resonant absorber comprises a cavity enclosing a mass of air, with a narrow opening to the outside – Fig 12. The air mass can effectively act as a 'spring' at the resonant frequency of the cavity and under those conditions absorb appreciable sound energy exciting the resonance.

The resonant frequency of such a 'cavity' or Helmholtz resonator is given by:

$$f_r = 85\,250 \sqrt{\frac{S}{LV}} \text{ Hz}$$

where S = cross section area of opening in sq inches
L = length of opening in inches
V = volume of cavity in cubic inches

$$= 55 \sqrt{\frac{S}{LV}} \text{ Hz}$$

where S is in m^2, L in m and V in m^3

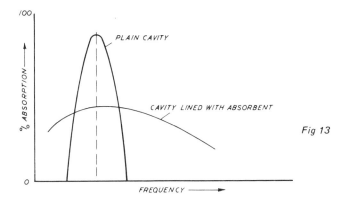

Fig 13

At the resonant frequency, such a resonator is capable of absorbing up to nearly 90% of sound impinging on the cavity (via the opening). The performance is modified considerably by lining the cavity with an absorbent material when, although the peak absorption is substantially reduced, absorption is spread over a wider range of frequencies — Fig 13.

Perforated panels backed by a sub-divided air space have the property of acting as multiple resonant absorbers. The resonant frequency of such a panel is given by

$$f_r = 200 \sqrt{\frac{p}{L(t + 0.8d)}} \text{ Hz}$$

where p = percentage of open area

L = depth of air space, inches

t = thickness of the panel, inches

d = diameter of the perforations inches

$$f_r = 5\,000 \sqrt{\frac{p}{L(t + 0.8d)}} \text{ Hz}$$

where L, d and t are in millimetres

This is only an approximate formula, but it will indicate optimum values of p, L and t to provide maximum attenuation at a specific frequency. The actual performance of such a panel may be considerably modified by the introduction of absorbent material behind it, when the percentage of open area will largely govern the absorption coefficient achieved at higher frequencies.

If the perforated panel is thin, and employed primarily as a protective cover for an absorbent material, the principal effect is again to reduce the absorption at higher frequencies in inverse proportion to the percentage of open area. The frequency at which this reduction is likely to become apparent can be estimated from the formula:-

$$f = \frac{40p}{d} \text{ Hz (approximately)}$$

It follows that a large number of small diameter holes giving a specific open area are beneficial in delaying the loss of absorption with increasing frequency.

See also chapters on *Acoustic Materials, Acoustic Enclosures, Noise in Domestic Buildings.*

Acoustic Materials

MATERIALS USED for sound reduction treatment may be sound *absorbent* or sound *insulating* (or have a combination of both properties). Sound *absorbent* materials are generally lightweight and of fibrous or porous nature, with the characteristic of reducing *reflection* of sounds. Sound *insulating* materials are barriers to the *transmission* of sound. They are dense, heavyweight materials, their effectiveness as a barrier being directly related to their superficial weight (mass per unit area).

Sound absorbent materials can, however, also contribute to sound *insulation* by reducing the sound level generated within a room or enclosure. Thus, the sound level transmitted through the walls is lower than it would be with no treatment. This does not justify them being called insulating materials and the distinction between the two basic types of materials should be appreciated.

Materials are listed in alphabetical order for convenience of reference, regardless of whether they are insulating materials and/or absorbents. Also included are a number of low density materials which are primarily *thermal* insulators rather than sound attenuators and are specifically used as such. At the same time, however, they may also have sound absorbent properties in varying degrees.

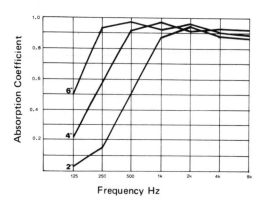

Fig 1 Effect of thickness on sound absorption of 80 pores-per-inch reticulated polyurethane foam. (After Scott acoustical laboratory test).

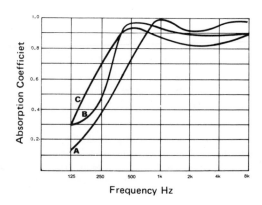

Fig 2 Acoustical properties of flexible polyurethane foam.
A — 1" thick foam with no airspace.
B — 1" foam with ½" airspace.
C — 1" foam with 1" airspace.

Acoustic Foams

Acoustic foams are sponge-like materials and have an ability to absorb sound energy by changing it into heat. The capacity of such a material to absorb sound energy is dependent primarily on the permeability and thickness of the cellular structure or foam and, to a lesser extent, on the pore size and surface treatment. The rigidity of the foam has little or no effect on acoustic performance.

In general, the higher the permeability the greater the sound absorption of a foam, although there is an optimum permeability for a particular thickness. The thinner the foam the higher the flow resistivity (and thus the lower the permeability) needs to be for optimum performance. This affects particularly the lower frequency performance — see Fig 1. Other parameters being the same, the smaller the pore size (*ie* the greater the number of pores per unit area), the greater the absorption capability. Thus, with control over all these parameters, foams can be 'tailored' during manufacture to provide optimum acoustic properties.

Acoustic foams are particularly suitable for middle and high frequency sound treatment. For low frequency applications performance can be improved by spacing away from a hard surface — see Fig 2.

Acoustic foams can be made from any 'expandable' material, but the main type used is flexible polyurethane foam of open pore structure. Apart from being inert and a fire-retardant material (except that it may fume) polyurethane foam can also be dyed or painted for decor effect without appreciably affecting its sound absorbent properties. Polyurethane foams are also produced with a polymer film surface (to resist scuffing and abrasion), bonded to open weave fabrics, sheet vinyl, sheet lead, and other materials.

Flexible foams are:

(a) Light in weight (b) Self-supporting (c) Of continuous construction
(d) Easy to bend and shape (e) Easy to fix (f) Efficient as absorbers
(g) Inexpensive (h) Capable of many design variations (i) Clean and hygienic

There are a number of foam chamber lining constructions available depending upon the performance required and the overall budget.

Rigid acoustic foams are particularly suitable for acoustic wedges in sizes up to 36 in long for fixing direct to walls (or ceilings) with a suitable adhesive. Because of their light weight they have minimal edge drop.

Asbestos

With asbestos now widely regarded as a health-hazard material, its use and application is largely now discontinued. It is, however, still used for heat-and-fire resistant applications, including non-combustible asbestos/cement mouldings, insulation boards and other products. Such materials may also have good acoustic properties. Foamed asbestos is produced from asbestos fibre and kaolin in sheet form (usually with aluminium foil backing) as a thermal insulant. It can also be used for sound and vibration. Typical sound absorption properties of low density 10 kg/m^3 (0.63 lb/ft^3) foamed asbestos are:

Frequency (Hz)	125	250	500	1 000	2 000	4 000
Sabines absorption	0.36	0.51	0.94	0.80	0.65	0.80

TABLE I – SOUND TRANSMISSION LOSS THROUGH 2 IN THICK LONG-GRAIN BALSA, DENSITY 7–8 lb/ft^3

Frequency Hz	160	200	250	315	500	630	1 000	1 600	2 000	3 150
Transmission loss dB	17	13	10	11	12	14	21	25	26	29

Balsa

Balsa is the lightest available commercial wood with a density that may range from 95 kg/m^3 (6 lb/ft^3) to 300 kg/m^3 (18 lb/ft^3).

. The possibilities of balsawood as a sound absorbent/sound insulating material have only recently attracted interest. Some early test figures of sound transmission loss for typical balsa panels of 110–130 kg/m^3 (7–8 lb/ft^3) density and 2 in thickness are given in Table I. These indicate a mean transmission loss of 19 dB over a frequency range of 100–3 000 Hz or 21 dB over a frequency range of 125–4 000 Hz. These figures are in general agreement with theoretical values predicted by the mass law.

Specifically, however, these data would appear to relate to sound incident on long-grain balsa panels — *ie* panels constructed from conventional cut block. The majority of balsa application for structural cores and thermal insulation panels is in end-grain configuration. Here, blocks of long-grain balsa are cut into 'slices' which are glued up into rigid panels, the face areas of which are thus end grain.

Recent (limited) laboratory tests on the sound transmission loss of end-grain balsa panels yielded figures substantially higher than the mass law would predict, namely:

Panel thickness	Balsa density	Sound reduction
50 mm	160–190 kg/m^3 (10–12 lb/ft^3)	27 dB(A)
50 mm	95–110 kg/m^3 6–7 lb/ft^3	26 dB(A)
25 mm	160–190 kg/m^3 10–12 lb/ft^3	25 dB(A)

Sound Absorption Properties of Balsa

One apparent further advantage of using end-grain balsa for an enclosure, or inner lining of an enclosure is that, being a low density 'porous' material, it should act as a sound absorber. Here, more detailed performance figures are available and at first sight are not particularly attractive (Table II). Figures for end-grain balsa are further compared with other sound absorbing materials in Table III.

This, however, is not necessarily the complete story. It is possible to increase the sound absorption coefficient of a solid material by grooving or blind-drilling holes *etc*, and balsa is a material

TABLE II – MEASURED SOUND ABSORPTION COEFFICIENT OF 2 IN THICK END-GRAIN BALSA

Frequency Hz	125	160	200	250	315	400	500	630	800	1K	1.25K	1.6K	2K	2.5K	3.15K	4K	5K	6.3K	8K
Balsa Density: 6–7 lb/ft^3		0.02	0.01	0.01	0.02	0.04	0.10	0.20	0.24	0.27	0.24	0.18	0.15	0.21	0.25	0.22	0.25	0.28	0.46
10–12 lb/ft^3	0.01	0.03	0.03	0.02	0.03	0.12	0.15	0.23	0.20	0.24	0.23	0.24	0.18	0.19	0.16	0.19	0.28	0.28	0.43

ACOUSTIC MATERIALS

TABLE III — TYPICAL SOUND INSULATION VALUES OF WALLS AND PARTITIONS

Construction	Approximate Weights		Average Sound Reduction (100–3200 Hz)
	kg/m^2	lb/ft^2	dB
Double 112.5 mm brick with 50 mm cavity, thin wire ties. 13 mm plaster both sides	480	98	53
225 mm brick with 13 mm plaster both sides	480	98	50
112.5 mm brick with 13 mm plaster both sides	270	55	45
100 mm clinker block with 13 mm plaster both sides	185	38	43
75 mm clinker block with 13 mm plaster both sides	150	31	41
9.5 mm plasterboard and 13 mm plaster on both sides of stud frame	65	13	35
9.5 mm plasterboard on both sides of stud frame	20	4	30
3 mm hardboard on both sides of stud frame	6	1.2	23

TABLE IV — SOUND ABSORPTION OF SOME BUILDING MATERIALS

Material	Sound-absorption coefficients Hz					
	125	250	500	1000	2000	4000
Walls						
Brick, unglazed	0.03	0.03	0.03	0.04	0.05	0.07
Concrete block, painted, sealed	0.10	0.05	0.06	0.07	0.09	0.08
Concrete block, unpainted	0.36	0.44	0.31	0.29	0.39	0.25
Glass, heavy plate	0.18	0.06	0.04	0.03	0.02	0.02
Plaster on concrete block	0.12	0.09	0.07	0.05	0.05	0.04
Plywood, 3/8 in	0.28	0.22	0.17	0.09	0.10	0.11
Draperies						
Medium weight 14 oz/yd^2 draped to half area	0.07	0.31	0.49	0.75	0.70	0.60
Floors						
Concrete or terrazzo	0.01	0.01	0.02	0.02	0.02	0.02
Tile, asphalt, linoleum, or concrete	0.02	0.03	0.03	0.03	0.03	0.02
Wood	0.15	0.11	0.10	0.07	0.06	0.07
Carpet, loop pile, medium heavy jute back over hair pad	0.06	0.21	0.67	0.55	0.56	0.60
Carpet, indoor outdoor, foam back	0.02	0.02	0.15	0.39	0.74	0.77
Ceilings						
Concrete	0.01	0.01	0.02	0.02	0.02	0.02
Gypsum board, ½ in	0.29	0.10	0.05	0.04	0.07	0.09
Plywood, 3/8 in thick	0.28	0.22	0.17	0.09	0.10	0.11
Suspended, acoustical panels	Consult manufacturers' literature					
Suspended, non-acoustical panels	0.14	0.15	0.15	0.15	0.13	0.15
Metal, fluted roof deck			0.10			

which can be very easily worked in this manner. Such grooving, *etc*, could be taken to half thickness without materially affecting the rigidity of the panel (although it would, of course, reduce the sound absorption of such a panel).

Brickwork and Building Materials

Brickwork is a good insulator against random sound, particularly if plastered on one (or both) side(s) — see Table III. Particular requirements are set by Building Regulations and Building Codes for various types of solid and cavity walls.

Most common building materials also have a useful performance as sound absorbents, some specific data being given in Table IV.

Clinker Block

Clinker block is a lightweight aggregate concrete block using furnace residue, containing no more than 10—25% unburnt carbonaceous material. It is a standard building material for cavity walls and partitions, with a sound reduction performance, when plastered, almost directly comparable with a similar thickness of brick.

Other materials in the same category, generally classified as lightweight aggregate concerned with different types of lightweight concrete are:

(i) Foamed slag — an aggregate used for heavier concrete and foundation work.
(ii) Expanded clay.
(iii) Expanded pulverized fuel ash.
(iv) Expanded slate.
(v) Pumice.
(vi) Kieselguhr — for the production of diatomaceous concrete.
(vii) Vermiculite — yielding lightweight cavities in the density range 300—900 kg/m^3
    (19—51 lb/ft^3) which can be cut, sawn, nailed and screwed.

Cork

Insulation corkboard is produced in the density range 100—320 kg/m^3 (6—20 lb/ft^3). The higher densities 80—100 kg/m^3 (5—6 lb/ft^3) have good sound absorption properties in the middle frequency range (see Fig 3), Slightly heavier densities are used for thermal insulation, and densities of about 175 kg/m^3 (11 lb/ft^3) upwards for anti-vibration treatment, also for insulation against impact or percussive noise.

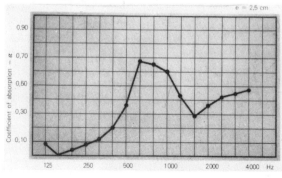

Fig 3 Typical sound insulation properties of cork board 80/100 kg/m^3.

Concrete

Concrete rates as a good sound insulator for random sound (see Table IV), but will generally transmit impact sound unless combined with some other form of insulation or isolation treatment. Being hard, concrete surfaces will also readily reflect sound (*ie* concrete is a poor sound absorber). See also chapter on *Noise in Domestic Buildings.*

Lightweight concrete materials (aerated concrete or gas concrete) developed as thermal insulants have appreciably lower sound insulation than dense concrete, a typical figure being 42—45 dB for a density of 640 kg/m^3 (40 lb/ft^3).

A new application of gas concrete in Britain is for the inner leaves of cavity brick walls and also for the construction of roofing slabs.

Fibre Metal

Fibre metal consisting of random interlocked structures of metal fibres in a variety of alloys and low densities have attractive sound absorbing properties when used as cavity fillings with a hard surface backing at a 'tuned' depth (approximately ¼ of the wavelength to be absorbed). Resulting absorption performance is broad band and bandwidth can be further increased by providing a variety of cavity depths, or using multiple layers of fibre metal — see also Fig 4.

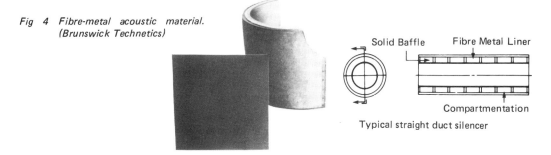

Fig 4 Fibre-metal acoustic material. (Brunswick Technetics)

Typical straight duct silencer

Foam Glass

Foamed glass is an insert, impenetrable low density material produced as a thermal insulant. It is used as a pipe insulant and also for thermal insulation in flat roof constructions. Since the cells are non-interconnecting its sound absorption properties are probably low, but no specific figures are available.

Glass

Glass is highly reflective to sound (*ie* a very poor sound absorber); but an excellent sound insulator. Thickness for thickness, glass is about as good as medium-density concrete in this respect. Its performance as a sound insulator is, however, limited by the practical thicknesses of glass which can be employed due to economic and structural load considerations. Thus, practical single glass panels seldom offer a sound reduction better than 35 dB. See chapter on *Acoustic Glazing.*

Glass Fibre

Glass fibre wool is both an excellent thermal insulant and sound absorber. It is widely used for the manufacture of acoustic tiles, boards and panels; and for plastics. General building materials based on gypsum are insulating wallboard, ceiling panels and impregnated wood-wool slabs.

Insulating wallboards consist of a core of aerated gypsum plaster faced on both sides by mill board. One surface is also usually covered with polished aluminium foil to act as a reflective surface facing the side of the cavity where the board is attached to timber studding.

Impregnated wood-wool slabs comprise compressed wood fibres impregnated under pressure with gypsum cement. They are widely used as a building material for roofing and/or lightweight walling, providing both sound and thermal insulation.

Honeycomb Panels

Brazed-up all metal honeycomb panels consisting of a perforated face sheet and plain backing sheet with a honeycomb core (Fig 5) have a particular application for sound absorbent linings in absorptive type silencers carrying high temperature gas flows (*eg* gas turbine exhausts). Basically such a construction, with the perforated surface inward-facing, presents a multiplicity of Helmholtz resonators.

Acoustic attenuation (sound absorbing) panels of this type can be flat, or with single or double curvature, or be surfaces of revolution such as a jet pipe bullet.

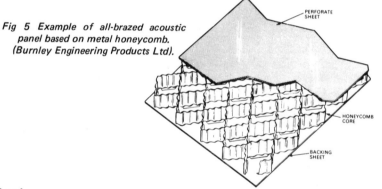

Fig 5 Example of all-brazed acoustic panel based on metal honeycomb. (Burnley Engineering Products Ltd).

Lead

Lead sheet is an exceptionally good sound insulating material, not only because of its high superficial weight but also because of its natural limpness eliminating coincidence effects. This means that a sound barrier incorporating lead can be lighter and thinner than other sound barriers offering a similar performance — *eg* see Fig 6. Until recently the main limitation in the use of lead was the high cost of rolling the slab to the thin gauges required. This has been overcome by the development of continuous cast thin sheet lead.

Lead sheet is specified by superficial weight (kg/m^2 or lb/ft^2). One square foot of 1/64 in thick sheet lead weighs 1 lb, the usual sizes available being 1 lb (1/64 in thick), 2 lb (1/32 in thick) and 3 lb (3/64 in thick). Thin sheet lead has the advantage that it can be easily cut (*eg* with scissors), wrapped or draped around a structure; nailed, stapled, *etc;* and can readily be bonded to other surfaces.

Lead materials for sound insulation also include the following:

Lead/foam sandwich — sheet lead laminated between polyurethane foam. The lead is usually a single layer of 1 lb sheet, with foam thicknesses ranging from 6 mm (¼ in) to 50 mm (2 in). A particular advantage of this material is that it can readily be cut and shaped by hand and bonded to another surface, the foam providing isolation of the lead maintaining its desirable limpness for optimum acoustic performance.

ACOUSTIC MATERIALS

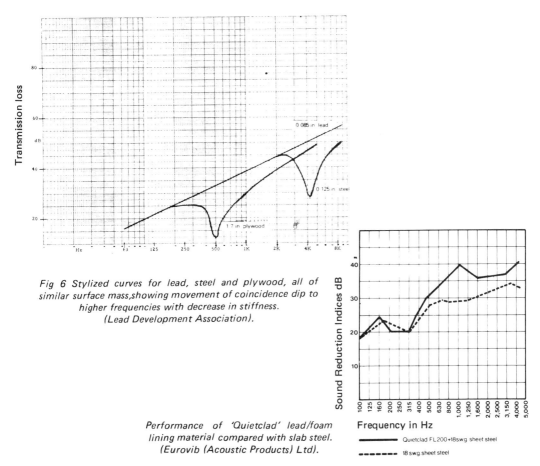

Fig 6 Stylized curves for lead, steel and plywood, all of similar surface mass, showing movement of coincidence dip to higher frequencies with decrease in stiffness.
(Lead Development Association).

Performance of 'Quietclad' lead/foam lining material compared with slab steel.
(Eurovib (Acoustic Products) Ltd).

Lead-loaded plastic sheet — usually vinyl or neoprene sheet loaded with powdered lead to yield a flexible curtain or blanket sound barrier. Fabric reinforcement may also be incorporated in the sheet. This type of material is free from the one limitation of sheet lead used as a drape material — *ie* its tendency to cold flow.

Leaded plastics — usually lead-loaded epoxy in the form of 'damping' compounds for application by trowelling; casting compounds for patching or filling voids; or in the form of mould tiles (damping tiles).

Mastic Coatings

Polymer based catalytic-curing acoustic mastics are tough, durable and having the consistency of thick plaster can be applied to irregular as well as regular surfaces. They normally have maximum effectiveness as a sound insulant when applied over 25–50 mm (1–2 inch) thicknesses of 64–128 kg/m^2 (4–8 lb) density glass fibre insulation.

Metal Jacketing

A proprietary example of acoustical jacketing is Muffl-Jac© which is a special composite of high quality lead, laminated to an outer protective layer of type 3003 or 5005 high purity aluminium.

The product is specifically designed to reduce sound generated by, or within, piping and equipment, with an acoustical effectiveness superior to that achieved by the separate applications of lead and aluminium, and is available in rolls, sheets and prefabricated forms.

A special 3/16 in corrugation is incorporated into standard Muffl-Jac© to add stiffness (for the attenuation of lower frequencies) and strength, reduce glare, allow for expansion and eliminate 'coil breaks' over small size piping.

For maximum acoustical effectiveness, a separation should exist between the radiating surfaces and the acoustical jacketing — eg the jacketing can benefit from being applied over 1—2 in thicknesses of glass fibre or mineral fibre insulation.

Mass-Loaded Vinyl

Heavyweight vinyl sheet — or mass-loaded vinyl — is produced in standard weights of 2.4—4.9 kg/m^2 (0.5—1.0 lb/ft^2), with a performance as a sound barrier material directly comparable with that of lead sheet of the same superficial weight. It is produced by high temperature fusion using non-lead fillers (which do not normally include asbestos).

Like lead it is a limp material, but much tougher and more resilient. It is also resistant to water, oils, weak acids and alkali, and fungi. For the same superficial weight it is thicker than lead — eg for 1 lb/ft^2 surface weight, mass-vinyl has a thickness of 2.00 mm, compared with 0.4 mm for lead.

Mass-vinyl is produced in plain and fabric reinforced forms, and also in composite construction — eg film-foam laminates for rollaway curtains. Sound reduction performance, in all applications, can be evaluated as that of sheet lead of the same superficial weight.

Mineral Wool

Mineral wool is manufactured from a volcanic deposit called diabase by melting the rock (together with limestone and coke) and converting it into fibres by a spinning process. The quality and size

TABLE V — TYPICAL FORM OF ROCKWOOL ACOUSTIC INSULATION

Form	Density kg/m^3	Thickness mm	Applications
Acoustic quilts	48	25—50	Acoustic insulation (absorbers) behind perforated linings; Sound absorbing layer in lightweight partitions: duct insulation
Acoustic pads	16—32	25—50	Sound absorbing layer in perforated ceilings
Acoustic slabs	32—48	25—50	Sound absorbing layer(s) behind perforated linings; sound absorbing layer in perforated ceilings
Rockwool slabs	20—144	25—110	Acoustic insulation of panels, partitions, walls, ceilings: duct insulation (in rigid and semi-rigid forms)
Resin-bonded Building slabs (load bearing)	64—96	12.9—28	Impact sound insulation; sound insulation of electrically heated concrete floors
Resin-bonded Building slabs (non load-bearing)	48	20	Mainly thermal insulation of cavity or solid brick or block walls, etc.

ACOUSTIC MATERIALS

of the fibres is determined both by the selection of raw material and control of the manufacturing process to produce a base material with consistent properties. This material is then further processed to produce a variety of products used for sound treatment, *eg* slabs, pads, mats, quilts, *etc*, as well as acoustic panels of all types and rigid moulded sections — see Table V.

Rockwool is very widely used in the building industry as a thermal insulant with fire retardent properties. Certain products may combine both good thermal and sound absorption properties. Others are produced specifically for acoustic or thermal insulation.

Phenolic Foam

Phenolic foams can be produced in densities ranging from 8—160 kg/m^3 (0.5—10 lb/ft^3). They are rigid materials which may have a high proportion of interconnecting cells, in which case they can be effective sound absorbers. The main advantage of phenolic foam is that it can withstand higher temperatures than most other types of plastic foams (130°C maximum). It also has a very low co-efficient of thermal expansion.

Phenolic foam is used in the building industry as a core material for structural panels; also in block form for (thermal) insulation of concrete roofs.

Plasterboard

Examples of the sound insulation performance of separating wall constructions using plasterboard are given below:

Basic construction	Range of average sound reduction index (dependent on specific construction)
Timber-framed partitions with 75 x 50 mm studs at 600 mm centres lined both sides	34—42 dB
Gyproc metal stud partitions with studs at 600 mm centres	35—48 dB
Gyproc demountable metal stud partitions with studs at 600 mm centres	36—45 dB
Gyproc laminated plasterboard partitions	31—48 dB
Plasterboard cellular core partitions	29—47 dB

See also chapter on *Sound Reduction in Domestic Buildings*.

Steel

Steel is a convenient material for the construction of strong, rigid enclosures for noise machines, *etc*, although its performance is frequency-dependent (*ie* exhibits coincidence effects) and additional damping treatment is usually necessary or desirable. The effectiveness of steel (and other metals) as a sound barrier is directly proportional to the superficial weight (*ie* follows the mass-law for sound insulation).

Vinyl

PVC thick film 48 kg/m^3 (3 lb/ft^3) average in weight, can be used for transparent noise barriers (*eg* flexible curtains), acoustic performances following the limp-mass principle. Such a barrier may be capable of providing up to 25 dB sound reduction, or normally 20 dB overall. Transmission loss decreases markedly with frequencies below 1 000 Hz — *ie* thick PVC films are most effective for middle and higher frequency treatment.

Vinyl foams (expanded PVC) are employed mainly as core materials and as thermal insulants. They are the strongest of the cellular plastics. Interconnected cell foams have good sound absorption properties.

Other vinyl products specifically developed for sound treatment include lead-loaded vinyl and fibre-loaded vinyl. Both are limp, flexible materials with better sound reduction properties than plain fibre. Greater sound reduction can be achieved using multi-layers of such materials. Lead-loaded vinyl is produced in various grades ranging from 0.75–7.5 kg/m^3 (0.15–1.5 lb/ft^3). It is also used with polyurethane foam facings, and with aluminium foil facing. The latter is used to provide both thermal and acoustic insulation for ducts and tiles.

Vermiculite

Exfoliated vermiculite is a widely used loose-fill thermal insulant, but rather less effective in this form as a sound absorbent. Bonded foliated vermiculite is produced in the form of lightweight rigid boards which have both good thermal insulation and sound absorption properties.

Wood Fibre Boards

Insulation board is a processed wood fibre board with a density of less than 160 kg/m^3 (10 lb/ft^3). Boards of this type are also produced with perforations as acoustic boards — see also Table VI.

Particle board or chipboard is made from solid fragments of wood rather than fibrous pulp, bonded with a synthetic resin binder. Densities range from 160–320 kg/m^3 (10–20 lb/ft^3). Both are standard building construction materials the sound reduction properties of which depend on the form of construction, cementing behind the board and any infill used.

TABLE VI — SOUND ABSORPTION COEFFICIENTS OF WOOD FIBRE PRODUCTS

Type of board	Thickness (mm)	Cavity behind board (mm)	Absorption Co-efficient Frequency (Hz)					
			125	250	500	1000	2000	4000
Plain insulating board	12.7	152.4	0.13	0.56	0.15	0.13	0.17	0.20
Acoustic boards								
Regular perforations	12.7	152.4	0.09	0.67	0.47	0.61	0.73	0.80
Regular perforations	12.7	25.4	0.20	0.30	0.35	0.55	0.70	0.70
Regular perforations	12.7	Solid	0.10	0.20	0.40	0.50	0.45	0.50
Micro-perforations	12.7	19.1	0.19	0.68	0.46	0.65	0.73	0.61
Micro-perforations	12.7	Solid	0.11	0.28	0.67	0.65	0.71	0.59
Regular perforations	19.1	25.4	0.20	0.50	0.70	0.85	0.75	0.65
Irregular perforations	19.1	25.4	0.20	0.40	0.49	0.51	0.80	0.86
Cross grooves	19.1	50.8	0.08	0.13	0.34	0.50	0.67	0.80

Acoustic Enclosures

COMPLETE ENCLOSURE of the noise source is usually the best — and possibly the only suitable — treatment when the overall noise reduction required is 15 dB(A) or greater. Basic rules to follow in the design and construction of a close fitting enclosure are:

(i) Overall sound reduction index of the structure must be as high as possible.
(ii) Maximum absorption at all frequencies of interest must be provided inside the enclosure.
(iii) The integrity of the enclosure must be high (*ie* it must be 'leakproof' as regards sound).
(iv) Mechanical isolation between the enclosure structure and the machine must be as complete as possible.

Basically the sound reduction index (SR1) of any given panel is directly related to its superficial weight, *ie*:

Sound Reduction Index = 20 log Mf − 43 dB ....(1)

where M is the superficial weight of the panel in kg/m²

f is the frequency considered, Hz

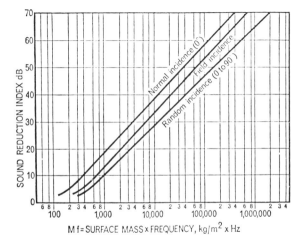

Fig 1 The mass laws for partitions.

This is consistent with the sound impinging on the panel from a direction at right angles to it (normal incidence). In practice the panel may also receive sound reflected on to it at random incidence, modifying performance somewhat. In a typical case the effective sound reduction index probably lies between the two — see Fig 1. Relating the sound reduction index required to the field incidence line then gives the mass frequency value required. From this can be calculated the superficial weight of panel required to provide this level of attenuation at all frequencies of interest. A final figure for superficial weight is then chosen on this basis. Specifically, if frequencies up to and including 250 Hz are the directing factors, estimates for the full superficial weight are calculated for the lowest frequency. If the most significant frequencies are 500 Hz or higher, about 75% of this value should be sufficient.

Various other materials can also be used for modular panel construction, some of which may offer distinct advantages over sheet steel (*eg* easier construction and assembly). The limiting factor is the density of the material or composite which must be high enough to avoid an unacceptable panel thickness. Such materials, however, may have disadvantages, in varying degrees of susceptibility to damage, lack of structural strength and difficulty of sealing.

Loss of Performance

If the enclosure could be made as a completely uniform one-piece box, then the overall sound reduction index as used in the noise reduction formula (equation (1)) would be virtually the same as that of the basic panel from which the box is constructed. For very small items of plant coverable by a 'lift-on' box, and where no ventilation or access is required, such is indeed the case.

Real enclosures however inevitably suffer from a number of potentially weak areas, which unless carefully designed and assembled will have the effect of reducing the overall effective sound reduction to something rather less than the potential of the basic panel construction.

Thus in the case of a non-homogeneous panel there will be a loss of sound insulation in 'weak' areas, with the average sound reduction index for the whole panel being given by:

$$R_{av} = 10 \log_{10} \frac{S1 + S2 + S3 + \ldots}{(S_1 t_1) + (S_2 t_2) + (S_3 t_3) + \ldots} \qquad \ldots \ldots (2)$$

where S1, S2, S3 are the respective surface areas and $t_1, t_2, t_3 \ldots$ are the respective sound transmission coefficients.

Thus if suffix 1 in the average transmission coefficient calculation is taken as applying to the basic panel construction, and suffixes 2, 3, *etc,* apply to other areas of the enclosure such as doors, windows, ventilation openings, or even openings between panel edges, all of which will, more often than not, have a higher transmission coefficient than the basic panel, the average transmission coefficient will also be higher than that of the panel, *ie* the average sound reduction will be lower.

The calculation is best carried out graphically and one convenient representation of the fundamental expression is shown in Fig 2. To calculate the average for panels consisting of more than two areas of different sound reduction index, first use the chart to evaluate the average between any two, the main panel and one weak area would be best. Then evaluate the combination of a third area with the average of the first two, and so on. Obviously, the weakening effect of areas of different construction will be much less if the individual sound reduction index of each of the other areas can be designed to be as close as possible to that of the basic construction panel. Designs for achieving this will now be considered for the 'weak' areas likely to be necessary in practical enclosures.

ACOUSTIC ENCLOSURES (A)

For machinery: **MAFUND® pads**

an elastomeric cushioning material designed for isolation of shock, vibration and structure-borne sound

EICHLER

Austria:
A-1101 Vienna, Pernerstorfergasse 5, ☎ (0222) 64 91 81/63, Telex 131105 eicha

USA: LORD CORPORATION Allforce Acoustics, Box 1067, Erie, PA 16512, ☎ 814:838-7691

GB: **Designed for Sound Ltd.**, 59/61 High Street, Wivenhoe, Essex, ☎ 020622/4477, TX: 987637

ANTIVIBRATION
with insta plant

Insta Plant is the modern method of fixing industrial machinery to any solid floor. No more bolting or floor damage, with Insta Plant, layout changes can be made in a matter of hours.

For further information contact:
Bury Cooper Whitehead Ltd
BCW
Hudcar Mills, Bury
Lancashire. BL9 6HD
Telephone 061 764 2262

RICH MULLER
can now supply from U.K. stock

PERFORATED METAL
Can now be obtained from our new stockholding operation in the United Kingdom.
Available for immediate despatch, we offer:

Galv. Steel 1250 x 2500 mm
(3 mm holes)
Mild Steel 1000 x 2000 mm
(various)
Stainless 1000 x 2000 mm
(various)

TRY US for an efficient service and keen prices

RICH. MULLER (UK) LTD
London House,
77 High Street
Sevenoaks, Kent TN13 1LD
Tel: 0732 459595 Telex: 95658

ACOUSTIC ENCLOSURES (B)

The new Modular System for Machinery Protection

A compact monitoring package of proven electronics including vibration, displacement, temperature, speed and other modules.

Designed to meet most recognised specifications including: BASEEFA, Lloyds, C.E.G.B. and C.S.A.

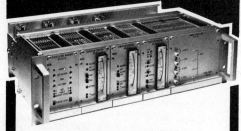

WEIR PUMPS LTD.
Newlands Road, Cathcart,
Glasgow G44 4EX.
Tel: 041-637 7141
Telex: 77161/2

W37

SOUND ADVICE

CRL 2.35
Sound Measuring System

Automatic Memory
Type 1 Precision Accuracy
Leq
Peak Capability
Dose
Frequency Analysis

Also available:
Industrial Sound Level Meters
Audiometers
Automatic Noise Alarms

Send for details from:-
CIRRUS RESEARCH LIMITED
1/2 York Place, Scarborough, YO11 2NP.

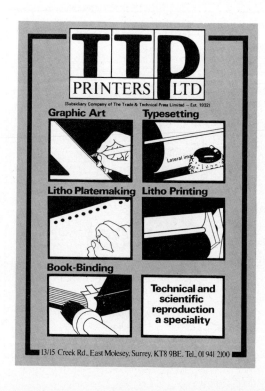

ACOUSTIC ENCLOSURES

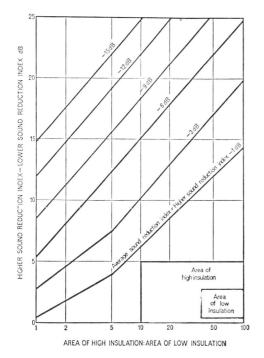

Fig 2 Loss of sound insulation due to 'weak' areas.

Two acoustically protected carbins and a control room in a strip aluminium rolling plant.
(ACE).

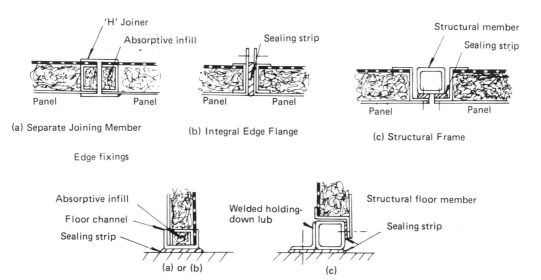

(a) Separate Joining Member

(b) Integral Edge Flange

(c) Structural Frame

Edge fixings

(a) or (b)

(c)

Floor or Masonry Wall Fixing

Fig 3 Methods of joining panels.

Practical acoustic enclosures are, almost without exception, constructed from individual panels joined together at their edges to form the whole 'box' — the size of individual panels being fixed by the overall size of the surface of which it forms part, the size of the opening or viewing area required, or simply the available size of the basic sheet material of which it is constructed.

The prime requirement of joints between panels is that they must allow absolutely no leakage of acoustic energy through. A good practical guide to the design of suitable joints is to regard them as having to be airtight. This in turn will almost certainly require the use of some form of soft flexible sealing strip in rubber or neoprene.

In current commercial designs, one of three methods is commonly used: edge joining by integral flange; edge joining by separate 'H' or channel strip; or complete panel support by separate frame. Fig 3 shows an example of each of these, together with suitable methods for sealing the edges of wall or roof panels to concrete floors and masonry walls.

Doors

Of all the facilities required in an enclosure, access is probably the most called for. It is of course perfectly feasible to provide this by making one or more complete panels removable, although this would demand a construction system along the lines of Fig 3(c).

It is more usual however — especially in cases where frequent access for maintenance, work loading, or even direct operation of the machine is required — for one or more panels to be attached by hinges to form a door.

Acoustic doors are in fact of fundamental importance in the field of noise control. There is very little to be added in the specific context of designing doors for close fitting acoustic enclosures, except perhaps to emphasize the importance of constructing the door leaf to the same specification as the basic enclosure panel wherever possible, and of ensuring that the construction of the enclosure in the region of the door is adequately stiff to prevent progressive 'sag' or other distortion of the door frame, leading to the appearance of gaps around the edges when closed. Enclosure doors must of course be provided with the same degree of sealing as is employed on doors in the more conventional masonry surroundings, and furniture such as roller catches and espagnolette bolts must be of equivalent standard.

Sliding doors can be used to good effect on prefabricated enclosures, but without positive seal compression on closure, their effectiveness is probably limited to say 15–20 dB(A) noise reduction.

Ventilation

Where ventilation of the enclosure is necessary, noise leakage can be minimised by fitting a directtive silencer to both intake and discharge openings. Alternatively, it may be possible to bring air into and out of the enclosure via an acoustic louvre. This is a louvre performing all the architectural non-vision and weather-proofing functions of the conventional article, but having sound absorbing vanes. The basic elements of its construction are shown in Fig 4 together with typical performance figures, although these may be expected to vary by up to 3 dB with different manufacturers' designs.

Work Feed

In many cases, the machine being enclosed is one involved directly in production, such as woodworking or automatic pressing from strip feed, as opposed to 'service' machinery such

ACOUSTIC ENCLOSURES

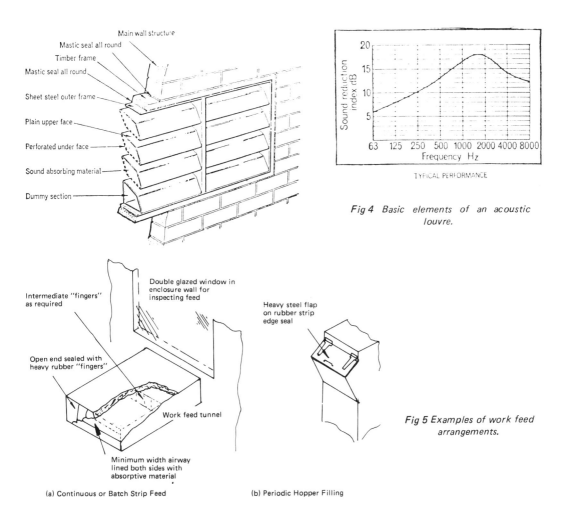

Fig 4 Basic elements of an acoustic louvre.

Fig 5 Examples of work feed arrangements.

(a) Continuous or Batch Strip Feed
(b) Periodic Hopper Filling

as electrical power generators or air compressors. Whatever other features may be required in the design of enclosures for such machines, openings for feeding in and taking out work are vital.

As with any other opening, the overriding requirement is to preserve the acoustic integrity of the panel, in other words to provide an 'attenuated' opening. For machines which require long or continuous feeds such as the two examples just mentioned, the usual solution is to provide an acoustically lined tunnel through which the work has to pass, as indicated in Fig 5.

Other machines require a supply of material only periodically, the continuous feed coming from their own hopper. Granulators for recycling scrap plastic, or hammer mills are good examples here. In such cases it is often sufficient to provide a flap in the enclosure, hinged rather than sliding, so that a good seal can be ensured when the flap is closed. If the flap is horizontal or inclined, its weight will probably be sufficient to seal, otherwise a quick release wedge catch will be required.

IAC acoustic enclosure in foundry provides a combined noise and dust control solution for the fettling area.

IAC modular acoustic enclosure for high speed punch press at Automotive Products Ltd in Leamington Spa.

Windows

Enclosures frequently require a means of inspecting the interior without necessarily opening access doors, for both safety and production operations. It is cheaper for example to leave gauges and meters indicating machine performance, attached to the machine than to extend them outside. Operators feeding enclosed machinery from outside with long materials such as timber also of course require visual access.

As with acoustic doors, the design of windows for high noise reduction, is of universal importance in buildings. Again there is little to add in the special context of acoustic enclosures, except to point out the obvious limitations in design flexibility of which the most important is the restricted airspace that can be accommodated between panes of double glazing. To some extent the resulting loss of attenuation can be offset by using heavy panes, but more likely if the eventual sound reduction index for the panel as a whole is not to be seriously reduced, areas of inspection windows should be kept as low as possible.

Service Entries

The problem here is to bring the various conduits and pipes feeding power, water or fuel supplies, *etc* through the enclosure walls without introducing leakage points. If designing for a new installation, services can be brought in under the wall *via* a trough set in the foundations. If the trough is covered over a length of two metres or so and absorptive material laid in under the cover to fill the gaps around pipes, *etc,* attenuation through the trough should be adequate, (see Fig 6). Alternatively, the trough may be grouted with mortar or mastic joins

ACOUSTIC ENCLOSURES

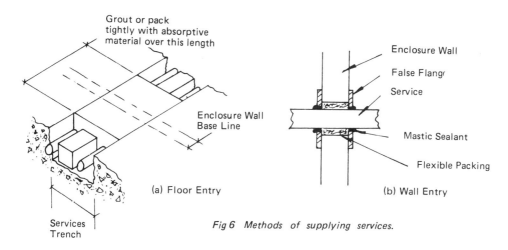

Fig 6 Methods of supplying services.

where it passes under the enclosure wall. One or two services which can be transported in fairly flexible pipes, may be passed through the airways of ventilation air silencers. With engine exhaust gas in particular, if the gas is released inside the enclosure just at the entrance to the air discharge silencer where it is purged by the vent air fans this does have the advantage of removing any need for secondary silencing, the primary silencer also being enclosed with the engine. It does of course require suitably corrosion resistant construction for the discharge air silencer, and is only permissible with an outside installation where the fume-carrying ventilation air is not likely to affect anyone.

More usually, services have to be brought through the actual enclosure wall. Here we have two options. One is to provide a very heavy (up to 6 mm) plate panel which carries connectors on both sides, so that the service is brought up and connected to the outside, and a separate section joins the inside connector to the machine. The alternative is to provide a clearance hole in the wall panel, and pass the service through, sealing by, for example, the method indicated in Fig 6.

Of the two, a specific services panel is to be preferred, particularly if more than one service is to be introduced, as each must be sealed separately if taken through the wall.

SPL in Rooms

Where the noise source is fully enclosed within a room the reduction in reverberant sound pressure level at a point in the room due to enclosure of the source, is given approximately by:

$$SPL_1 - SPL_2 = R - 10 \log S_E + 10 \log A_E \text{ dB} \qquad \ldots\ldots\ldots(3)$$

where SPL_1 and SPL_2 are respectively the reverberant sound pressure levels in the room before and after enclosure of the source, dB

- R is the sound reduction index of the enclosure wall, dB
- S_E is the amount of surface area of the enclosure radiating into the room, m^2
- A_E is the total absorption inside the enclosure, $S_E \bar{\alpha}_E$, m^2 units.
- $\bar{\alpha}_E$ is the average absorption coefficient inside the enclosure.

In terms of parameters which the noise control engineer can use to maximize the room sound pressure level reduction, the overall S_E can be more or less discounted. Whether the enclosure is 'walk-in' or contained, the difference in overall surface area likely to occur in practice has a small effect on the overall performance, and the area term can be regarded as being more or less constant for the particular machine being enclosed.

The design parameter offering most scope for achieving a given target reduction is R, the sound reduction index of the structure. There is indeed some variation possible in internal absorption, but beyond a certain range of engineering methods of providing the absorption, scope for further increase is limited.

Reducing Reverberant Sound

In the case of an enclosure, because the sound reduction index of the panel remains constant, being a property of the enclosure construction, sound pressure level outside the enclosure will vary directly as sound pressure inside. If, therefore, the inside sound level can be reduced, a corresponding noise reduction will be achieved outside. This reduction is obtained by providing the maximum possible amount of acoustic absorption inside the enclosure.

Since almost invariably the absorption required will be broad band in nature, the treatment most often to be found in close fitting enclosures is the blanket type porous absorber. Even for machines which produce strong discrete frequency radiation, it will be found that practical engineering limitations on the design of a practical modular panel will dictate the use of such material.

The only significant performance parameter which the designer of acoustic panels has at his disposal is the thickness of lining, which of course means the thickness of the panel. Broadly, the design rule is — the lower the frequency, the thicker the absorptive material required. Traditionally, panel thickness, (at least for prefabricated modular enclosures) is either 50 mm for average reduction of up to 35 dB(A) or 100 mm to give up to perhaps 45 dB(A) average reduction. Sometimes only 75 mm of the thicker panel is taken up with absorptive lining, the rest being used for the mass infill referred to earlier in the discussion on maximizing sound reduction index.

Once the thickness of the absorptive material has been chosen, it is true to say that within the range of engineering materials available, there is little to choose between one type and another. The main design parameter is density, and this should be in the range from 30 to 100 kg/m^3, a significantly lighter material being unlikely to have the required absorption properties, while a denser material would be wasteful.

The Room as an Acoustic Enclosure

Another class of problem frequently encountered is where the noisy machinery is in one room, and people affected are in adjacent rooms, or outside the building.

In such cases, the engineering expressions, which can be used to estimate sound pressure level in the receiving space, are of somewhat different form, *viz:*

Room to room

$$L_2 = L_1 - R_{av} + 10 \log (S_p/A_2) \text{ dB} \qquad \qquad \ldots\ldots\ldots(4)$$

ACOUSTIC ENCLOSURES

Room to outside

(a) observer near wall

$$L_2 = L_1 - R - 6 \text{ dB} \qquad \qquad \ldots\ldots\ldots(5)$$

(b) observer some distance from wall

$$L_2 = L_1 - R_{av} + 10 \log S_p - 20 \log r - 14 \text{ dB} \qquad \qquad \ldots\ldots\ldots(6)$$

where

L_2	is the sound pressure level at the observer, either in the adjacent room or in the space outside.	
L_1	is the reverberant sound pressure level in the source room	dB
R	is the sound reduction index of the intervening partition	dB
S_p	is the area of partition	m^2
A_2	is the total absorption in the receiving room	m^2 units
r	is the distance of an outside observer from the wall of the source room	m

Although as stated these are of different form from the original equation for noise reduction in the same room due to enclosure, it will be immediately apparent that the requirement to maximize sound reduction index remains. It should be noticed here too that for room transmission or the case of the outside observer some distance away from the wall, average sound reduction index must be used as before. Noise reaching an outside observer immediately next to the source room wall, will however be determined only by the sound reduction index of the material immediately next to the observation position.

It is assumed that the partition area S_p, absorption in the receiving room, and distance r from the outside wall are all fixed by architectural or operational considerations and may, therefore, be regarded as constant; the only other parameter the acoustic designer has at his disposal is sound pressure level inside the source room.

To reduce this, there is a choice of:

(a) reducing strength of the source machine, a measure dealt with in some detail in various other chapters, but out of context in this one.

(b) providing a close fitting acoustic enclosure over the machine, as covered in detail in the foregoing sections of this chapter.

(c) providing the maximum amount possible of acoustic absorption inside the source room.

In other words, exactly the same three basic design rules enumerated earlier for close fitting prefabricated enclosures, have to be applied to the building structure for an acoustic room.

See also chapters on *Sound Insulation and Absorption* and *Acoustic Materials.*

Sound Barriers

ANY VERTICAL surface will act as a barrier to airborne sound waves, the effectiveness of such a barrier depending on the properties of the material concerned and the physical form of the barrier. Hard, dense materials will *reflect* sound, whereas porous, low density materials will *absorb* sound. The ultimate amount of sound transmission will, therefore, depend on the reflecting (insulating) and/or absorbent properties of the barrier, together with its effective height and length. There can also be flanking transmission through the solid media concerned — Fig 1.

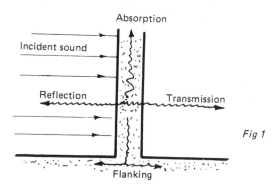

Fig 1

To act as an effective sound *insulator* the barrier material should have a superficial weight or surface *mass* of at least 7 kg/m^2. A number of common building materials meet this requirement — see Table I. Building debris, sand bags, soil (mounds of earth) or barriers built from old tyres, *etc* are other possibilities for permanent barriers. Topsoil can often be used to 'seal' such barriers. For purposemade barriers, wood-wool slabs fixed to timber posts can form a very effective, durable and low cost barrier.

On an open site, buildings, stored materials and other solid obstructions can all act as sound barriers. Earth embankments can also act as sound barriers. However, unless suitably sited, such barriers may merely transfer a noise problem from one area to another.

To be most effective a noise barrier needs to be placed as close as possible either to the noise source or the receiving position. There should be no gaps or joints in the barrier through which sound would 'leak', and ideally the length of the barrier should be at least ten times its height. Alternatively, if this is not possible, the barrier should enclose the noise source.

Rigid steel acoustical panel barriers offer good sound attenuation, capable of providing 15 dB(A) or greater noise reduction. Their main disadvantage is that they are heavy and relatively

SOUND BARRIERS

91 m long x 12 m high sound barrier formed from a structural steel frame housing 100 mm thick composite acoustic panels consisting of sound absorbent material sandwiched between faces of solid and perforated sheet steel on opposite sides of the wall.
(ICI Limited)

TABLE I — SOUND INSULATION CHARACTERISTICS OF COMMON BUILDING MATERIALS
(BS5228:1975)

Material	Thickness	Surface mass	Mean sound reduction index (100 Hz to 3150 Hz)
	mm	kg/cm^2	dB(A)
Asbestos cement boards	6	12	26
Brickwork	113	220	35 to 40
Chipboard	18	12	26
Clinker blocks	75	100	23
Fibreboard (insulation board)	12	4	18
Compressed straw	50	17	28
Plaster board	13	12	26
Plywood	6	4	21
Woodwool cement slabs both faces plastered with 13 mm	76	70	35

costly, as well as being difficult to erect and dismantle. Thick plywood is a lighter material for manipulation, but does not offer the same acoustical performance.

Sound *absorbent* materials are generally poor sounding insulators in that they readily transmit sound through them. Their usefulness is as linings to reduce the reflections of sound incident on particular surfaces. Particular use of sound absorbing materials is made for the *lining* of screens or semi-enclosures where the object is both to reduce the sound generated by a source, such as a noisy machine or tool, and at the same time to reduce the overall noise level in the vicinity of the source which would otherwise be reflected by a sound barrier and thus add to the discomfort of the machine or tool operator.

Some typical sound absorbing materials are listed in Table II, together with average absorption coefficients. The absorption coefficient, multiplied by 100, represents the *percentage* of sound absorbed by the material.

In general, a simple sound barrier can give an anticipated reduction of up to 15 dB(A), although in the majority of practical cases a figure of 10 dB(A) is regarded as more usual.

Sound Curtains

Specifically a barrier is taken to be a rigid screen, commonly constructed from modular acoustic

TABLE II — SOUND ABSORBING MATERIALS FOR LINING OF COVERS AND ENCLOSURES

Material thickness		Average absorption coefficient
mm		Between 125 Hz and 4000 Hz
Glass fibre	50	0.7
Mineral wool	50	0.8
Straw slabs	50	0.4
Woodwool slabs	50	0.6

Suspended noise absorbers installed in a print shop. (Acousticabs Ltd).

panels. An alternative is the non-rigid acoustical *curtain* comprised of high density flexible material (*eg* lead-loaded vinyl or similar).

Thin acoustical curtains provide easy maintenance and operational flexibility but are quite limited as to actual noise reduction, particularly if they are designed as barriers to contain sound, and not absorbers. Curtain performance can be enhanced by adding absorptive foam to the interior surface, in which case the level of noise absorption achieved depends on the foam thickness and the area covered. This presents certain practical problems.

Thinner foams are poor absorbers but allow for complete curtain coverage. Thicker foams have good absorption properties but generally need to be applied in strips or segments, reducing their overall effectiveness. Exposed foam surfaces are subject to abrasion and other damage, but this particular limitation can be overcome by a protective outer covering. Another point which has to be considered with foams is that they may present fire and toxic smoke hazards.

A number of proprietary materials have been developed to overcome such limitations inherent in foam-lined curtains or absorber/barriers. Lead is an obvious choice for sound insulation, but is unsuitable as a curtain material. This limitation is overcome with composite constructions employing a lead septum as a central element with layers of glass fibre or similar materials on both sides to act as a sound absorber. The outer surfaces of such a composite may be protected with a further layer of aluminized glass fibre scrim or cloth.

Such composition curtains, and those which are absorbent rather than insulating (reflecting) barriers may be finished with a quilted surface to provide an attractive outer surface. Quilting also has the property of providing acoustic decoupling of noise impacting and penetrating the curtain.

A disadvantage of quilting is that the presence of stitch holes means that the sub-strata are not fully protected against harsh environments and so plain surface finished would normally be used in such conditions.

Gullfiber's mobile acoustic screen can be wheeled to the required position and pulled out to a length of up to 11 m (36ft) to wrap around a noise source. (Industriakustik Ltd).

Modular acoustic screens arranged to make individual grinding and welding booths. (Industriakustik Ltd).

An alternative material for sound insulating curtains is mass-loaded vinyl. In equivalent weights it offers similar noise transmission loss with the advantages of lower cost, easier handling and installation, greater durability and freedom from the toxicity hazard associated with lead.

Originally thick PVC film in weights up to 3 lb/ft^2 was developed as a sound barrier material, proving most effective for sound absorption in the middle and upper frequency ranges, but inferior to lead overall. Much better performance was achieved by mass-loading, the first of these materials being lead-filled vinyl. Today's heavyweight vinyl sheeting is produced by high temperature fusion using non-lead fibrous fillers, and generally excluding asbestos (another material held to be hazardous to health).

Acousticurtain in a factory, installed on sliding gear to provide an instantly moveable acoustic partition. (Acousticabs Ltd).

On a weight-for-weight basis (theoretically any limp barrier material should have the same sound reduction value if the mass per unit area is the same), 2 mm thick mass-loaded vinyl is directly comparable with 0.39 mm (1/64 in) thick sheet lead, both weighing 1 lb/ft^2. In terms of *thickness* of insulation material, mass-loaded vinyl needs to be about five times as thick as sheet lead for similar performance as a sound insulator.

The practical advantages are that the thicker mass-loaded vinyl is tougher, resistant to tearing or ripping during handling, and does not take a set when bent. Like thin sheet lead it is also easily cut with a knife or scissors and established application techniques used with lead systems are generally applicable. Further, for the same weight of material, the price of mass-loaded vinyl is stable and less than that of sheet lead.

Other advantages offered by mass-loaded vinyl are that it is weather resistant and unaffected by oils, weak acids and alkalis. Being a non-metallic material it cannot generate electrolytic corrosion in contact with metals. There is also the practical advantage where large surface areas are to be treated that it is generally available in 54 inch wide rolls, whereas the maximum width available in lead sheet is usually 48 inches.

Mass-loaded vinyl is produced in standard weights from 0.5 to 1.0 lb/ft^2, giving STC ratings from 19 to 27 (see Fig 2). By comparison the usual sizes available in sheet lead are 1 lb/ft^2 (1/64 inch thick), 2 lb/ft^2 (1/32 in thick) and 3 lb/ft^2 (3/64 in thick). To provide comparable performance with 1/32 in or 3/64 in thick lead, therefore, multiple layers of mass-loaded vinyl would have to be used, although material cost should still be favourable to the vinyl material.

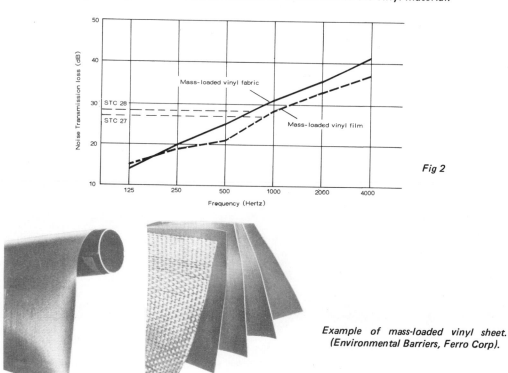

Fig 2

Example of mass-loaded vinyl sheet. *(Environmental Barriers, Ferro Corp).*

Installation cost would, however, be increased by the necessity of applying two or more layers. Against this is the fact that vinyl sheeting required to provide a similar STC rating to sheet lead should not prove an embarrassment, except in certain specialized applications.

Mass-loaded vinyl is available in both fabric-reinforced and non-reinforced forms. Various other forms of filled and non-filled vinyl sheeting are also available as sound insulation materials — *eg* curtain materials comprising one or more sheets of vinyl encased in foam, with a tough outer film to protect the foam from mechanical or chemical attack. These curtain materials can provide sound reduction directly comparable to other constructions containing equivalent weights of lead.

Mass-loaded vinyl fabric is suitable for sound insulation on machinery, pumps, ducting, housings and enclosures and transportation applications.

Mass-loaded vinyl film, with or without fabric reinforcement is a flexible noise barrier material suitable for rollaway curtains, *etc*; also for pipe and duct noise laggings, machinery covers, ceiling noise barriers, crosstalk barriers, wall and door septums, *etc*.

Semi-Enclosures

Semi-enclosures, also known as *acoustic sheds* to distinguish them from machine enclosures, are structures erected around a source of noise, but not fully enclosing the source, leaving space for an

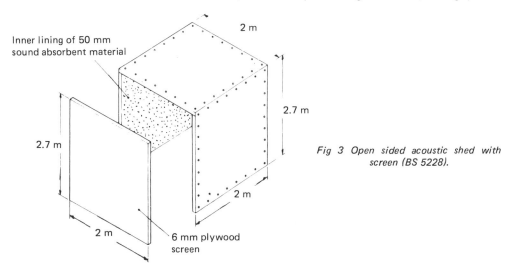

Fig 3 Open sided acoustic shed with screen (BS 5228).

operator to work a noisy tool, *etc* — Fig 3. The main structure may or may not be of sound insulating material, but will have an inner lining of sound absorbent material to reduce sound reflection inside the acoustic shed — see also Table III.

Fairly lightweight construction is used in the case of portable acoustic sheds, *eg* typically plywood on timber framing. It is important that there should be no gaps at the joints or corners. For permanent acoustic sheds blockwork or breezeblock is to be preferred. Again all joints should be properly made. Any gaps between the sides and ground can, if necessary, be sealed with a flap of suitable heavy grade flexible material.

The thickness recommended for the sound absorbent lining is 50 mm; or if mounted on battens 25 mm. Glassfibre or mineral wool linings generally require a wire mesh or perforated screen covering to ensure that they stay in place.

TABLE III — MEASURED SOUND REDUCTION GIVEN BY TYPES OF PARTIAL ENCLOSURE (BS5228 : 1975)

Type of Enclosure	REDUCTION		
	Facing the opening(s)	Sideways	Facing the rear of shed
	dB(A)	dB(A)	dB(A)
Open-sided shed lined with absorbent; no screen	1	9	14
Open-sided shed lined with absorbent; with reflecting screen in front	10	6	8
Open-sided shed lined with absorbent; with absorbent screen in front	10	10	10

Sound will, of course, be radiated directly through the open end of the shed, but this can be stopped by an acoustic screen — see Fig 4. This screen may or may not be lined with sound absorbent material on the operator's side. Usually it would not be, unless special requirements dictated otherwise.

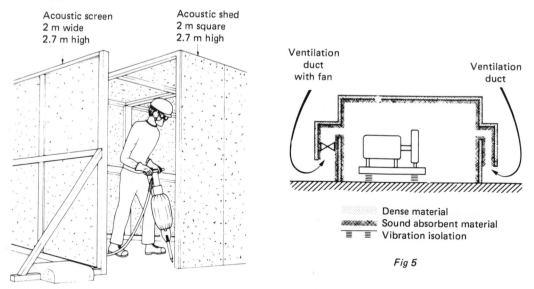

Fig 4 Acoustic shed BS5228.

Fig 5

Semi-enclosures for machines have openings either for access or ventilation and take the general form shown in Fig 5. The main structure should be of sound insulating material with a suitably high surface mass (10 kg/cm^2 or higher), with a sound absorbent lining at least 28 mm thick (although this thickness can be reduced if any high frequency noise is present). Again the favoured materials for sound absorbent linings are glassfibre of rock wool behind wire mesh or a perforated screen; or acoustic tiles.

See also chapters on *Sound Insulation and Absorption, Acoustic Materials* and *Acoustic Enclosures*.

Acoustic Treatment of Floors and Ceilings

TO BE effective as sound insulation, flooring needs to provide a desirable degree of attenuation of airborne sound and also to minimize impact noises caused by footsteps, vibrations from cleaning equipment, and other sources of vibration. In general *airborne sound reduction* can be achieved by a suitable *mass* of floor structure; and vibration and/or *impact sound insulation* is best provided by isolating the floor surface from the structural unit. The latter is the so-called 'floating floor' where the necessary floor mass is carried on a resilient insulating cushion or quilt or acoustic isolation pads.

A floating floor, as its name implies, is an internal floor slab or screed separated from the structural floor by a layer of resilient material forming an impedance barrier to transmission of energy to the slab forming the ceiling of the room below. One example of a simple floating floor is shown in Fig 1.

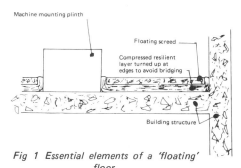

Fig 1 Essential elements of a 'floating' floor.

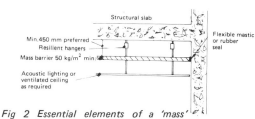

Fig 2 Essential elements of a 'mass' ceiling.

A number of proprietary floating floors are also available, one system for example consisting basically of prefabricated boards to form the base for a cast screed, and incorporating a number of integral anti-vibration pads. Such floor construction can give a very significant — at least 10 dB — improvement whether in fact the source is located in the plant room or the room below.

In some cases, particularly where the problem is with an existing installation, it is clearly impracticable to lay a floating floor. Then the only solution is to provide a resiliently suspended mass barrier as a false ceiling in the room below, as indicated in Fig 2. Note here that a heavy ceiling and a good seal all round are essential. Normal suspended lightweight 'acoustic' ceilings give virtually no improvement at all. Although Fig 2 shows the specific case of a mass ceiling under a plant room, the greatly improved noise reduction can again be expected for transmission in either direction.

TABLE I — CONCRETE FLOORS

Thickness		Weight lb/ft^2	Average Sound Transmission Loss dB
in	(mm)		
2	(50)	25	42
4	(100)	50	47
8	(200)	100	52
16	(400)	200	57
32	(800)	400	62
64	(1 600)	800	67

TABLE II — CONCRETE FLOORS

Construction	Average sound reduction (100–3 200 Hz) dB	Slope	Sound Insulation Grading in Dwellings		
			Airborne	Impact	Overall
Concrete floor (reinforced concrete or hollow pot slab weighing not less than 45 lb per sq ft) with hard floor finish	45	B	Grade II	4 dB worse than Grade II	—
Concrete floor with floor finish of wood boards or ¼inch thick linoleum or cork tiles	45	B	Grade II	Grade II	Grade II
Concrete floor with floor finish of thick cork tiles or of rubber on sponge rubber underlay	45	B	Grade II	Probably Grade I	Grade II
Concrete floor with floating concrete screed and any surface finish	50	B	Grade I	Grade I	Grade I
Concrete floor with floating wood raft	50	B	Grade I	Grade I	Grade I
Concrete floor with suspended ceiling and hard floor finish	48	B	Probably Grade I	2 dB worse than Grade II	—
Concrete floor with suspended ceiling and wood board floor finish	48	B	Probably Grade I	Grade II	Grade II
Concrete floor with suspended ceiling and floor finish of thick cork tiles or rubber on sponge-rubber underlay	48	B	Probably Grade I	Grade I	Probably Grade I
Concrete floor with 2inch lightweight concrete screed and hard floor finish	48	B	Probably Grade I	4 dB worse than Grade II	—
Concrete floor with 2inch lightweight concrete screed and floor finish of thick cork tiles or of rubber on sponge-rubber underlay	48	B	Probably Grade I	Probably Grade I	Probably Grade I
Concrete floor weighing not less than 75 lb per sq ft (reinforced concrete slab 6–7inch thick) with hard floor finish	48	B	Grade I	4 dB worse than Grade II	—
Concrete floor weighing not less than 75 lb per sq ft with floor finish of thick cork tiles or of rubber on sponge-rubber underlay	48	B	Grade I	Grade I	Grade I

ACOUSTIC TREATMENT OF FLOORS AND CEILINGS

Concrete Floors

In terms of airborne sound reduction, a 2 inch (50 mm) solid concrete floor will have an average sound transmission loss of about 42 dB. Since sound absorption follows the mass law, each doubling of the floor thickness will result in an improvement in sound reduction of only 5 dB. It follows that the sound reduction possible is limited by the practical consideration of the weight of concrete involved — see Tables I and II.

In practice a solid concrete floor slab of approximately 4 inches (100 mm) thick will give an air sound insulation performance of Grade II standard and an impact sound insulation figure inferior to Grade II requirements. Impact sound insulation, however, can be improved by a soft surface layer — *eg* cork tiles, or carpet or rubber flooring on a soft underlay. A minimum of 6 inches (150 mm) of solid concrete slab would provide Grade I standard airborne sound insulation, but worse than Grade II standard impact sound insulation unless again finished with a soft surface (Table III).

Performance to Grade II requirements in both cases can be provided by a 4 inch (100 mm) thick concrete slab with a floating screen applied over an acoustic quilt or slab fibrous material which need be no more than ½ inch (12.5 mm) thick. Screen thickness required is usually of the order of 2–3 inches (50–75 mm). This can substantially reduce the weight of the concrete floor.

TABLE III – SOUND INSULATED FLOOR CONSTRUCTIONS

Structural Floor Slab	Floating Floor	Insulation	Sound Insulation Airborne	Sound Insulation Impact	Remarks
4 inch concrete (45 lb/in^2)	2½ inch (65 mm) Screed (min)	½ inch (12.5 mm) Quilt	Grade I	Grade I	Quilt protected by bituminous paper and one mesh blow screed
4 inch concrete	Timber raft	½ inch (12.5 mm) Quilt (min)	Grade II possibly Grade I	Grade I	Raft of floorboards raised on 2 x 2 timbers at 16 inch centres
4 inch concrete	Chipboard raft	½ inch (12.5 mm) Quilt	Grade II possibly Grade I	Grade I	18 mm chipboard flooring on 2 x 2 battens, Polythene sheeting over quilt
4 inch concrete	Chipboard	½ inch (12.5 mm) min building slab	Grade II possibly Grade I	Grade II possibly Grade I	Performance largely dependent on choice of slab material
Timber joist		Pugging applied between joists	Grade II or less	Grade II or less	Mineral wool or sand pugging
Timber joist		Pugging plus acoustic quilt	Grade II or better	Grade II or better	Grade I possible with 2 inch depth sand pugging
Timber joist	Timber raft	Pugging plus acoustic quilt	Grade II or better	Grade II or better	Floorboards on battens, quilt under. Grade I, possible with sand pugging

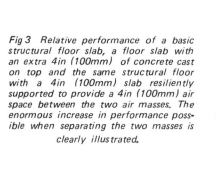

Fig 3 Relative performance of a basic structural floor slab, a floor slab with an extra 4in (100mm) of concrete cast on top and the same structural floor with a 4in (100mm) slab resiliently supported to provide a 4in (100mm) air space between the two air masses. The enormous increase in performance possible when separating the two masses is clearly illustrated.

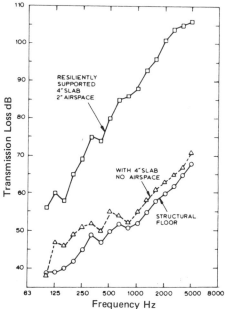

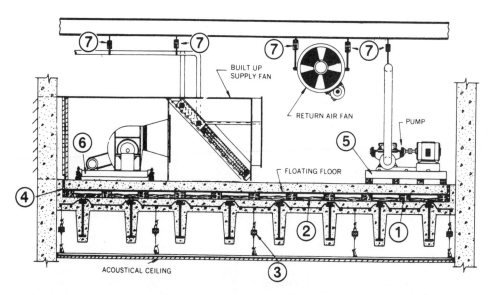

Fig 4 Three-step approach to floating floor design. This consists of a floated floor, high deflection spring isolators on all equipment, and a resiliently suspended ceiling of a dense impervious material. The system components are: 1. Isolation panel consisting of fibreglass isolation pads bonded to exterior grade plywood and supporting the floating floor; 2. Low-density acoustical fibreglass blanket; 3. Fibreglass ceiling isolation hangers; 4. Fibreglass perimeter isolation board with mastic seal; 5. Fibreglass-isolated inertia base for supporting pumps; 6. Spring-isolated inertia base for supporting fans; and 7. Spring and fibreglass hangers for isolating fans and piping.

Other types of floating floor constructions using a quilt or blanket acoustic insulation intermediate layer are detailed in Table III — (see also Table IV for wood joist floors).

An alternative approach is greater separation of floor slab and basic structural slab on resilient supports, including a substantial air space between them. This can provide markedly improved attenuation compared with a similar overall thickness of concrete. Fig 3, for example, shows the performance of a 4 inch (100 mm) concrete structural floor with a 4 inch (100 mm) floor slab cast directly on top; and with a 4 inch (100 mm) floor slab resiliently supported above the same structural floor with a 2 inch (50 mm) air gap. A possible disadvantage with this type of floating floor construction is the increased overall depth of floor provided.

This particular form of floating floor lends itself to further treatment for critical applications when virtually complete freedom from transmitted vibration is required. The floor slab can be floated on isolation mounts and the intervening spaces filled with acoustic blanket or quilt material. Machines or equipment carried on the floor are mounted on high deflection spring isolators. Finally an acoustic ceiling can be suspended from the structural floor slab.

An example of this more elaborate type of treatment is shown in Fig 4.

TABLE IV — WOOD JOIST FLOORS

Construction	Average sound reduction (100–3 200 c/s) dB	Slope	Sound Insulation Grading in Dwellings		
			Airborne	Impact	Overall
Plain joist floor with plasterboard and single-coat plaster ceiling (no pugging)					
Thin Walls	34	C	8 dB worse than Grade II	8 dB worse than Grade II	—
Thick Walls	36	C	4 dB worse than Grade II	5 dB worse than Grade II	—
Plain joist floor with plasterboard and single-coat plaster ceiling and 3 lb per sq ft pugging on ceiling					
Thin Walls	39	C	4 dB worse than Grade II	5 dB worse than Grade II	
Thick Walls	44	C	Possibly Grade II	Possibly Grade II	Possibly Grade II
Plain joist floor with ¾ inch lath-and-plaster ceiling (no pugging)					
Thin Walls	40	C	Probably 4 dB worse than Grade II	Probably 6 dB worse than Grade II	—
Thick Walls	45	C	Grade II	Grade II	Grade II
Plain joist floor with ¾ inch lath-and-plaster ceiling and 17 lb per sq ft pugging on ceiling					
Thin Walls	45	B	Grade II	Grade II	Grade II
Thick Walls	48	B	Grade II or possibly Grade I	Grade II	Grade II

cont...

TABLE IV — WOOD JOIST FLOORS (contd) ...

Construction	Average sound reduction (100–3 200 c/s) dB	Slope	Sound Insulation Grading in Dwellings		
			Airborne	Impact	Overall
Floating floor with plasterboard and single-coat plaster ceiling (no pugging)					
Thin Walls	39	C	4 dB worse than Grade II	3 dB worse than Grade II	—
Thick Walls	44	C	Possibly Grade II	Possibly Grade II	Possibly Grade II
Floating floor with plasterboard and single-coat plaster ceiling and 3 lb per sq ft pugging on ceiling					
Thin Walls	43	C	2 dB worse than Grade II or	2 dB worse than Grade II or	—
Thick Walls	48	C	possibly Grade I	possibly Grade I	Grade II or Grade I
Floating floor with ¾inch lath-and-plaster ceiling (no pugging)					
Thin Walls	43	C	2 dB worse than Grade II	Grade II	—
Thick Walls	48	C	Grade II or Grade I	Grade I	Grade II or Grade I
Floating floor with ¾inch lath-and-plaster ceiling and 3 lb per sq ft pugging on ceiling					
Thin Walls	45	C	Possibly Grade II	Grade II	Possibly Grade II
Thick Walls	48	C	Grade II or Grade I	Grade I	Grade II or Grade I
Floating floor with ¾inch lath-and-plaster ceiling and 17 lb per sq ft pugging on ceiling					
Thin Walls	49	B	Probably Grade I	Probably Grade I	Probably Grade I
Thick Walls	50	B	Grade I	Grade I	Grade I

Resilient Seatings

Resilient seatings in elastomeric load bearing materials are similar to resilient bearings (see chapter on *Resilient Mounting of Structures*), except that they are designed and employed primarily for static load distribution. At the same time, however, they can promote control of noise and vibration.

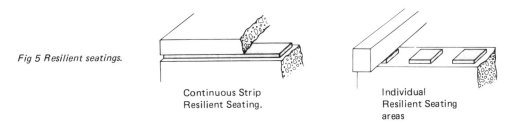

Fig 5 Resilient seatings.

Continuous Strip Resilient Seating.

Individual Resilient Seating areas

Resilient seatings are designed for use on all types of structures in areas where a separating member is desirable — Fig 5. They can be employed between any combination of materials to eliminate high stresses and support edges, thus providing an economic and effective means of avoiding cracks developing through shrinkage, rotation differential settlement, creep and thermal movements within the structure. They can also be used to absorb shock loads, prevent spalling of concrete; fretting corrosion between steel and concrete seatings; and eliminate electrolytic action between dissimilar metals in contact.

Where it is necessary for a resilient seating to accommodate large structural movements it may be necessary to provide a sliding surface in conjunction with a resilient seating as this can allow for considerable movement in any plane. This type of seating is similar in concept and working to a bridge bearing.

There are a considerable number of resilient seating products available designed to cater for the full range of bearing loads found in structural applications. The majority are elastomeric compounds, either plain or reinforced. The type of material chosen, and thickness used, will depend entirely on the degree of horizontal and rotational movement required, together with the degree of unevenness of surface or lack of parallelism between the two mating structures which has to be accommodated. Products which do not readily deflect with a minimum of lateral spread may have surface reliefs in order to allow the required displacement

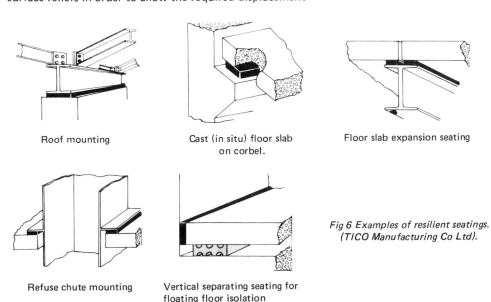

Roof mounting

Cast (in situ) floor slab on corbel.

Floor slab expansion seating

Refuse chute mounting

Vertical separating seating for floating floor isolation

Fig 6 Examples of resilient seatings. (TICO Manufacturing Co Ltd).

Some examples of resilient seating applications are shown in Fig 6. A general application is to load bearing vertical and horizontal areas. Specific applications of major importance include theatre and domestic flooring applications to allow relative structural movement and reduce transmission of impact; and resiliently supported floors in tower blocks situated in areas subject to high temperature variations. In the latter case because of the high temperature movement, standard concrete-to-concrete connection may not be possible — *eg* see Fig 7.

A new and unusual application of resilient seatings is to prevent self-excited oscillation of factory chimneys due to wind forces. This can be a particular problem with modern welded steel chimneys which have very low inherent structural damping. Mounting the chimney on a resilient layer (Fig 8) is cheaper and usually more effective than fitting helical spoilers or vortex breakers to the top section of the chimney. Alternatively a double-seating arrangement is the most effective method.

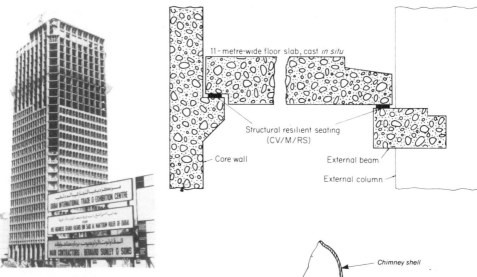

Fig 7 Dubai International Trade & Exhibition Centre shown in course of construction has upper floor slabs simply supported on TICO resilient seatings.

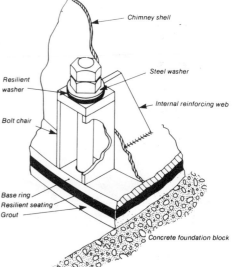

Fig 8 Resilient seating at base of 30 m high chimney. (TICO Manufacturing Co Ltd).

Acoustic Glazing

THE MAIN significance of windows and glazed areas in buildings is their permeability to outside sounds. External sound will normally be random incident on windows, so there can be marked spread of sound through a window area, further diffused by reflections from interior walls. Thus the ratio of window area to facade area is significant. The most marked loss occurs within the window/facade area ratio up to 20%, after which increasing the window area has a much smaller effect. The 'turning point', area 20%, is significant in that the insulation of the whole facade is not very much better than that of the window area itself — see Fig 1.

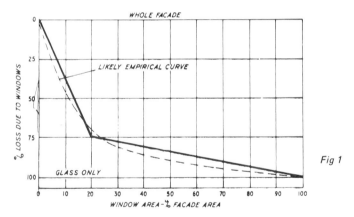

Fig 1

This leads to the following ground rules:

(a) Small windows are unlikely to be effective in maintaining the sound insulation of the whole facade.

(b) Altering the shape of the windows will have little appreciable effect on sound insulation as it is window area which is significant.

(c) Large window areas do not automatically mean additional 'transparency' to sound. The damage has largely been done in this respect once the window area exceeds 20% of the total area.

(d) The insulation of the whole facade can only be maintained by improving the sound insulation offered by the windows. The size and shape of the windows can be chosen from considerations of the daylight, aesthetic, view or heating requirements.

(e) An alternative, or additional possibility exists, namely reduction of the noise incident upon a window. This may be achieved by siting (*ie* increasing the distance from the noise source), the use of barriers and buildings as noise screens, locating noise tolerant rooms on the noisy facades, and rooms requiring silence on the quieter sides of the building. This can be effective in reducing the sound insulation required of the windows.

The sound insulation values of window glass can be predicted with reasonable accuracy from the Mass Law curve (Fig 2). In other words with single glazing glass thickness (or more accurately glass weight) is the significant parameter. Sound insulation is, in fact, increased by about 5dB for each doubling of the mass per unit area of the glazing. A typical figure for average constructions is a sound insulation figure of about 25 dB or rather less, depending on how well the window seals.

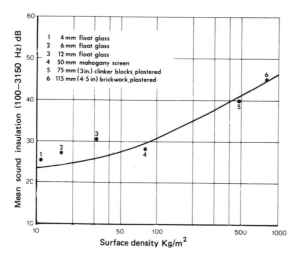

Fig 2 Sound insulation values of window glass and other materials related to Mass Law.

As a general rule glass has rising characteristics of sound insulation with increasing frequency, although this is modified by resonance effects. As a result, increasing the thickness of glass does not automatically ensure that the overall effect will be that much better. The thicker the glass the lower the coincident frequency. Thus 12 mm float glass with its initial frequency of 1 kHz has sufficient mass to be a good barrier to traffic noise. Doubling the glass thickness would give an overall improvement in sound reduction of 5 dB but reduce the critical frequency to 500 Hz, at which frequency typical traffic noise has strong components.

The critical frequency of glass can be calculated as:

$$\text{Critical frequency} = \frac{12\,000}{t} \text{ Hz}$$

where t is the glass thickness in millimetres

The real acoustic performance of single, monolithic glasses are displayed in Fig 3, which shows the results of a statistical analysis of data originating from laboratories all over the world. While the frequencies at which wave coincidence occurs can be predicted with accuracy, the actual fall in sound insulation and bandwidth over which it operates depend on the acoustic damping of the glass.

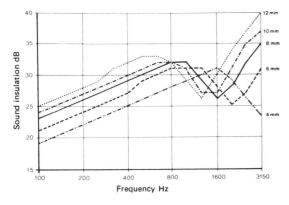

Fig 3 Acoustic performance of single monolithic glasses.

Double Glazing

Double windows, employing two glasses with an intervening air space (generally known as double glazing) offer a direct solution to increasing the mass per unit of a window without necessarily going to excessive glass thickness, combined with further insulation offered by separating the glass into two leaves with an airspace between. To be effective as a *sound* insulator, however, the air space must be generous. A minimum airspace of 100 mm is normally necessary for any realistic improvement in sound insulation, with an optimum maximum of about 300 mm. With greater separation there is very little improvement in sound insulation. For practical (and aesthetic) reasons, the normal maximum adopted is 200 mm. Airspaces of less than 100 mm may also be used; but less than 50 mm airspace gives sound insulation directly comparable to the same (total) weight of single glazing.

The sound insulation performance of double glazing can be improved by incorporating absorbent material in the sides. This is widely used to offset the less than optimum performance realized with smaller air gaps. Ultimately, the overall performance will depend very largely on the integrity of the panel and frame seals. Typical performances of good double glazing using differing thickness of glass are shown in Fig 4.

Sealed double glazing units do not normally offer really high sound insulation because: (a) the relatively narrow airspace is insufficient to decouple the movements of the component glasses and (b) sound energy is transferred *via* the peripheral sealing strip. Generally, they offer an

Fig 4 Sound insulation of double glazing. (Pilkington Environmental Advisory Service).

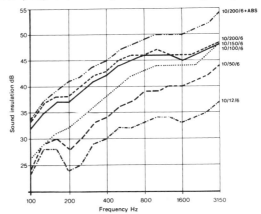

improvement of up to 3 dB in sound insulation over that provided by the thicker of the component glasses alone. Nevertheless, by using a thick glass in such a unit, an acceptable compromise is often achieved between reasonably high sound insulation, coupled with the benefits of thermal insulation, and compactness.

As a general principle, it is usual to specify glasses of differing thicknesses so that sympathetic resonances are avoided. In other words, the coincidence frequencies of the two glasses should be displaced so that while one is resonating (and consequently providing lower sound insulation than the norm), the other is unaffected, and vice versa. By this means, smoother overall sound insulation is obtained; a thickness difference of about 30% is usually sufficient to secure this. Thus, when a total glass thickness of 16 mm is to be installed, it is better to avoid using two leaves of 8 mm glass, 10 mm and 6 mm glasses (or 12 mm and 4 mm glass) being preferred. Even greater control is possible using laminated glass or triple glazing.

Resonance Effects

Double glazing exhibits two further types of resonance — see Fig 5. The mass-air-mass resonance, where the enclosed air acts like a spring in the energy transfer between the glasses, depends on the thickness of the two glasses and on their separation distance, according to the relationship:

$$f_{mam} = 120 \left(\frac{1\,000}{(M_1 + M_2)\,d} \right)^{1/2}$$

where M_1 and M_2 are the masses (kg/m^2) of the glasses

and d = the separation in millimetres

Additional high frequency resonances occur because of standing waves in the airspace separating the two glasses. The dominant Cavity Resonance is at a frequency of

$$f = \frac{V}{2d} \text{ kHz}$$

where V = velocity of sound in air (340 m/s)

d = glass separation, in millimetres

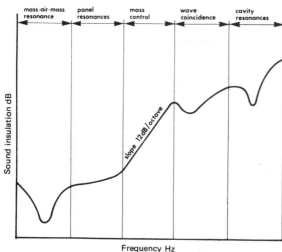

Fig 5 Typical double-glazed window characteristics.

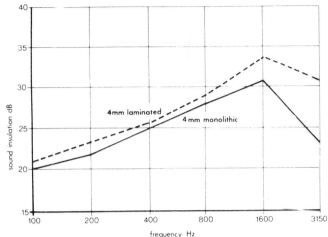

Fig 6 Comparison of monolithic and laminated glasses.

Laminated Glasses

Laminated glasses, two or more glasses bonded together by transparent plastic interlayers (usually of polyvinyl butyral), offer definite advantages in sound insulation. Fig 6 shows that a single, solid glass has a dip in the sound insulation curve at the coincidence frequency, which, for common glass thickness, occurs in the 1 000 Hz to 2 000 Hz region. This dip is smoothed out by using a laminated glass of the same overall thickness as the single glass. There is thus a significant improvement at the coincidence frequency, but at other frequencies the very small improvement achieved must be weighed against the cost.

Sealed double glazing units, incorporating laminated glass, are sometimes used where space limitations do not permit the use of dissimilar glass thicknesses to suppress coincidence effects. Here two panes of the thinner substance may be employed, one being laminated. Little further improvement is gained from laminating the second glass also. Laminated glass offers no acoustic advantage in wide airspaced double windows, because in these arrangements it is mainly the decoupling effect of the airspace itself which secures the resulting high sound insulation.

Special Window Designs

Virtually any required sound reduction can be achieved by suitable design of glass windows and/or glass screens, following the general principles outlined above, but due consideration must also be given to other factors such as space restrictions, load bearing capacity of structures and economics, etc. Thus there is unlikely to be any 'ideal' design for a particular application. It has to be an acceptable compromise. The value of the special window designs also depends on the magnitude of flanking transmissions — eg the performance of the window may be greatly improved, but actual improvement in sound insulation may only be marginal because sound is entering the insulated space by other paths.

Glasses of different thicknesses are not difficult to fix and may be considered for building applications, but setting one glass at an angle to the other and the use of flexible mountings are devices that are usually reserved for the very special observation windows of broadcasting studios, audiology laboratories and the like.

The sound insulation required of a facade is best found by subtracting, at each frequency, the acceptable indoor noise from the existing outdoor noise. This gives a criterion curve against which the properties of various constructions can be matched and a selection made. A simplification can be used for selecting the windows when traffic noise is the dominant problem.

Acoustic Doors

SOUND REDUCING doors or *acoustic doors* are produced in a variety of configurations to suit the method of opening best suited to the application, the amount of noise reduction required, the most effective method of sealing and suitable choice of door furniture. The overall design must also take into account measures which may be necessary to eliminate potential flanking paths. Finish of both door faces and frames may also be significant.

The acoustic performance of sound reducing doors is specified in terms of sound reduction index plotted against frequency. Obviously with the same basic type of door, performance can vary with different constructions — *eg* see Fig 1. A point to watch is that the specified acoustic performance as determined by tests will relate to the door(s) being mounted in a specific type of wall, which is not necessarily the same construction as the wall into which the door is finally fitted. This difference can modify the overall performance achieved.

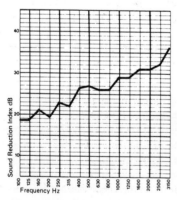
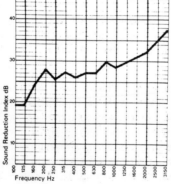
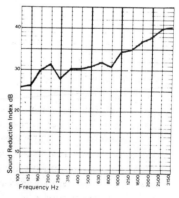

Door Construction:
Solid core of laminated Cedar rails, faced with 4mm crossbanding, hardwood veneered or laminate faced. Hardwood lipped on long edges. Overall thickness 1.5/8 in (41 mm).

Frame and Threshold Construction:
Frame is 4 in x 2 in (100mm x 50mm) in hardwood or softwood with integral or planted stops, fitted with seals and closing on a stepped threshold.

Door Construction:
Solid core of laminated Cedar rails, faced with 4mm crossbanding; hardwood veneered or laminate faced. Hardwood lipped on long edges. Overall thickness 1.3/4 in (44mm).

Frame and Threshold Construction:
Frame is 4 in x 2 in (100mm x 50mm) hardwood or softwood with integral or planted stops and fitted with seals, closing on a splayed threshold.

Door Construction:
Special construction consisting of a solid core of laminated Cedar rails with an acoustic membrane on the centre line. Faced with 4mm crossbanding hardwood veneered or laminate faced. Hardwood lipped on all edges. Overall thickness 2.1/8 in (54 mm).

Frame and Threshold Construction:
As for other assemblies.

Fig 1 Comparative examples of all wood sound-insulating doors. (Shephard & Peller Ltd).

ACOUSTIC DOORS (A)

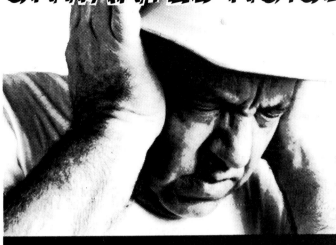

UNWANTED NOISE HURTS

When Health, Safety and Productivity are at risk quietly effective CLARK ACOUSTICAL Doors will isolate the noise. Manual or power operation to suit traffic.

High attenuation designs available for Radio and TV studios.

Send for details.

ACOUSTICAL SLIDING DOORS

Clark Door Limited
Willow Holme, Carlisle CA2 5RR
Tel: 0228-22321 Telex: 64131

CLARK DOOR

MADE IN BRITAIN

HODGSON & HODGSON LTD.

Manufacturers of sound control components from glass and mineral fibres.

Acoustic Screens — Booths — Enclosures.
Spatial Absorbers — Sound absorbent components for all types of equipment.

Crown Industrial Estate,
Anglesey Road,
Burton-on-Trent,
DE14 3PA.

Tel: (0283) 64772
Telex: 342250

IF THIS IS HOW YOU'RE HEARING YOU SHOULD GET YOUR NOISE TESTED.

Noise-induced hearing loss is an insidious process... whose results are unfortunately permanent. Partial or total deafness cannot be cured.
There is usually an acceptable solution to the problem of noise, at a price that could be a pleasant surprise.
The appropriate acoustic enclosure or operator's refuge will attenuate the noise effectively — and the health and safety of shop-floor workers is safeguarded permanently, at an acceptable cost.
Just stop and consider a few current facts... the size and number of compensation claims in hearing-loss cases are escalating rapidly. Settlements now average £8,000 per claim, with some individuals receiving as much as £20,000 to £30,000. The insurance business is watching developments with increasing concern. It follows therefore, that acoustic protection is a viable economic investment when offset against the likelihood of increased premiums.
Sound Solutions Ltd. offer an unrivalled range of enclosures, attenuators louvres and refuges that provide effective and economic protection
Our custom-design facilities enable us to protect operators without compromising machine access — yet often our acoustic measures present valuable energy-conservation opportunities as a bonus for you.
The Factory Inspectorate do not consider ear defenders to be an adequate protection.
Let Sound Solutions help you stay within the Code of Practice, and improve labour relations.

Sound Solutions Ltd.
Victor Works, Bolton Hall Road,
Bolton Woods, Bradford BD2 1BQ
Telephone 0274-597273

Noise Pollution needs a Sound Solution

Frank Driver Noise Reducing Cabinets

For Generators, Compressors, Pumps, etc., up to 4 ft. x 3 ft. x 2 ft. or custom made for larger sizes. Ask for details.

Frank Driver Generators
206, Elgar Road, Reading, Berks. RG2 ODF
Telephone (0734) 752666. Telex 848193

S3210

AVOSHIELD
Lead barrier sheet for sound insulation

The use of thin lead sheet in sound control has become well established, with a diverse range of applications. AVOSHIELD lead sheet is available in sheets, or coils in varying lengths, widths and thicknesses from:

AVOMET LTD
Unit 14,
Rassau Industrial Estate,
Ebbw Vale,
Gwent NP3 5SD
Telephone: 0495 305457

Absorba ACOUSTIC SCREENS
...Protect Your Work Force

"Absorba Screens are for isolating noisy machines or motors, and these screens can be supplied either in semi-portable form for erection by your own personnel, or permanent booths and complete enclosures, installed by us.
We can offer a wide range of systems to meet most conditions, and noise surveys can be undertaken.

Please send for full details

HEAFIELD INDUSTRIES LTD.,
CONNAUGHT ST., OLDHAM OL8 1DE
PHONE: 061-624 8331

ACOUSTIC DOORS

IAC Noiselock acoustic doors installed into masonry structure for the Test Cells at Rover "Range Rover" plant in Solihull, Warwickshire.

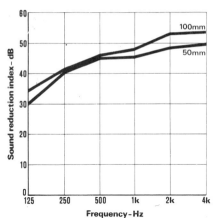

Acoustic performance of standard construction single-leaf sound-reducing door mounted in 225 mm brick wall (ICI Acoustics).

Hinged Doors

Hinged doors normally provide the simplest and cheapest solution, provided they are insulated and achieve an airtight seal when closed. There are a wide variety of proprietary designs, performance characteristics of which are specified by the manufacturers. They are the usual answer to 'standard' size openings not exceeding 1 metre. One possible disadvantage with hinged doors is that they necessarily take up floor space for opening. Swinging them back against the wall might solve these problems but both wall and floor space is lost. Double hinged doors are designed to reduce the size of the opening circle and to lighten the weight on the hinges.

Additional problems can arise with larger hinged doors as sound insulation tends to make such doors very heavy and if the hinges are too tight the doors sag and drag on the floor.

Basically hinged doors are generally advantageous provided they are used for pedestrians and do not exceed a width of 100 cm. If used in narrow corridors, 180° open doors are preferred. Double hinged doors are not always desirable as the seal in the middle can prove unreliable.

Sealing gaskets have to be fitted to the outer edges of double hinged doors, and this makes them vulnerable to damage if goods are being moved through these larger openings. Sealing the bottom of hinged doors requires a raised sill to be fitted to the floor so that the bottom gasket fits the sill in the same way as the top.

This can be a nuisance when moving goods in or out as the sill has to be negotiated each time. Some manufacturers eliminate the raised sill by fitting 'rising' hinges, these operate so that the door rises from the floor when it is opened. In this way the bottom gaskets are spared much wear and tear.

Designs to be avoided are those doors fitted with normal hinges and a very flexible rubber 'sweeper' gasket on the bottom. These doors open and close with difficulty due to the friction of the rubber gasket on the floor. Of course, after a while the friction disappears, but so will the rubber gasket.

Sliding Doors

Sliding doors can prove advantageous when openings wider than 1 to 2 metres have to be closed, provided sufficient sliding space is available. Sliding doors also offer a more flexible choice as they can slide either to the right or left and be fitted on the inside or outside of the opening. They have the inevitable limitation, however, of being a more costly solution because of the more complicated structure needed.

With careful planning, sliding doors can be located in almost any opening, of virtually any width, using two, three or more separate doors, although the greater the number of doors the greater the inter-door sealing problems. For positive sealing the design may incorporate a lever mechanism which presses the door(s) onto the frame and sill when closed, and disengages the door(s) from wall and frame to facilitate opening on rollers. The design of the track and rolling system is extremely important to provide equal distribution of door weight and minimal rolling resistance.

Vertical sliding acoustic doors are comparatively rare, but have been used with considerable success in special applications. They have the advantage of providing a completely unobstructed floor space when open.

Concertina topped acoustic door spanning 12 metre width and height of 6 metres from floor level at entrance to engine test bay. Noise reduction achieved 30 dB(A). (ICI Acoustics).

General Construction

Lightweight doors may be of built-up wood construction, with or without internal sound absorbent material. Most industrial acoustic doors are of steel panel construction with internal bracing and layers of acoustic damping and absorbent materials to give an overall thickness of the order of 50–100 mm. They are normally fitted into mild steel frames with compression seals fitted to the periphery of the doors.

Door Furniture

Particular factors which affect the design and selection of suitable door hinges and fasteners are:
- (i) The most vulnerable part of an acoustic enclosure for allowing sound to break out must be the places where joints and apertures appear.
- (ii) It follows that unless these apertures are treated with the same degree of care and expertise as the rest of the chamber, the final acoustic result will not be as anticipated.

ACOUSTIC DOORS

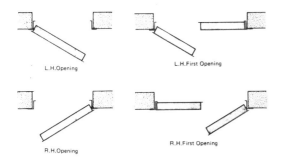

Examples of single-hinged doors and sealing points.

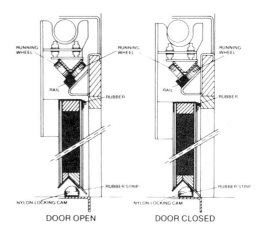

Section through Markus AM2-55 sound door for studios, etc, capable of achieving sound reduction of 55 dB between 125 Hz and 4 000 Hz. (Markus Hermetic Doors Ltd).

(iii) Of necessity some of these apertures will be doors and although the acoustic value of the door itself can be the same as the rest of the chamber, the results actually achieved depend entirely upon the effectiveness of the seal between the door and the door frame.

(iv) To achieve a satisfactory seal it is necessary to ensure that the method of retaining the door in the frame, together with the method of construction, are both compatible and efficient.

(v) To retain the door it is usual, although not essential, to use a door fastening device or devices on one side of the door and hinges on the other; therefore the hinges are not just a means of swinging the door open. They have an important role to play when the door is closed.

(vi) Unless hinges of appropriate design and manufacturing tolerances are fitted to the frame and door with similar accuracy, then the required seal will not be achieved on the hinge side and the acoustic exercise is valueless.

In the case of iron or cast steel door furniture, the manufacturing tolerance issue is an important one for acoustic steel doors. First, having drilled and tapped both the door and door frame to receive the hinge and back plate assembly, it is important that the hinge pins are in alignment with each other and it is essential in hinge manufacture, that all the fixing holes and hinge pin holes are drilled, after heat treatment of the casting, to the same manufacturing tolerances. The shrinkage rate of the casting does not then affect the position of alignment of the fixing holes and hinge pin holes and the likelihood of obtaining an effective door seal is greatly increased. In the event of a hinge breakage, it is not easy to plug and re-drill a steel door and any replacement article should be to the same manufacturing tolerance as the original article. This cannot be so with a hinge where the holes are cast in as no one has control over the annealing shrinkage rate. Most commercial refrigeration door furniture for example is manufactured with the holes cast in and therefore is not really suited for steel door use.

See also chapter on *Acoustic Enclosures*.

Fan and Air Duct Silencers

THE USE of fans to distribute ventilating or conditioned air throughout buildings will almost certainly produce excessive sound pressure levels in the areas served if some form of fan silencer is not used. Equally, industrial fans used to supply process or cooling air, or fans located out of doors in such installations as cooling towers or furnace draught systems, all produce sound power which in many cases is transmitted away from the fan, perhaps in the form of simple airborne energy if the fan is open running. This will produce a sound pressure level at nearby operators' positions or residential property perhaps, which, when compared with the appropriate criterion of acceptability, is found to be excessive.

In some cases it may be possible to achieve the required reduction by redesign of the fan installation geometry. Again, in the case of fans located out of doors, relocation or the use of simple screens and barriers may suffice. In the majority of cases though, it will be found necessary to provide a device which will reduce, not necessarily the amount of sound power actually produced by the fan, but certainly the amount which is allowed to leave the fan and be dispersed through the system. This requires a knowledge of the sound power level of the fan and the characteristics of the transmission path so that excess sound pressure can be removed by a silencer.

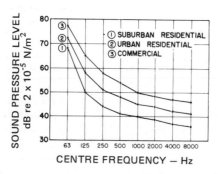 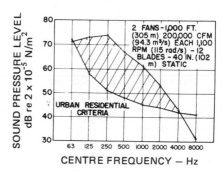

Fig 1 Desirable maxima fan noise levels (left) and forced draught inlet noise levels (right). (Environmental Elements Corporation).

Absorptive Silencers

The precise mechanism by which a silencer extracts sound energy from the flow depends upon whether it is of the absorptive type. In the current state of the art fan silencers are of the absorptive type, having a characteristically very wide frequency band performance, similar to the frequency spectrum of sound power produced by the majority of the present generation of industrial fans. There are some special cases such as systems employing multi-stage axial fans or high pressure centrifugal blowers whose noise is characterized by strong discrete frequency components, where the possible use of narrow band reactive silencers of the lined or plain expansion chamber type should not be discounted.

Absorptive silencers, as the name implies, attenuate acoustic energy by a process of progressive absorption as the sound field passes between boundaries (the duct walls) which are faced with acoustically absorptive material. Fig 2 indicates diagrammatically how the outer edges of plane waves, the modes which usually carry most of the energy, are immersed in the lining material. This portion of a wave which is moving through the relatively high flow resistance presented by the acoustic material, has literally to work much harder to overcome the frictional forces exerted on the air contained inside the pores and interstices of the lining, than have the centre regions of the wave moving through relatively 'free' air. Continuous working of the ends of the wave in transforming vibrational energy into heat energy in this manner, results in a correspondingly progressive 'bleeding' of replacement energy from the centre regions of the wave. The overall result then is a general reduction in the amplitude of molecular vibration, or less sound power at the end of the lined section.

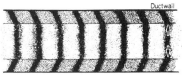

Fig 2 Dissipation of energy at ends of plane waves in a lined duct.

Performance of Ducted Silencers

The performance of a ducted silencer is described in terms of its Insertion Loss (IL). An insertion loss generally is the difference in sound pressure level observed at some fixed point before and after the silencer is fitted. If the insertion loss of the silencer is known, then for any input fan sound power level, the sound power level to the duct system after the silencer is simply the fan SWL minus the silencer IL.

A simplified expression for insertion loss in a lined duct is given by King. The resulting form for the attenuation in dB per unit length of lined duct run is:

$$\text{Attenuation} = \frac{P}{S} \cdot f(z)$$

where

P is the 'wetted' perimeter of the duct section covered by the lining

S is the duct cross sectional area

f(z) is some function of the complex acoustic impedance of the duct lining material

Analysis of the value of P/S for some representative duct shapes will show that its value, and hence the amount of attenuation, increases for air passages whose width w is small compared to their length H (see Fig 3). In the limit it can be shown that the value of P/S approximates to 2/w for either circular or rectangular configurations. This means that for a given lining material (ie fixed f(z)), the narrower the airway between linings the higher the attenuation. There is obviously

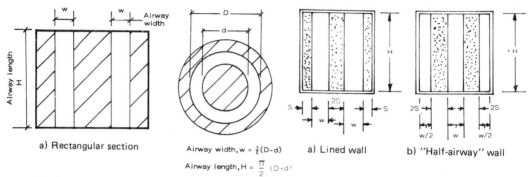

Fig 3 Airway geometry.

Fig 4 Combination of lined ducts to form splitter silencers.

a limit here though, if the duct is to carry any flow. In practice a reasonable free flow area is maintained by placing two or more channels together to form a so-called 'splitter' silencer as indicated in Fig 4. Note here how the central 'splitter' is really two linings located back-to-back. Because symmetry of distribution of acoustic energy can be assumed in each of the airways, it is not necessary to have a physical barrier between the two linings. By the same reasoning it can be shown to be unnecessary to have a partition down the centre of an airway when two 'half-airway' silencers are joined.

The exact form of the function f(z) remains open to question. For engineering design estimates, simplified formulas are normally used, *eg*:

$$\text{Attenuation} = \frac{2700}{w} \cdot \alpha \quad \text{dB/metre}$$

where

α is the random incidence absorption coefficient of the lining material
w is the airspace in mm

An alternative formula often widely quoted is:

$$\text{Attenuation} = \frac{1800}{w} \cdot \alpha^{1.4} \quad \text{dB/metre}$$

The simplification of the expression for attenuation to a straightforward dependence on absorption coefficient means of course that the characteristic behaviour of the silencer is very similar at the low frequency end of the spectrum to that of conventional porous-type acoustically absorptive materials. To maximize performance then, it is necessary to look for a lining material with the highest absorption coefficient practicable. Within the range of engineering materials currently available there is not too much to choose between them. Glass fibre for example will give much the same result as slag wool, other mineral wool, or even expanded polyurethane foam. Even some variation in density does not appear to produce marked differences in performance, provided it lies in the range 30 to 100 kg/m^3.

Below this the material becomes too loosely packed to offer effective flow resistance to molecular vibrations in the elemental volumes of air carrying acoustic energy, whilst at very high

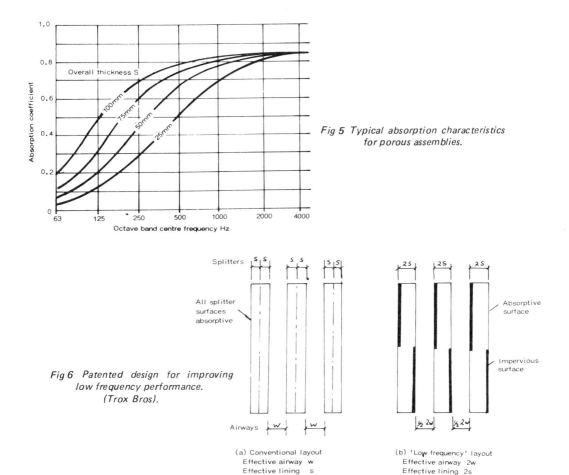

Fig 5 Typical absorption characteristics for porous assemblies.

Fig 6 Patented design for improving low frequency performance. (Trox Bros).

temperature (flue-dilution or gas turbine exhaust systems for example), there is a significant reduction in gas density requiring smaller interstitial passages inside the material to give the same amount of flow resistance and hence energy transfer. Material densities of perhaps 150 kg/m^3 are sometimes used in such applications.

By far the most significant parameter determining absorption coefficient is thickness of material. Fig 5 shows somewhat idealized curves of absorption coefficient over the frequency bands for various material thicknesses. The need for a good depth of material to obtain high performance at low frequencies is evident. On the other hand, there are obvious engineering limitations on the practical thickness of material that can be used to line ducts or form splitters — flow resistance being one already mentioned.

Most commercial manufacturers of packaged splitter type silencers seem to have opted for a lining thickness of around 100 mm (splitter thickness of 200 mm), as the best compromise. One enterprising company has attempted to resolve the conflicting requirements of attenuation and flow resistance by employing a more or less standard module (splitter thickness + airway width), but facing one side of the splitter with acoustically impervious sheet, so that only its other face is open to the acoustic energy in the airway on that side. Instead, therefore, of having the conventional arrangement of one airway width, say, w, lined on either side with material of thickness s, the module is transformed into one of airway width 1 lined on one side only with material of thickness 2s. Because of the properties of symmetry that can be assumed for the sound field, which were mentioned earlier, this is acoustically equivalent to an airway of width 2w lined on both sides with material of thickness 2s (see Fig 6).

The object of the exercise is to get the benefit of the considerably enhanced low frequency absorption coefficient of the effectively double thickness material, as indicated for example in Fig 5, and indeed a corresponding improvement in attenuation is to be expected. However, the improvement must be offset by the loss of attenuation due to effectively doubling the airway width. Which effect will eventually dominate clearly depends upon the exact role of absorption in the process of energy conversion. Simplified formulas do not give this with sufficient accuracy.

The performance to be expected from the more conventional splitter configurations (in this case the specific instance of a 1 200 mm silencer length, lining thickness 100 mm, splitter thickness 200 mm, and varying airway width) is shown in Fig 7.

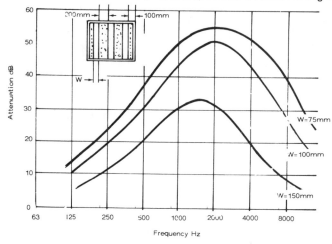

Fig 7 Typical attenuation from 1200mm long packaged splitter silencer.

Frequency Dependency

At any particular frequency attenuation increases with reducing airway as is to be expected. More interesting perhaps is the shape of the individual curves which are quite characteristic for absorptive silencers. At frequencies up to around 500 Hz the shape of the attenuation/frequency curve is very similar to the characteristic shape of absorption coefficient of the lining material (Fig 5). After reaching its peak value the silencer performance declines and attenuation falls off instead of remaining at that level for further increases of frequency as would be expected from prediction based on material absorption coefficient. The reason for this is to be found in the geometry of the system rather than in the absorption properties of the acoustic material.

At wavelengths of the duct sound field which are small (that is, high frequency sound) compared with passage width, most of the energy is 'beamed' along the centre of the airway. Consequently the portion of the wave actually immersed in the wall lining, and upon which the energy transformation mechanism is operating, is of relatively lower value, and the rate of exchange, or attenuation, is correspondingly reduced. As a design guide, peak attenuation frequencies beyond which performance will start to drop again are those whose wavelengths are some

FAN AND AIR DUCT SILENCERS

TABLE I – ATTENUATION FROM 200 mm THICK SPLITTERS (100 mm THICK LININGS)

Length mm	Air Passage width mm	Attenuation in dB in octave bands Hz							
		63	125	250	500	1000	2000	4000	8000
900	50	8	16	27	45	55	55	55	50
1200	50	10	20	36	55	55	55	55	55
1500	50	13	24	42	55	55	55	55	55
1800	50	15	30	51	55	55	55	55	55
2100	50	17	34	55	55	55	55	55	55
2400	50	19	38	55	55	55	55	55	55
900	75	6	11	19	34	45	45	39	28
1200	75	7	14	26	46	55	55	52	38
1500	75	9	17	30	48	55	55	55	42
1800	75	10	20	34	50	55	55	55	46
2100	75	12	23	40	55	55	55	55	55
2400	75	13	26	45	55	55	55	55	55
900	100	5	9	16	30	39	39	31	26
1200	100	6	12	23	40	51	51	41	29
1500	100	8	15	26	43	53	53	45	32
1800	100	9	17	30	47	55	55	49	36
2100	100	11	20	35	55	55	55	55	43
2400	100	12	23	40	55	55	55	55	47
900	125	4	7	13	25	32	32	23	15
1200	125	5	9	19	33	42	42	42	30
1500	125	7	12	22	38	47	47	34	20
1800	125	8	14	26	43	52	52	39	23
2100	125	9	17	30	50	55	55	46	28
2400	125	10	19	34	55	55	55	52	32
900	150	3	6	11	20	25	25	15	8
1200	150	4	7	15	26	33	33	19	11
1500	150	5	9	18	33	41	41	24	13
1800	150	6	11	22	39	49	49	29	16
2100	150	7	13	26	45	55	55	34	19
2400	150	8	15	29	52	55	55	39	21
900	175	2	5	9	17	21	21	13	5
1200	175	3	6	13	22	28	28	16	7
1500	175	4	8	15	28	35	35	21	8
1800	175	5	9	19	33	42	42	25	9
2100	175	6	11	22	39	49	49	29	11
2400	175	7	13	25	45	55	55	33	11
900	200	1	4	8	15	19	19	11	3
1200	200	2	5	11	20	25	25	14	4
1500	200	3	7	13	25	31	31	18	5
1800	200	4	8	17	29	37	37	22	7
2100	200	5	10	20	34	43	43	25	7
2400	200	6	11	22	39	49	49	29	8

1.75 times the width of the airway. Unfortunately none of the simplified formulas predict performance after the maximum, but again for initial purposes only, an assumption of a decay rate of 10 dB/m/octave for frequencies above that of maximum attenuation, should not prove to be too inaccurate.

Fig 7 was specifically for one length and one airway. Table I shows the performance to be expected from 100 mm liners (200 mm splitters) for a range of lengths and airways. Note here that for a given material and lining thickness, the only factors determining attenuation are silencer length and airway. Overall width and height of the silencer cross section affect only aerodynamic resistance as discussed later.

Performances for other splitter thicknesses can be obtained from individual manufacturers, but the number of decibels attenuation provided by a given airway and length combination with say, 200 mm splitters can be expected to increase or decrease approximately in direct proportion to the change of absorption resulting from the change of splitter thickness. Change of absorption coefficient in turn can be estimated to a sufficient degree of accuracy by the curves in Fig 5. It is important to note here however that the relevant curve to be used is the one for the appropriate *lining* thickness, which is of course *half* the splitter thickness.

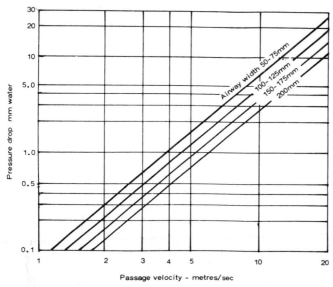

Fig 8 Typical pressure loss from splitter silencers.

Flow Resistance

Except in the case of the simple lining of plain rectangular or circular ducts, silencers invariably involve some blockage of the flow path and therefore present some degree of aerodynamic resistance leading in turn to a pressure drop across the silencer. Fig 8 shows the order of magnitude to be expected for the pressure drop across silencers with 200 mm thick splitters and varying airway widths as a function of flow velocity in the air passage. This is given by:

$$\text{Passage velocity} = \frac{\text{Volume flow}}{\text{number of airways} \times \text{airway width} \times \text{airway height}}$$

With volume flow in m^3/sec, and airway width and height both in m, passage velocity is then in m/sec.

FAN AND AIR DUCT SILENCERS

The values shown here are typical for splitters with flat ends. Considerable improvements can be achieved by fairing the ends both at inlet and outlet to the silencer. It is best to consult individual manufacturers for figures appertaining to their particular designs, but it should be possible to obtain improvements varying from 10% with simple angle fairings up to perhaps 20% or more with fully aerodynamically shaped ends.

Secondary Noise (Regeneration)

Secondary noise will be generated at both intake and discharge from the silencer. Invariably though, it is the discharge side of a silencer located on the downstream side of the fan which will require the most careful examination for the effects of its regenerated noise. Noise generated at the intake to a silencer in the same position relative to the fan will almost certainly be much lower than the fan noise at that point. Since the silencer is attenuating the fan noise by a sufficient amount (by definition) it can clearly deal equally well with its own intake generated noise.

Similar remarks apply to silencer discharge noise when the silencer is located upstream of the fan. Noise produced at the intake of an upstream silencer has the potential, it is true, of travelling unimpeded through the upstream duct system. On the other hand the amount of regenerated sound power actually leaving the intake of an upstream silencer is somewhat lower than Table II would indicate, because the actual generating mechanism takes place some little distance inside the silencer airway. Then, because of the highly absorptive surroundings, the whole process of producing acoustic energy is rather less efficient.

TABLE II — OVERALL SOUND POWER LEVEL OF SELF-NOISE FROM SPLITTER SILENCERS

$$SWL = 55 \log_{10} V + 10 \log_{10} N + 10 \log_{10} H - 45 \text{ dB}$$

where, V is the local velocity in the splitter airway m/sec
N is the number of airways
H is the airway height (rectangular) or circumference (cylindrical) mm

Distribution of the energy over the frequency spectrum can be obtained by subtracting the following from the overall sound power level:

Octave band centre frequency	Hz	63	125	250	500	1000	2000	4000	8000
Correction for octave band sound power level	dB	−4	−4	−6	−8	−13	−18	−23	−28

Optimum Silencer Position

While, by and large, a duct silencer will attenuate whatever sound power passes through it wherever it is located in the transmission system, there are some very important points to be checked if its potential performance is not to be seriously impaired. These may be summarized as:

(i) Does the silencer cause any maldistributed or turbulent flow into the fan?
(ii) Does the air passage 'jet' flow impinge upon any other duct element?
(iii) Is the flow into the silencer evenly distributed over all its free area?
(iv) Can 'unsilenced' sound power in the run of ducting between fan and silencer break out through the duct wall into noise sensitive areas?

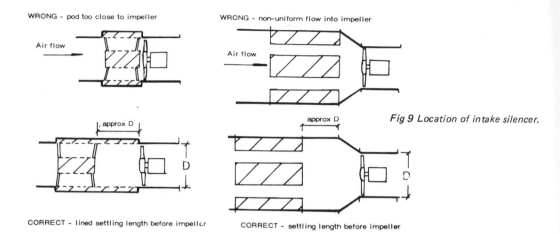

Fig 9 Location of intake silencer.

(i) Location Relative to the Fan

Located close to the fan impeller which carries the distribution of sources the acoustic lining of the silencer will be exposed to energy contained in many more duct modes than the simple plane wave which is predominant at more than a few fan diameters downstream. The majority of these modes involve impingement of energy onto the absorptive material at oblique angles rather than the grazing angle of the ends of the plane waves, and for this the absorption mechanism is much more efficient.

For this reason insertion losses quoted by most manufacturers, which are based on the silencer being close-coupled to the fan, may show figures considerably in excess of what the unit would be expected to produce if used as a simple lined duct located some way down the system (see Table III).

Equally, considerable care has to be taken over location of silencers relative to the fan intake to avoid generating a turbulent flow in the intake. While the straight through silencer will cause no problem in this respect, the addition of a pod to turn it into a cylindrical splitter also adds a wake, and introduces just the type of flow into the fan which should be avoided at all costs. Indeed the paradoxical situation can almost be reached when a 5 dB silencer fixed direct to the upstream flange of the fan causes an extra 10 dB in sound power level to be generated by the fan, resulting in a net 5 dB more noise than occurred before the silencer was installed.

TABLE III — INSERTION LOSS FROM 600 mm DIAMETER 600 mm LONG LINED DUCT
(75 mm WALL LINING THICKNESS)

Frequency Hz	63	125	250	500	1000	2000	4000	8000
Used as straight-through silencer direct coupled to fan	1	4	7	12	12	8	7	6
Used as simple lined duct section remote from fan	1	1	2	3	3	1	1	0

FAN AND AIR DUCT SILENCERS

Besides a turbulent wake from the central pod causing broad-band noise generation by the fan, there is also the possibility of extra discrete frequency noise generation from the aerodynamic interaction between pod support arms and fan impeller. The noise producing potential of a rectangular splitter placed too close to the fan intake is equally obvious.

Fortunately the solution requires only the inclusion of a plain duct section, of length equal to not less than one fan intake diameter, between silencer and fan. Besides avoiding extra sound power generation by the fan, there is also a small acoustic performance bonus to be gained if the length of plain duct can be acoustically lined.

(ii) 'Jet' effect

In conventional heating and ventilating systems, it is normal to design for silencer passage velocities of not more than about 20 m/sec, a limit determined by both acceptable flow resistance and regenerated noise. In some industrial applications, particularly where high temperatures are involved such as in gas turbine exhausts, velocities may rise to around 30 m/sec. Even at the velocities usually worked to of some 15 m/sec however, their potential for generating acoustic energy by means other than self-noise in the silencer (as previously discussed), must be recognized and avoided.

Discharge of air from a silencer passage in the centre regions of the duct and parallel to its axis will not normally generate, even at velocities of around 20 m/sec, significant jet sound power levels compared with 'silenced' fan noise at that point — such velocities being relatively small in 'free jet' terms. Problems are very likely to arise though if the jet is allowed to impinge upon any solid surface located near the silencer discharge. It is in fact well known that noise from a free jet is considerably increased (perhaps by as much as 10 to 15 dB) if the jet is allowed to interact with an adjacent solid surface. For this reason the type of installation giving rise to such situations, as for example indicated in Fig 10, should be avoided at all costs.

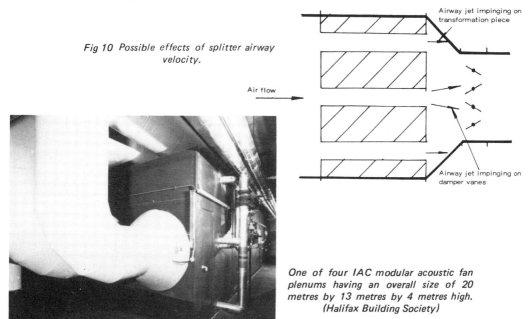

Fig 10 Possible effects of splitter airway velocity.

One of four IAC modular acoustic fan plenums having an overall size of 20 metres by 13 metres by 4 metres high. (Halifax Building Society)

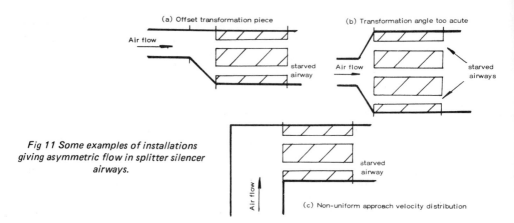

Fig 11 Some examples of installations giving asymmetric flow in splitter silencer airways.

(iii) Full Flow

Rectangular silencers are often 'built up' of a number of single air passages to achieve sufficient free flow area for a reasonable aerodynamic resistance. By definition it is important that the flow through the silencer should be equally distributed over all its free area. Some installation designs however set up entry conditions which effectively prevent this happening, and some examples of these are shown in Fig 11. Clearly, attempting to force more than its allocated proportion of air through an airway will result in excessive velocity in the passage, with the inevitable result of more resistance and self-generated noise than were intended.

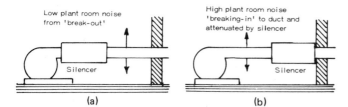

Fig 12 Effect of plant room silencer positioning.

(iv) Breakout

As a general rule it is advantageous to have the attenuating device as close to the source as possible since this minimizes the physical space that 'unsilenced' sound can occupy. In the context of duct silencers, long lengths of ducting carrying out the full sound power always present a risk of energy breaking out through the duct wall which, if the duct passes through a noise sensitive area, can clearly result in excessive sound pressure in that region. In buildings served by a ducted system, the main fan silencer is nearly always located in the plant room. Whether it is best placed near to the fan there, or immediately next to the point where the ducting passes through the wall depends upon whether one is trying to achieve low plant room noise levels (Fig 12(a)), or whether the object is to prevent high noise levels, which may well be acceptable if confined to the plant room itself, from being transmitted along the duct system to the served areas after having broken in through the duct walls (Fig 12(b)).

Silencer Construction

A vital component in any silencer is the choice of acoustic infill material, which must be capable of withstanding a wide range of adverse conditions.

Glass and mineral wools which are resistant to the majority of chemical components likely to be encountered, and which can operate in gases at temperatures up to 500°C, are readily available in the range of densities specified earlier, in either quilt form for cylindrical silencers, or as semi-rigid resin-bonded slabs for rectangular splitters.

Open-celled expanded polyurethane foam is sometimes used with excellent acoustic results, and has distinct engineering advantages in being substantially dust free and easy to support on internal duct surfaces by normal industrial adhesives. The material is somewhat limited however in respect of its resistance to fire and to some chemical compounds, notably gas flows high in solvent content.

Problems are often encountered with the durability of acoustic material under prolonged (sometimes years) exposure to airflow and the associated turbulence and mechanical vibration. Adequate mechanical support and protection of the surfaces of the material under such conditions, which at the same time allows it to do its acoustic work, is a design consideration of prime importance. The expanded foams are reasonably resistant to abrasion and scrubbing arising from continuous airflow next to their absorptive surfaces, and being substantially free of dust, can for most applications be installed with unfaced surfaces.

The mineral fibres on the other hand need a protective facing to prevent dust and fibre carry-over in the short term and complete erosion in the long term. Protective facings for slabs and quilts are offered by a number of manufacturers, and these range from the 'soft' type such as non-woven glass cloth, glass thread mesh, treated scrim cloth, and sprayed PVC, through to the full positive protection of perforated sheet steel. Whilst the soft facings have been used for a number of years, and still are offered in some standard products, they are very susceptible to mechanical damage in manufacture and handling prior to installation. Also, under the constant fretting action of air turbulence, small lesions in the facings will almost certainly develop to the point where the facing material is almost completely removed, along with probably a fair amount of the infill material itself. Silencers exposed to atmospheric air, sometimes with high moisture content, are particularly susceptible to this.

The hard facings virtually remove all risks of degeneration of the material. The only acoustic requirement for a hard facing is that the net percentage open area of the facing material should be not less than 20% to preserve the broad frequency band absorption characteristic of the infill, and that hole size should not exceed some 6 mm unless there is additional support for the material behind. Hard facings can in fact be manufactured from almost any material, the most common currently in use being perforated sheet steel, 'expanded' sheet steel, welded mesh, or even chicken wire.

Even hard facings however will not prevent the ingress into the infill of water, grease, fine dust and other liquids or solid particulate matter carried by the flow. The method most commonly employed to prevent this is to wrap the infill behind its hard facing with one of the very thin, probably 12 micron, polyester or similar impervious films which are on the market. While this does result in some loss of attenuation in the mid and high frequency bands, the important low frequency bands are generally unaffected. If the higher frequency attenuations are important, individual manufacturers should be consulted for the precise effect of the particular films they use upon their standard performances.

Mechanical Considerations

The outer case of the silencer and the internal construction forming splitter supports and infill facings, can in most cases be manufactured in whatever material is used to engineer the rest of the system, without affecting acoustic performance.

In ventilating and air conditioning systems for example, light gauge sheet steel, probably pre-galvanized, is the norm.

For more arduous duties such as some industrial process air or dust extraction systems, fume scrubbers, shipboard and mining applications, much heavier gauges of steel are necessary. Plate thicknesses of up to 6 mm are employed, with exposed surfaces protected by one or other of the bitumastic or chlorinated rubber paints if any significant chemical corrosion hazards exist.

For very serious corrosion hazards, the entire silencer is frequently manufactured in PVC or GRP, providing the gas being handled is at normal temperature.

For combinations of corrosive gases and high temperatures, the most widely employed solution is to manufacture either the complete silencer, or at least the components directly exposed to the flow, in stainless steel sheet. This material is often demanded too in the food and pharmaceutical industries where frequent sterilization is mandatory.

Where weight is critical, such as in the case of roof extract units mounted on lightweight structures, the silencer may be manufactured in aluminium sheet.

See also chapters on *Fan Noise* and *Air Distribution Systems*.

Industrial Silencers

PURE DISSIPATIVE silencers are based on the use of flow-resistive materials, normally in the form of porous acoustic linings. Such materials may be used to cover the interior surface — as in lined ducts — or be in the form of parallel baffles. The latter arrangement may be considered as splitting the volume involved into a series of lined ducts. This effectively increases the absorptive length of the silencer without increasing its actual length, although its diameter may have to be increased.

The simplest form of absorptive silencer is a duct made from sound-absorbing material, or lined with such a material. In the latter case a perforated inner liner may be incorporated to hold the absorbent material in place. The performance of such a silencer is given by:

$$TL = 4.2\alpha^{1.4} L/d$$

where α = absorption coefficient of material
L = length of duct
d = diameter of duct

The transmission loss achieved with lined ducts is primarily due to the attenuation provided by the absorbent lining. However, when the total cross-sectional area of the lining approaches the same value as that of the full passage through the duct, reflected waves are produced, yielding a further transmission loss. This condition can also be promoted by marked changes in area at the ends of the duct, or by an input which is not plane axial.

Example of acoustic enclosure assembled from standard modules. (Burgess Industrial Silencers Ltd).

A plane axial input comprises a sound wave travelling parallel to the longitudinal axis of the duct or silencer. In practice if the wavelength is less than the geometric dimension of the duct it will normally have axial as well as longitudinal components. Equally a typical noise input may have components in the radial plane as well as the longitudinal plane — Fig 1. Where these components are substantially equal in all three planes, the input is said to be *random incident*.

The transmission loss of lined ducts is commonly calculated for plane axial input. For approximate working the Sabine formula for attenuation can be used:

$$\text{TL per foot} = 12.6 \frac{P}{A} \alpha^{1.4} \text{ dB}$$

where P = perimeter of lined section, inches
A = cross sectional area, square inches
α = absorption coefficient of lining

This formula only holds true for a relatively low range of frequencies (*ie* 250 Hz to 2 000 Hz) and becomes increasingly inaccurate as the ratio of duct length to wavelength increases. It is also dependent on the value of α being between 0.2 and 0.4, and the ratio of the major to minor dimension of a rectangular duct being between 1 and 2.

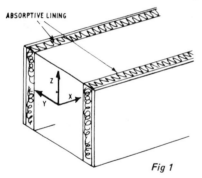

Fig 1

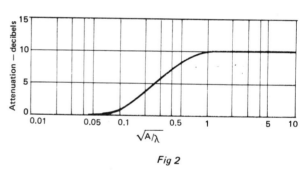

Fig 2

It will be appreciated that the performance determined for plane axial input needs correction where random incident input is involved. Mathematical analysis in this respect is complex and solutions are dependent on both the wavelength of the sound and the length of the duct. However, a simple form of correction can be developed from empirical data, based only on duct area (A) and wavelength of the sound (λ). This is shown in graphical form in Fig 2, with the random incidence correction rendered as a dedendum, in dB, related to $\sqrt{A/\lambda}$.

Industrial Absorptive Silencers

Industrial absorptive silencers normally consist of sections of rectangular ductwork containing 'splitters' comprised of acoustically absorbent infill or circular-section units, usually containing an absorbent lined 'pad' concentric with the axis of the main casing. Rectangular section silencers lend themselves most readily to insulation in connection with rectangular ductwork of equivalent dimensions. Circular-section silencers are often directly coupled to axial-flow fan installation spiral-wrap ductwork. Other more specialist types of attenuator include 'splitter-bends' where the performance of a lined mitre bend is combined with that of the splitter attenuator, or 'splitter-louvres' where inlet louvres to plant rooms are acoustically lined to give a degree of noise control on the inlet side of the fan systems. Cross-talk attenuators can involve a wide choice of configuration of relatively small cross-section ducts, often integrated with the inlet or discharge louvre of, for example, bathroom extract and supply systems in hotels and flats, where a severe cross-talk problem can exist from the propagation of speech along the common riser duct.

The choice of the acoustically absorbent materials for the construction of dissipative silencers will be limited by the need to meet specific design criteria. They must be non-flammable, dust-free, unaffected by water, vermin, *etc,* and must not erode in the high flow velocities likely to be encountered in the air-ways. Both fibreglass and mineral wool comply with most of the aforementioned requirements and to prevent any erosion of this fibrous material the splitter elements are usually faced, either with fibreglass or cellulose tissue materials, or are contained in cotton scrim behind perforated metal facings. When the silencers are required to deal with grease-laden atmospheres, or must be steam sterilized (as for instance when used in air-conditioning of operating theatres, *etc*) the permeable facings may be replaced by a thin plastic film which itself must be protected from erosion by either bonding it directly to the fibrous material or by placing it behind a perforated metal facing. Such a configuration is likely to have a reduced performance at middle

Chemical Process silencers for the inlet and discharge ducts (500 mm diameter) of fans handling process gas containing corrosive elements. The absorptive 'pod' type silencers utilized stainless steel and high nickel alloys in their construction and a specially developed acoustic packing material. The acoustic packing was in fact a fibre made from 'Propathene' polypropylene, encapsulated in 'Propafilm' polypropylene film and retained in a polypropylene mesh bag. The packing was retained behind perforated metal in the construction of the silencer.
(ICI Acoustics)

A high capacity gas exhaust system comprising four large bore pipes leading to a silencer, generated noise levels of up to 124 dB(A) at 5 metres and caused complaints from residents living about two miles away. The silencer was operating efficiently but noise was "breaking out" through the pipe walls up-stream of the silencer. Acoustic lagging specified and fitted by ICI Acoustics reduced the noise levels at 5 meters by 14 dB(A), stopped the complaints and also improved the working conditions of people in the area. This type of treatment is effective in reducing noise from many types of pipe or duct systems.

and high frequencies, but the low frequency performance (which invariably contains the critical bandwidth which determines the overall design performance of the attenuator) will be unchanged, or even increased.

Reactive Silencers

The simplest form of reactive silencer is a plenum chamber or expansion chamber — Fig 3. The chamber is normally cylindrical and may or may not contain baffles. Industrial silencers of this type commonly have inlet and discharge ducts offset to minimize the direct transmission of sound across the chamber. The characteristic attenuation is a period function of $\frac{2\pi}{\lambda} \times L$ where L is the length of the chamber and λ is the wavelength of the sound, repeating every 180 degrees. The actual ratio of attenuation achieved at the peaks is a function of the square of the area ratio, or ratio of chamber cross section, to inlet (pipe) cross section.

The principal advantages offered by reactive silencers are relatively small size and high mechanical stability (ie they contain no 'soft' materials which may disintegrate and/or migrate under continual vibration). Construction, however, tends to become increasingly complex, and more critical, if a silencer of this type is designed to provide large attenuation over a wide frequency range. A simpler solution, in such cases, is often that of using a reactive silencer in series with an absorptive type silencer.

Plenum chambers can provide high attenuation at low frequencies, but, when used for industrial silencers, tend to suffer from the basic disadvantage of needing to be of large volume.

Fig 3

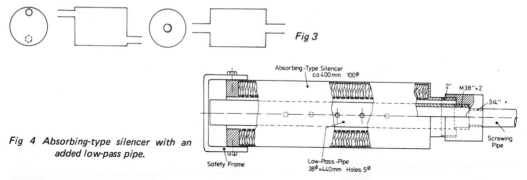

Fig 4 Absorbing-type silencer with an added low-pass pipe.

Reactive-Absorptive Silencers

A basic-absorptive silencer is an expansion chamber lined with absorbent material. With suitable design the overall attenuation achieved can be greater than the sum of the attenuation of the tuned volume and that of the lining. (Fig 4).

Inserts

The use of splitters has already been described. The thickness of the splitter is a significant parameter for low frequency attenuation when the absorption is dependent on the thickness of the absorptive material. To avoid restricting the flow passages, the bulk of the silencer may have to be increased appreciably. Another method of treating low frequency sound is the use of side branches or absorbing elements on the sides of silencers.

For the treatment of high frequencies, *baffles* may be inserted on the flow path through rectangular silencers. In the case of circular silencers, *centre bodies* may be used for the same purpose.

See also chapters on *Air Distribution Systems* and *Fan and Air Duct Silencers.*

INDUSTRIAL SILENCERS

IAC Noishield Acoustic Louvres provide acoustic solution for roof-top cooling towers and plant at the headquarters of ATV in Birmingham.

IAC 5S Quiet Duct Silencers provide intake noise control for forced draft boiler fans. Completed acoustic installation provided a reduction from an anticipated 105 dBA to an actual 73 dBA when four of the five boilers were operating.

Examples of custom-built enclosure constructed on site to control noise and dust from a casting fettling operation. (Burgess Industrial Silencers Ltd).

Silencing Gas Turbines

THE THREE main sources of gas turbine noise are the inlet, exhaust and the turbine casing (with which are associated various ancillaries such as gears, pumps, coolers and ventilators). Intake noise levels are typically high frequency and sound power levels may be as high as 145 dB for larger units. Exhaust power levels may also reach 130 dB, although here the noise spectrum is generally shifted downwards to a lower frequency range. Casing levels associated with gas turbines are generally lower than intake or exhaust levels, but are nevertheless of sufficient magnitude to require consideration as a major source of noise. Fig 1 shows the three basic noise contents of a typical unsilenced gas turbine, with noise level measured at a distance of 400 feet.

There is also the fact that in general applications the gas turbine drives another machine, which itself will contribute noise to the environment. Figure 2, for example, analyzes the various sound power levels generated in a typical 10 000 horsepower gas turbine driver compressor.

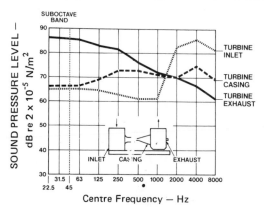

Fig 1 Typical unsilenced gas turbine noise levels at 400 ft. (Environmental Elements Corp).

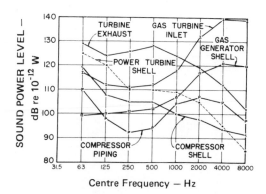

Fig 2 Typical 10 000 hp (7457 kW) aircraft gas turbine driven compressor noises. (Environmental Elements Corp)

Intake Silencing

Intake noise is at maximum intensity at a frequency equal to the product of the axial compressor rotating speed and number of blades (blade passing frequency). Noise is reduced by the use of silencers and elbows, and sometimes even by pointing the intake upwards, so that a favourable directivity affects results.

SILENCING GAS TURBINES

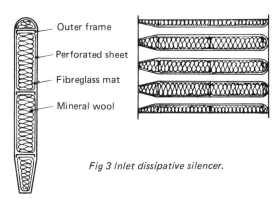

Fig 3 Inlet dissipative silencer.

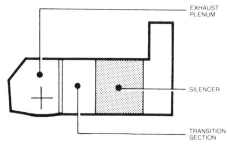

Fig 4 Lateral exhaust silencer.

Dissipative type (absorbent) silencers are used, normally consisting of superimposed parallel baffles (Fig 3), the ideal baffle thickness being equal to one half of the wavelength to be treated. In practice they are generally made slightly thicker. Overall geometry is important to avoid self-generated noise, and the silencer is often split by vibration-isolating joints to minimize noise spread through the shell. The silencer is also normally coated to limit noise radiated from its walls.

Exhaust Silencing

The noise spectrum at exhaust is characterized by high intensities at low frequencies. This, plus the fact that gas temperatures reach up to 500–550°C (resulting in greater wavelengths) means that much thicker baffles (up to 800 mm) must be used.

Since the noise at higher frequencies is also considerable, it is often necessary to have to use two separate silencers of different thicknesses (one for high and one for low frequencies). Whenever very low residual noise levels are required, it is more important to limit self-generated noise at exhaust rather than at inlet, due to the fact that exhaust flow rates are higher and much more irregular.

The problem of self-generated noise can be overcome by exhaust systems such as in Fig 4. The silencer in this solution was located before the elbow so that the incident waves are planar, and thus the elbow provides maximum attenuation and cuts self-generated noise from silencer flow.

One of eight IAC acoustic enclosures housing gas turbine sets at Greenwich Power Station of the L.T.E. The intake silencer and lagged exhaust silencer are also visible.

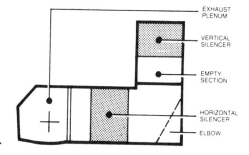

Fig 5 Lateral exhaust system.

The configuration shown in Fig 5 where there is one silencer before and another after the elbow, may be used when even greater attenuation is desired. It allows maximizing of both elbow efficiency and the efficiency of the vertical silencer which, coming after the elbow, is hit by random-incident waves. Evidently, in the second silencer, velocity must be reduced and distribution regularized as much as possible. This requires the use of large passage areas, turning vanes (made of perforated sheets so that acoustical effectiveness is not impaired), and a hollow duct right after the elbow to even out velocity.

Mechanical factors are much more important in exhaust silencers than in inlet silencers due to their great temperature variations (ranging around 500 deg C) and, above all, to the different temperature constants of the exhaust components. To overcome this, the silencer baffles are designed so that they can freely dilate with respect to the ducts they are housed in. Also, the panels themselves are assembled so that they can move internally to some extent, thereby preventing stresses that could result in damage and even breakage.

A cutaway view of a typical silencer panel is shown in Fig 6. The layers, consisting of a screen, high temperature blanket, and fibreglass mat, serve to provide greater protection to the rock wool and guarantee silencer efficiency throughout the life of the turbine.

Recovery boilers used for gas heat recovery systems are also effective as exhaust silencers and can reduce the requirements for adequate exhaust silencing.

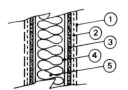

1. Perforated sheet
2. Stainless steel screen
3. High temperature mat
4. Fibreglass mat
5. Mineral wool

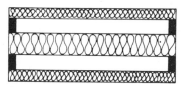

Section of typical turbine exhaust silencer

Fig 6 Exhaust silencer panel.

Self-generated and Flanking Noise

Self-generated noise originates from turbulence of air passing through the silencer and can reach surprisingly high levels. A formula for estimating the amount of self-generated noise for each octave in the spectrum is:

$$PWL_{OCT} \approx 131 + 55 \log M_F + 10 \log A_F - 45 \log \frac{P}{100} + 7.5 \log \frac{273 + T}{293}$$

where M_F = V_F/C with V_F = face velocity referring to A_F (m/sec)
A_F = face area
P = % open area
T = temperature of air or gas (°C)

Formula values may be reduced by approximately 5 dB(A) using panels with trailing edges tapered approximately 7°. The formula is valid only when velocity is evenly distributed as it is in inlet silencers, as opposed to exhaust silencers, where appreciable variances can be found.

Silencer Materials

Materials must ensure proper operation and durability from both mechanical and acoustical standpoints. The insulating material must have the right flow resistance value to be acoustically effective, and be such that its characteristics remain unchanged in time, even under the severe conditions it is subjected to inside the exhaust duct.

A material commonly used is rock wool, where the flow resistance can be calculated from:

$$R_1 = \frac{K \rho m^{1.53}}{d^2}$$

where $K = 3.18 \times 10^3$
ρ = air density
m = density (kg/m^3)
d = fibre diameter (microns)

High density materials are normally chosen as having better resistance to high temperatures as well as being less prone to being squashed. A representative choice is a density of around 90 kg/m^3 with a fibre diameter of about 6 μm. Metal fibre is an alternative material.

Lagging and Enclosure

In addition to using silencers to reduce duct noise, laggings and complete acoustic enclosure are also important treatments. Fully lagged and fully enclosed, very substantial reductions in overall noise levels can be achieved. The effect of such treatment is enhanced if the enclosure is set up on its own separate foundations rather than the turbine baseplate, which in turn is mounted on anti-vibration mounts.

Typically enclosure panels consist of an outer plate, an internal perforated plate and an insulating material of the kind used in silencers, whose thickness depends on the amount of noise reduction desired. In some types, there is an intermediate layer which may be either a steel plate or a layer of bituminous material.

See also chapter on *Aircraft and Airport Noise*.

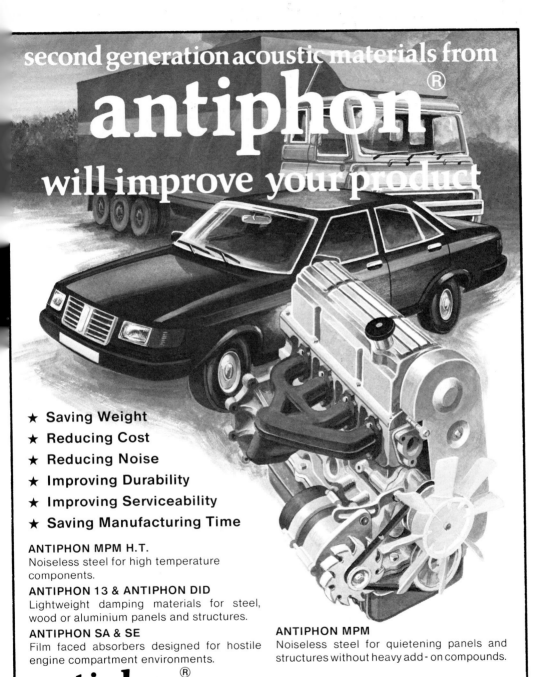

Rubber in engineering

With its unique properties, Natural Rubber can make an important contribution to many advanced engineering applications – protecting against impact, shock and vibration, misalignments and slow movements, aggressive environments, seawater, low temperatures. In addition, Natural Malaysian Rubber is backed by an unsurpassed range of services and publications:

Technical advisory service
Technical advice on design and manufacture of components and performance in service.

Engineering design with Natural Rubber
This free 48-page NR technical bulletin shows how performance of rubber components relates to shape, size, temperature, mode of deformation and the compound used.

Natural Rubber engineering data sheets
Each EDS gives extensive test data on a specific engineering rubber. Introductory sheets deal with test methods, properties and the effect of compounding. Free on request.

Engineering films
Details of films, including a 30-min. colour film 'Engineering with rubber' will gladly be supplied.

Articles and publications on engineering subjects
Reprints of articles and papers dealing with various uses of rubber in engineering, and a wide range of publications covering all aspects of NR cultivation, processing, technological advances and engineering applications are freely available. Please send for details.

Two Research Laboratories
Two great laboratories, responsible for basic research and practical developments, both always ready to co-operate in solving new problems: The Rubber Research Institute of Malaysia's Technology Center, Sungei Buloh, Selangor, Malaysia, and the Malaysian Rubber Producers Research Association's Tun Abdul Razak Laboratory, Brickendonbury, England.

MALAYSIAN RUBBER RESEARCH AND DEVELOPMENT BOARD

Tun Abdul Razak Laboratory,
Brickendonbury, Hertford SG13 8NL.
Tel: Hertford (0992) 54966. Telex: 817449.

SECTION 5b

Machine Balance and Vibration

IN THE case of any rotating element any unbalance in mass generates an acceleration force proportional to the square of the rotational speed, resulting in vibration. The fundamental frequency of this vibration is rev/min/60 Hz, but higher harmonies may also be generated. Standard treatment is to achieve a suitable degree of balance so that the actual vibration levels generated are within acceptable limits. Minimizing the mass of the rotatory elements will also minimize vibration losses, although usually this is a secondary consideration.

A distinction is drawn between static and dynamic balance. Considering a basic rotating machine element as an equivalent rotor mounted on a shaft, *static* unbalance will generate vibrations in a plane at right angles to the shaft (A–A in Fig 1). Any *dynamic* unbalance present will generate additional vibrations in random-axial planes, which can be represented by resultants in two diagonal planes at 45 degrees to the shaft axis (B–B and B'–B' in Fig 1). It also follows that static balance does not ensure vibration-force running, *ie* simple balance of the rotor in plane A–A is not enough.

Fig 1   *Fig 2*

Provided the system is rigid (*eg* the shaft does not deflect) both static and dynamic balancing can be achieved by adding (or removing) weights in two balancing planes A–A and B–B (or B'–B'). Adjustment of weight in plane A–A compensates for static unbalance and this can be done by simple static tests with the assembly mounted horizontally on knife edges. Adjustment of weight in plane B'–B' provides a couple compensating for the couple produced by dynamic unbalance. In this case suitable balance can only be arrived at under dynamic conditions. This can be done by trial-and-error methods, although they can be tedious, time-consuming and not necessarily entirely satisfactory. Balancing machines are much to be preferred as these can detect vibrations of the order of a few microinches only and thus it is readily possible to arrive at any required degree of unbalance. A further advantage offered by balancing machines is that the combined effect of static and dynamic unbalance can be treated as one, using two weights as shown in Fig 2.

Practical Balancing Requirements

The centrifugal force generated by vibration on a rotating system is directly related to the vibration amplitude and the rotating mass. Fig 3a shows this relationship for 10 mils (0.004 mm)

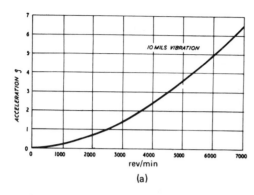
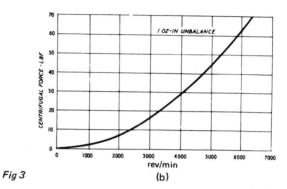

Fig 3

peak-to-peak vibration amplitude expressed in terms of g. Fig 3b shows the same relationship rendered in more measurable units — *ie* centrifugal force generated by specific unbalance 'couple' (mass-distance). It is seen from the latter that an apparently small degree of unbalance can produce a substantial centrifugal force, increasing as the square of the rotational speed. This vibration is potentially damaging in stress-cycling the rotating assembly as well as increasing the loads on bearings. It can also be felt by the operators of machines, *etc,* in addition to being a generator of noise.

The ability of the person to sense vibration by touch is dependent on the frequency of vibration. At speeds of about 8 000 rev/min and above, a vibration amplitude of 1/10 mil (4 micron) can be detected by the fingertips. At a speed of 1 000 rev/min, at least four times this amplitude of vibration must be present to be detected by touch, and even greater amplitudes at lower frequencies. This threshold of vibration is shown in Fig 4, together with an arbitrary curve drawn at a higher level defining the acceptable level for vibration in industry. The latter represents vibration levels below which a machine operator, for example, would disregard vibration.

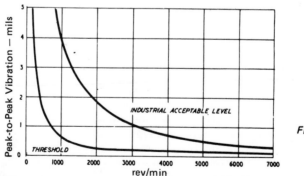

Fig 4

This subjective evaluation has no realistic engineering meaning, but can be expanded and rendered in a subjective/objective relationship as shown in Fig 5. This defines various degrees of subjective assessment of vibration levels — very smooth, smooth, *etc* — in terms of vibration amplitude and rotational speed.

This can serve only as a very general guide, but can be used to establish the degree of balancing required on a rotational machine. Thus, if the machine has an operating speed of 1 200 rev/min and is to be 'smooth running', the balance must be such that the vibration amplitude does not exceed about 1.3 mils (0.0005 mm). This can be checked by measurement with the assembly

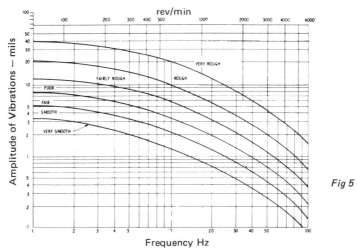

Fig 5

running in its own bearings, and balance adjusted as necessary. Balancing the rotating unit to the required vibration level on a balancing machine will not necessarily give the desired result since the vibration amplitude may well increase when the rotating unit is run in its own bearings. The standard of balance required can, however, be established by *in situ* testing and adjustment of balance as necessary, and then transferring to the balancing machine to measure the vibration amplitude present without bearings. The latter figure will set the balance requirements for the particular design.

This system is still too general for specific recommendations as the interpretation of what is 'smooth' or 'rough' running is capable of different subjective translations. It can serve as a guide for 'spotting' requirements, or for plotting further curves by interpolation to calculate the difference in balance requirements if the design of the rotating machines involves a speed change.

If, for example, a particular rotating machine with a speed of 500 rev/min proves satisfactory in operation with a vibration amplitude of 2 mils, increasing the operating speed to 1 000 rev/min with the same degree of balance would change the characteristics from 'smooth' running to 'fair' — Fig 6. To maintain the same degree of smooth running as before, the balance must be improved to reduce the vibration amplitude to a little over 1 mil (0.0004 mm), as indicated by the interpolated curve.

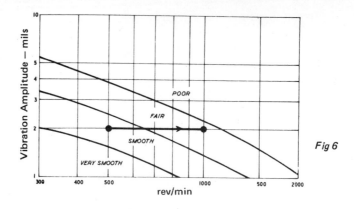

Fig 6

TABLE I — ISO RECOMMENDATIONS NO 1940 (DRAFT)

Quality Grade G	eω mm/sec	Rotor types — General examples
G 4 000	4 000	Crankshaft-drives of rigidly mounted slow marine diesel engines with uneven number of cylinders.
G 1 600	1 600	Crankshaft-drives of rigidly mounted large two-cycle engines.
G 630	630	Crankshaft-drives of rigidly mounted large four-cycle engines. Crankshaft-drives of elastically mounted marine diesel engines.
G 250	250	Crankshaft-drives of rigidly mounted fast four-cylinder diesel engines.
G 100	100	Crankshaft-drives of fast diesel engines with six and more cylinders. Complete engines petrol or diesel for cars, trucks and locomotives.
G 40	40	Car wheels, wheel rims, wheel sets, drive shafts. Crankshaft-drives of elastically mounted fast four-cycle engines (gasoline or diesel) with six and more cylinders. Crankshaft-drives for engines of cars, trucks and locomotives.
G 16	16	Drive shafts (propeller shafts, cardan shafts) with special requirements. Parts of crushing machinery. Parts of agricultural machinery. Individual components of engines (petrol or diesel) for cars, trucks and locomotives. Crankshaft-drives of engines with six and more cylinders under special requirements.
G 6.3	6.3	Parts of process plant machines. Marine main turbine gears (merchant service). Centrifuge drums. Fans. Assembled aircraft gas turbine rotors. Fly wheels. Pump impellers. Machine-tool and general machinery parts. Normal electrical armatures. Individual components of engines under special requirements.
G 2.5	2.5	Gas and steam turbines, including marine main turbines (merchant service). Rigid turbo-generator rotors. Rotors. Turbo-compressors. Machine-tool drives. Medium and large electrical armatures with special requirements. Small electrical armatures. Turbine-driven pumps.
G 1	1	Tape recorder and phonograph (gramophone) drives. Grinding-machine drives. Small electrical armatures with special requirements.
G 0.4	0.4	Spindles, discs, and armatures of precision grinders. Gyroscopes.

MACHINE BALANCE (A)

REGO
ANTIVIBRATION MOUNTS

▲ MACHINE BASE STABILITY
▲ OPTIMUM VIBRATION REDUCTION
▲ UP TO 90% ELIMINATION OF STRUCTURAL NOISE
▲ SIMPLICITY OF SELECTION

BESTOBELL PROTECTION

P.O. BOX 6,
135 FARNHAM ROAD,
SLOUGH,
BERKSHIRE SL1 4UY.
TEL: 0753 23921 TELEX: 848107

SHAKE RATTLE & ROLL

WITH

NOVIBRA

the natural choice
ANTI-VIBRATION MOUNTS
for efficient, controlled isolation

NOVIBRA rubber to metal bonded
Anti-vibration mounts eliminate
harmful vibrations and
greatly reduce structure borne noise.

For sound advice move to NOVIBRA by

TRELLEBORG

90 Somers Road, Rugby, Warwickshire CV22 7ED.
Telephone (0788) 62711. Telex 311144

T71/83

ANTI-VIBRATION MOUNTINGS

MANUFACTURERS OF ANTI-VIBRATION
MOUNTINGS AND RUBBER BONDED
TO METAL PRODUCTS

 LIMITED

THAMES WORKS
  LOWER TEDDINGTON ROAD
  HAMPTON WICK
  KINGSTON-UPON-THAMES
  SURREY KT1 4HA
  TELEPHONE: 01 977 2201
  TELEX NO. 262284 REF. NO. 3533

LORD Mechanical Group

THE INTERNATIONAL EXPERTS

- **VIBRATION ISOLATION**

- **TRANSMISSION COUPLINGS**

- **SHOCK MOUNTS**

- **ELASTOMERIC BEARINGS**

- **DAMPING MATERIALS**

LORD CORPORATION (U.K.) LIMITED
The Brook Trading Estate, Deadbrook Lane,
ALDERSHOT, Hampshire. GU12 4XB.
Tel: Aldershot (0252) 26225 Telex: 859484

It is more convenient to deal with residual unbalance as a specific quality, rather than analyze the system involved in degrees of balance, although the implication is the same. Thus residual unbalance is, effectively, a measure of the degree of balance. It can be expressed directly in terms of the vibration amplitude resulting, or as a force couple. The degree of balance specified in terms of vibration amplitude is usually more readily measured. However, force-compensating balancing machines measure the force couple present, so balance requirements can be determined in such units directly without calibration. Both methods are, therefore, in use.

Most specific recommendations for balance requirements can be rendered in terms of the force couple present due to unbalance. A further requirement may be that the velocity amplitude of the periodic motion of the centre of gravity of the rotating units should be constant by adopting *normalized unbalance* as the working criterion, or the force couple unbalance divided by the weight of the rotating unit. This normalized unbalance must be inversely proportional to rotational speed. Different degrees of unbalance can then be given a *quality number* or *quality grade*.

This method of treatment of unbalance is adopted as the basis of ISO Recommendations for balancing machines related to acceptable residual unbalance per unit of rotor mass in $\frac{g \cdot mm}{kg}$ or centre of gravity displacement in μm — see Table I and Fig 7.

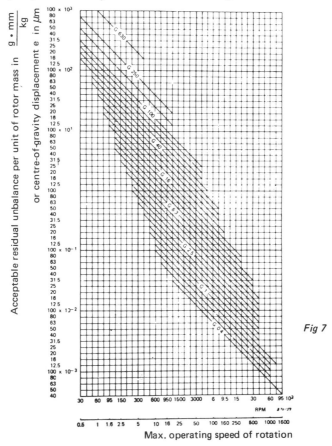

Fig 7

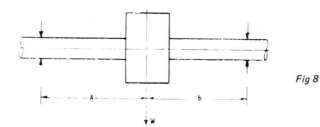

Fig 8

Critical Speeds

Even in a balanced rotating system there will be a rotational speed which corresponds to the resonant frequency of the system, exciting a bending mode because no such system can be fully rigid (*ie* there will be some static shaft deflection).

In the case of a weightless shaft carrying a rotor of weight W supported between two bearings, (Fig 8) static deflection (Y_s) will be responsible for a centrifugal force (F) under rotational conditions where:

$$F = \frac{W}{g}(Y_s \omega^2)$$

If eccentricity or dynamic unbalance (e) is also present, this formula will be modified to:

$$F = \frac{W}{g}(Y_s + e)\omega^2$$

where
  ω is the angular velocity

The actual deflection resulting can be expressed as:

$$Y = \frac{W}{g}(Y_s + e)\omega^2$$

$$\alpha = \frac{W}{g}\omega^2$$

where
  α is the force required to produce unit deflection for a given type of load and shaft support; or numerically

$$\alpha = \frac{3EI(a+b)}{a^2 + b^2}$$

where
  E = modulus of elasticity of shaft material
  I = moment of inertia of shaft section

MACHINE BALANCE AND VIBRATION

For critical conditions it follows that the deflection will become infinite, whence

$$\omega^* = \sqrt{\frac{\alpha g}{W}}$$

where
ω^* = critical angular velocity

or in terms of critical speed (N^*)

$$N^* = \frac{30\omega^*}{\pi} = 9.56\omega^* \text{ rev/sec}$$

In the case of horizontal shafts the assumption of a critical speed implies a specific static deflection due to the load W, *viz:*

$$Y_s = \frac{W}{\alpha} = \frac{Wg}{\omega^2 w} = \frac{g}{\omega^{*2}}$$

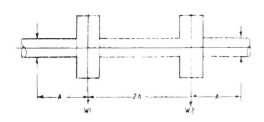
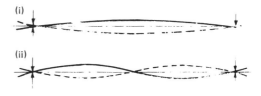

Fig 9

Where two rotors are mounted on the same (weightless) shaft at equal distances from end bearings (Fig 9), there are two possible deflection curves, as shown. Corresponding values for α deflection and critical velocity are:

Case (i)
$$\alpha_1 = \frac{3EI}{4a^3}$$
$$Y_1 = \frac{4Fa^3}{3EI}$$
$$\omega^*_1 = \sqrt{\frac{\alpha g}{W}}$$

Case (ii)
$$\alpha_2 = \frac{6EI}{a^3}$$
$$Y_2 = \frac{Fa^3}{6EI}$$
$$\omega^*_2 = \sqrt{\frac{\alpha g}{W}}$$

This means that there are two possible critical speeds, the second critical speed being in the ratio:

$$\sqrt{\frac{\alpha_2}{\alpha_1}}$$

or 2.83 times the first critical speed.

Similarly for M rotors there will be M critical speeds. In all cases the greatest value of Y (*ie* greatest shaft deflection) will be that corresponding to the first mode of deflection and thus to the lowest critical speed.

For a rough approximation, the first critical speed can be calculated from:

$$W_1{}^* = C \frac{g}{Y_s}$$

where

Y_s is the static deflection

C is an empirical coefficient, dependent on the manner of loading (and typically lying between 1.0 and 1.268 for shafts with two bearings).

In the case of non-uniform loading, static deflection is best determined graphically.

For more complete calculations, the following relationship may be employed:

$$\frac{1}{\omega^*} = \frac{1}{\omega_s{}^2} + \frac{1}{\omega_1{}^2} + \frac{1}{\omega_2{}^2} + \ldots\ldots$$

where

ω_s = critical velocity of the shaft
ω_1 = critical velocity of a weightless shaft with a concentrated load 1
ω_2 = critical velocity of a weightless shaft with a concentrated load 2 and so on

If the diameter of the shaft is not constant, then the equivalent diameter can be determined from:

$$\text{equivalent diameter } d = \frac{d_1 l_1 + d_2 l_2 + d_3 l_3 + \ldots\ldots}{L}$$

where

d_1 = diameter of length l_1
d_2 = diameter of length l_2, *etc*
L = complete length of shaft between bearings

Alternatively, critical length may be derived graphically.

In general, the shaft should be stiff enough (*ie* static deflection low enough) for the operating speed to remain always below the first critical speed (*eg* maximum operating speed should not be greater than 80° of the first critical speed). It is also desirable to avoid fractions of the critical speed for continuous operation — *eg* 1/2n, 1/3n, and 1/4n. Thus the optimum operating speed range is from 0.5 to 9.8 times the critical speed. If flexible elements are introduced, maximum operating speed should not exceed 0.7 times the critical speed.

Super-Whirling

Shafts which have to be operated above the first critical speed should again avoid its multiples and must be regarded as flexible rather than rigid elements. This considerably complicates the methods required to ensure satisfactory dynamic balance. Except in particular circumstances it is always advisable to stiffen the shaft for higher speed running so that it is still operating below the first critical speed, to minimize vibrational troubles and make dynamic balancing easier. This also avoids the inevitable vibration mode which would otherwise have to be introduced in accelerating the shaft through the first critical speed to achieve operational speed.

A particular trouble which may occur with smaller shafts run at high speeds, and thus above the frist critical speed, is that severe synchronous whirling may develop over a range of speeds well above the first critical speed. This is generally known as 'super-whirling' and can lead to severe vibration and wear. The onset of super-whirling, however, depends on a certain degree of un-balance being present, so that a perfectly dynamically balanced shaft is free from this effect (although perfect balance is virtually impossible to obtain). The degree of vibration present is also dependent on the clearance space in bearings. Although super-whirling may be present, if the clearance space is sufficiently small (or the bearing is a resilient packing) no vibration is noticeable. The presence of super-whirl, however, can promote rapid wear on the bearing, increasing the clearance and allowing vibration and further rapid increase in wear to develop.

Reciprocating Unbalance

In machines with reciprocating elements, reciprocating unbalance is the result of counter force generated by the acceleration and deceleration of the reciprocating elements. The resulting change in momentum of the complete system along a given axis is then equal to the resultant of all similarly directed internal forces, which in turn accelerate or decelerate the support frame in a corresponding mode. With conventional reciprocating movements the position is complicated by the fact that the reciprocating motion is not sinusoidal, resulting in frequencies which are the fundamental and higher harmonics of the rational speed. Specifically, with a crank mechanism, the second harmonic is the most prominent, its magnitude being dependent on the ratio of the connecting rod length to the crank radius.

Torsional Vibration

Torsional vibrations may be excited by motion forces resulting from acceleration or deceleration, or from torque loading and are a particular feature of reciprocating machines and internal combustion engines.

In the case of i/c engines, the predominant torsional vibration exciting forces are gas pressure torques. In the case of a four-stroke engine, harmonics of significance are $n/2$, n, $3/n2$, where n is the engine torsional speed (crankshaft speed). Higher order harmonics can generally be ignored. Representative values can be interpolated, or extracted, from empirical data and analyzed in terms of a Fourier series.

The total exciting torque is derived by adding the inertia torques resulting from unbalanced reciprocating masses, perfect balance of these masses being impossible because of the non-harmonic motions involved. Inertia effects arise in the form of alternating torques generated about the crankshaft axis. Again, only the fundamental and two harmonics are usually significant.

Typical first-mode natural frequencies are of the order of 250 to 350 Hz for 4-cylinder in-line engines, 220 to 310 Hz for 6-cylinder in-line engines, and 300 to 350 Hz for V-8 engines.

Critical speeds are rotational speeds at which the frequency coincides with the natural frequency of the shaft, when resonance will occur. Certain harmonic orders are of major significance ('major orders'), because all the cylinders are in phase. This occurs for orders m, so that m/N = ½, 1, 1½, 2, etc, where N is the number of cylinders. Major orders are thus associated with major critical speeds, defined as:

$$\text{major critical speeds} = \frac{f_n \times 60}{m}$$

Thus, in the case of a 4-cylinder in-line engine, m = 2, 4, 6, 8, etc, major critical speeds would thus be $30f_n$, $15f_n$, $10f_n$, etc. Assuming that a typical first-mode natural frequency of 300 Hz applies, corresponding crankshaft speeds would be 9 000, 4 500, 3 000 rev/min, etc.

The significance, or more specifically, the severity of the effect of the major critical speeds is independent of the firing order. The relative effect of other orders can, however, be changed by modifying the firing order. Thus no modification of firing order, based on equal firing intervals, can eliminate the effect of major orders. As a consequence some form of torsional vibration damping or absorption, generally known as a harmonic balancer, is almost invariably employed on automobile engines and similar drivers. This has the effect of substantially reducing the displacement or amplitude of crankshaft torsional vibrations.

Inertial Torque Formula

The fundamental frequency of the gas-pressure torque is one half the rotational speed of the crankshaft (in the case of 4-stroke engines). In terms of the angular crankshaft velocity (ω), this yields harmonics of:

$\omega/2, \omega, 3\omega/2, 2\omega$, etc

A formula for the inertial torque developed is:

$$Q_I \; 0.5 \; M_{rec} - \omega^2 r \left(\frac{r}{2l} \cdot \sin(\omega t) - \sin(2\omega t) - \frac{3r}{l} \sin(3\omega t) \right)$$

where

M_{rec} = equivalent reciprocating mass
mass of piston plus approximately one third of the mass of the connecting rod

r = radius of crankpin

l = length of connecting rod

The total exciting torque can thus be determined by adding these quantities for various harmonics, or orders, usually starting from ω and generally referred to as a number, starting from 1. Major orders are then defined by the fact that the vectors for all cylinders are in phase, *viz*

order number/number of cylinders = 0.5, 1.0, 1.5, 2.0, *etc*.

The speeds at which the frequencies of these major orders coincide with the natural frequency of the shaft are then referred to as major critical speeds. Resonant effects at such speeds are normally countered by the fitting of torsional dampers or harmonic balancers, tuned to the major critical speeds. Certain minor critical speeds may also occur, but the relative severity of these can often be adjusted by other means, such as by an alteration of the firing order of the cylinders.

MACHINE BALANCE AND VIBRATION

TABLE II – CAUSES OF MACHINE VIBRATION

Immediate	Predominant Frequency	Long Term
Unbalanced rotating parts	1 x rev/min	Increasing with (i) wear (ii) age (iii) corrosion (iv) non-elastic behaviour
Unbalanced reciprocating parts	1 x rev/min	Increasing with (i) wear (ii) age (iii) corrosion (iv) non-elastic behaviour
Fluctuating loads or forces	1 x rev/min	As (i), (ii) and (iii) above, plus (v) environmental changes (vi) accumulation of foreign matter (vii) damage (viii) inadequate lubrication
Misalignment	1 x rev/min	As (i), (ii), (iv), (v), (vi), (vii) and (viii) above
Loose mountings	1 x rev/min	As (i), (ii), (v) and (vii) above.
Rolling bearing noise	n x rev/min	As (i), (ii), (iii) and (iv) above.
Gear noise	number of teeth x rev/min	As (i), (ii), (iii) and (iv) above.
Oil whirl	½ x rev/min	Temperature changes
Environmental		Pressure changes Humidity changes
Energy transfer		Energy transfer

TABLE III – VIBRATION FROM MACHINE ELEMENTS

Component	Frequency of Vibration			Remarks
	Order of Shaft rev/min	High	Impact	
Rotating parts	✓			Depends on degree of unbalance.
Eccentric shafts	✓			Depends on degree of eccentricity.
Misalignment	✓			Can produce marked axial vibration.
Mounts	✓			If loose, can also be resonant.
Belts and pulleys	✓			
Bearings		✓		Broad band frequencies generated.
Gears		✓	✓	Depends very largely on accuracy of manufacture and gear type.
Slipping clutches	✓	✓	✓	
Loose or broken parts	✓		✓	
Lubricant	✓*	✓**		*Due to oil whirl. **Due to lack of lubricant

Summary Tables

Basic causes of vibration in machines are summarized in Table II. The column 'immediate' refers to parameters inherent in the design and/or construction of the machine, with effects apparent when the machine is first put into service. The column 'long term' describes changes which may occur in service to affect the original vibration levels generated by industrial sources.

Table III presents a check list of fundamental frequencies likely to be present in machine vibrations.

Table IV is a simple check and indicates suitable treatment for different sources of vibration and the field of attenuation concerned. See also chapters on *Machine Noise*, *Vibration Isolation*, *Dynamic Analysis of Vibration* and *Machinery Health Monitoring*.

TABLE IV – NOISE AND VIBRATION TREATMENT FOR MACHINES

	Action	Vibration Reduction	Noise Reduction	Remarks
Directly generated vibration	(i) reduce at source	✓	✓*	*Noise may not be significant.
	(ii) use isolation mounts		✓	Eliminate structural-borne noise.
	(iii) fit enclosure		✓	Reduce airborne noise.
Coupled vibration	(i) reduce cause	✓	✓*	*Noise may not be significant.
	(ii) isolate	✓	✓	
	(iii) change speed or frequency	✓	✓*	*Noise may not be significant.
	(iv) add damping	✓		
	(v) detune	✓	✓	
	(vi) change mass	✓	✓	
	(vii) change stiffness	✓	✓*	*Noise may not be significant.
Torsional vibrations	(i) use flexible coupling	✓		Noise levels usually insignificant.
	(ii) fit torsion damper	✓		
	(iii) adjust flywheel mass	✓		
Directly generated noise	(i) reduce at source		✓	
	(ii) fit silencer		✓	
	(iii) fit enclosure		✓	Reduce airborne noise.
	(iv) isolation mount		✓	Reduce structural-borne noise.
	(v) resonant absorber	✓		
	(vi) acoustic shielding		✓*	*In preferred direction
Resonance	(i) detune	✓	✓	
	(ii) change coupling	✓	✓	
	(iii) change stiffness	✓	✓	
	(iv) change mass	✓	✓	
	(v) reduce radiating area		✓	

Vibration Isolation

VIRTUALLY ALL practical dynamical systems (*eg* machines) are generators of unwanted forces (such as centrifugal force in unbalanced motors) which in turn generate vibrations (*ie* force-excited vibrations). These are transmitted from the system to any supporting structure, unless the force generation (the machine itself) is supported on some system of vibration-isolation. Equally, vibrations can be generated by motion and these motion-excited vibrations transmitted to a coupled mass, again unless there is some form of vibration-isolation between the two.

The ideal system for vibration isolation is where the vibration generator is separated from the associated mass by free space, when no transmission of vibration is possible from one to the other. This defines a perfect vibration isolation system with a *transmissibility* of zero. At the same time this is a purely theoretical concept since — short of levitation — there is no means of one part of the complete system supporting the other in free space. Practical supporting systems therefore carry one of the masses on a resilient mounting which inevitably have a finite value of transmissibility, defined as:

$$\text{Transmissibility (T)} = \frac{1}{1 - (f_f/f_n)^2}$$

where

f_f is the forcing frequency of force excited or motion excited vibrations present

f_n is the natural frequency of the resilient system

A transmissibility of less than 1 implies some degree of isolation; the smaller the value the greater the effectiveness of the system.

As f_f/f_n approaches 1, transmissibility also approaches ∞, *ie* there is no isolation; and when $f_f = f_n$, transmissibility reaches a (theoretically) infinite value. The critical parameter is thus the ratio f_f/f_n, the overall relationship between which and transmissibility is shown in Fig 1. Here it is seen that transmissibility reaches 1 when $f_f/f_n = \sqrt{2}$, rising steeply to a fully resonant condition at 1.4142 $f_f/f_n = 1$. Thus with f_f/f_n ratios of less than $\sqrt{2}$, vibration is actually *amplified* rather than reduced. It follows, therefore, that any isolation system must achieve a resonant natural frequency substantially lower than the lowest exciting frequency present. At this point it can also be made clear that isolators control the effect rather than the cause of vibration. In other words, they do not reduce the amplitude or frequency of the original source of vibration, but merely reduce the amount of *transmission* of this vibration from one point of the system to the other. Isolators do, however, change the *phase angle* between a sinusoidal vibratory input and the vibratory response of the supported body by 180 degrees in an undamped system.

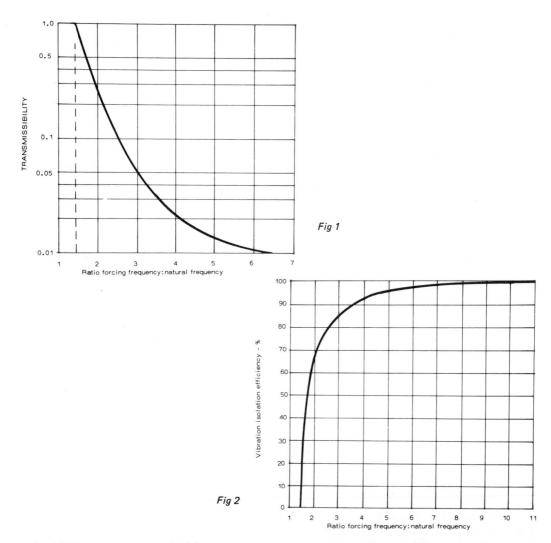

Fig 1

Fig 2

In general, satisfactory vibration isolation is usually obtained where a frequency ratio of 2.5 or greater is present. If the supporting structure itself is resilient, however, a considerably higher ratio may be necessary. For practical elimination of vibration (*ie* an efficiency of over 90%) a ratio of 4 or more is necessary. The particular practical difficulty which can arise in achieving high frequency ratios is the excessive amount of spring deflection that may be required.

The effectiveness of a vibration-isolation system can also be defined in terms of *isolation efficiency*, viz:

$$\text{isolation efficiency (\%)} = 100 \left(1 - \frac{1}{1 - (f_f/f_n)^2}\right)$$

This is shown graphically in Fig 2.

VIBRATION ISOLATION

Since all vibration-isolators are essentially springs, and the natural frequency of a spring is a simple function of its static deflection under load, transmissibility or isolation efficiency can also be plotted directly against spring deflection — Fig 3. This enables the amount of spring deflection to be read directly for a required transmissibility or isolation efficiency for any forcing frequency. Note here that these two diagrams specify forcing frequency in cycles per minute consistent with conventional machine practice, but isolator spring performance is normally stated in Hz (cycles per second). Fig 4 further extends this basic performance analysis with both f_f and f_n expressed in Hz.

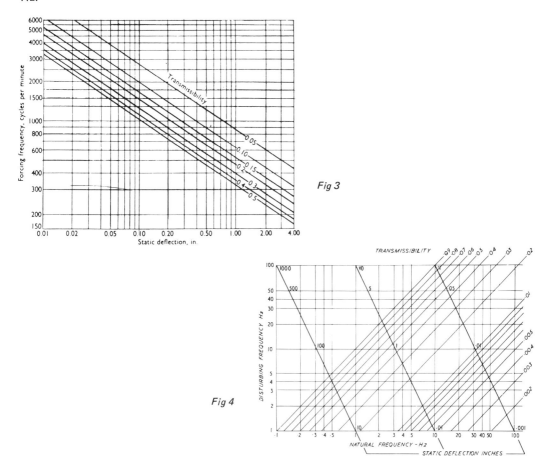

Fig 3

Fig 4

(See chapter on *Anti-Vibration Mounts* for further description of how springs work as vibration-isolators).

It is important to appreciate that the calculation of the natural frequency of isolators on the basis of the static modulus of the material can often lead to misleading results, and that it is the *dynamic* modulus which should be used in the case of all high hysteresis materials.

Isolators may be used to mount a machine directly on to a baseplate, floor or foundation, or to a substantial inertia block (usually a concrete block) which is itself isolated. Either method of mounting or suspension may have up to six degrees of freedom, each mode of vibration having its own natural frequency. The various modes may be uncoupled or coupled, depending on the geometry and mass distribution of the system, and the stiffness of the isolators. In the case of coupled modes, vibration in one mode will excite vibration in the coupled mode, with corresponding movement in both modes.

The particular value of an inertia block is that it can be proportioned to bring the combined centre of gravity of the system in line with unbalanced forces, thus giving the greatest possible moment of inertia to counteract any vibration couples. In other words, it can help dissipate some of the generated forces before they reach the isolators. They are also effective for limiting movement of the system whilst passing through resonant speed (*eg* starting up and slowing down), as well as at normal operating speeds.

The optimum weight of an inertia block can be determined by calculation based on the unbalanced forces, desired limit of vibration amplitude and isolation efficiency. The actual shape may also be important to support all components in proper alignment. In general, inertia blocks in concrete need to be at least 150 mm (6 in) thick, the ratio of inertia block weight to weight of supported equipment varying between 1.5:1 and 8:1, depending on the type of supported equipment (the high figure being typical for reciprocating machines).

Degrees of Freedom

For simplified analysis a system with a single degree of freedom is usually taken, corresponding to a single mass supported on a spring and with one plane of symmetry related to the principal axes of inertia of the unit.

The suspended mass then has only one natural frequency given by:

$$f_n = \frac{1}{2\pi} \sqrt{\frac{Kg}{W}} \quad Hz$$

where K = stiffness of spring or spring rate
W = weight supported
g = acceleration of gravity

If the spring system is damped, the natural frequency becomes

$$f_n = \frac{1}{2\pi} \sqrt{\frac{Kg}{W}} \, (1 - \frac{C}{C_c})^2$$

where C = the damping coefficient
C_c = critical value of $C = 2\sqrt{\frac{Kg}{W}}$

See also chapter on *Principles of Vibration*.

To segregate the six modes of vibration (six degrees of freedom) which may actually be present in a practical system, three planes of symmetry would have to be selected, with considerably more complicated mathematics worked — and probably several unknowns. In most cases involving six modes of vibration, in fact, reasonably close results can normally be obtained working with two planes of symmetry.

VIBRATION ISOLATION

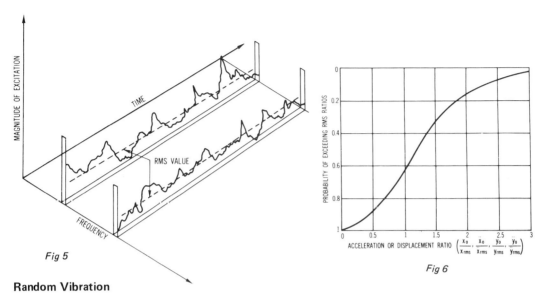

Fig 5

Fig 6

Random Vibration

Four parameters can be used to describe random vibration — excitation, magnitude, frequency, and time. At any frequency interval, the signal has some RMS value (Fig 5). To simplify analysis, the assumption is that the excitation is a stationary process — *ie* the RMS value stays constant with respect to time for any discrete frequency band. This implies that the time origin of the excitation is irrelevant — that is, that the magnitude of the RMS value does not vary with time.

Maximum excitation, of course, damages equipment the most. The probability that it will exceed certain ratios of the RMS value is predictable. For example, the probability that peak vaues will exceed three times the RMS value is only 0.02 (Fig 6). A plot of mean squared acceleration *vs* frequency could be used to define random excitation, but then the magnitude would depend on and vary with the bandpass of the measuring instrument. The use of the mean squared acceleration density, or power spectral density, avoids this limitation. This method is much more precise, as the density is defined as the mean squared acceleration passed by the filter with a one Hz bandpass width.

Among other forms, the power spectral density can be shown as a constant function of frequency, combined with superimposed discrete spectra (if the source or path of vibration energy has resonant effects), or as a plot of varying amplitude. However, this method makes it impossible to reconstruct the signals' original time history, since the signals are averaged.

Through a series of successive integrations, an equation for calculating the RMS acceleration can be derived:

$$\ddot{x}_{rms} = \sqrt{S_f(f_b - f_a)} \tag{1}$$

An expression for the RMS displacement can also be derived. For the case where f_a is small relative to f_b (*eg* 5 Hz and 2 000 Hz) an approximation may be used:

$$x_{rms} \approx \frac{5.66\, \ddot{x}_{rms}}{\sqrt{f_a^3 f_b}} \tag{2}$$

A simple, single-degree-of-freedom system responds to random vibration by oscillating at its natural frequency.

The magnitude of the response acceleration depends on the system's natural frequency, system resonant transmissibility, and the power spectral density of the excitation. It is given by:

$$\ddot{y}_{rms} = \sqrt{(\pi/2)f_n S_f T_r} \tag{3}$$

Equation 3 is the most important relationship the design engineer has to work with, and is plotted in Fig 7 as a nomogram.

If the system is lightly damped ($T_r \geqslant 2$), as is usually the case, only the power spectral density in the region of system resonance need be considered. In isolated instances, such as for a heavily damped system and variable power spectral density input, Equation 3 cannot always be used.

A simple system responds to a 'white-noise' random excitation by oscillating at its natural frequency — only the magnitude of the response varies randomly. Again, the probability that a peak value will exceed some ratio of the RMS response (as calculated from Equation 3) can be predicted from Fig 6.

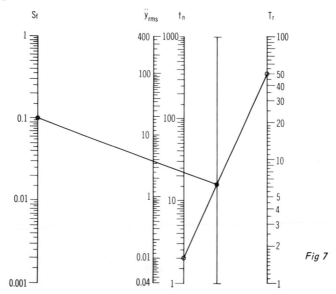

Fig 7

For a simple system oscillating at its natural frequency, response displacement can be calculated from:

$$y_{rms} = 9.8\, \ddot{y}_{rms}/f_n^2 \tag{4}$$

Figure 8 plots this expression as a nomogram for design use.

The typical response of a structure of multi-degree of freedom system to random inputs is shown in Fig 9. Since most real structures are not simple systems and random excitations in actual service are usually not constant over broad frequency bands, the response of a structure to a random input cannot be determined simply.

VIBRATION ISOLATION

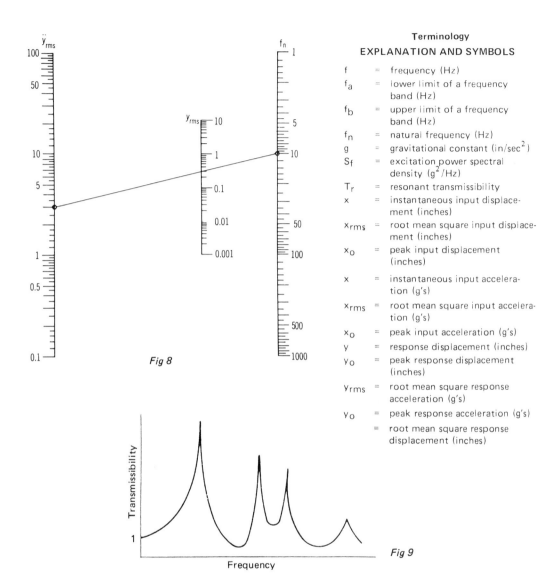

Fig 8

Fig 9

Terminology
EXPLANATION AND SYMBOLS

f	=	frequency (Hz)
f_a	=	lower limit of a frequency band (Hz)
f_b	=	upper limit of a frequency band (Hz)
f_n	=	natural frequency (Hz)
g	=	gravitational constant (in/sec^2)
S_f	=	excitation power spectral density (g^2/Hz)
T_r	=	resonant transmissibility
x	=	instantaneous input displacement (inches)
x_{rms}	=	root mean square input displacement (inches)
x_o	=	peak input displacement (inches)
$\ddot{x}$	=	instantaneous input acceleration (g's)
$\ddot{x}_{rms}$	=	root mean square input acceleration (g's)
$\ddot{x}_o$	=	peak input acceleration (g's)
y	=	response displacement (inches)
y_o	=	peak response displacement (inches)
$\ddot{y}_{rms}$	=	root mean square response acceleration (g's)
$\ddot{y}_o$	=	peak response acceleration (g's)
	=	root mean square response displacement (inches)

Mathematically, the normal modes of a structure can be determined from transmissibility *vs* frequency data. The modes are expressed as equivalent single-degree-of-freedom systems with given natural frequencies and damping characteristics. Superimposing these simple equivalent systems in the correct proportion approaches the response of the more complex structure.

The response of this kind of system to random excitations can be computed by just such an analytic approach. The answer for the response of the complex system will contain all frequencies in the excitation spectrum, each at a different RMS amplitude. This procedure usually takes too much time, however, and a modification of Equation 3 can be used as a shortcut:

See also chapter on *Anti-Vibration Mounts*.

Anti-Vibration Mounts

THE BASIC principle of vibration isolation is support of the mass of a vibrating machine on a resilient mount (or series of mounts) having a lower natural frequency than that of the lowest generated vibration. Considering vibration as a series of individual impacts or blows on the mount, the first blow causes the resilient mount to deflect downwards. When the force of the blow is spent the mount starts to return at its own frequency. Since this is less than the frequency of the blows it is still returning when it receives the next blow, then offering a 'cancelling' force. The greater the ratio of the disturbing frequency to the natural frequency of the mount, the greater the amount of vibration energy absorbed in this way. The *efficiency* of such a mount is thus given by:

$$\text{Efficiency (\%)} = 100 \times \left(1 - \frac{1}{(R^2 - 1)}\right)$$

where R is the ratio of forcing frequency to natural frequency.

Actual efficiency achieved is, however, also dependent on the rigidity of the structure supporting the system. No anti-vibration mount is 100% efficient, so some vibration is transmitted and the resonant frequency of the base can also be significant.

Specifically, therefore, an anti-vibration mount is some form of spring (or spring system). However, if a pure spring it will have no self-damping properties, so whilst it may inhibit transmission of vibration, it will not modify the original vibration. To do this, the spring must have damping properties, or be associated with a separate damper. Thus vibration-isolation and vibration-damping may need separate consideration.

The most essential requirement of a vibration-isolation material is that it must be truly resilient so that it has the ability to return to its original height when loads or forces are removed. Provided such a material is loaded within its elastic limit it will then have a long effective life.

To be effective in *reducing* vibration, the material must also have good damping characteristics.

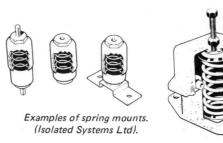

*Examples of spring mounts.
(Isolated Systems Ltd).*

ANTI-VIBRATION MOUNTS (A)

VIBRATION  ISOLATION

MACHINE INSTALLATION
Machine Tools — Power Presses — Generators
Moulding Machines — Fans — Compressors

MARINE — OFFSHORE
Exhaust and Flare Stacks — Auxiliaries —
Deckhouses and Modules — Boat Fenders

AEROSPACE — DEFENCE
Electronics — Helicopter Rotor Dampers
Gun Recoil Absorbers — Packaging

STOP-CHOC LTD 710 Banbury Avenue, SLOUGH, Berks SL1 4LH
Telephone: Slough (0753) 33223 Telex: 848620

DAVIES LIGHTING FITTINGS

For use on overhead Cranes and other Heavy Machinery, where vibration is a problem, a full range of lighting fittings incorporating anti-vibration assemblies and alternative types of mounting brackets. Control gear available.

Suitable for GES Lamps:-
- TUNGSTEN
- HPS/SON
- MBFU
- MBFRU
- MBTF

A. DAVIES & SONS LTD.
Alpha Works, Ashton Road, Bredbury, Stockport.
Tel: 061—430 5297. Telex: 666501

DESIGN FOR A QUIET LIFE

**Foam-lead-foam sound barrier.
Soundmat LF Embossed
can achieve 40-50 decibel reductions.**

This is just one of the Soundcoat range of Noise Control Materials which are available from Stock. We also die-cut these materials to your drawings. A first-class range of pressure sensitive adhesives makes the application of parts a clean and simple process.

For information on noise control materials contact us NOW.

Ferguson & Timpson Limited
Gasket Cutting Division
5 Atholl Avenue, Hillington
Glasgow G52 4UA
Tel: 041-882 4691
Telex: 77108

Branch Offices at:
London, Birmingham, Hull, Liverpool.

ANTI-VIBRATION MOUNTS (B)

NOISE

Trubros Acoustics specialise in controlling noise problems in a wide variety of situations.

Acoustic enclosures to keep noise in, and to keep noise out.

Acoustic screens to isolate noise from small and large process areas.

Portable acoustic air conditioned cabins, for personnel in uncomfortable working conditions.

Acoustic cladding, in decorative style, for office and factory areas.

TRUBROS ACOUSTICS LTD
Imperial Works   Station Road Kegworth Derby DE7 2FR
Telephone Kegworth (05097) 2104

TICO
pads for noise and vibration control

TICO LF/PA low frequency mounting pads are part of a comprehensive series of TICO Pad materials specifically designed to reduce the transmission of noise and vibration throughout modern industry.

The extensive load range of TICO Pad materials covers a wide variety of mounting applications including standard workshop machinery, heavy duty Power Hammer Anvils and Foundations, and the complete isolation of plant room air conditioning equipment, including pipework.

Brochures detailing static and dynamic properties of TICO Pad materials are available upon request.

Please make use of our free consultative service.

JAMES WALKER & CO LTD
LION WORKS WOKING
SURREY GU22 8AP
ENGLAND
Telephone: Woking 5951
Telex: 859221

VIBRATION REDUCTION.

WHEN THE GOING GETS TOUGH

ALL METAL ISOLATORS

Stainless Steel knitted cushions for damping to give the dynamic property of rubber, combined with the durability of steel, resulting in a product capable of taking on any environmental foe, with increased life expectancy.

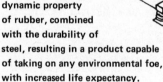

C.M.T. WELLS KELO LTD.,
Kingsland, Holyhead, Gwynedd LL65 2SN
Telephone: 0407 2391/5 Telex 61518 Welkel G

Natural Frequency and Deflection

The natural frequency of a spring is directly related to its static deflection (d) under load, *viz:*

$$f_n = \frac{188}{\sqrt{d}} \qquad\qquad f_n = \frac{947}{\sqrt{d}}$$

where d is in inches　　　　　　where d is in millimetres

Its performance as an isolator is expressed by its *transmissibility,* which is a function of the ratio of the disturbing frequency to the natural frequency of the spring, *viz:*

$$\text{transmissibility (T)} = \frac{1}{1 - R^2}$$

where　R = ratio of disturbing frequency to the natural frequency of the mounted assembly.

Isolation *efficiency* can then often be expressed directly in terms of transmissibility as 100 (1−T) per cent. (See also Table I).

TABLE I — VIBRATION ISOLATION PERFORMANCE

Forcing Frequency / Mount Frequency	Transmissibility	Isolation Efficiency	Result
1	amplified	—	worse than rigid mount
$\sqrt{2}$	1.0	0	same as rigid mount
1.5	0.80	20%	very poor
2	0.33	67%	fair to good
2.5	0.19	81%	good
3	0.125	87.5%	very good
4	0.06	94%	excellent
5	0.04	96%	—
6	0.02	98%	—
10	0.01	99%	virtually vibration free mounting

Types of Vibration Isolators

Basic *methods* of vibration-isolation are summarized in Table II. Unit mounts, more commonly called anti-vibration mounts or flexible mounts, include numerous individual types, as summarized in Table III.

TABLE II — METHODS OF ISOLATION

Method	Materials		Remarks
Unit mounting	(i) (ii) (iii) (iv)	Metal springs Moulded rubber Air springs Hydraulic 'springs'	Secured between machine bedplate and mounting base; or as industrial mounts between machine foot and bearer or mounting base
Pad mounting	(i) (ii) (iii) (iv)	Rubber Cork Felt Rubberized fabric	Simple machine mounts, may or may not be bonded in place
Suspension systems	(i) (ii)	Helical coil springs Leaf springs	For the isolation of concrete foundation blocks
Area mountings	(i) (ii) (iii) (iv) (v)	Rubber Cork Rubberized fabric Ridged rubber mats Felt	Laid in the form of a carpet under concrete blocks

TABLE III — ANTI-VIBRATION MOUNTS

Diagrammatic	Type	Construction	Features and Applications
	Simple pad	Rubber, rubberized fabric, rubberized cork, felt.	Simple, inexpensive mounts for stationary machines — may or may not be bonded in place.
	Area mount	Rubber, rubberized fabric, rubberized cork, cork, felt, ridged or contoured rubber mats	Carpet type mount laid under stationary machines over concrete floor or concrete blocks
	Spring	Helical steel spring	Excellent vibration isolation and can be tuned over a wide range of natural frequencies. Also good as a shock absorber, but has very little damping. Can be designed for very heavy loads. Suitable for high temperature surroundings.
	Damped spring	Helical steel spring with integral damping device	All the advantages of a spring for vibration isolation and steel absorption with large damping. Damper may be of friction, viscous or elastomeric (hysteresis) type
	Miniature rubber	Basically a rubber grommet with or without metal mounting plates	Typical natural frequency 15 Hz. Typical maximum load 10–12 lb. Available in a variety of rubbers with different damping characteristics

cont...

TABLE III — ANTI-VIBRATION MOUNTS (contd.)

Diagrammatic	Type	Construction	Features and Applications
	Simple rubber	Cylindrical, square or contoured rubber block with bonded-in studs	Simple, inexpensive vibration isolation mount. Typical natural frequency 5–20 Hz. Maximum load about 300 lb. Moderate damping characteristics.
	'Captive' rubber	Rubber block with bonded-in stud, captive in metal base	Excellent general-purpose flexible mount for reciprocating engines and machines up to about 3000 lb weight. Typical natural frequency 9–12 Hz. Large variety of proprietary designs
	Suspension systems	Helical coil or leaf springs	Used for the isolation of concrete foundation blocks for stationary machines
	Hydraulic	Load damping rubber bellows with interconnecting orifice	Superior damping and shock absorbing performance to captive rubber mounts. Recently developed for automobile engine mounting.
	Premature Pneumatic (air spring)	Heavy duty sealed bellows interally pressurized at 20–100 bar	Typical natural frequency 2–5 Hz. Loads up to 20 000 lb. Excellent performance as isolation mount and shock absorber
	Pneumatic (self-levelling)	Externally pressurized air spring	For vibration-free fixed level isolation of precision machine tools, etc. Natural frequency may be as low as 1–1.5 Hz.
	Free-standing mounts	Various, but commonly metal spring with elastomeric damping and incorporating height adjustment	Mounts for stationary machines and attached only to the machines
Proprietary	Isolating mounts	Numerous, including designs and constructions for bolted down and free-standing mounts	Designed as vibration isolation mounts, or combined isolation mounts and shock absorbers
	Woven mesh (cushions)	Crimped stainless steel mesh rolled into a cylindrical cushion	Particularly suitable as a damping material used in conjunction with metal spring isolators for high temperature applications. Also useable on its own as an isolator.

In general metal springs become preferable for unit mounts where the required static deflections exceed 12.5 mm (½ inch), and are capable of carrying many heavy loads, but can be rivalled by air springs. For smaller deflection elastomeric mounts are often preferred with the advantage of also providing vibration damping. (Air springs also provide damping).

Metal springs (usually steel) have a wide application since they approach closely the ideal spring performance and lend themselves to being produced with a wide range of characteristics (*ie* a wide range of natural frequencies, dependent on the static deflection under load). They can also be used at temperatures far in excess of that permissible with non-metallic resilient materials, and in sizes to carry the heaviest loads. Their main limitation is their very low damping characteristics which, in particular applications, may require the addition of separate dampers of the viscous or friction type. A wide variety of vibration and shock-absorbing mounts are produced on this basis, employing a composite construction of springs and dampers. Without the use of separate dampers, the damping provided by a metal spring is of the order of 0.1% of the actual damping.

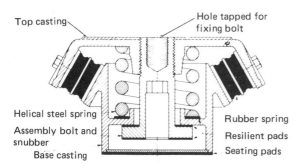

Combined spring and rubber in shear/compression mount.

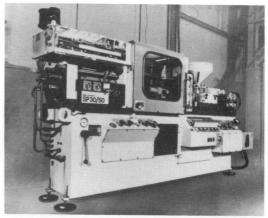

Heavy machine mounted on TICO S/AD Adjustamounts. (James Walker & Co Ltd).

Metal springs are readily capable of accommodating large deflections and thus can be tuned to very low frequencies. In this respect they are superior to all other isolating materials — see Fig 1 — provided the necessary deflection can be accommodated, or the suspension does not become unstable laterally. For a given geometry, it is possible to establish a stable working region — see Fig 2. Low frequency metal springs do, however, have the unfortunate characteristic of readily transmitting higher frequencies, and although, of course, a spring can easily be stiffened to provide isolation at high frequencies only, it would then pass lower frequencies.

ANTI-VIBRATION MOUNTS

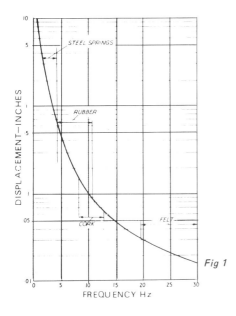

Fig 1

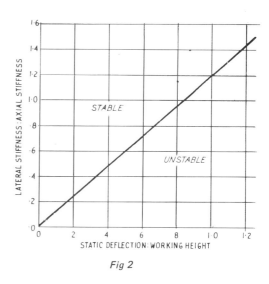

Fig 2

Spring design and performance are well established and production can be closely controlled to yield predicted performance figures.

Metal springs form a complete, and separate, class of isolators in which the resilient element is:
(a) helical compression springs under compression;
(b) helical compression springs under tension;
(c) leaf springs.

The last named have more limited, and specialized, application, but have one specific advantage over helical coil springs in that a fair degree of damping is provided by interleaf function. They are used mainly for complete suspension systems (*eg* the isolation of concrete mounting blocks). Helical coil springs may also be used for a similar purpose (with the addition of discrete dampers as necessary). Metal springs of composite construction are normally produced as unit mounts.

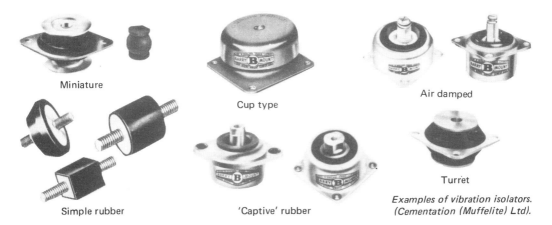

Examples of vibration isolators. (Cementation (Muffelite) Ltd).

TABLE IV — MINIMUM OUTSIDE DIAMETERS FOR STABLE SPRINGS*

Defl. Min. o.d. Load	¾"	0.76 to 1.0	1.1 to 1.5	1.6 to 2.0	2.1 to 2.5	2.6 to 3.0	3.1 to 3.5	3.6 to 4.0	4.1 to 4.5	4.6 to 5.0	5.1 to 5.5
250	2½"	2½	2.7/8	3½	4½	5½	5½	7	7	7	7
251– 500	2½	2.7/8	2.7/8	4½	4½	5½	7	7	7	7	8½
501–1000	2½	2.7/8	4½	4½	5½	7	7	7	7	8½	8½
1001–1500	2½	2.7/8	4½	5½	7	7	7	7	7	8½	8½
1501–2000		2.7/8	4½	5½	7	7	7	7	7	8½	8½
2001–2500		2.7/8	4½	5½	7	7	7	7	7	8½	10
2501–3000				5½	7	7	7	7	8½	8½	10
3001–3500	USE MULTIPLE ELEMENT SPRINGS					7	7	8½	8½	8½	10
3501–4000						7	7	8½	8½	10	10
4001–4500						8½	8½	8½	8½	10	10
4501–5000							8½	8½	8½	10	10

*Vibration Eliminator Co.Inc.

A guide as to the minimum spring outside diameter required to ensure stable springs for various loads and static deflections, is given in Table IV.

An important point to note with steel springs is that they provide a direct transmission path for *noise*. Thus although they are effective for vibration isolation it is necessary to introduce an elastomeric pad (or similar treatment) to break the direct sound path and minimize sound transmission.

Elastomeric Mounts

A number of resilient materials are employed for isolators and can show advantages over metal springs in that these are often more readily mounted and that inherent damping characteristics are also present, due to absorption of energy within the material through molecular deformations and internal spring combinations. Bonding, for example, is often a perfectly practical form of mounting with resilient material isolators. Elastomers also have sound-absorbing properties, are free from the need for adjustment, and are generally more economic than metal springs where static deflections required do not exceed 12.5 mm (½ inch). Elastomeric mounts can be designed for larger deflections, but at the expense of increased volume and cost. Both natural and synthetic rubbers may be used; also sponge rubber (although this has limited application).

Elastomeric materials may be used as anti-vibration mounts in a number of ways:

(i) *Unit Mounts* — fabricated for direct attachment to the baseplate of machines, or to mounting blocks, *etc*.

(ii) *Pad Mounts* — simple cut or formed shapes on which the mass to be mounted rests and may or may not be bonded in place.

(iii) *Area Mounts* — laid in the form of a complete 'carpet' (usually under an inertia block). This method has now been largely superseded by the use of pad mounts interspaced with lighter 'crushable' material, such as polyurethane foam, to accommodate the shape factor necessary with rubber pads.

Basic Characteristics of Rubber

The stiffness of rubber as a spring material can be determined completely in terms of the modulus of the material and the geometry of the section. The modulus can be modified by compounding, so that a wide variety of rubbers are available with different characteristics. The dynamic modulus differs appreciably from the static modulus, although this is largely influenced by the hardness of the rubber. Natural rubbers also tend to have more resilience than synthetic rubbers and can thus provide lower natural frequencies in similar geometry. The use of natural rubber is, however, limited by compatibility — *eg* natural rubbers cannot be used where they may come in contact with oils or other hydrocarbons.

The hardness of the rubber also affects the inherent damping properties. Soft rubbers have low hysteresis and relatively little internal damping. Harder rubbers, in general, tend to provide higher damping. Typical damping rates ($C_r = C/C_c$) range from under 2% for soft rubbers to about 15% or more for rubbers of 80 degrees hardness. The hysteresis of a rubber can also be modified by compounding.

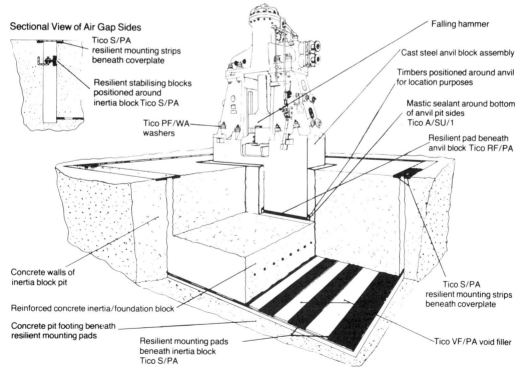

Layout of a typical hammer installation with a resiliently mounted foundation block.
(James Walker & Co Ltd).

A general characteristic of rubbers is that they are incompressible. That is to say, under compressive loading they can only deflect if free to expand laterally. If constrained, they will remain stiff. Thus, when employed in compression, sufficient clearance must be allowed for lateral expansion, or a ribbed periphery incorporated to accommodate expansion. Anything that restricts lateral expansion, such as bonding to a surface, will tend to increase the stiffness of the material under compression. The free area available for lateral expansion will in effect, govern the stiffness of the rubber per unit of load-carrying area.

This incompressibility characteristic of rubber limits the degree of deflection which can be achieved with solid rubber pads, *etc*, particularly when used as large area mats. A solution is to mould the rubber in the form of studded or grooved mats which will materially increase the deflections possible without requiring lateral clearance, and thus lower the minimum natural frequency which can be realized with the material. The degree of grooving, *etc*, which can be achieved is limited by the stability required; alternatively, if high deflections are required from a rubber mat, discrete dampers may also have to be added to provide the necessary stability.

Elastomeric Unit Mounts

Basic methods of stressing rubber are in compression, tension or shear, (Fig 3). For a given load, deflection is least in compression and greatest in shear, for a given volume of rubber. Loading in tension is seldom used except for relatively light loaded hangers, since the integrity of such a mount is reliant on the strength of the rubber-to-metal bond in a practical unit, which is relatively low (*eg* typically 45–70 lb/in^2 or 3–5 kg/cm^2). Loading in compression thus provides the greatest load capacity, and loading in shear the greatest flexibility (deflection) and lowest natural frequency.

COMPRESSION   TENSION   SHEAR

Fig 3   *Fig 4*

Combined loading in compression and shear is provided by an angled mount when both unit load capacity and deflection depend on the angle selected for the mount. To provide greater stability with compression/shear mounts they are commonly produced in back-to-back or *conical* configuration – Fig 4.

When strained in shear a rectangular rubber pad will assume a parallelogram form, the relation between force and deflection being:

$$\text{deflection (inches)} = \frac{FGt}{S}$$

where F = force in pounds
       G = shear modulus of rubber, lb/in^2
       t = thickness of rubber, inches
       S = area of rubber, square inches

The value of the modulus G is dependent on strain. It remains appreciably constant for small strains (small deflections) but then decreases with increasing strain. As a general rule the static strain defined by d/t should not exceed 0.5.

Franking machines at the Inland Revenue mounted on TICO LF/PA/10 material for reduction of noise and vibration to the building structure. (James Walker & Co Ltd).

When used as a compression pad, performance is generally determined empirically, or from semi-empirical charts or formulas, taking into account the free movement available for expansion and whether or not the pad is bonded. Stiffness is, however, also directly related to the length/width ratio of the pad.

When used in shear compression a rubber isolator is normally designed to provide its primary (isolation) performance stressed in shear. The free movement available is, however, restricted, so that compressive loading provides a snubbing action restricting the movement or drift of the pad. Loading in shear compression is generally considered to provide the best compromise between load/deflection characteristics and rubber life.

It should be appreciated that the performance of rubber-like materials cannot be predicted accurately from the static modulus of the material and basic first principles since the degree of damping inherent in the material will modify the transmissibility. Also non-linear load/deflection characteristics may be present. Standard transmissibility/frequency ratio curves and standard deflection test data can, therefore, only be used as a guide to natural frequency and the amount of compression likely to be experienced under static stress, respectively. True values can only be obtained by empirical tests on the materials concerned, taking into account the damping present and dynamic characteristics of the material.

Air Springs and Isolators

Air springs are the most efficient form of vibration-isolator, capable of achieving efficiencies of 99% or better in specific cases. They can also be designed to cover a wide range of load capacities, typically from 100 lb to over 50 000 lb. Another extremely favourable characteristic is that they can be designed with natural frequencies as low as 5 Hz (300 cycles per minute) and even very much lower still with an auxiliary air volume (air reservoir).

Most air springs, too, can show high damping ratios ranging from about 0.01 for consolidated air springs up to 0.1 for rolling diaphragms. Controlling and varying the damping can be accomplished by various forms of orifice damping.

Pad and Area Mounts

Materials used for pad and area mounts include the following:

(i) *Configurated Rubber* — (a) studded pads;
                                          (b) ribbed pads;
                                          (c) ribbed or honeycombed sections with band facings.

(ii) *Rubberized Fabrics*
(iii) *Reinforced Rubber*   — (a) metallic reinforcement;
                                                 (b) non-metallic reinforcement.
(iv) *Cork*   — (a) agglomerated cork;
                                                 (b) rubber bonded cork.
(v) *Felt and Bonded Felt*
(vi) *Foamed Materials*
(vii) *Steel Mesh*

The principal properties of interest in the case of resilient materials are:

static stiffness
dynamic stiffness   governing natural frequency;

damping ratio   governing damping;

permissible unit loading
effect of steady static load   governing load-carrying capabilities.
effect of dynamic strain amplitude

Other properties of significance are:
  (a) static creep rate;   (b) compatibility with ambience;

  (c) effect of ageing on dynamic properties;   (d) tolerances of dynamic properties under operating conditions.

Configurated Rubber

Configurated rubber may be adopted for mounting pads, the object being to increase the static deflection (and thus reduce the natural frequency) of the material. The properties are otherwise similar to those of solid rubber.

Common forms of configuration are studs or ribbing of one or both surfaces; this provides the simplest and cheapest solution. If concrete is to be poured over such a configurated surface, however, it is necessary to lay rigid sheet material (usually steel) over the pad in order to prevent the concrete filling the 'open' spaces, which would destroy the object of configuration.

Alternative forms of configuration which overcome this limitation are:
  (i)   accommodating the 'open' spaces within the pad section.
  (ii) covering the configurated section with a hard face.

Reinforced Rubber

A variety of rubber mounts of composite construction are described as reinforced rubber isolators where the metallic component is effective in providing stability or in modifying the physical performance of the unit. Rubberized fabric is also a type of reinforced rubber. This classification is, however, more properly reserved for rubber pad materials with non-metallic reinforcement, designed to provide very high load bearing qualities whilst still retaining good resilience. Such materials are perhaps more specifically described as resilient bearing materials, load bearing being a primary function whilst, at the same time, providing isolation and damping of vibrations which may be generated in, or excited in, the supported system.

ANTI-VIBRATION MOUNTS

A typical non-metallic reinforcing material is asbestos, used in conjunction with suitable 'age-proof' rubber such as neoprene. The compounding, fabrication, moulding pressure and vulcanization can be adjusted, as necessary, to meet specific requirements for static deflection, shear deflection and stress. Significant parameters, as far as dynamic properties are concerned, are:

(i) the dynamic modulus of elasticity (E), from which the dynamic stiffness can be obtained using the formula:

$$K = E\frac{A}{t}$$

where A is the area of the pad and
t is its thickness

(ii) the shape factor(s), which is defined by the ratio of loaded area to free-to-bulge area. Laminated constructions make a pad far less sensitive to shape factor than homogeneous rubber pads of the same overall dimensions.

(iii) the loss factor (N), defined as the tangent of the phase angle by which the damping force produced by the pad lags behind the stiffness or spring force. This is a more convenient term to use than the damping ratio since the latter is a property of the complete vibratory system whereas the loss factor is a property of the pad only. At resonance, however,

$$N = 2 \times \text{damping ratio}$$

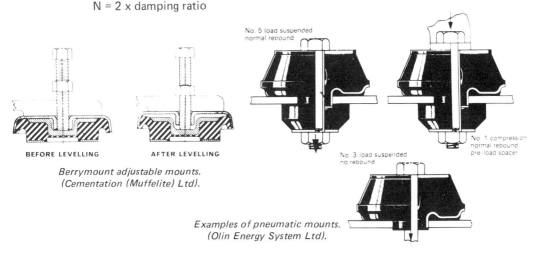

Berrymount adjustable mounts.
(Cementation (Muffelite) Ltd).

Examples of pneumatic mounts.
(Olin Energy System Ltd).

Cork

The simple form of cork used for isolation mounts is usually compressed cork particles, bonded under high pressure into a slab material. The degree of subsequent compressibility is controlled by the density of the slab (or basically its porosity). Unlike rubber it is a compressible material and may accommodate deflections of up to 30% without lateral expansion. Its natural frequency depends on density (increasing with density) and loading (decreasing with loading) although a practical limit is placed on the minimum possible resonant frequency for any density and thickness by the maximum load which the material can safely support without becoming overstressed and permanently deformed.

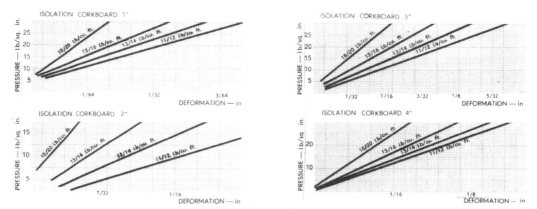

Fig 5 Load/deformation of typical cork isolation material.

The generally recommended loadings for optimum performance are 5 to 25 lb/in^2 corresponding to a density range from 0.18 to 0.34 (see also Fig 5). Equally, cork isolators must be subject to reasonable loading in order to perform effectively since deflection increases rapidly with both very low and high loads — *ie* the optimum operating range is based on a reasonably linear proportion of the load-deflection curve.

Cork isolators are generally used in compression, when the damping provided is normally of the order of 6% and independent of density. In some applications cork is used in shear-compression.

Agglomerated cork pads or mounts can have a number of disadvantages. Under repeated stress cycling the material may crumble. Also if water-soluble binder is used, the particles may become disassociated under unfavourable conditions — *eg* dampness. Cork itself is also an absorbent material and can become waterlogged if subject to continual immersion. Some of these objections can be overcome by using a water-resistant binder, whilst physical strength can be enhanced by steel-binding. This can take the form of strap-type steel binding around the edge of the cork or bonded metal facings.

Despite the physical limitations of cork as an isolation material, it is relatively cheap, and has useful enough properties for it to continue to be used as an isolator, particularly as regards its compressibility characteristics.

Rubber-Bonded Cork

Materials in this category comprise cork particles bonded with rubber. To a large extent this enables the favourable characteristics of both materials to be realized without the individual disadvantages. Thus, rubber-bonded cork can have the same compressibility characteristics as cork, *ie* it does not require 'clearance' to permit the shape factor necessary for deflection of rubber alone. At the same time the rubber bond gives the same compatibility as the rubber to the composition and overcomes the relatively fragile characteristics of the cork alone. Rubber-bonded cork is, therefore, an extremely useful general purpose material for isolation pads. It is generally rated as suitable for medium or high frequency isolation.

Felt

Felt is a 'traditional' vibration-isolation material and its use continues, although it has the disadvantage of being absorbent. Its desirable properties would be nullified should it become waterlogged or soak up oil, *etc.* Felt is most effective as a vibration isolator in small area pads cut from

ANTI-VIBRATION MOUNTS

the softest felt in 6 mm or 25 mm (¼ inch or 1 inch) thicknesses. These pads can be held in place with cement if vibration is high. The force-deflection curve for felt is substantially linear for deflections up to 25% of the thickness, after which stiffness increases rapidly with increasing deflection. In general, therefore, the loading of the felt should not exceed that giving a maximum of 25% deflection. Typical vibration absorption characteristics are shown in Fig 6.

The natural frequency is determined primarily by the thickness of pad rather than its area and static loading, with a minimum value of around 20 Hz for a 25 mm (1 inch) thick pad. The relationship to thickness is not in direct proportion, however. With denser felts the natural frequency is increased and also becomes more dependent on static load. In general, 30 Hz is representative of a practical minimum natural *frequency* for felt pads, thus restricting their application to the isolation of frequencies above $\sqrt{2 \times 30}$, or say, above 40 Hz. Damping, however, is quite high, which is a useful feature for reducing amplitude of vibration at resonance.

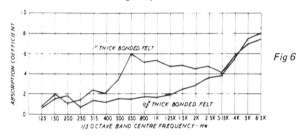

Fig 6

Foamed Materials

Foamed materials are also used as mounting pads although here again the primary virtues of such materials are as acoustic absorbents rather than vibration isolators. Their main limitation as mounts is their load-carrying capacity. Within such limitations they can be useful for low frequency treatment because of their high compressibility, although lack of stiffness can present stability problems.

A particular application of foamed material pads is as intermediate spacers used with rubber mounting pads, as previously mentioned.

Whilst steel mesh can be used in the form of pads, load carrying capabilities are strictly limited in this form. It is more generally employed in the form of inserts (with a coil spring to promote return action) for heavy duty applications where the primary requirement is shock absorption rather than continuous vibration isolation.

Basically steel mesh cushions manufactured by crimping and rolling into a cylindrical cushion and then compressed to beyond its elastic limit act as a variable rate, variable damped steel spring.

Crimped stainless steel mesh incorporated in vibration isolators. (CMT Wells Kelo Ltd).

When such cushions are incorporated in a vibration isolation mounting, apart from their obvious freedom from response to environmental conditions, they have a minimum deflection under increasing load due to the variable rate characteristic and when suitably selected also maintain substantially constant natural frequency over a wide range of loads. The non-linear damping characteristic which (under vibrating conditions) varies with displacement has the advantage that at high frequencies and low amplitudes damping is at a desirably low level, while under conditions of low frequency excitation, where greater amplitudes are experienced, damping increases, resulting in minimum transmissibility.

Both spring rate and damping rate characteristics can be varied within a substantial range by the adoption of different mesh structures and by varying other parameters covering the fabrication of the cushions.

Miscellaneous Materials

Various other materials with vibration-isolation properties include earth, sawdust, balsa wood and other soft timbers, and rigid plastic foams.

Balsa wood is a lightweight material which, in end grain configuration, is capable of sustaining a substantial unit load, at the same time providing good vibration isolation with damping. The load can also be distributed using metal facing plates on a balsa mat.

The load bearing capacity of balsa (*ie* its compression strength) varies both with grain direction and balsa density. The latter can range from 6 lb/ft^3 to 16 lb/ft^3. Strength figures can be estimated for the following formulas giving the limit of proportionality:

Long grain compression (lb/in^2) lies between 9–13 x density (lb/ft^2)

Long grain compression (kg/m^2) lies between 0.057–0.04 x density (kg/m^3)

End grain compression (lb/in^2) = 183 x density (lb/ft^3) − 650

End grain compression (kg/m^2) = 0.8 x density (kg/m^3) − 45

Expanded polystyrene is also a low density material, but with substantially lower mechanical strength. As a vibration-isolation material, therefore, the main application is in floating floor construction rather than anti-vibration mounts.

Expanded polystyrene is made in a number of grades each capable of withstanding a range of 4–150 kN/m^2 (6–22 lb/in^2). The basic grades available are shown in Table V.

See also chapters on *Vibration Isolation, Damping* and *Resilient Mounting of Structures*.

TABLE V – ISO GRADE POLYSTYRENE BOARD

		Expanded Polystyrene – Physical Properties			
	Colour Code	Compressive Stress for 1%* Deformation		Compressive Stress for 10%† Deformation	
		kN/m^2	lb/in^2	kN/m^2	lb/in^2
Impact sound duty	Yellow/violet	14.4	2.08	53	7.68
Standard duty	Yellow	26	3.77	86	12.47
High duty	Black	51	7.39	128	18.56
Extra high duty	Green	78	11.31	178	25.82

* Corresponds approximately to a resonant frequency of 32 Hz when used with 25 mm (1 in) thick polystyrene.

† Corresponds approximately to a resonant frequency of 10 Hz when used with 25 mm (1 in) thick polystyrene.

Damping

DAMPING IN the general sense is accelerated decay of the amplitude of vibrations or oscillations. More specifically it is a means of controlling resonant response to vibration, the greater the damping present the less the vibrational movement at resonance. It is thus a basic method of controlling vibration, shock or noise. In the case of mechanical systems the three basic methods available are dry friction damping, viscous damping and hysteresis damping. Lesser use is made of magnetic damping and air damping — see Table I.

TABLE I — TYPES OF DAMPING AND THEIR CHARACTERISTICS

Type	Characteristics	Applications	Remarks
Viscous	1. Velocity sensitive 2. Temperature sensitive 3. High cost	In general best suited to low frequency applications.	Has advantage of high returnability. May present sealing problems
Friction	1. Constant force 2. Not temperature dependent 3. Low cost	Applications where maximum amplitude is fixed. Particularly suitable for operating over a wide temperature range.	May present wear problems
Hysteresis	1. Displacement sensitive 2. Temperature sensitive 3. Frequency sensitive 4. Small size and low weight 5. Low to moderate cost	Broad range of applications and particularly for high frequencies	Damping and elasticity is an ambient property of the material used (elastomers)
Magnetic	1. Velocity sensitive	Specialized	Damping is proportional to velocity-generated induced current
Air			Damping regulated by orifice flow from bellows type enclosure

Materials themselves may also possess *inherent* damping as the result of inter-molecular reactions in transferring mechanical energy into heat. Rigid materials (*eg* metals and structural materials) as a general rule have little internal damping whilst others, broadly classified as viscoelastic, have relatively high internal damping. With suitable combinations of metals or structural materials and viscoelastic materials composite structures can be produced with good damping properties in structural designs.

Dry Friction Damping (Constant Damping)

The principle of dry friction damping is shown in Fig 1. Movement of the spring-mass system generates rubbing friction between a stationary element and one attached to the mass. This provides a damping force which is constant and independent of displacement and velocity. However the damper must be designed to accommodate the maximum extent of displacement anticipated. Ideally the input amplitude should not exceed $4F/\pi K$.

Transmissibility at resonance is low but at some higher frequency ratio the transmissibility reaches a constant value rather than continuing to decrease. Also for each friction damper there is a critical amplitude which, if exceeded, will result in 'runaway' resonant conditions.

Transmissibility characteristics can be improved by adding springs in series with the damper, producing an amplitude-dependent system. At higher frequencies all the deflection then occurs in the springs with the damper itself no longer sliding.

Fig 1

Fig 2

Viscous Damping

Viscous damping applied to a spring-mass system is shown in simple diagrammatic form in Fig 2. The damper produces a force (F) which is proportional to the velocity of the system. If the motion is sinusoidal and C is the damping constant:

$$F = C\omega X_o \cos \omega t$$

It follows that the damping force is a function of the exciting frequency and the amplitude of motion across the damper.

Critical damping (C_o) is given by $2\sqrt{Km}$. The amount of damping in any practical system is normally defined as a percentage of the critical damping.

Since damping forces are proportional to velocity, viscous dampers are particularly suitable for systems requiring high retunability.

Hysteresis Damping

Hysteresis damping — shown diagrammatically in Fig 3 — produces a force which is in phase with the velocity and proportional to the displacement, resulting in a linear equation of motion.

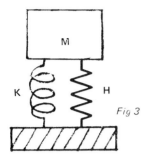

Fig 3

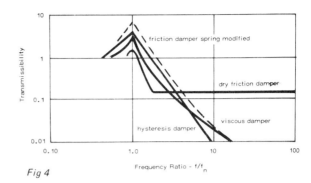

Fig 4

Damping force $(F) = i\,h\,X_O$

where X_O is the displacement

h is the coefficient of hysteresis damping

i is the complex operator indicating a phase angle of $90°$

Transmissibility at resonance is equal to $1/\eta$, where η is the loss factor of the material. Response characteristics are similar to viscous damping except that increasing frequency has no effect on a hysteresis damper. It is thus more attractive than a viscous damper for high frequency applications. Comparative transmissibility curves are shown in Fig 4, but see also Table I.

Layer Damping

The basis of layer damping is a coating or insertion of viscoelastic material in structural members (usually a beam or plate). As the base structure vibrates the layer(s) of viscoelastic material undergoes stretching and compression, dissipating a large proportion of mechanical energy. To be effective in damping, however, the viscoelastic layer must be capable of storing a large amount of energy before dissipating it, which in practice means that an optimum layer stiffness exists for maximum energy dissipation. This in turn is a function of the type of structure, the stiffness of the elastic layer(s) and the boundary conditions.

Layer damping may be *constrained* or *unconstrained*. The following description is extracted from ASME publication 64-WA/RP-8.

A constrained layer of damping material consists of a composite laminate made up of a damping core material sandwiched between skins of structural material in a three- or multi-ply layer. A typical three-ply construction is shown in Fig 5. The damping action is a direct result of shearing of the viscoelastic core material as the laminate is flexed. The amount of damping obtained is a function of the material and properties of both the skin and damping media, the geometric properties such as size and shape, and other factors such as excitation, frequency and temperature.

By proper design techniques and choice of damping materials, it is possible to design constrained-layer damping treatments which preserve the structural integrity of the composite yet have predictable response characteristics over a broad frequency range. Practically speaking, damping materials are available which will limit structural resonant amplitudes to less than 10 to 1 over a frequency range in excess of 1 decade. By proper choice of the damping material, it is possible to maintain and in some instances even increase the effective dynamic stiffness of a laminate. Design techniques are available which will allow the accurate, reliable prediction of such composite-laminate properties.

INTEGRAL DAMPING:
applying damping material between the skins of plates or beams.
(Referred to as constrained layer damping material.)

Fig 5

ADDITIVE DAMPING:
applying damping material to the surface of a beam or plate.
(Referred to as unconstrained layer damping material.)

Fig 6

An unconstrained or free-layer damping treatment consists of a simple two-ply construction, Fig 6. In this form of treatment, a highly damped material is applied directly to the structure to be controlled. Composite damping is obtained by alternate extension and compression of the damping layer as the structure is flexed. As with the constrained-layer treatment, the effectiveness of the damping treatment depends upon forcing the damping material to move and dissipate energy.

Damping Materials

The choice of damping materials is controlled by the type of construction and how efficiently the material will be used. For constrained-layer construction, many properties must be considered. The dynamic sheer elastic (G') and damping (G'') moduli are of prime importance because the damping action is basically shear. A typical effective polymeric material compound has an elastic shear modulus ranging from 100 lb/in^2 to 10 000 lb/in^2, or even higher for some specialized applications, and a shear damping modulus of the same magnitude. The properties of such materials are relatively stable with respect to frequency but are usually extremely sensitive to temperature. It is most desirable to have as broad a temperature range as possible, particularly for military applications.

The final choice of a constrained-layer damping material will depend upon the other environmental requirements, such as humidity, salt spray, and solvent resistance. Some design flexibility is possible by varying core thickness and polymer stiffness to obtain the most effective response characteristics. A typical loss-factor versus exciting-frequency curve for a simple three-ply laminate is shown in Fig 7.

The material choice for unconstrained-layer damping treatments is somewhat more limited than for constrained-layer treatments but follows similar basic rules. Since the damping mechanism is extension and compression of the damping layer, the dynamic-elastic and dynamic-damping components of Young's modulus become important. Since any unconstrained-layer damping treatment should control the composite structural response effectively, it is reasonable that this

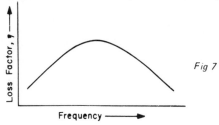

Fig 7

material should be very stiff and exceedingly well damped. A soft material, even though exceedingly well damped, would not act as an effective unconstrained treatment. The explicit expression for the damping of a composite structure is directly proportional to the loss factor of the damping material (η_2), the ratio of the Young's modulus of the viscoelastic damping material to the Young's modulus of the structural material (e_2), and the relative thickness, the ratio of the thickness of the damping layer to the thickness of the structure (h_2). The composite loss factor (η) then is equal to the quantity:

$$\eta = \eta_2\, e_2\, h_2 \left[\frac{3 + 6h_2 + 4h_2^2}{1 + e_2\, h_2\, (3 + 6h_2 + 4h_2^2)} \right]$$

This expression forcefully defines the requirements for the unconstrained-layer material. It is most apparent that a stiff, highly damped material is essential for effective damping treatment. A typical effective material would have an E' of 10^6 and a loss factor in excess of 1. The effect of material stiffness is clearly shown by comparing the composite loss factor obtained from a material with an E' of 10 000 and a loss factor of 1 to a material with an E' of 1 000 000 and a loss factor of 1. The composite loss factors would be in the order of 0.2 for the E' of 10^6 and 0.02 for the material with E' of 10^4.

Spaced Damping

The stiffer the structure the more difficult it becomes to dampen. A significant parameter in this respect is the distance between the neutral axis of the damping layer and the neutral axis of the structure being damped. This is known as the *spaced damping distance*. The greater this distance the greater the damping effect.

An example of how this can be accomplished by modification of the design is shown in Fig 8 which considers the basic case of a damping layer applied to a hat-shaped structural section (a). A double-thickness, half-width layer affords some immediate improvement by increasing the spaced damping distance (b). Solution (c), which involves a change in the choice of hat section, is even better.

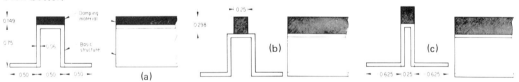

Fig 8 Three ways to apply damping to a hat-shaped structural section. It can be simple layer on, (a) but a double-thickness, half-width layer, (b) improves effectiveness. A slight structural change, (c) helps even more. (Dimensions are in inches).

An alternative method of measuring the spaced damping distance is through the use of a spacing layer — Fig 9. The spacer material should be lightweight, but have very high shear stiffness to maintain the coupling between damping layer and base material. At the same time it must have low extension stiffness, although this can be provided by gaps in the spacer material if necessary. In this case the length of each spacer should be less than the wavelength of the bending waves present in the structure — see also Fig 10.

A comparison between the performance of maximized space damping and conventional layer damping based on the form of structure shown in Fig 9 is shown in Fig 11. In general all space

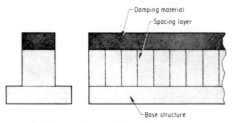

Fig 9 General spaced damping treatment, using segmented spacer layer to increase distance H_{31}.

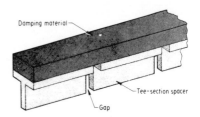

Fig 10 Tee-section spaced damping treatment. In this example, 1 inch tee sections, 2 inches long, are separated by a gap of ¼ inch.

damping treatments have the same type of composite loss factor. At lower frequencies the composite loss factor can be calculated from the original equation given. At higher frequencies the damping layer becomes increasingly uncoupled from the base and the loss factor starts to fail.

Assuming that the structure involved is a beam (or beam-like) the original equation can be re-written:

$$\eta_C = \frac{\eta_3 \left[\dfrac{W_3 H_3^3}{12} + W_3 H_3 H_{31}^2 \right]}{\left[\dfrac{E_1}{E_3} I_1 \dfrac{W_3 H_3^2}{12} \right] + W_3 H_3 H_{31}^2}$$

where η_3 = loss factor of damping material
 W_3 = width of damping layer
 E_1 and E_3 = the respective elastic moduli of the materials
 H_3 = thickness of damping layer
 H_{31} = distance between neutral axes of layers

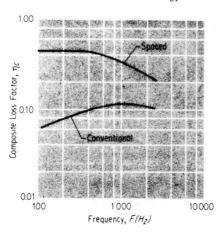

Fig 11 Comparison of spaced and conventional damping.

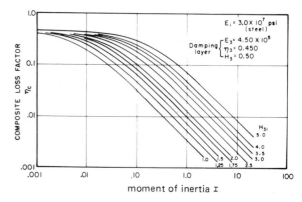

Fig 12 Composite loss factor v.s moment of inertia of structure being damped for various values of H_{31} (distance between neutral axis of layers).

Figure 12 shows composite loss factor (η_c) vs moment of inertia of the base structure (I_1) for various values of H_{31} constructed from the aforementioned equation. For the construction of this curve, the damping layer width was taken to be the same as the base and 0.50 in thick. Damping material properties of $E_3 = 4.50 \times 10^5$ and $\eta_3 = 0.45$ were assumed with a steel-based structure.

Figure 12 may be interpreted as follows:

Taking a 4 inch square tube as an example, the moment of inertia of this structure is 1.17 in^3. (This is a per unit width value). If we look at the figure, we can find the composite loss factor associated with the values of H_{31}. It is known that the neutral axis of bending of a square tube is at its centre. This says that for a spaced damping treatment of the beam, H_{31} is equal to one-half the thickness of the beam plus one-half the thickness of the damping layer, plus the total thickness of the spacer, or in this case:

$$H_{31} = 2.5 \text{ in} + H_s$$
$$H_s = \text{thickness of spacer}$$

Thus, if a spacer 1 inch thick is used, H_{31} is 3.5 in and the composite loss factor is 0.035. The type of plot shown may be developed for any damping material and any configuration of the damping layer. The plot will give a designer an idea of what magnitude of loss factor he can hope to achieve.

Panel Damping

When a solid panel of structure is excited into vibration, a proportion of the mechanical energy generated by the vibration is radiated in the form of airborne noise. In the case of metals and similar solids the proportion of radiated energy is high, since such materials possess little internal damping. Thus the only practical way of reducing vibration-excited radiated sound is to dissipate a high proportion of the vibrational energy in the form of heat in a damping layer (or damping device) attached to the solid. The performance of such a damper can be computed on the basis of the rate of decrease of squared amplitude of free bending vibrations on an equivalent plate (D_t), when:

$$D_t = 27.3 \, \eta f_r \text{ dB per second}$$

where η = damping factor
f_r = resonant frequency of the plate, Hz

In practice, the effectiveness of any damping material is normally expressed directly in terms of D_t, ie in dB per second.

Damping may also be important in attenuating the free bending wave in the panel, when a comparable parameter is:

$$D_1 = 13.6\eta/\lambda \text{ dB per unit length}$$

where $D\lambda$ = one wavelength

or

$$D\lambda = 13.6\eta \text{ dB per wavelength}$$

where λ = wavelength (in compatible units)

Attenuation of the free bending wave can be of paramount importance in critical structures or those carrying rigidly mounted sensitive devices which might be harmed by excessive vibration.

The majority of applications of sound deadening materials, however, is concerned with reduction in airborne noise resulting from resonant vibration of panels.

Some typical treatments for vibration damping are summarized in Table II. Basically, the simplest form of treatment involves the addition of a high-hysteresis material as a surface layer. If this layer is continuously bonded to the surface, energy is dissipated mainly by the shear motion of the damping layer and resulting internal friction, given by the shear modulus of the material. This accounts for the temperature dependence of many surface layer materials where the shear modulus varies appreciably with temperature.

Performance also differs when the surface layer is unconstrained (as in the case of a simply applied layer), or constrained (as in the case of a surface layer covered with another material, such as metal foil or thin rigid sheet material). In the former case the damping factor is proportional to the square of the weight of the material and, when constrained, more or less linearly proportional to the weight. However, for a given weight of damping material, performance realized is likely to be about the same for weights of between 10% and 20% of the base panel weight: but above 20% the weight of the unconstrained layer becomes increasingly effective.

TABLE II – VIBRATION DAMPING TREATMENTS

Treatment	Characteristics	Remarks
Solid structures	Inherent property of materials and jointing.	Self-damping generally low. Materials with high internal damping may be used for special purposes.
Surface layers: (i) Mastics	Widely used. Damping layer usually applied to one side only	Performance varies with temperature and also with composition.
(ii) Other materials		See Table on Vibration Damping Materials
Sandwich construction	Damping material incorporated as a core	Damping of up to 80% of that of the damping layer itself can be achieved
Blankets: (i) Papers	Pleated or corrugated forms.	Lightweight and inexpensive.
(ii) Felts	Fabricated forms.	Performance depends on construction and weight.
(iii) Fibrous	Fabricated forms.	Lightweight material.
(iv) Multi-ply	Fabricated forms.	Usually alternating waved and plain layers.
(v) Septum loaded	Very high performance possible.	Usually mounted on rubber plugs.
Diaphragm deadeners	Tuned for maximum performance at specific frequencies.	Lightweight treatment for low frequency damping.
Dynamic adampers (Absorbers)	Spring-mounted mass systems tuned to provide damping by out-of-phase response.	Mainly applicable to rotating machines.
Fluid dampers	Provide energy dissipation by 'dashpot' action.	Mainly applied to sliding movement with vibration damping as a secondary characteristic.

The choice of material and treatment must also take into account the resonant frequency and temperature (the significance of the latter already having been mentioned). The effectiveness of the treatment may depend very largely on the response at resonant frequency, as the degree of damping provided at other frequencies may be negligible. Thus Fig 13 shows typical results on a bare and treated panel excited into steady state vibration, the treatment in this case giving a reduction of 20 dB in resonant sound. The downward shift of the peak frequency is due to the increase in composite mass with the stiffness remaining substantially unaltered. The use of stiffeners alone is not a method of providing damping, but only shifts the resonant frequency. In practical panels a considerable number of resonant frequencies are likely to be involved, but each will normally have a different relative loudness as far as radiated sound is concerned. Higher frequencies can readily be damped by increasing the mass of the panel or structure, but the proper use of sound deadening treatment not only allows for possible constructional cost saving (and, in specific applications, even more significant weight reduction), but also a direct approach to attenuation of lower frequencies.

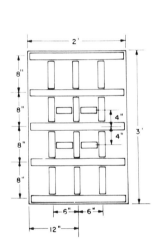

Fig 13 Two-dimensional spaced damping treatment.

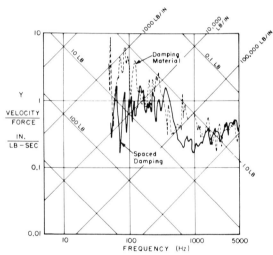

Fig 14 Driving point admittance of spaced damping treatment as compared to complete damping material coverage.

Damping Materials

Typical vibration damping materials are summarized in Table III. Of these, mastics are probably the most widely known as they are extensively used in the automobile industry, largely because of their low cost, suitability for application by spray, good adhesion and the fact that they can withstand baking temperatures. The latter is important since the efficiency of asphalt-base mastic deadeners is directly related to their 'drying' time which can be accelerated by baking. It should be noted, however, that the majority of these products are very specially formulated for the automobile industry and their sound deadening properties may vary appreciably with different formulations. In many cases some performance as a sound deadening material is sacrificed in order to enhance other properties required. Some mastics, in fact, may have relatively poor sound deadening properties, *eg* those formulated primarily as abrasion and corrosion resistant coatings for undersealing.

TABLE III — VIBRATION DAMPING MATERIALS

Material	Typical Decay Rates Achieved	Temperature Dependent	Proportionality to thickness(t) or weight(w)	Advantages	Limitations
Papers:					
(i) Pleated	Depends on form			Simple and lightweight	Limited performance
(ii) Crepe Blanket	5 dB to 60 dB per sec		$\propto t$ and $\propto w^2$	Low cost, light weight. Also sound-absorbing medium	
(iii) Laminated Asbestos	1 dB to 6 dB per sec		$\propto t$	Fire resistant	Fairly low performance
Fibre Board			$\propto t$	Standard building material	Very low performance
Sheet Asphalt	Similar to mastics			Ease of fixing	Limited performance
Laminated Mica	Up to 40 dB per sec		$\propto t$	Fireproof	Performance deteriorates with age
Sandwich Plates	Depends on thickness and properties of damping layer		$\propto t$	Rigid construction	
Sponge Rubber	Dependent on rubber composition	Yes		Simple material, also good vibration isolator	Needs special high hysteresis rubbers for good performance
Mastics	2 dB–32 dB per sec but varies widely with formulation of mastic	Yes	$\propto w^2$	Ease of application. Low cost. Good adhesion	Limited performance unless specially formulated. Limited performance at higher temperatures
Asphalted Felts:					
(i) 0.2–0.7 lb/ft^2	1 dB–25 dB per sec	Much less than mastics	$\propto w$ (and form)	Simple fixing	Performance specific to construction and form
(ii) Multi-ply (alternate indented and plain)	6 dB–40 dB per sec	Much less than mastics	$\propto w$ (and form)	Simple fixing	Performance specific to construction form
(iii) Septum loaded	Up to 400 dB per sec	No	Not significant	Very high damping possible	Isolation of septum important calling for special fastenings
Blankets:					
(i) Plain	Varies with material, density, fibre, form, etc	No	$\propto w$	Also good thermal insulation and sound-absorptive material	Not very effective dampers

cont...

TABLE III – VIBRATION DAMPING MATERIALS (contd.)

Material	Typical Decay Rates Achieved	Temperature Dependent	Proportionality to thickness (t) or weight (w)	Advantages	Limitations
(ii) Septum loaded	Up to 300 dB per sec	No	∝ t	Also good thermal insulation and sound-absorptive material	
Tuned Diaphragm Deadeners	Very high at selected frequency	No	∝ w² (attached mass)	Ease of fitting and removal. Particularly suitable for low frequency damping and lightweight treatment	May need to be combined with other treatment

Another characteristic of mastic materials is their considerable variation in sound deadening properties with temperature. They can normally be formulated to give optimum properties at specific temperatures between about −17.8 and 32°C (0–90°F), but can seldom achieve decay rates above 10 dB per second at higher temperatures. A change in temperature of only 20 degrees can also reduce the decay rate, compared with that for the optimum temperature, by as much as 25%.

Their chief advantages, however, are the readiness with which they will adhere to metals, simplicity of application and relatively low cost. They are difficult to remove, however, except for water-soluble types which generally have a somewhat lower performance.

Blankets are an attractive alternative to surface layers since they can be fabricated in a wide variety of forms, thicknesses and weights and secured by simple fastenings. It is generally an advantage, in fact, to employ 'spot' fastening so that movement and friction can develop between the surface and the blanket, increasing the rate of energy absorption of the combination. The majority of blankets, too, possess sound-absorption properties and can provide thermal insulation.

Composite or multi-ply constructions, applied particularly to asphalted felts, provide a high degree of internal friction and may be finished with corrugated or indented surfaces. These can take advantage of surface loading to provide increased flexure and damping, or, if cemented in place, should be only spot-cemented with a minimum of adhesive in order to preserve the voids which encourage flexure of the material. Similar treatment applies in the case of soft fibrous blankets, although these in themselves are not particularly effective as sound deadening materials since the bulk of the material is difficult to excite into relative motion.

Quite dramatic improvements can be realized, particularly with lightweight blankets, if the material is surface-loaded with a rigid sheet of relatively high mass (septum loaded) provided the septum itself is not rigidly connected to the base panel. They can be fastened with rubber 'rivets' or dashpads. The septum then acts as a pure inertial (surface) load, promoting high relative movement within the blanket or composite material and thus high frictional losses. Decay rates of up to 400 dB per second can be achieved with septum loaded blankets and composites, depending on the choice of blanket material and the inertia of the septum. Sheet metal is the most common septum material, but other materials (such as hard rubber) may be used, where it is required that the blanket should be mouldable to conform to curved surfaces, *etc.* Provided the septum is relatively massive (in a comparative sense), the effectiveness of a septum-loaded blanket in any specific material is more or less directly proportional to its thickness.

Damping in Structures

In many applications it is necessary to provide damping in structures — *eg* plates — that have bending waves travelling in two directions. Similar principles apply, Fig 14 showing typical treatment applied to a steel plate. The strips of spaced damping are applied in a checkerboard pattern on approximately 8 in centres. The damping treatment used is the same as that normally applied to a T-section. Figs 15 and 16 show the driving point admittance and corner-to-corner transfer admittance for the configuration shown in Fig 14 as compared with data from the same plate covered with 4.5 lb/ft^2 of Class II Navy damping tiles. As can be seen, the checkerboard pattern is at least equivalent to the complete coverage treatment and in many areas is better. The spaced damping treatment of this plate actually weighs less than the all damping material treatment (about 4.0 lb/ft^2). Spacing layers that are optimized to work in two directions may also be built.

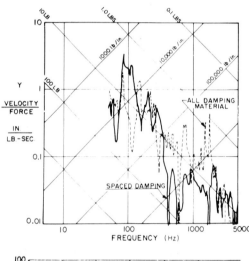

Fig 15 Transfer admittance of spaced damping treatment compared to complete damping material coverage.

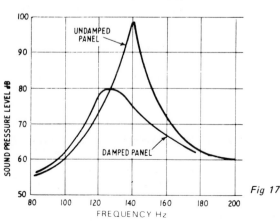

Fig 17

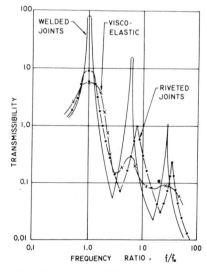

Fig 16 Transmissibilities for various methods of joint fabrication.

Damping at Joints

Damping in a built-up structure is largely dependent on the energy that can be dissipated by slip at the joint. To be effective in practical application, however, any joint slip in a basic mechanical assembly (*eg* in a bolted joint) can both degrade the stiffness of the structure and also produce wear and/or fretting corrosion. Nevertheless this does not necessarily rule out bolted or riveted joints as a source of high damping, compared with welded joints — see Fig 17.

In a joint capable of slip the magnitude of the friction force is given by:

$$F = \eta N$$

where

η is the coefficient of sliding friction for the two surface materials

N is the normal force across the interface

The damping force is a non-linear function of relative velocity, (see friction damping). In practice the response to sinusoidal excitation can be expected to be sinusoidal when the equivalent viscous damping coefficient (C_{eq}) is given by:

$$C_{eq} = \frac{4\eta N}{\pi \Omega X}$$

where

Ω is the forcing circular frequency and,

X is the slip amplitude

The energy dissipated per cycle is $E = 4\,NX$

N and X can be expected to show a substantially linear relationship, *viz*

$$X = -(X_F/N_L)\,N + X_F \text{ (approx)}$$

where

X_F = slip amplitude of free joint

N_L = normal force required to lock the joint

Maximum energy dissipation occurs when $N = N_L/2$, thus:

$$E_{max} = N_L\,X_F$$

$$= F_L\,X_F$$

where

F_L is the friction force required to lock the joint against slip

In practice this means optimum damping from a bolted joint is normally realized when the preload applied to the bolt(s) is one half the value required to lock the joint. Damping can be increased by interface treatment (*eg* insertion of polymer films or metal foil). Introduction of an interface lubricant may have a similar effect, but not invariably so.

Impact Damping

Transient vibrations excited by impacts tend to generate resonant vibration in a panel, or a characteristic ringing tone or impulse sound, often generally described as 'tinniness' because of the predominance of high frequency components. The latter is (mistakenly) often taken as a measure of the quality of the construction. For this reason damping treatment applied to the surface may be purely for subjective reasons — *ie* to provide a high decay rate for ringing tones and some (usually small) reduction in the loudness of the initial sound. Thus damping treatment applied to a relatively thin metal structure, such as a metal cabinet door, can give the impression of a heavy, 'quality' construction when the door is slammed or rapped with the knuckles.

Impact damping is, however, important in a design where shock excitation is repetitive. This applies particularly in automobile design to minimize the noise of rain on the body roof, shock noise due to door slamming (as well as adding to 'quality' sound), deadening of road rumble, *etc*: it also applies to office machine covers on typewriters, calculating machines, *etc*. Treatment involves similar considerations as for the damping of steady or continuous vibrations.

See also chapter on *Acoustic Materials*.

Resilient Mounting of Structures

SINCE THE early 1960s there has been increasing awareness of the need for, and advantages of, supporting building structures on some form of anti-vibration mount, isolating the building from ground vibrations such as might be generated by traffic or railways alongside (or underground), heavy machinery alongside, *etc.* Where the vibrations exist in the ground and cannot readily be reduced, the principle is to construct the building on resilient bearings in order to effectively reduce a range of noise and vibration frequencies transmitted to the building, by isolating it from its foundations; and by ensuring that the vibrations are not 'short circuited' into the building in any other way.

The method started in the US in 1915 and later in Canada using alternate layers of lead and asbestos fibre, but early results were disappointing due to stiffening of lead under constant static load. Also bearings cannot isolate frequencies below 50 Hz, though train vibrations in the ground are usually of the order of from 15–30 Hz, and on soft clay can be as low as 5½ Hz. The 'train rumble' noise given off by the 'drumming' of building walls and floors is much higher, sometimes extending to 600 Hz.

The first really successful treatment of buildings, in fact, was the employment of bridge-type bearing mounts within the building structure, the first example in the UK being the Albany Court building, a block of flats erected directly over the underground railway at St.James's Park Station. Whilst this was highly successful, bridge-type mounts were originally designed to accommodate shear movement which is not necessary in a building mount. This led to the development of a new design of building bearing mount in the form of a synthetic rubber/fabric reinforced resilient load bearing material. In this form the rubber provides the 'spring' for vibration isolation and the reinforcement restricts lateral flow and creep under compression. Compounding the rubber with a cellular material provides both damping and internal voids into which the rubber may flow when compressed.

Types of Bearings

Resilient bearings fall into five main types, as designated in BS6177:1982 — see Table I. All types are usually rectangular in shape, but can have detail variations (*eg* some may incorporate corrugations or other forms of voids).

For most applications it is sufficient for the mechanical properties of the material to be described by:

(i) Static load-deflection curves under normal loading (generally compression).

(ii) Long term creep behaviour — *eg* pad deflection against time under constant static (and if necessary dynamic) loading.

TABLE I — TYPES OF RESILIENT BEARINGS (BS6177 : 1982)
(See also Fig 1)

Type	Construction	Remarks
Plain elastomer bearing	Homogeneous vulcanized natural or synthetic rubber.	Very limited application. Needs location to prevent lateral spread.
Plain composite bearing	Rubber with homogeneously dispersed cellular particles.	Many possible combinations and physical properties.
Elastomeric sandwich bearings	Plain elastomeric material bonded between two parallel steel plates.	Many possible combinations and physical properties.
Multiple elastomeric sandwich bearing	Multiple layers of elastomeric material alternating with metallic or non-metallic reinforcement layers.	Many possible combinations and physical properties.
Multiple composite sandwich bearing	Multiple layers of composite elastomeric material with metallic or non-metallic reinforcement layers.	Many possible combinations and physical properties.

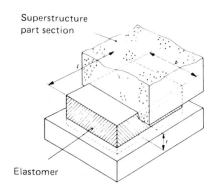

(a) Plain elastomer pad

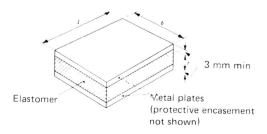

(b) Sandwich mounting

Note:
l = length of elastomer layer
b = breadth of elastomer layer
t = thickness of elastomer layer

$$\text{shape factor} = \frac{\text{load area}}{\text{bulge area}} = \frac{lb}{2(l+b)t}$$

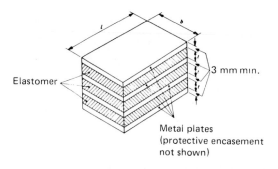

(c) Multiple sandwich mounting

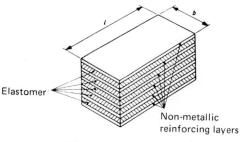

(d) Composite mounting

Fig 1

RESILIENT MOUNTING OF STRUCTURES

(iii) Dynamic modulus and loss factor at different frequencies.

Here it is important to appreciate that the dynamic properties of resilient bearings can differ under varying operating conditions and are affected by:

(a) static pressure;
(b) shape factor;
(c) operating frequency;
(d) strain amplitude;
(e) operating temperature;
(f) flexibility of the bearing supports and back-up structure;
(g) age and previous history of the bearing.

It is therefore essential that the installation of a resilient bearing has to be conducted at an early design stage to enable an economic, efficient and safe system to be engineered.

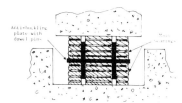

Double unit rubber/steel sandwich bearing with dowel anti-buckling device.

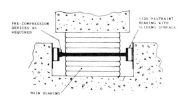

Side restraint anti-buckling system for use with multiple composite sandwich bearings.

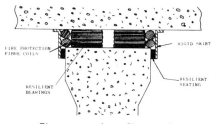

Fire protection fibre coils surrounding bearings, with isolated cover plate.

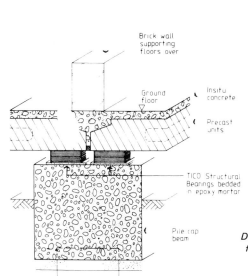

Diagrammatic view showing the ground floor construction at West End Sidings.

454   RESILIENT MOUNTING OF STRUCTURES

Irish Centre Community Hall is situated directly over the main railway line from St. Pancras Station — a classic example of when resilient bearings can provide noise and vibration isolation under extreme site conditions.

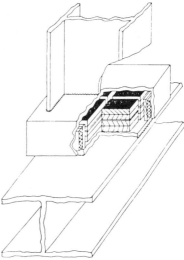

Typical example of module bearing housing.

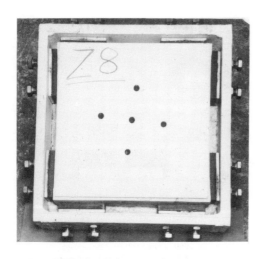

Example of a small accommodation module mounting with wind/shear restraint, (viewed from underside).

Resilient bearings at base of column with steel location dowels and capped resilient isolating collars.

RESILIENT MOUNTING OF STRUCTURES

Design Procedure.

The following stages give a good idea of the design and supervision procedure:

1. Site vibration survey including one for all types of trains — high speed passenger, local, goods, heavy freight, *etc,* recording several examples in each case. The recordings must be vibrograms of velocity or displacement and the natural frequency of the pick-up must be lower than the lowest frequency of ground wave, or be d.c. Meter readings from 'high frequency' accelerometers do not provide information about the low frequency waves, nor how many cycles they contain; neither do they provide adequate information about peak transients, all of which are required for design of insulation.

2. The quantity of vibration reduction needed (for living units it would be to about the 'boundary of perception') is next calculated, and the natural frequency of the sprung building is determined, based on the dynamic, not the static, elasticity of the bearings. The wind sway calculations of the building are based on the static stiffness.

3. The accurate dead weight of the building is found, and its mass centre. If the latter is eccentric for a rigid building the number of bearings must be adjusted to keep the structure upright and to prevent cracking, *etc.*

4. As the springs are quite resilient compared with the much more rigid ordinary foundations, the structure must often be made stiffer to accept wind loads without undue distortion, which might otherwise seriously damage lightweight partitions, external cladding or windows.

5. All rubber bearings must be protected against fire in vulnerable places.

6. A structure that is carried on resilient bearings should be designed on fail-safe principles so that in the event of a failure, partial failure or damage to the bearings, it will remain adequately supported and retain its safety and general serviceability under the design load.

7. Long term side stability of the sprung building is maintained by side acting resilient members which also take over the function of wind braces (Fig 2).

8. Finally, all variations of stiffness in the springs shown by these tests are compared with the designed stiffness of the building members, so that excess variations can be adjusted, *eg* see Figs 3 and 4.

   Specific design requirements are detailed in BS6177:1982.

The bearing characteristics can be 'tailored' over a very wide range. Compounding the rubber mix, fabrication, moulding pressure and vulcanization, plus bearing sizes, can be adjusted to meet particular requirements of static deflection, shear deflection and stress. Where large column loads/ areas of bearings are concerned it is advisable to employ a number of smaller bearing units in modular form to ensure equal dynamic and static properties throughout the structure.

The significant parameters of dynamic behaviour can be stated as:

(i) Dynamic stiffness — a function of the dynamic modulus of elasticity, pad area and thickness. This can be very different from the static stiffness (Fig 5).

(ii) Shape factor — laminated construction makes the bearing far less susceptible to excess internal stresses, long term fatigue, and lateral spread compared with homogeneous rubber bearings of the same overall dimensions.

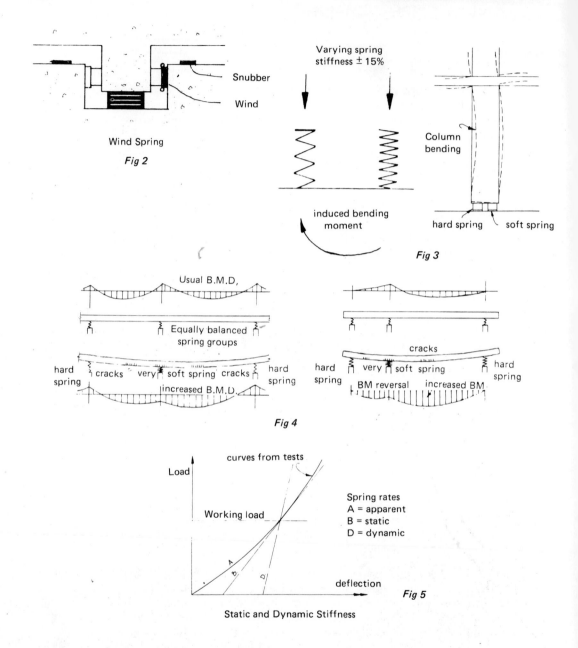

Fig 2 — Wind Spring

Fig 3 — Column bending

Fig 4

Fig 5 — Static and Dynamic Stiffness

(iii) Loss factor — a measure of the damping ratio of the pad, defined as the tangent of the phase angle by which the damping force produced by the pad lags behind the stiffness or spring force.

(iv) Model factor — test (usually laboratory) model springs are substantially less stiff dynamically than the dimensionally larger, though similarly proportioned, civil engineering resilient bearings.

Leak Paths

It is essential to appreciate that no form of isolation mounting remains effective if it is 'short-circuited' by a rigid connnection between the resiliently mounted structure and its surrounds, thus allowing vibration to 'leak' across this path. Also, clearances around a bearing should be such as to permit movement to its loaded profile without restriction. Care must be taken to ensure that restraint to a bearing is not caused through interference either by structural parts or by accumulation of surplus grout or other debris in its vicinity.

Service connections to a mounted structure should either be sufficiently flexible to act as isolators themselves, or incorporate flexible couplings to give the safe effect (*eg* elastomeric bellows fitted to drainpipes).

Where it is necessary to seal any small gaps between the mounted structure and its surrounds (*ie* its foundation or an adjacent unmounted structure) the sealing must be done with a material of adequate flexibility to accept any probable movement. Due allowance must also be made in the design for any consequent effects on the efficiency of the mounted structure.

To ensure that a mounted structure remains completely isolated from any adjoining structure, sufficient space should be left between them to allow for lateral movement. Such movements will include those from temperature, ordinary deflection and sway under load, foundation settlement, long term creep and deflections of mountings, including possible damage caused by fire or structural overload. Normally building tolerances and clearances should also be taken into account.

Examples of Resilient Bearings

TICO CV/CA is a high stress resilient bearing material of vulcanized laminate construction comprising a reinforcement of alternating layers of high tensile strength fabric bonded to plies of polychloroprene base rubber modified by the inclusion of cellular particles. The face layers are of slightly different formulation to facilitate bonding. It is dimensionally stable under widely varying atmospheric conditions and is rated for a maximum working stress of 7 000 kN/m^2 (1 000 lb/in^2) and can achieve natural frequencies down to 10 Hz without the need for additional horizontal stability.

Principal applications are the isolation from outside sources of vibration of tower office blocks, flats and theatres, and heavy duty oil rig accommodation modules, *etc.*

TICO CV/CA bearings with rigid fail safe support with vertical load capacity equal to the three resilient bearings

Luxury flats in Ebury Street, London. This 24 000 ton structure is isolated at first floor level on 1410 TICO CV/CA bearings 103 m thick, with basement flats constructed on 'box-within-box' principle, separated from the foundation by 351 TICO CV/M bearings. It is Europe's largest resiliently mounted building.

TICO CV/M is a medium load structural bearing material formulated from highest quality neoprene modified by the inclusion of cellular particles. It is capable of being used in a variety of ways from modular form bearings on pile caps to continuous strip footings; and may be employed in any thickness (in multiples of 25 mm) necessary to provide the required natural frequency. Maximum recommended stress is 1 400 kN/m^2 (200 lb/in^2).

Particular applications are for isolating partial or total structures from their foundations on sites subject to medium and low frequency noise and vibration (*eg* near railway lines, isolation of domestic housing, studio control rooms, floating floors, *etc;* and also lightweight oil rig accommodation and laboratory modules).

Sections of the bearings employed during construction of isolated barrier block at West End Sidings in close proximity to main line trains; a total of 1950 TICO CV/M bearings were used.

Fully floating isolated anechoic chamber carried on five layer TICO CV/LF/K bearings providing a natural frequency of approximately 62 Hz at 350 kN/m^2 stress.

Example of TICO CV/LF low frequency bearing using five layers.

TICO CV/LF material incorporates a moulded configurated rubber layer bonded between flat face plies of rubber composite material. It has been developed as a low load, low frequency structural bearing capable of a relatively high spring rate. It is produced in three grades catering for loads up to a maximum of 700 kN/m^2 (100 lb/in^2) with natural frequencies down to 7 Hz (two layers), or lower depending on the number of layers employed.

CV/LF materials are designed for total or partial support of all types of internal structures such as anechoic chambers, recording studios, floating floors, theatre stages, false roofs, *etc.*

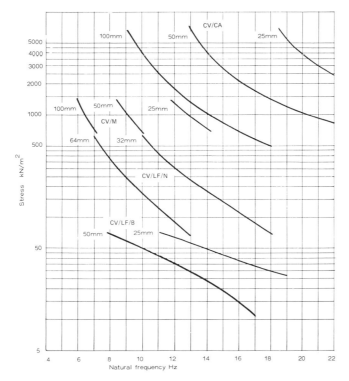

*Fig 6 Properties of structural bearings.
(TICO Ltd).*

Figure 6 shows recommended working stress ranges and natural frequencies of these materials. All can be bonded to concrete, steel or wood, using the appropriate adhesive.

See also section on *Resilient Seatings* in chapter *Acoustic Treatment of Floors and Ceilings.*

SECTION 6

Legislation

IN THE UK environmental noise did not become treatable as a statutory nuisance until the passing of the Noise Abatement Act 1960, although health, safety and welfare at work were contained in numerous Acts (eg the Mines and Quarries Act 1955; the Agriculture (Safety, Health and Welfare Provisions Act 1956); the Factories Act 1961; and the Offices, Shops and Railway Premises Act, 1963). From the appearance of the Noise Abatement Act the UK government has remained firmly committed to reducing noise nuisance, responsible departments being:-

Department of Environment (DoE) — with an overall watching brief, and certain specific responsibilities, although in the case of environmental noise from fixed sources control lies largely in the hands of local authorities.
Department of Employment and *Health and Safety Commission* — responsible for noise at workplaces.
Department of Transport and Trade — traffic noise and noise from civil aircraft.

Other departments involved in other aspects of noise policy include the Departments of Industry, Energy, Health and Social Security; the Ministry of Defence, the Home Office, the Scottish Development Department and the Welsh Office.

Specifically in the UK the Control of Pollution Act 1974 (replacing the original Noise Abatement Act) carries noise nuisance away from individual fixed sources. Surface vehicle noise is covered by The Motor Vehicles (Construction and Use) Regulations of 1968, amended in 1980. Aircraft noise is dealt with under the Air Navigation (Noise Certification) Act 1979 (but excludes helicopters).

The Health and Safety at Work Act 1974 remains the major vehicle covering the ramifications of noise in workplaces and employers' responsibilities, together with various Codes of Practice. On the public side, the Land Compensation Act 1973 confers a right for compensation for noise caused by public works.

The UK, France, the Federal Republic of Germany and the Netherlands are the only Member States of the Community that so far have comprehensive legislation on noise and in all cases legislation is fairly recent. All provide for limits to be set to control noise at source, and have provisions to control behaviour and exposure to noise. All four countries provide some form of noise standards for the construction of dwellings, but only the Netherlands provides for mandatory exterior noise limits at the facades of dwellings. In France and Germany non-mandatory guide values have been set for certain zones (residential, industrial). In the Netherlands mandatory limits apply in certain zones although exceptions can be made. The noise abatement zone concept of measuring existing noise levels and then not allowing deterioration is unique to the United Kingdom.

Under the Treaty of Rome, Articles 84 and 100, the European Commission is also directly concerned with the regulation of environmental noise in the UK and other member states on the basis that member states cannot freely set their own limits.

Directives issued and proposed by the EEC to date cover the following subjects:

Motor Vehicles (adopted) — reduction of noise levels incorporated in Construction and Use Regulations 1980, operating from 1st April 1983.

Motor Cycles (adopted) — noise levels and measurement methods incorporated in Construction and Use Regulation 1980, operating from 1st October 1982.

Mopeds — still under consideration.

Tractors (adopted) — revised noise limits, incorporated in Construction and Use Regulations, operating from 1st October 1982.

Railway Noise — not yet finalized as a Commission proposal.

Noise at the Workplace — preliminary draft 1982.

Subsonic Aircraft (adopted) — setting noise levels for civil aircraft.

Helicopters (proposal) — to establish noise limits for helicopters.

Household Appliances (proposal) — methods of measurement and checking noise emissions from various domestic appliances.

Construction and Plant Equipment (awaiting resolution) — general directives, also covering methods of measuring noise from construction machinery. Various other related directives are in proposal form (1982) covering tower cranes, current generators, air compressors, bulldozers, loaders and excavators, excavation loads and mobile cranes.

Powered Industrial Trucks (proposal) — to approximate the laws of member states.

Lawnmowers (proposal) — to set limits for noise emission from new lawnmowers.

SECTION 7
Buyers' Guide

Sub-sections   (a)   Trade Names Index 467

                    (b)   Classified Index of Manufacturers producing all
types of Noise and Vibration Equipment 470

                    (c)   Alphabetical List of Manufacturers with
addresses, telephone and telex numbers of
Head Office, Works and Branches 483

Sub-section (a)

TRADE NAMES INDEX

ADCO — Industrial intercom — Atkinson Dynamics, Division of Guy F. Atkinson Company.
ACOUSTIBOOTH — Acoustic enclosure — Acousticabs Ltd.
ACOUSTICAB — Acoustic enclosure — Acousticabs Ltd.
ACOUSTICURTAIN — Flexible acoustic barrier — Acousticabs Ltd.
ACOUSTILOUVRE — Flexible acoustic barrier — Acousticabs Ltd.
ACOUSTISCREEN — Sound absorption and insulating slabs — Acousticabs Ltd.
ACOUSTISLAB — Sound absorption and insulating slabs — Acousticabs Ltd.
ACCUMODE® — Signal conditioner — Endevco UK Ltd.
A.M.I. SARL — Microphones — Cirrus Research Ltd.
ANTIPHON — MPM (Metal-Plastic-Metal) constrained layer damping systems, second generation absorbing materials, first generation absorbing foams and barrier materials — Antiphon Ltd.
AVOSHIELD — Acoustic lead — Avomet Ltd

B & K — Instruments and accessories for noise and vibration analysis — Bruel & Kjaer (UK) Ltd.
BARAFOAM — Sound absorbing polyurethane flame-retardant foam — Kay Metzeler Ltd.

CEL INSTRUMENTS — Acoustic and vibration measuring instruments — Computer Engineering Ltd.
CIRRUS RESEARCH LTD — Sound level meters and associates equipments — Cirrus Research Ltd.
COBRA ACOUSTICS LTD — Acoustic materials and panels, compressors, consultants and noise control installations — Cobra Acoustics Ltd.
CONIC FLOW — Duct and fan silencer — Industrial Acoustics Company Ltd.
COUSTIBELL — Acoustic barrier curtain — Bestobell Protection.
COUSTILAM — Barrier and absorption composite — Bestobell Protection.
COUSTIPLATE — Laminated steel and aluminium sheet — Bestobell Protection.

DEDNER — Water-based bitumen emulsion — Supra Chemicals & Paints Ltd.
DEDSEAL — Solvent-based corrosion-resistant compound — Supra Chemicals & Paints Ltd.
DEDSHETE — Vibration damping pads — Supra Chemicals & Paints Ltd.
DELTA-SOUND-ABSORBER — Sound absorbers — Ewald Dorken AG.
DEMPISON — Damping materials — Bestobell Protection.
DUMAC — Machine condition monitoring equipment — Scientific Atlanta Ltd.

DUMOR – Acid resistant gloves and gauntlets – Heafield Industries Ltd.
DYNA-MONITOR® – Signal conditioner – Endevco UK Ltd.
DYNA-QUIP – Ball valves, quiet release hose couplings – IMI Norgren Enots Ltd.

ECONOCRUISE – Torsional vibration measurement equipment – Econocruise Ltd.
ENDEVCO® – Endevco UK Ltd.
ENOTS – Pneumatic cylinders and valves, tube fittings, tubing pneumatic accessories – IMI Norgren Enots Ltd.
ENTRAN – Semiconductor pressure transducers, accelerometers and load cells – Entran Inc.
EQUIFLEX – Anti-vibration mount – C.M.T. Wells-Kelo Ltd (Dynamics Division).

FORMAT – Polymeric heavy layer – Supra Chemicals & Paints Ltd.

GENRAD – Noise and vibration instrumentation – Gracey & Associates.
GRACEY (BRE) – Reverberation meter – Gracey & Associates.

illmant – Pipe insulation – Illbruck GmbH Schaumstofftechnik
illmod – Building seal tape – Illbruck GmbH Schaumstofftechnik
illsonic – Contoured acoustic panel – Ilbruck GmbH Schaumstofftechnik
INSTAPLANT – Anti-vibration felt – Bury Cooper Whitehead Ltd.
INTEW – Acoustic intensity system – GenRad European HQ
ISAP – Interactive signal analysis package – GenRad European HQ
ISOBASE® – Accelerometer – Endevco UK Ltd.
ISOBOX™ – Environmental isolation box – Endevco UK Ltd.
ISOSHEAR™ – Accelerometer – Endevco UK Ltd.

KURTOSIS – Bearing damage detection – Condition Monitoring Ltd.

MAFUND® **PAD** – Anti-vibration bearing pad – Eichler KG.
MARKUS – Sound reducing sliding and hinged doors – Envisodoor Markus Ltd.
MASCOLITE – Anti-vibration felt – Bury Cooper Whitehead Ltd.
MICROMODAL – Modal analysis sytem – GenRad European HQ
MICROTRAC™ – Digital tracking filter – Endevco UK Ltd.
MODULINE – Acoustic panel system – Industrial Acoustics Company Ltd.

NAGRA/KUDELSKI – Tape recorders – Hayden Laboratories Ltd.
NICOLET – Real time spectrum analyzers, signal processors, signal averagers, spectrometers, digital oscilloscopes, logic analyzers, digital plotters – Nicolet Instruments Ltd.
NOISELOCK – Acoustic panel – Industrial Acoustics Company Ltd.
NOISHIELD – Acoustic panel – Industrial Acoustics Company Ltd.
NORGREN – Compressed air processing equipment silencers (pneumatic coalescing lubrication systems) – IMI Norgren Enots Ltd.
NORWEGIAN ELECTRONICS (NEAS) – Noise and vibration instrumentaion – Gracey & Associates.
NOVIBRA – Anti-vibration mounts – Trelleborg Ltd.

ONO SOKKI — FFT, single-, dual- and multi-channel analyzers — Computer Engineering Ltd.

PICOMIN® — Accelerometer — Endevco UK Ltd.
PIEZITE® — Piezoelectric material — Endevco UK Ltd.
PIXIE® — Semi-conductor strain gauge — Endevco UK Ltd.
PRESTO — Foams — Ferguson & Timpson Ltd.
PULSAR INSTRUMENTS — Sound level meters — Cirrus Research Ltd.

QUIET DUCT — Duct silencer — Industrial Acoustics Company Ltd.
QUIET FLOW — Fan plenum — Industrial Acoustics Company Ltd.
QUIET VENT — Cross-talk silencer — Industrial Acoustics Company Ltd.

REDFLEX — Rubberized G.P. gloves — Heafield Industries Ltd.
REGO A.V. MOUNTS — Anti-vibration mounts — Bestobell Protection
RION — Sound and vibration measuring equipment — Rion Co Ltd.
RIONET — Hearing aids — Rion Co Ltd.

SENNHEISER — Microphones, headphones test equipment — Hayden Laboratories Ltd.
SENTINEL — Vibration switch — Condition Monitoring Ltd.
SIGNAP — Rotating machining signature analysis package — GenRad European HQ
SOUNDCOAT — Noise control materials — Ferguson & Timpson Ltd.
250 SOUND SHELTER — Audiometric chamber — Industrial Acoustics Company Ltd.
SUPER NOISELOCK — Acoustic panel — Industrial Acoustics Company Ltd.
SUPRA FR FOAM — Polyether flame-retardant acoustic foam — Supra Chemicals & Paints Ltd.
SUPRAMAT — Urethane floor mats and mouldings — Supra Chemicals & Paints Ltd.

THERMASON — Thermal insulator and panel vibration dampener — Supra Chemicals & Paints Ltd.
THERMOS — Thermal and acoustic insulating materials — Hodgson & Hodgson Ltd.
TICO — Structural resilient bearings, resilient seating and sliding bearings for civil construction applications — TICO Manufacturing Ltd
TICO — Anti-vibration materials for machinery and pipe mounting, bellows — James Walker & Company Ltd.
TRACKWALL — Movable acoustic room divider — Industrial Acoustics Company Ltd.
TSL — Time series language — GenRad European HQ

VIBREX — Track and balance systems for helicopter and aircraft — Environmental Equipments Ltd.
VYON — Pneumatic silencers, filter media — Porvair Ltd.

willseal — Building seal tape — Illbruck GmbH Schaumstofftechnik

Sub-section (b)

CLASSIFIED INDEX OF MANUFACTURERS PRODUCING ALL TYPES OF NOISE AND VIBRATION EQUIPMENT

Absorbers
Acoustic Consultants & Engineers
Acousticabs Ltd
Bestobell Protection
Bury Cooper Whitehead Ltd
Ewald Dörken AG
Ferguson & Timpson Ltd
Hodgson & Hodgson Ltd
Illbruck GmbH Schaumstofftechnik
Industrial Acoustics Company Ltd
Stop-Choc Ltd
Tantalic Acoustical Engineering Ltd
Trelleborg Ltd
Trubros Acoustics Ltd
James Walker & Co Ltd

Accelerometers
BBN Instruments Corporation
Bruel & Kjaer (UK) Ltd
Castle-Microair Ltd
Cirrus Research Ltd
Computer Engineering Ltd
Condition Monitoring Ltd
Endevco UK Ltd
Entran Devices Inc
Environmental Equipments Ltd
Gracey & Associates
PCB Piezotronics Inc
Rion & Co Ltd
Scientific Atlanta Ltd
Vibro-Meter AG
Weir Pumps Ltd

Acoustic Emission Equipment
Endevco UK Ltd

Acoustic-Emission Measurement Systems
Bruel & Kjaer (UK) Ltd

Acoustic Emission — Testing Service
Endevco UK Ltd

Acoustic Intensity Equipment
GenRad European HQ

Acoustic Materials — Absorption
Acoustic Consultants & Engineers
Acousticabs Ltd
Antiphon Ltd
Avomet Ltd
Bestobell Protection
Burgess Industrial Silencing Ltd
Castle-Microair Ltd
Cobra Acoustics Ltd
Bury Cooper Whitehead Ltd
Ewald Dörken AG
Ferguson & Timpson Ltd
Harrison & Jones (Acoustic Div) Ltd
Heafield Industries Ltd
Hodgson & Hodgson Ltd
Illbruck GmbH Schaumstofftechnik
Industrial Acoustics Co Ltd
Kay-Metzeler Ltd
Sound Solutions Ltd
Supra Chemicals & Paints Ltd
Trelleborg Ltd
Trubros Acoustics Ltd

Acoustic Materials — Damping
Antiphon Ltd

Acoustic Materials — Isolation
Acoustic Consultants & Engineers
Acousticabs Ltd
Antiphon Ltd

Bestobell Protection
Burgess Industrial Silencing Ltd
Bury Cooper Whitehead Ltd
Castle-Microair Ltd
Harrison & Jones (Acoustic Div) Ltd
Illbruck GmbH Schaumstofftechnik
Industrial Acoustics Company Ltd
Supra Chemicals & Paints Ltd
TICO Manufacturing Ltd
Trelleborg Ltd
Trubros Acoustics Ltd

Acoustic Offices
Industrial Acoustics Company Ltd

Acoustic Panels
Acoustic Consultants & Engineers
Acousticabs Ltd
Bestobell Protection
Burgess Industrial Silencing Ltd
Castle Microair Ltd
W. Christie & Grey Ltd
Cobra Acoustics Ltd
Ewald Dörken AG
Frank Driver
Ferguson & Timpson Ltd
Harrison & Jones (Acoustic Div) Ltd
Heafield Industries Ltd
Hodgson & Hodgson Ltd
Illbruck GmbH Schaumstofftechnik
Industrial Acoustics Company Ltd
Sound Solutions Ltd
Tantalic Acoustical Engineering Ltd
Trubros Acoustics Ltd

Acoustic Rooms
Acoustic Consultants & Engineers
Acousticabs Ltd
Burgess Industrial Silencing Ltd
Castle Microair Ltd
W. Christie & Grey Ltd
Ferguson & Timpson Ltd
Harrison & Jones (Acoustic Div) Ltd
Hodgson & Hodgson Ltd
Industrial Acoustics Company Ltd
Sound Solutions Ltd
Tantalic Acoustical Engineering Ltd
Trubros Acoustics Ltd

Acoustic Screens
Acoustic Consultants & Engineers
Acousticabs Ltd
Bestobell Protection
Burgess Industrial Silencing Ltd
Castle Microair Ltd

W. Christie & Grey Ltd
Frank Driver
Ferguson & Timpson Ltd
Harrison & Jones (Acoustic Div) Ltd
Heafield Industries Ltd
Hodgson & Hodgson Ltd
Illbruck GmbH Schaumstofftechnik
Industrial Acoustics Company Ltd
Sound Solutions Ltd
Tantalic Acoustical Engineering Ltd
Trubros Acoustics Ltd

Adhesives
Ferguson & Timpson Ltd

All-Weather Monitoring Equipment
Computer Engineering Ltd

Aluminium Sheet, Sound Dampened
Antiphon Ltd
Bestobell Protection
Castle Microair Ltd

Amplifiers
Bruel & Kjaer (UK) Ltd
Condition Monitoring Ltd
Endevco UK Ltd
Environmental Equipments Ltd
Gracey & Associates
Rion Co Ltd
Vibro-Meter AG

Amplifiers, Change
Castle-Microair Ltd

Amplifiers, Power — for vibrators
Castle-Microair Ltd

Analyzers, Drop Test
Genrad European HQ

Analyzers, Frequency
Bruel & Kjaer (UK) Ltd
Castle-Microair Ltd

Computer Engineering Ltd
Condition Monitoring Ltd
GenRad European HQ
Gracey & Associates
Hakuto International (UK) Ltd
Nicolet Instruments Ltd
Rion Co Ltd
Scientific Atlanta Ltd
Vibro-Meter AG

Analyzers, Octave Band
Bruel & Kjaer (UK) Ltd
Castle-Microair Ltd
Cirrus Research Ltd
Computer Engineering Ltd
GenRad European HQ
Gracey & Associates
Hakuto International (UK) Ltd
Nicolet Instruments Ltd
Rion Co Ltd
Scientific Atlanta Ltd
Weir Pumps Ltd

Analyzers, Real Time
Bruel & Kjaer (UK) Ltd
Computer Engineering Ltd
Condition Monitoring Ltd
GenRad European HQ
Gracey & Associates
Hakuto International (UK) Ltd
Nicolet Instruments Ltd
Rion Co Ltd
Scientific Atlanta Ltd
Vibro-Meter AG

Analyzers, Seismic
Genrad European HQ

Analyzers, Third Octave Band
Bruel & Kjaer (UK) Ltd
Castle-Microair Ltd
Cirrus Research Ltd
Computer Engineering Ltd
GenRad European HQ
Gracey & Associates
Hakuto International (UK) Ltd
Rion Co Ltd
Scientific Atlanta Ltd

Anechoic Acoustic Foam Wedges
Ferguson & Timpson Ltd
Harrison & Jones (Acoustic Div) Ltd
Illbruck GmbH Schaumstofftechnik
Industrial Acoustics Company Ltd
Kay-Metzeler Ltd
Trubros Acoustics Ltd

Anechoic Chambers
Bruel & Kjaer (UK) Ltd
Burgess Industrial Silencing Ltd
Harrison & Jones (Acoustic Div) Ltd
Industrial Acoustics Company Ltd
Kay-Metzeler Ltd
Rion Co Ltd
Trubros Acoustics Ltd

Anti-Vibration Mountings
Acoustic Consultants & Engineers
A.V.A. Ltd
Bestobell Protection
Bury Cooper Whitehead Ltd
Castle-Microair Ltd
W. Christie & Grey Ltd
C.M.T. Wells-Kelo Ltd (Dynamics Division)
Eichler KG
Industrial Acoustics Company Ltd
Lord Corporation (UK) Ltd
Stop-Choc Ltd
Tantalic Acoustical Engineering Ltd
Trelleborg Ltd
James Walker & Co Ltd

Anti-Vibration Structural Bearings
Eichler KG
Stop-Choc Ltd
TICO Manufacturing Ltd
Trelleborg Ltd

Audiology Chambers
Industrial Acoustics Company Ltd

Audiometric Equipment
Acoustic Consultants & Engineers
Bruel & Kjaer (UK) Ltd
Cirrus Research Ltd
Computer Engineering Ltd
Industrial Acoustics Company Ltd
Rion Co Ltd

Balancing Equipment
Bestobell Protection
Bruel & Kjaer (UK) Ltd
Endevco UK Ltd
Gracey & Associates
Vibro-Meter AG
Weir Pumps Ltd

Blankets, Acoustic
Acousticabs Ltd
Avomet Ltd
Bestobell Protection
Castle-Microair Ltd
Ferguson & Timpson Ltd
Hodgson & Hodgson Ltd

Booth Ventilation Kits
Burgess Industrial Silencing Ltd
Castle-Microair Ltd
W. Christie & Grey Ltd
Industrial Acoustics Company Ltd
Trubros Acoustics Ltd

Booths, Sound Attenuating

Acoustic Consultants & Engineers
Acousticabs Ltd
Bestobell Protection
Burgess Industrial Silencing Ltd
Castle-Microair Ltd
W. Christie & Grey Ltd
Ferguson & Timpson Ltd
Hodgson & Hodgson Ltd
Industrial Acoustics Company Ltd
Rion Co Ltd
Tantalic Acoustical Engineering Ltd
Trubros Acoustics Ltd

Booth Ventilation Kits

W. Christie & Grey Ltd

Bump Test Machines

Castle-Microair Ltd

Burner Silencers

Industrial Acoustics Company Ltd
Tantalic Acoustical Engineering Ltd

Calibration Equipment

Bruel & Kjaer (UK) Ltd
Castle-Microair Ltd
Cirrus Research Ltd
Computer Engineering Ltd
Condition Monitoring Ltd
Environmental Equipments Ltd
GenRad European HQ
Gracey & Associates
PCB Piezotronics Inc
Rion Co Ltd

Ceilings, Acoustic

Acoustic Consultants & Engineers
Acousticabs Ltd
Castle-Microair Ltd
W. Christie & Grey Ltd
Ewald Dörken AG
Illbruck GmbH Schaumstofftechnik
Industrial Acoustics Company Ltd
Tantalic Acoustical Engineering Ltd
Trubros Acoustics Ltd

Compressors, Silenced

Cobra Acoustics Ltd
IMI Fluidair Ltd
Supra Chemicals & Paints Ltd
Tantalic Acoustical Engineering Ltd
Trubros Acoustics Ltd

Computer Aided Test Systems

GenRad European HQ

Computer and Word Processor Covers

Acoustic Consultants & Engineers

Computer Based Monitoring Systems

Condition Monitoring Ltd

Condition Monitoring Equipment

Bruel & Kjaer (UK) Ltd
Computer Engineering Ltd
Condition Monitoring Ltd
Endevco UK Ltd
GenRad European HQ
Gracey & Associates
Scientific Atlanta Ltd
Vibro-Meter AG
Weir Pumps Ltd

Consultants

Acoustic Consultants & Engineers
Acoustic & Vibration Technology
Acousticabs Ltd
Castle-Microair Ltd
Cirrus Research Ltd
Cobra Acoustics Ltd
GenRad European HQ
Gracey & Associates
Illbruck GmbH Schaumstofftechnik
Industrial Acoustics Company Ltd
Malaysian Rubber Producers' Research Association
Scientific Atlanta Ltd
Weir Pumps Ltd

Correlators

Nicolet Instruments Ltd

Dampeners

Castle-Microair Ltd
Supra Chemicals & Paints Ltd
Trubros Acoustics Ltd
James Walker & Co Ltd

Dampers

C.M.T. Wells-Kelo Ltd (Dynamics Division)
Econocruise Ltd
Lord Corporation (UK) Ltd
Stop-Choc Ltd

Dampers (Viscous)

W. Christie & Grey Ltd
Econocruise Ltd
Supra Chemicals & Paints Ltd

Data Memory Systems

Gracey & Associates
Data Memory Systems

Data Acquisition Systems

Condition Monitoring Ltd

Deadeners
Antiphon Ltd
Castle-Microair Ltd
Ferguson & Timpson Ltd
Illbruck GmbH Schaumstofftechnik
Supra Chemicals & Paints Ltd

Doors, Sound Resisting
Acoustic Consultants & Engineers
Acousticabs Ltd
Burgess Industrial Silencing Ltd
Castle Microair Ltd
W. Christie & Grey Ltd
Clark Door Ltd
Envirodoor Markus Ltd
Ferguson & Timpson Ltd
Industrial Acoustics Company Ltd
Tantalic Acoustical Engineering Ltd
Trubros Acoustics Ltd

Dosimeters
Bruel & Kjaer (UK) Ltd
Castle Microair Ltd
Cirrus Research Ltd
Computer Engineering Ltd
Gracey & Associates
Rion Co Ltd

Double Glazing
Selectaglaze Ltd
Trubros Acoustics Ltd

Ducting, Sound Insulated
Acoustic Consultants & Engineers
Burgess Industrial Silencing Ltd
W. Christie & Grey Ltd
Harrison & Jones (Acoustic Division) Ltd
Tantalic Acoustical Engineering Ltd
Trubros Acoustics Ltd

Duct Connectors, Sound Reducing
James Walker & Co Ltd

Dynamic Alignment Equipment
Scientific Atlanta Ltd

Electric Generators, Sound Insulated
Trubros Acoustics Ltd

Electric Motor Hoods
Trubros Acoustics Ltd

Enclosures, Acoustic
Acoustic Consultants & Engineers
Acousticabs Ltd
Bestobell Protection
Burgess Industrial Silencing Ltd
Castle-Microair Ltd
W. Christie & Grey Ltd
Ferguson & Timpson Ltd
Harrison & Jones (Acoustic Division) Ltd
Heafield Industries Ltd
Hodgson & Hodgson Ltd
Industrial Acoustics Company Ltd
Tantalic Acoustical Engineering Ltd
Trubros Acoustics Ltd

Energy Absorbers
Acousticabs Ltd
C.M.T. Wells-Kelo Ltd (Dynamics Division)
James Walker & Co Ltd

Environmental Noise Monitors
Computer Engineering Ltd

Exciters
Bruel & Kjaer (UK) Ltd
Castle Microair Ltd
Endevco UK Ltd
Rion Co Ltd

Fabricators, Installations
Acoustic Consultants & Engineers
Acousticabs Ltd
Bestobell Protection
Ferguson & Timpson Ltd
Tantalic Acoustical Engineering Ltd
Trubros Acoustics Ltd

Fan Jackets, Acoustic
Bestobell Protection
Ferguson & Timpson Ltd
Tantalic Acoustical Engineering Ltd

Fan Silencers
Acoustic Consultants & Engineers
Acousticabs Ltd
Bestobell Protection
Burgess Industrial Silencing Ltd
W. Christie & Grey Ltd
Ferguson & Timpson Ltd
Harrison & Jones (Acoustic Division) Ltd
Industrial Acoustics Company Ltd
Tantalic Acoustical Engineering Ltd
Trubros Acoustics Ltd

Fatigue Testing Analysis
GenRad European HQ

Filters, Acoustic
Castle Microair Ltd
Cirrus Research Ltd

Computer Engineering Ltd
Gravey & Associates
Harrison & Jones (Acoustic Division) Ltd
Rion Co Ltd
Tantalic Acoustical Engineering Ltd

Filters, Bandpass
Bruel & Kjaer (UK) Ltd
Castle-Microair Ltd
Gracey & Associates
Rion Co Ltd
Vibro-Meter AG
Weir Pumps Ltd

Flexible Couplings
James Walker & Co Ltd

Gas Turbine Silencers
Burgess Industrial Silencing Ltd
Industrial Acoustics Company Ltd
Tantalic Acoustical Engineering Ltd

GRP Constrained Damping
Antiphon Ltd

Hearing Aids
Rion Co Ltd

Hearing Conservation
Bruel & Kjaer (UK) Ltd
Castle-Microair Ltd
Computer Engineering Ltd
Gracey & Associates
Industrial Acoustics Company Ltd
Rion Co Ltd

Hearing Protection Devices (HPD)
Castle-Microair Ltd

Hearing Tests, Personal
Bruel & Kjaer (UK) Ltd
Rion Co Ltd

Impact Absorbtion Devices
James Walker & Co Ltd

Instrument Hire
Castle Microair Ltd
Cirrus Research Ltd
Gracey & Associates

Insulation
Avomet Ltd

Intercommunications Equipment, Industrial
Atkinson Dynamics, Division of Guy F. Atkinson Company

Lead Sheet, Acoustic
Avomet Ltd
Bestobell Protection
Supra Chemicals & Paints Ltd

Light Fittings, Anti-Vibration
A. Davies & Sons (Electrical Engineers) Ltd
Stop-Choc Ltd

Line Silencers
Acousticabs Ltd
Industrial Acoustics Company Ltd
Tantalic Acoustical Engineering Ltd

Linings, Acoustic
Acoustic Consultants & Engineers
Acousticabs Ltd
Avomet Ltd
Bestobell Protection
Burgess Industrial Silencing Ltd
Castle-Microair Ltd
Ferguson & Timpson Ltd
Harrison & Jones (Acoustic Division) Ltd
Heafield Industries Ltd
Hodgson & Hodgson Ltd
Illbruck GmbH Schaumstofftechnik
Kay-Metzeler Ltd
Tantalic Acoustical Engineering Ltd
Trubros Acoustics Ltd

Load Cells
Entran Devices Inc

Louvres, Acoustic
Acoustic Consultants & Engineers
Acousticabs Ltd
Bestobell Protection
Burgess Industrial Silencing Ltd
W. Christie & Grey Ltd
Ferguson & Timpson Ltd
Heafield Industries Ltd
Industrial Acoustics Company Ltd
Tantalic Acoustical Engineering Ltd
Trubros Acoustics Ltd

Machinery Mounts
Acoustic Consultants & Engineers
A.V.A. Ltd
Bestobell Protection
Castle-Microair Ltd

C.M.T. Wells-Kelo Ltd (Dynamics Division)
Environmental Equipments Ltd
Industrial Acoustics Company Ltd
Lord Corporation (UK) Ltd
Stop-Choc Ltd
Trelleborg Ltd
James Walker & Co Ltd

Microphones

Bruel & Kjaer (UK) Ltd
Cirrus Research Ltd
Computer Engineering Ltd
Gracey & Associates
Hayden Laboratories Ltd
PCB Piezotronics Inc
Rion Co Ltd

Microphones, High Intensity Noise

Endevco UK Ltd

Modal Analysis Systems

Endevco UK Ltd
GenRad European HQ
Gracey & Associates
Nicolet Instruments Ltd
Rion Co Ltd
Scientific Atlanta Ltd

Music Noise Control

Castle Microair Ltd
Cirrus Research Ltd
Computer Engineering Ltd
Illbruck GmbH Schaumstofftechnik
Industrial Acoustics Company Ltd

Music Practice Rooms, Acoustic

Castle Microair Ltd
Industrial Acoustics Company Ltd

Noise Analyzer — Community

Castle-Microair Ltd

Noise Average Meters

Castle Microair Ltd
Computer Engineering Ltd
Gracey & Associates
Rion Co Ltd

Noise Cancelling Headsets and Communicating Equipment

Castle-Microair Ltd
Hayden Laboratories Ltd

Noise Control Installations

Acoustic Consultants & Engineers

Acousticabs Ltd
Bruel & Kjaer (UK) Ltd
Castle Microair Ltd
W. Christie & Grey Ltd
Cobra Acoustic Ltd
Ferguson & Timpson Ltd
Frank Driver
Industrial Acoustics Company Ltd
Rion Co Ltd
Sound Solutions Ltd
Tantalic Acoustical Engineering Ltd
Trubros Acoustics Ltd
James Walker & Co Ltd

Noise Exposure Monitors

BBN Instruments Corporation
Bruel & Kjaer (UK) Ltd
GenRad European HQ
Gracey & Associates
Rion Co Ltd

Noise Generators

Bruel & Kjaer (UK) Ltd
Computer Engineering Ltd
Gracey & Associates
Rion Co Ltd

Noise Meters

Bruel & Kjaer (UK) Ltd
Castle Microair Ltd
Cirrus Research Ltd
Computer Engineering Ltd
GenRad European HQ
Gracey & Associates
Rion Co Ltd

Noise Exposure Monitors

Castle Microair Ltd
Computer Engineering Ltd

Oscillators

Bruel & Kjaer (UK) Ltd

Oscilloscopes

Bruel & Kjaer (UK) Ltd
Nicolet Instruments Ltd

Partitions, Acoustic

Acoustic Consultants & Engineers
Acousticabs Ltd
Bestobell Protection
Burgess Industrial Silencing Ltd
Castle-Microair Ltd
W. Christié & Grey Ltd
Ferguson & Timpson Ltd
Harrison & Jones (Acoustic Division) Ltd

Heafield Industries Ltd
Hodgson & Hodgson Ltd
Industrial Acoustics Company Ltd
Tantalic Acoustical Engineering Ltd
Trubros Acoustics Ltd

Peak Hold Meters

Bruel & Kjaer (UK) Ltd
Castle-Microair Ltd
Computer Engineering Ltd
GenRad European HQ
Gracey & Associates
PCB Piezotronics Inc
Vibro-Meter AG

Peak Noise Level Meters

Bruel & Kjaer (UK) Ltd
Castle Microair Ltd
Computer Engineering Ltd
GenRad European HQ
Gracey & Associates

Perforated Metal

Bestobell Protection
Rich, Muller (UK) Ltd
Trubros Acoustics Ltd

Pickups

Castle-Microair Ltd
Rion Co Ltd
Scientific Atlanta Ltd
Weir Pumps Ltd

Piezoelectric Devices

BBN Instruments Corporation
Castle-Microair Ltd
Computer Engineering Ltd
Condition Monitoring Ltd
Endevco UK Ltd
Environmental Equipments Ltd
Gracey & Associates
Rion Co Ltd
Scientific Atlanta Ltd
Vibro-Meter AG
Weir Pumps Ltd

Pressure Meters, Digital

Entran Devices Inc

Pressure Monitors

Condition Monitoring Ltd

Pressure Transmitters

Entran Devices Inc

Printers, Digital

Castle Microair Ltd
Gracey & Associates
Harrison & Jones (Acoustic Division) Ltd

Recorders, Digital Cassette

Bruel & Kjaer (UK) Ltd
Computer Engineering Ltd

Recorders, Digital event

Bruel & Kjaer (UK) Ltd
Castle Microair Ltd
Condition Monitoring Ltd
GenRad European HQ

Recorders, Graphic Level

Bruel & Kjaer (UK) Ltd
Castle Microair Ltd
Computer Engineering Ltd
Gracey & Associates
Nicolet Instruments Ltd
Rion Co Ltd

Recorders, Oscillograph

Bruel & Kjaer (UK) Ltd
Computer Engineering Ltd

Recorders, Tape (Instrumentation)

Bruel & Kjaer (UK) Ltd

Recorders, XY

Bruel & Kjaer (UK) Ltd
Castle Microair Ltd
Gracey & Associates
Hakuto International (UK) Ltd
Nicolet Instruments Ltd

Recorders, XYl

Castle Microair Ltd
Nicolet Instruments Ltd

Recorders, XYT

Castle Microair Ltd
Nicolet Instruments Ltd

Refuges, Acoustic

Acoustic Consultants & Engineers
Acousticabs Ltd
Bestobell Protection
Burgess Industrial Silencing Ltd
Castle Microair Ltd
W. Christie & Grey Ltd
Ferguson & Timpson Ltd
Industrial Acoustics Company Ltd
Tantalic Acoustical Engineering Ltd
Trubros Acoustics Ltd

Resilient Materials
Heafield Industries Ltd
TICO Manufacturing Ltd

Resins, Sound Deadening
Antiphon Ltd
Castle-Microair Ltd

Reverberation Chambers
Burgess Industrial Silencing Ltd
Castle-Microair Ltd
Industrial Acoustics Company Ltd
Rion Co Ltd

Rubber Bonded-to-Metal Products
A.V.A.Ltd
Castle-Microair Ltd
W. Christie & Grey Ltd
Lord Corporation (UK) Ltd
Stop-Choc Ltd
Trelleborg Ltd
James Walker & Co Ltd

Reverberation Meters
Gracey & Associates

Rubber Constructional Bearings
Eichler KG
TICO Manufacturing Ltd
Trelleborg Ltd
James Walker & Co Ltd

Rubber Mouldings
Ferguson & Timpson Ltd
TICO Manufacturing Ltd
Trelleborg Ltd
James Walker & Co Ltd

Sealing Systems
James Walker & Co Ltd

Servo Analysis
GenRad European HQ

Shaker (Vibrator) Control Equipment
Bruel & Kjaer (UK) Ltd
Castle-Microair Ltd
Condition Monitoring Ltd
Environmental Equipments Ltd
GenRad European HQ
Instron Environmental Ltd

Shock Absorbers
Castle-Microair Ltd
C.M.T. Wells-Kelo (Dynamics Division)
Environmental Equipments Ltd
Stop-Choc Ltd
James Walker & Co Ltd

Shock Analyzers
BBN Instruments Corporation
Castle-Microair Ltd
Computer Engineering Ltd
GenRad European HQ
Gracey & Associates
Scientific Atlanta Ltd

Shock Mounts
Castle Microair Ltd
Eichler KG
Lord Corporation (UK) Ltd
Stop-Choc Ltd
Trelleborg Ltd
James Walker & Co Ltd

Shock Test Equipment
Endevco UK Ltd

Signal Processing Equipment
Castle Microair Ltd
Condition Monitoring Ltd
Endevco UK Ltd
GenRad European HQ
Gracey & Associates
Nicolet Instruments Ltd
Scientific Atlanta Ltd
Vibro-Meter AG

Signature Analysis
GenRad European HQ

Silencers
Acousticabs Ltd
Burgess Industrial Silencing Ltd
Castle Microair Ltd
W. Christie & Grey Ltd
IMI Norgren Enots Ltd
Industrial Acoustics Company Ltd
Porvair Ltd
Tantalic Acoustical Engineering Ltd
Turbros Acoustics Ltd

Silencers, Exhaust
Burgess Industrial Silencing Ltd
W. Christie & Grey Ltd
Ferguson & Timpson Ltd
IMI Norgren Enots Ltd
Industrial Acoustics Company Ltd
Porvair Ltd
Tantalic Acoustical Engineering Ltd

BUYERS' GUIDE — CLASSIFIED INDEX

Silencers, Intake
Burgess Industrial Silencing Ltd
W. Christie & Grey Ltd
Ferguson & Timpson Ltd
Industrial Acoustics Company Ltd
Porvair Ltd
Tantalic Acoustical Engineering Ltd
Trubros Acoustics Ltd

Silencers, Pipeline
Burgess Industrial Silencing Ltd
Castle Microair Ltd
Industrial Acoustics Company Ltd
Tantalic Acoustical Engineering Ltd

Silencers, Pneumatic
Burgess Industrial Silencing Ltd
IMI Norgren Enots Ltd
Industrial Acoustics Company Ltd
Porvair Ltd
Tantalic Acoustical Engineering Ltd

Silencers, Pneumatic Exhaust (Coalescing)
IMI Norgren Enots Ltd

Slip Tables
Instron Environmental Ltd

Snubbers
Burgess Industrial Silencing Ltd
IMI Norgren Enots Ltd
Stop-Choc Ltd
Trelleborg Ltd

Sound Barriers
Avomet Ltd
Supra Chemicals & Paints Ltd

Sound Insulating Walls
Acoustic Consultants & Engineers
Acousticabs Ltd
Burgess Industrial Silencing Ltd
Castle Microair Ltd
Ferguson & Timpson Ltd
Heafield Industries Ltd
Hodgson & Hodgson Ltd
Tantalic Acoustical Engineering Ltd
Trubros Acoustics Ltd

Sound Intensity Meters
Bruel & Kjaer (UK) Ltd

Sound Level Calibrators
Bruel & Kjaer (UK) Ltd
Castle Microair Ltd
Cirrus Research Ltd
Computer Engineering Ltd
GenRad European HQ
Gracey & Associates
Rion Co Ltd

Sound Level Indicators
Bruel & Kjaer (UK) Ltd
Castle-Microair Ltd
Cirrus Research Ltd
Computer Engineering Ltd
GenRad European HQ
Rion Co Ltd

Sound Level Meters
Bruel & Kjaer (UK) Ltd
Castle-Microair Ltd
Cirrus Research Ltd
Computer Engineering Ltd
GenRad European HQ
Gracey & Associates
Hayden Laboratories Ltd
Rion Co Ltd

Sound Level Meters, Impulse
Bruel & Kjaer (UK) Ltd
Castle-Microair Ltd
Cirrus Research Ltd
Computers Engineering Ltd
GenRad European HQ
Gracey & Associates
Rion Co Ltd

Sound Level Meters, Integrating
Bruel & Kjaer (UK) Ltd
Castle Microair Ltd
Cirrus Research Ltd
Computer Engineering Ltd
GenRad European HQ
Gracey & Associates
Rion Co Ltd

Sound Level Meters, Precision
Bruel & Kjaer (UK) Ltd
Castle Microair Ltd
Cirrus Research Ltd
Computer Engineering Ltd
GenRad European HQ
Gracey & Associates
Rion Co Ltd

Sound Monitoring Equipment
Bruel & Kjaer (UK) Ltd
Castle Microair Ltd
Cirrus Research Ltd
Computer Engineering Ltd
GenRåd European HQ

Gracey & Associates
Nicolet Instruments Ltd
Rion Co Ltd

Soundproof Cabins
Acoustic Consultants & Engineers
Acousticabs Ltd
Bestobell Protection
Burgess Industrial Silencing Ltd
Castle Microair Ltd
W. Christie & Grey Ltd
Ferguson & Timpson Ltd
Industrial Acoustics Company Ltd
Rion Co Ltd
Sound Solutions Ltd
Tantalic Acoustical Engineering Ltd
Trubros Acoustics Ltd

Soundproofed Rooms (Prefabricated)
Acoustic Consultants & Engineers
Acousticabs Ltd
Burgess Industrial Silencing Ltd
Castle-Microair Ltd
W. Christie & Grey Ltd
Ferguson & Timpson Ltd
Industrial Acoustics Company Ltd
Rion Co Ltd
Tantalic Acoustical Engineering Ltd
Trubros Acoustics Ltd

Spectrum Analyzers
Bruel & Kjaer (UK) Ltd
Castle Microair Ltd
Computer Engineering Ltd
Condition Monitoring Ltd
Environmental Equipments Ltd
GenRad European HQ
Gracey & Associates
Hakuto International (UK) Ltd
Nicolet Instruments Ltd
Rion Co Ltd
Scientific Atlanta Ltd

Spring Isolated Table
W. Christie & Grey Ltd

Stainless Steel Sheet, Sound Dampened
Antiphon Ltd

Statistical Level Meters
Bruel & Kjaer (UK) Ltd
Castle Microair Ltd
Computer Engineering Ltd
Gracey & Associates
Rion Co Ltd

Steel Sheet, Sound Dampened
Antiphon Ltd

Strain Measuring Equipment
Bruel & Kjaer (UK) Ltd
Condition Monitoring Ltd
Entran Devices Inc
Instron Environmental Ltd
Vibro-Meter AG

Stroboscopes, Electronic
Bruel & Kjaer (UK) Ltd
Castle-Microair Ltd
Condition Monitoring Ltd
Environmental Equipments Ltd
Gracey & Associates
Scientific Atlanta Ltd

Structural Analysis
GenRad European HQ

Tacho Systems
Condition Monitoring Ltd

Tape Recorders
Computer Engineering Ltd

Telex Covers, Acoustic
Acoustic Consultants & Engineers
Harrison & Jones (Acoustic Division) Ltd
Tantalic Acoustical Engineering Ltd

Telephone Hoods, Acoustic
Bestobell Protection
W. Christie & Grey Ltd
Harrison & Jones (Acoustic Division) Ltd
Tantalic Acoustical Engineering Ltd

Telex Muffs
Harrison & Jones (Acoustic Division) Ltd

Test Cells
Acoustic Consultants & Engineers
Acousticabs Ltd
Burgess Industrial Silencing Ltd
Castle-Microair Ltd
Industrial Acoustics Company Ltd
Scientific Atlanta Ltd

Test Facilities
Castle Microair Ltd
Gracey & Associates
Industrial Acoustics Company Ltd

BUYERS' GUIDE — CLASSIFIED INDEX

Transducers
BBN Instruments Corporation
Bruel & Kjaer (UK) Ltd
Castle-Microair Ltd
Computer Engineering Ltd
Condition Monitoring Ltd
Endevco UK Ltd
Entran Devices Inc
Gracey & Associates
Hakuto International (UK) Ltd
Hayden Laboratories Ltd
Rion Co Ltd
Scientific Atlanta Ltd
Vibro-Meter AG
Weir Pumps Ltd

Transducers, Displacement
Bruel & Kjaer (UK) Ltd
Castle-Microair Ltd
Computer Engineering Ltd
Condition Monitoring Ltd
Rion Co Ltd
Vibro-Meter AG
Weir Pumps Ltd

Transducers, Displacement, Velocity & Acceleration
Scientific Atlanta Ltd

Transducers, Force
Bruel & Kjaer (UK) Ltd
Castle-Microair Ltd
Entran Devices Inc
PCB Piezotronics Inc
Rion Co Ltd
Vibro-Meter AG

Transducers, Load
Vibro-Meter AG

Transducers, Pressure
BBN Instruments Corporation
Bruel & Kjaer (UK) Ltd
Castle-Microair Ltd
Endevco UK Ltd
Entran Devices Inc
Gracey & Associates
PCB Piezotronics Inc
Vibro-Meter AG

Turnkey Noise Control Systems
Industrial Acoustics Company Ltd

Typewriter Covers, Acoustic
Harrison & Jones (Acoustic Division) Ltd

Ultrasonic Leak Detectors
Cirrus Research Ltd

Ventilating Systems
Acoustic Consultants & Engineers
Burgess Industrial Silencing Ltd
Trubros Acoustics Ltd

Vent Silencers
Burgess Industrial Silencing Ltd
Castle-Microair Ltd
W. Christie & Grey Ltd
Industrial Acoustics Company Ltd
Tantalic Acoustical Engineering Ltd
Trubros Acoustics Ltd

Vibration Attenuation Materials
Antiphon Ltd
Castle-Microair Ltd
W. Christie & Grey Ltd
C.M.T. Wells-Kelo Ltd (Dynamics Division)
Ferguson & Timpson Ltd
Harrison & Jones (Acoustic Division) Ltd
Illbruck GmbH Schaumstofftechnik
Industrial Acoustics Company Ltd
Tantalic Acoustical Engineering Ltd
TICO Manufacturing Ltd
James Walker & Co Ltd

Vibration Control Systems
Bruel & Kjaer (UK) Ltd
Castle-Microair Ltd
W. Christie & Grey Ltd
Condition Monitoring Ltd
GenRad European HQ
Gracey & Associates
Industrial Acoustics Company Ltd
Instron Environmental Ltd
Rion Co Ltd
Scientific Atlanta Ltd
Stop-Choc Ltd
James Walker & Co Ltd

Vibration Damping Materials
Antiphon Ltd
Bestobell Protection
Castle-Microair Ltd
C.M.T. Wells-Kelo Ltd (Dynamics Division)
Ferguson & Timpson Ltd
Harrison & Jones (Acoustic Division) Ltd
Illbruck GmbH Schaumstofftechnik
Industrial Acoustics Company Ltd
Stop-Choc Ltd
Supra Chemicals & Paints Ltd
TICO Manufacturing Ltd
Trubros Acoustics Ltd
James Walker & Co Ltd

Vibration Deadeners
Antiphon Ltd
Bestobell Protection
Castle Microair Ltd
Ferguson & Timpson Ltd
Harrison & Jones (Acoustic Division) Ltd
Industrial Acoustics Company Ltd
Stop-Choc Ltd
Supra Chemicals & Paints Ltd
James Walker & Co Ltd

Vibration Generators
Bruel & Kjaer (UK) Ltd
Castle Microair Ltd
Condition Monitoring Ltd
Endevco UK Ltd
Gracey & Associates
Inston Environmental Ltd

Vibration Isolators
A.V.A. Ltd
Bestobell Protection
Bury Cooper Whitehead Ltd
Castle Microair Ltd
W. Christie & Grey Ltd
C.M.T. Wells-Kelo Ltd (Dynamics Division)
Eichler KG
Environmental Equipments Ltd
Industrial Acoustics Company Ltd
Stop-Choc Ltd
Tantalic Acoustical Engineering Ltd
James Walker & Co Ltd

Vibration Measuring & Monitoring Equipment
BBN Instruments Corporation
Bruel & Kjaer (UK) Ltd
Castle Microair Ltd
Cirrus Research Ltd
Computer Engineering Ltd
Condition Monitoring Ltd
DISA
Econocruise Ltd
Endevco UK Ltd
Entran Devices Inc
GenRad European HQ
Gracey & Associates
Hakuto International (UK) Ltd
Nicolet Instruments Ltd
PCB Piezotronics Inc
Rion Co Ltd
Scientific Atlanta Ltd
Vibro-Meter AG
Weir Pumps Ltd

Vibration, Data Aquisition
Vibro-Meter AG

Vibrators
Bruel & Kjaer (UK) Ltd
Castle Microair Ltd
Condition Monitoring Ltd
Endevco UK Ltd
Environmental Equipments Ltd
Instron Environmental Ltd

Vibrometer, Laser Dappler
DISA

Windscreens
Computer Engineering Ltd
Gracey & Associates
Rion Co Ltd

Windscreens (Microphone)
Bruel & Kjaer (UK) Ltd

Windscreens for Sound Level Meters
Castle-Microair Ltd

Wood Constrained Damping
Antiphon Ltd

Sub-section (c)

ALPHABETICAL LIST OF MANUFACTURERS WITH ADDRESSES, TELEPHONE AND TELEX NUMBERS OF HEAD OFFICE AND WORKS

ACOUSTIC CONSULTANTS & ENGINEERS, St. Anne's House, North Street, Radcliffe, Manchester M26 9RN
    Telephone: 061 724 9511   Telex: 677407
ACOUSTIC & VIBRATION TECHNOLOGY, 27 Bramhall Lane South, Bramhall, Stockport, Cheshire.
    Telephone: 061 440 9392   Telex: 669028
ACOUSTICABS LIMITED, Stonebow House, The Stonebow, York.
    Telephone: 0904 36441   Telex: 57813
ANTIPHON LIMITED, 170 Park Road, Peterborough PE1 2UF.
    Telephone: 0733 49987   Telex: 32887
ATKINSON DYNAMICS, Division of Guy F. Atkinson Company, 10 West Orange Avenue, So. San Francisco, CA 94080, USA. Telephone: (415) 583 9845   Telex: 34297 Atkinson SGF
A.V.A. LIMITED, Thames Works, Lower Teddington Road, Hampton Wick, Nr. Kingston-upon-Thames, Surrey KT1 4HA. Telephone: 01 977 2201   Telex: No.262284 Ref.No.3533
AVOMET LTD, Rassau Industrial Estate, Ebbw Vale, Gwent NP3 5SD
    Telephone: 0495 305457   Telex: 497114
BBN INSTRUMENTS CORPORATION, 50 Moulton Street, Cambridge, MA 02238, USA.
    Telephone: (617) 491 0091   Telex: 92-1470
BESTOBELL PROTECTION, P.O.Box 6, 135 Farnham Road, Slough, SL1 4UY.
    Telephone: Slough 23921   Telex: 848107
    Works: South Nelson Road, Cramlington, Northumberland, NE23 9BL.
    Telephone: 067 071 6611   Telex: 53157
BRUEL & KJAER (UK) LTD., Cross Lances Road, Hounslow, TW3 2AE.
    Telephone: 01 570 7774   Telex: 934150
BURGESS INDUSTRIAL SILENCING LIMITED, Shaftesbury Avenue, Simonside Industrial Estate, South Shields, Tyne & Wear, NE34 9PH.
    Telephone: 566721   Telex: 537404
BURY COOPER WHITEHEAD LIMITED, Hudcar Mills, Bury, Lancashire, BL9 6HD.
    Telephone: 061 764 2262   Telex: 669000 BCW Ltd
CASTLE-MICROAIR LIMITED, 3 West Street, Coach Mews, St.Ives, Huntingdon, Cambs PE17 APL
    Sales Office: Unit 1 Hogwood Industrial Estate, Finchampstead, Wokingham, Berkshire RG11 4QW
    Telephone: 0734 730050   Telex: 858893 Fletel G
    Works: P.O.Box 20, Scarborough, North Yorkshire
    Telephone: 0723 66347 Telex: 527244 Castle G
W. CHRISTIE & GREY LTD., Sovereign Way, Tonbridge, Kent TN9 1RJ
    Telephone: 0732 366444   Telex: 957194   Grams: Vibrolator Tonbridge
CIRRUS RESEARCH LIMITED, 1/2 York Place, Scarborough, North Yorkshire YO11 2NP
    Telephone: 0723 71441

CLARK DOOR LIMITED, Willow Holme, Carlisle.
Telephone: 0228 22321   Telex: 64131
C.M.T. WELLS-KELO LTD (Dynamics Division), Progress Works, Kingsland, Holyhead, Anglesey, Gwynedd LL55 2SN.
Telephone: 0407 2391/2/3/4/5   Telex: 61518
COBRA ACOUSTICS LTD, Unit C1, Stafford Park 15, Telford, Shropshire.
Telephone: 0952 616109
COMPUTER ENGINEERING LIMITED, 14 Wallace Way, Hitchin, Herts SG4 OSE.
Telephone: Hitchin 52731/3   Telex: 826615
CONDITION MONITORING LTD, Units 2/3, Tavistock Industrial Estate, Twyford, Berks RG10 9NJ
Telephone: 0734 342636   Telex: 847151
A. DAVIES & SONS (Electrical Engineers) LTD, Alpha Works, Ashton Road, Bredbury, Stockport, Cheshire SK6 2QF.
Telephone: 061 430 5297   Telex: 666501
DISA, Techno House, Redcliffe Way, Bristol BS1 6NU
Telephone: 0272 291436   Telex: 449695   Grams: Disaworks Bristol
Works: Disa Elektronik A/S Mileparken 22, DK-2740, Skovlunde, Denmark
Telephone: 010 452 842211   Telex: 0063 35349
FRANK DRIVER, 206 Elgar Road, Reading, Berkshire
Telephone: 0734 752666   Telex: 848193
EWALD DORKEN AG, Wetterstr. 58, D 5804, Herdecke, West Germany
Telephone: 02330/631   Telex: 08239 428
ECONOCRUISE LIMITED, 180 Wood Street, Rugby CV21 2NP, Warwickshire.
Telephone: 0788 74431   Telex: 311331
EICHLER KG, A-1101 Wien, Pernerstorfergasse 5, Postfach 165
Telephone: 0222 649181   Telex: 131105 eich a
ENDEVCO U.K. LTD, Melbourn, Royston, Herts, SG8 6AQ
Telephone: Royston (0763) 61311   Telex: 81522
ENTRAN DEVICES, INC, 10 Washington Avenue, Fairfield, NJ 07006
Telephone: 201 227 1002   Telex: 130361
ENVIRODOOR MARKUS LIMITED, Great Gutter Lane, Willerby, Hull HU10 6DT, N.Humberside
Telephone: 0482 659375   Telex: 527088
ENVIRONMENTAL EQUIPMENTS LIMITED, Fleming Road, London Road Industrial Estate, Newbury, Berkshire.
Telephone: 0635 40240   Telex: 847868 Envair G.
FERGUSON & TIMPSON LIMITED, 5 Atholl Avenue, Hillington, Glasgow G52 4UA
Telephone: 041 882 4691   Telex: 77108   Grams: FT HO G
London Office: Thistle House, Selinas Lane, Dagenham, Essex RM8 1TB
Telephone: 01 593 7611   Telex: 23371   Grams: FT LDN G
GENRAD EUROPEAN HQ, Norreys Drive, Maidenhead, Berks.
Telephone: 0628 39181   Telex: 848321
Works: Santa Clara, 2855 Bowers Avenue, Santa Clara, California, USA.
Telephone: 408 727 4400   Telex: 910 338 0291
GRACEY & ASSOCIATES, Threeways, Chelveston, Northants NN9 6AJ
Telephone: 0933 624212   Telex: 312517
HAKUTO INTERNATIONAL (UK) LTD, 159a Chase Side, Enfield, Middlesex EN2 OPW
Telephone: 01 367 4633   Telex: 299288
HARRISON & JONES (ACOUSTIC DIVISION) LTD, Chaul End Lane, Luton, Beds LU4 8HB
Telephone: 0582 595151   Telex: 826613
HAYDEN LABORATORIES LIMITED, Hayden House, Chiltern Hill, Chalfton-St-Peter, Bucks SL9 9UG
Telephone: 02813 88447/89221   Telex: 949469 Haylab
HEAFIELD INDUSTRIES LIMITED, Connaught Street, Oldham OL8 1DE
Telephone: 061 624 8331/3   Telex: 667462 Comcam G Megson
HODGSON & HODGSON LIMITED, Crown Industrial Estate, Anglesey Road, Burton-on-Trent, Staffs DE14 3PA
Telephone: Burton (0283) 64772   Telex: 342250
ILLBRUCK GMBH SCHAUMSTOFFTECHNIK, Burscheider Strasse 454, 5090 Leverkusen 3
Telephone: 02171 391-0   Telex: 8515733 illb d

IMI FLUIDAIR LIMITED, Radcliffe, Manchester M26 OJB.
   Telephone: 061 723 2421   Telex: 667071   Grams: Fluidair, Radcliffe
IMI NORGREN ENOTS LIMITED, Shipston-on-Stour, Warwickshire
   Telephone: 0608 61676   Telex: 83208 (Norsec G)   Grams: Philsym, Shipston
INDUSTRIAL ACOUSTICS COMPANY LIMITED, Walton House, Central Trading Estate, Staines, Middlesex TW18 4XB
   Telephone: Staines 56251   Telex: 25518
INSTRON ENVIRONMENTAL LIMITED, Coronation Road, High Wycombe, Bucks
   Telephone: 0494 33333   Telex: 83222
KAY-METZELER LIMITED, New Mill, Park Road, Dukinfield, Cheshire
   Telephone: 061 330 7311
LORD CORPORATION (UK) LTD, Unit 8, The Brook Trading Estate, Deadbrook Lane, Aldershot, Hants GU12 4XB.
   Telephone: 0252 26225   Telex: 858260 Lorcog
MALAYSIAN RUBBER PRODUCERS' RESEARCH ASSOCIATION, Tun Abdul Razak Laboratory, Brickendonbury, Hertford SG13 8NL
   Telephone: Hertford 54966   Telex: 817449   Grams: Rubresearch, Hfrd.
RICH.MULLER (UK) LTD, 77 High Street, Sevenoaks, Kent
   Telephone: 0732 459595   Telex: 95658
NICOLET INSTRUMENTS LIMITED, Budbrooke Road, Warwick, Warwickshire CV34 5XH
   Telephone: 0926 494111   Telex: 311135
   Works: Nicolet Scientific Corporation, 245 Livingston Street, Northvale, New Jersey 07647, USA
   Telephone: 201 767 7100   Telex: 71099 19619
PCB PIEZOTRONICS INC, 3425 Walden Avenue
   Telephone: 716 684 0001   Telex: 710 263 1371
PORVAIR LIMITED, Estuary Road, Riverside Industrial Estate, King's Lynn, Norfolk.
   Telephone: 0553 61111   Telex: 817115
RION CO LTD, Ikeda Building, 7-7 Yoyogi 2-chome, Shibuya-ku, Tokyo 151, Japan
   Telephone: 03 379 3251   Telex: J28437   Grams: Rionet Tokyo
   Sales Office: 20-41 Higashimotomachi 3-chome, Kokubunji, Tokyo 185, Japan
   Telephone: 0423-22-1133
SCIENTIFIC ATLANTA LTD, Horton Manor, Horton, Slough SL3 9PA
   Telephone: 02812 3211   Telex: 849406
   Works: 25 Bury Mead Road, Hitchin, Herts SG5 1RT
   Telephone: 0462 31101   Telex: 826087
SELECTAGLAZE LTD, Sutton Road, St. Albans, Hertfordshire
   Telephone: St Albans (0727) 37271
SOUND SOLUTIONS LTD, Victor Works, Bolton Hall Road, Bolton Woods, Bradford BD2 1BQ
   Telephone: 0274 597273   Telex: 51449 Chacom G for Sound Solutions
STOP-CHOC LTD, 710 Banbury Avenue, Slough, Berks SL1 4LH
   Telephone: Slough 33223   Telex: 848620
SUPRA CHEMICALS & PAINTS LTD, Hainge Road, Tividale, Warley, West Midlands B69 2NF
   Telephone: 021 557 9361   Telex: 336238
   Works: Acoustics Division, Globe Street, Wednesbury, West Midlands WS10 ONN
   Telephone: 021 502 2857
TANTALIC ACOUSTICAL ENGINEERING LTD, Dee Cee House, Princes Road, Dartford, Kent
   Telephone: Dartford 7700   Telex: 8953600
TICO MANUFACTURING LIMITED, Tico Works, Hipley Street, Old Woking, Woking, Surrey GU22 9LL
   Telephone: Woking 62635/6   Telex: Group 85221   Grams: Ticodraft, Wok
TRELLEBORG LIMITED, 90 Somers Road, Rugby, Warwickshire CV22 7ED
   Telephone: 0788 62711   Telex: 31 11 44
TRUBROS ACOUSTICS LTD, Imperial Works, Station Road, Kegworth, Derby DE7 2FR
   Telephone: 05097 2104
VIBRO-METER AG, 1701 Fribourg, Switzerland
   Telephone: 037 821141   Telex: 36232
JAMES WALKER & CO.LTD, Lion Works, Woking, Surrey GU22 8AP
   Telephone: Woking 5951   Telex: 859221
WEIR PUMPS LTD, Newlands Road, Cathcart, Glasgow
   Telephone: 041 637 7141   Telex: 77161/2

Index

A

Absorption coefficient . . 39,40,287,315,324,333,
    . . . . 336,337,344,380—82
Absorption performance. 40
Acceleration. 47,57,58,61
Accelerometers 156
Acoustic absorptiveness 326
Acoustic attenuation panels 340
Acoustic chamber. 45
Acoustic control room 80
Acoustic curtains 356
Acoustic doors 348
Acoustic enclosure .216,317,345—53,391,397,399
    basic design and construction rules . . . 345
    commercial design 348
    doors. 348
    foundries. 350
    in rooms 351
    loss of performance. 346
    reducing reverberant sound 352
    rooms as . . . . .352—53
    service entries 350
    ventilation 348
    windows 350
    work feed 348
Acoustic energy. 231,247
Acoustic foams 335
Acoustic materials . . . . .334—44
Acoustic panels 352
Acoustic power 2,9
Acoustic power level 2
Acoustic room. 44,46
Acoustic screens. . . . . See Screens and Screening
Acoustic sheds. . . . . .359—60
Acoustic studio 45
Acoustic trauma. 75
Acoustic treatment 45
Acoustic tunnels 262
Acoustical jacketing 341
Air absorption. . . . . 9—10,41
Air conditioning system 282
Air conduction sensibility 87

Air distribution systems . . . . .231—45
    bends. 232,236
    branches 233
    cross-talk. 244
    dampers 239
    duct fittings 237
    duct rungs 236,242
    duct systems. 236
    ducted system to atmosphere 245
    fan running in open plant room 245
    flanking transmissions 241,244
    flexible duct connectors 244
    free jet noise. 240
    grilles and diffusers 239
    indirect transmission 241,244
    in-duct elements. 237
    induction units 240
    noise breakout. 242
    noise control. 231,235
    noise sources. 231
    noise transmission path. 232
    plain duct runs 232
    plenum chambers . . . . .233—34,236
    proprietary units 234
    secondary noise 237,240
    terminal units 235,238
    transmission to atmosphere 245
    turbulent flow. 238
    vibration isolation. 243
    See also Fan noise

Air-gapping 46
Air intake silencer. 210
Air Navigation (Noise Certification) Act 1979
   276,463
Air springs 431
Airborne noise. 200,217,295,444
Airborne sound coefficient 298
Airborne sound reduction 361
Aircraft noise 70,134,273,463
    bypass jet engine 277
    future patterns 280
    jet-aircraft . . . . .275,277—78,280

INDEX 487

jet noise suppression 279
jet-prop aircraft 275
legislation 276,280
noise footprints 279
piston-engined aircraft 273–75
turbofan 277
Aircraft noise measurement 19
Airport noise 273
Alphanumeric printer. 130
Amplitude effects. 98
Amplitude parameters 147
Anechoic chamber 45
Anechoic field. 16
Anechoic room 42,45
Anechoic wedges 44
Anti-alias filters 126
Anti-vibration mounts 422–36
   building structures 451
   efficiency 422,423
   elastomeric. 428–31
   pad and area materials 431–36
   types and applications 424–25
   See also. . . . . Resilient bearings
Articulation Index 81,284
Asbestos 335,433
Atmospheric attenuation. 13
Atmospheric effects 13
Atmospheric moisture effects 13
Attenuation . .9,13,45,81,87,93–94,291,379,380,
  382,383,391,392,414
Audiogram. 77,137
Audiology rooms 143
Audiology screening programmes 142
Audiometer booths. 142
Audiometer calibration. 140–42
Audiometer tone frequencies 140
Audiometers. 138–39
   diagnostic 139
   monitoring. 139
   research 139
   self-recording 139
   simple 139
   types of 139
Audiometric enclosures 143
Audiometric facility 142,144
Audiometric programmes 144
Audiometry 82,137–44
Auditoria. 301–7
   acoustical defects. 304
   acoustics control 306
   basic requirements 306
   curved reflective ceiling and inclined S
      rating levels. 305
   recommended room volumes 304
   reflecting surfaces. 305
   shapes and acoustic properties. 304
Auditory threshold 137

Average noise measurement 107
Averaging modes 128

B

Background noise. . . 73,81,257,282–84,286,292,
  301,305
Baffles 260,261,394
Balancing. . . . . .See Machine balance
Balsa wood. 336,436
Beams 65
Bearing housing 208
Bearing mounts 207
Bearings 63,202–8
   conical 207
   cylindrical 206
   elastomeric. 205–6
   metallic. 207
   mountings 207
   non-metallic 203,204
   plain 202
      noise control. 203
      wineglassing 208
   plastic 203
   rolling 204–5
   sandwich. 206
   spherical 206
   vibration 202
   See also Resilient bearings
Bel 1
Bending waves. 448
Blankets 447
Bone conduction 88
Bone conduction sensitivity 87
Boundary conditions 66
Brickwork 338
Building materials. 338,355
Building Regulations 292,293,338
Building Research Establishment 120
Building Research Station 268
Building structures, anti-vibration mounts . . . 451

C

Carpeting effects 287,299
Cavity depth effects on sound absorption . . . 330
Cavity resonance 372
Ceiling height 42
Ceilings. 287
   curved reflective. 305
   suspended 288,361
Ceramic microphone 121
Chart recorder. 117
Chimneys 368
Chipboard 344
'Chirp' signal. 173
Civil Aviation Act 1978 276

Classrooms	307
Clinker block	338
Coincidence effect	316,317
Coincident frequency	319
Cold finger	101
Commercial buildings	*See* Offices
Community reaction	69,263
Complex shock	60
Complex wave	50
Composite panels	320–21
Compression-ignition engines	209–11
Compressors	221,250
gas turbine driven	396
vibration	222
Computer aided vibration monitoring system	186
Concrete	339
Concrete floors	362,363
Construction machines	223
Construction site analysis	230
Construction site equipment	221–30
sound power levels	224–26
Construction site noise sources	227–28
Continuous broad-band excitation	167
Control of Pollution Act 1974	463
Convolution integral	161
Cooling methods	218
Cooling towers	218
Cork	
isolation mounts	433–34
rubber-bonded	434
Corkboard	338
Corrected Noise Levels	71
Correction factors	302
Cremier formula	319
Critical frequency	316,317,319,370
Critical speeds	408,412
Cross-talk	244,393
Cut-off frequency	42,43

D

Damped single frequency transient vibration	53
Damping	422,426,437–50
air	437
applications	444
basic methods	437
critical	438
dry friction	438
free bending wave	443
friction	437
hysteresis	437,438
impact	450
joints	449
layer	439–40
magnetic	437
panel	443
performance	443

spaced	445,448
spaced distance	441
structures	448
treatment characteristics	444
viscous	437,438
Damping force	439
Damping materials	438,440,445–47
Damping ratio	53,54
Data logging	130
Data storage	130
dB(A) measurement	72,75
Dead finger	101
Dead room	44
Deafness	75
Decay rate	38,39
Decibel	1,3
Decrement	54
Degrees of freedom	54,55,56,60,64,418,420,421
Density effects	329
Department of Employment	463
Department of Environment	463
Department of Transport and Trade	463
Diesel engine noise	209,211–13
Diesel engine soundproof enclosure	216
Diesel engine vibration	213
Diesel generators	216
free standing	218–19
inside buildings	220
lift-off type of enclosure	219
permanently sited	219
Diesel locomotives	212
Digital signal processor (DSP)	186
Direct sound	37
Directivity factor	11,12,40
Directivity index	12,20–22
Discomfort level	75
Displacement	47,64
Domestic buildings	292–300
background noise	292
Deemed to Satisfy constructions	293
evaluation	295
external noise	299–300
floor coverings	298–99
independent timber frame method	297
insulation between houses	293
internal treatment	298
laminated plasterboard method	297,298
masonry wall insulation	296
multiple-occupancy	292
remedial treatment	297
standards	292
suspended ceiling/ partition constructions	293
Door furniture	376–77
Doors	348,374–77
general construction	376
hinged	375
sliding	375

INDEX 489

Double glazing. 371
        resonance effects 372
Double surfaced panels. 319
Drive motor noise. 242
Dual-purpose halls 306
Duct systems . . . . See Air distribution systems
Dynamic analysis 152,160–68
Dynamic balance 403
Dynamic frequency analyzers 125
Dynamic modulus. 417,433

E

Ear anatomy. 74
Ear defender head sets 87
Ear muffs 87–89, 91–92
Ear plugs. 81,87,89–90
     selective 91
Ear protectors . . . See Hearing protection devices
Earthwork operations 230
Eddy-current probe. 155,157
EEC. 464
Effective Perceived Noise Level (EPNL) 36
Elastomeric mounts. 428–31
Electro-acoustic systems 286
Electro-hydraulic servo-valves 180
Electro-hydraulic vibrator 179
Electro-magnetic vibrator 178–79
Energy attenuation constant. 41
Energy content 50
Environmental noise 73
Environmental noise analyzer 134
Environmental noise monitoring 132–36
     custom designed systems. 134
     hybrid systems 134
     low-cost system 133
     microphone system 133
     readout. 132
Equivalent continuous noise level (L_{eq}) . 27,73,85,
. . . . 118,259,260
Exhaust noise . 209–11,217,222,223,274,397,398
Expanded polystyrene 436
Expansion chamber. 223,394
External noise. 282,291
External sounds. 43,369

F

Factory noise 257–66
     background 257
     basic criterion 264
     categories of. 257
     community reaction 263
     escape to neighbourhood. 262
     L_{eq} measurement. 259
     noise sources. 262

Fan blades 200
Fan casing noise. 242
Fan design 235,378
Fan noise. . . . 231,232,235,236,241,246–54,378
    and windage 198–99
    corrections for installation design. . . . 250
    designing for minimum noise 252
    form of running 250
    general characteristics of 246–47
    guide vanes. 251
    intake turbulence 248–50,253
    interaction noise 249,254
    multi-staging. 251
    open plant room 245
    pressure–volume characteristic 253
    standby axial fan configuration 251
    turbulent flow 247–48,252,253
    vortex noise 247,252
Fan noise levels 378
Fan sound power levels. 245,246,247
Far field conception 10
Far field measurements. 105
Fast Fourier Transform (FFT) analysis techniques
    and tools. 166–168
Felts 447
    asphalted. 447
    vibration-isolation material 434
Fibre metal 339
Fibreglass 393
Flanking transmissions 317
Flexible foams. 335
Floating floors. 361
Floor coverings 298
Floor slab 364
Floors 287,361–68
Flow resistance 327,380,381,384
Fluctuating noise 85
Flutter echo 288
Foamed glass 339
Foamed materials as mounting pads 435–36
Forced vibration 47,55,98
Foundries 350
Fourier analysis 51,125,126,152,175
Fourier relationship. 50,51
Fourier series 50,161,411
Fourier spectrum 126
Fourier transform. 126,161,162,167,174
Free field conditions 9,16
Free field measurements 20,105
Free vibration 47
Frequency analysis 121–29
    comparison of linear and logarithmic. . 167
Frequency analyzer 134
Frequency displacement 47
Frequency domain 51,152,172
Frequency domain analysis 161
Frequency effects. . . 8,13,15,20,42,76,97,98,100
     121,147,200,248,287,310,316,318,326
     329,330,333,382,392,404,426

Frequency ranges 150,151
    sound intensity measurements. 112
Frequency ratio. 55,416
Frequency resolution. 172
Frequency response. 76,149
Frequency response analyzers, single-sine . . . 163
Frequency response function 161,171
Frequency spectrum 238
Friction noise 202

G

Gas turbine enclosure. 310
Gas turbines . . . . .396—99
    exhaust silencing 397
    intake silencing . . . . .396—97
    noise sources. 396
    self-generated and flanking noise 398
Gear manufacture inaccuracies 194,195
Gear noise 192
    effect of damping. 195
    general requirements 198
    sources of 198
Gear noise analysis 195
Gear noise parameters 199
General Fourier Series 162
Ghost noise 195
Glass 399,370
Glass fibre 359,380
    noise absorbers 261
Glass fibre wool 339
Glass screens. 373
Glazing. . . . . 287,369—73
    See also Double glazing

Granulators 349
Graphic recorders. 130,133
Ground absorption . . . . .13—14
Guildhall School of Music and Drama 458
Gypsum 339

H

Hammer mills 349
Hand drill 194
Handicap percentage 85
Hanning function. 127
Hanning weighting 128
Hanning window 167,172
Headsets 81
Health and safety aspects 74
Health and Safety at Work Act 1974 83,463
Health and Safety Commission 119,463
Health and Safety Executive. 83
Hearing conservation in industry . . . . . . .82—86
Hearing conservation programme 82,83
Hearing damage 74,82,138

Hearing efficiency. 137
Hearing handicap 75,85
Hearing impairment. 74
Hearing levels 86
Hearing loss 75,77,137,138,144
Hearing mechanism 76
Hearing protection 84,85
Hearing protection devices. . . . . . 80—81,87—94
    BS attenuation data. . . . . .93—94
    field performance. . . . . .88—89
    semi-aural devices. . . . . .90—91
    specified performance 89
Heat recovery boilers 398
Helmets 92
Helmholtz resonator 332
Honeycomb panels 340
Houses . . . . *See* Domestic buildings
Human ear. 74
Human figure model 98
Hydraulic tools 222

I

Impact damping. 450
Impact meter measurement 259
Impact noise. 200
Impact sound insulation 322,361
Impact testing 174
Impeller blading 249
Impregnated wood-wool slabs 340
Impulse response function 161
Impulse sound measurement. 25
Impulse sound meters. 120
Impulse sound pressure level 25
Impulsive machine noise 258
Impulsive noise measurement . . . . .108—9
Industrial noise . . . . *See* Factory noise
Inertial masss excitation systems 180
Inertial torque. 412
Infrasound effects. 20,97
Infrasound meters. 120
Instrumentation recorder. 130
Insulating wallboards 340
Insulation board. 344
Integrating sound level meters 117
Interference effects 38
Intermittent noise. 85
Internal combustion engines . . . . . . 209—20,270
    development of quieter structures . . . 216
    noise control techniques 213
Inverse square law. 10,11
Isolation efficiency . . . . .416—17

J

Jerk. 47
Jet effect. 387

INDEX

L

L_{10} Noise Index. 29
      measurement of. 269–70
L_{eq} . . . . . See Equivalent continuous noise level
Lagging. 399
Laminated glasses. 373
Land Compensation Act 1973. 463
Lead-filled vinyl. 357
Lead/foam sandwich 340
Lead-loaded plastic sheet. 341
Lead materials. 340
Lead sheet 340
Leaded plastics 341
Lecture halls. 306
Legislation. 82,83,463–64
Linear vibration. 63
Loss factor. 433
Loudness assessment 77
Loudness index 33
Loudness level. 31
Loudness scale. 31,32
Lubrication 202,204

M

Machine balance. 403–14
      critical speeds 408–10,412
      inertial torque. 412
      ISO Recommendations. 407
      normalized unbalance 407
      practical requirements 403–7
      quality number or grade 407
      reciprocating unbalance 411
      torsional vibration 411–12
Machine noise 193–201,257,258
      basic generators 193
      impulsive. 258
      reduction. 201
      treatment methods 414
Machine noise data 265–66
Machine noise measurement. 106
Machine noise radiators 200
Machine vibration
      basic causes 414
      fundamental frequencies. 414
      generators 193
      treatment methods 414
      See also Anti-vibration mounts; Damping;
      Resilient mounting of structures;
      Vibration isolation
Machinery health monitoring 181–90
      basic systems 183
      continuous or periodic. 184
      continuous equipment 185
      critical machines 181
      multiple noise sources 187–89

      non-critical machines. 183
      parameters monitored 181,184
      periodic equipment. 185–87
      semi-critical machines 181
      specific machine requirements. 184
      synchronous power spectra 189
Magnification factor 55
Mass controlled attenuator. 318
Mass-loaded vinyl 342,357–59
Mass per unit area law 318–19
Mastics 341,445,447
Mathematical analysis 63
Mean square pressure ratio. 3
Measurement . . . . See Noise measurement;
      Sound measurement
Metal jacketing 341
Microphone amplifiers 150
Microphone configurations 111
Microphone reciprocity calibrators 115
Microphones
      for environmental noise monitoring . . 133
      in sound level meters 114
Mineral wool. 342,359,380
Modal analysis. 170–76
Modes and mode configurations. 65–66
Monitoring
      application. 83
      audiometer. 139
      environmental. 132–36
      machinery health 181–90
      terminal 135,136
Motor vehicle sound level meter. 272
Motor vehicles. 70,267–72
      legislation 267
      See also Road Traffic Noise
Motor Vehicles (Construction and Use)
      Regulations 267,463
Mufflers 222,223,279
Muffl-Jac. 341,342
Multi-degree-of-freedom systems 54,55
Multi-rotor system 62
Music reception 301–2, 306
Musical and acoustic factors related 303

N

Narrow band analysis. 123
Natural frequency. . . . 52,53,60,64,418,420,422,
      .423,435
Near field conception. 10
Near field sound pressure level measurement . . 24
Neighbourhood noise. 70,262–64
Neoprene. 433
Noise from identical sources. 4
Noise Abatement Act 1960 463
Noise Abatement Requirements. 276

Noise Advisory Council 27
Noise and Number Index (NNI) 29
Noise annoyance 69,263
Noise averages 25
Noise certification 276
Noise criteria curves 35
Noise dosimeters 119
Noise energy 223
Noise energy measurement, time-varying . . . 117
Noise environment 119
Noise Exposure Forecast (NEF) 30
Noise exposure levels (L_{AX}) 73
Noise exposure patterns 84
Noise hazards warning 119
Noise Immission Level (NIL) 85
Noise indices 27
Noise Insulation Regulations 1973 269
Noise legislation 82,83,463–64
Noise level correction 71
Noise level reduction 70
Noise levels 17,83
      corrected 71
      damaging 75
      night time 264
Noise measurement 105–12
      average 107
      general 108
      impulsive 108–9
      machine 106
      office . . . . See Offices
      outdoor 105
      practical 112
      room 107
Noise monitoring 83
Noise Monitoring Terminal 135,136
Noise nuisance 463
Noise policy 463
Noise pollution level 30
Noise rating curves 35
Noise reduction coefficient 299
Noise Regulations 82
Noise scales 27
Noise sources 10–11,22,70
Non-linearity effects 167
Non-point source 10
Noys 31

O

Occupation noise 282
Octave band analysis 26,32,33,82,122,259
Octave band filters 122
Octave band sound levels 143
Octave band spectra corrections 246
Offices 282–91
      acceptable noise levels 283
      average noise levels 107,283
      background noise 282–84,286
      basic requirements 282
      external noise 282,291
      individual 289
      noise measurement 107
      noise sources 282
      occupation noise 282
      open-plan 288
      partitions 289–91
      reverberation time 286–87
      screens 288–89
      sound field 287
OmniScribe strip chart recorder 131
One third-band analysis 32
One-third octave synthesis 129
Opening effects 321
Oscillator/demodulator/amplifier 155
Outdoor noise 70
Outdoor noise measurement 105

P

Panel damping 443
Panel vibration 444
Parallel filter analyzers 124,125
Particle board 344
Partitions 289–91
Perceived Noise Level (PNL) 36
Percussive tools 221
Perforated facings 331
Perforated panels 329–33
Periodic noise signals 167
Periodic random noise 173
Permanent threshold shift 74,75,137
Personal monitoring 119
Personal sound exposure meters 119
Phase angle 50,56,415
Phase interlocking 150
Phase shift 56
Phenolic foam 343
Phon 31
Piezoelectric pick-up 151
Piling alternatives 229
Piles
      bearing 229
      retaining 229
Piston phone 116
Piston slap 214,215
Plasterboard 343
Plates 65,66
Plenum chamber 233–34,236,394
Pneumatic tools 221
Point source 9,11
Polar diagram 11
Polystyrene board 436
Polyurethane foam 380
Porosity 327

INDEX

Porous ceiling tiles 329
Porous materials. 327
    density effects. 329
Power spectral density 51,58
Preferred Noise Criterion (PNC). 36
Presbyacusis 74,75
Prescribed surface. 24
Pressure drop 327
Pressure loss. 384
Printer enclosure 286
Probability density 51,57
Proximity probe. 154
Pseudo random noise. 172
PVC film. 357

Q

Quiet Heavy Vehicle (QHV) Project 270
Quiet zone. 262

R

Radiation coefficient. 295
Random amplitude sine wave 58
Random aperiodic motion. 57
Random excitation 171
Random incident 392
Random vibration. 47,56,419–21
Readout 132
Real time analysis. 126,186
Reference distance 24–25
Reflected sound. 37
Resilient bearings 451–59
   design procedure 455–56
   examples of 457–58
   types of 452
Resilient materials 428
Resilient mounting of structures 451–59
   leak paths 457
Resilient seatings 366–68
Resonance effects. 372
Resonant absorbers. 332
Resonant frequency 318,319,321,333
Response displacement. 420
Reverberant field 16,22
Reverberant room. 38
Reverberation time 38–41,260,286–87,295,
  301,302,306
Reverberation time meter 120
Reverberation room 41
Road surface noise 270,271
Road traffic noise. 267–72
   assessment of 268
   comparison of UK and EEC noise limits.271
   L_{10} measurement 269–70
   statutory reductions 271
Rocksil density effects. 329

Rocksil products mounted over cavity and
  without facings 331
Rockwool acoustic insulation 342
Room acoustics 37
Room constant 39–41
Room diagonal 42,43
Room noise measurement 107
Root mean square. 50,57,58,114,419–21
Rotary tools. 223
Rubber. 451
   basic characteristics. 429–30
   configurated. 432
   reinforced 432
   shear compression. 431
   stressing methods 430
Rubber-bonded cork 434
Rubbing surfaces 202

S

Sabine formula 392
School rooms 307
Screens and screening 70,288,355,357,359,360,373
Semi-enclosures 359–60
Semi-reverberant field 16,23
Semi-reverberant room 38
Shadow zone 15
Shaft speed measurement 157
Shaft system. 62,63
Shaft whirling 411
Shape factors 433
Ship noise 308–11
   accommodation spaces. 308
   criteria guide. 309
   machinery spaces 309–11
   noise levels. 308
   passenger cabins. 311
Ship vibration transmission and suppression. . 311
Shock analysis. 60
Shock excitation 450
Shock phenomenon. 47,59
Shock response system 60
Shock spectrum. 60,61
Signal conditioning 130,185
Silencers 209,211,212,236,244,275,348,378
   absorptive 378,382,391–94,397
   breakout 388
   circular-section 392
   dissipative 391,397
   ducted 379
   exhaust. 397,398
   full flow 388
   industrial. 391–95
   infill materials. 388–89
   inlet. 398
   intake. 397
   jet effect 387

location of 386
material requirements 399
mechanical considerations . . . . . .389—90
optimum position. . . . . .385—88
performance. 391
reactive. 394
reactive-absorptive 394
rectangular section 392
secondary noise (regeneration) 385
splitter . . . . 380—85, 392,394
Simple harmonic motion. 47
Simple periodic vibration 47
Sine waves. 51
Single-degree-of-freedom systems. . 54—56, 60,64
Single event noise exposure (L_{AX}) 27
Single-sine analysis techniques and tools. . 162—66
Sinusoidal testing. 173
Sinusoidal vibration. 52
Skid resistance. 270
Slag wool. 380
Sliding surfaces 202
Slit effects 321
Sone 31,32
Sound 1
Sound absorbing materials. . . . 260,315,324,334,
. 355,356
disintegration 328
general characteristics of 325
general classification 325
surface finish 328
Sound absorption . . . . .46,301,315,322—24
basic principles . . . . .322—24
characteristics of 329
See also Absorption coefficient
Sound amplification system 305
Sound barriers. . . . . .14,315,334,354—60
effectiveness of 354
material requirements 354
Sound conditioning system 286
Sound curtains . . . . .355—59
Sound directivity 11
Sound exposure and L_{eq} relationship 118
Sound exposure meters. 118
personal 119
Sound insulating materials. . . . 315,316,334,355
Sound insulation . . . . .315—22
impact 322,361
of walls and partitions 337
Sound insulation panel 316
Sound intensity 6,8
Sound intensity level 1,2,6
Sound intensity measurement. . . . . . 110—112
frequency range of 112
Sound level calibrators 116
Sound level measurement 70
Sound level meters . . . . . . 113—21,147,148,268
additional facilities 116
calibration . . . . .114—16

calibrator types 115
circuitry . . . . .113—14
grades and standards for 113
impulse. 120
infrasound 120
integrating 117
microphones. 114
signal outputs 117
Sound levels 1,80
adding 3
percentage change in 5
subtracting. 4
Sound measurement . . . . .16—26
comparison of conditions 17
hemispherical 22
indoor 16
outdoor 16
quantitative 17
Sound power level . . . 1,2,5—6,9,12,17,19,20,23,
. 232,274
Sound pressure level . 2,3,5—6,9,17,26,38,40—42,
. . . . 77,78,138,279,295,315,328,351,378
Sound pressure level measurement 75
Sound propagation 7
characteristics of 11
modification of 13
Sound reduction 9,70
Sound reduction coefficient. 326
Sound reduction/frequency relationship. . . . 290
Sound reduction index (R) .109—10,242,290,315,
. . . . 316,318—20,345,346,352,353,374
Sound reflection 14
Sound spectra 19
Sound transmission loss 315
Sound velocity 7,47
Sound waves. 2,7,8,14
Spectral leakage. 167
Spectrum analysis. . . . . .122—29
See also Frequency analysis
Spectrum measurement 189
Speech, acoustic power. 305
Speech communication. . . . . .78—81
Speech intelligibility 284,301,305
Speech Interference Level (SIL). 79
Speech masking 284
Speech reception . . . . 301—2,306
Spring-mass system 52
Springs
air. 431
application. 426
low frequency 426
metal . . . . .423—28
minimum outside diameters 428
natural frequency 417,423
noise transmission. 428
performance. 427
types 427
unit mounts 426

INDEX

Spruance class destroyers 310
Standing waves 38
Static balance 403
Static deflections 423,426
Static modulus 417
Statistical analysis. 85
Steel 343
Step relaxation 174
Structural elements 65
Structure-borne noise. 216
Subjective noise parameters 31
Super-whirling 411
Suppressors 279
Surface finish 328
Suspended ceilings 288,361
Suspended noise absorbers 260–62
Swept-sine vibration testing 165
Swept spectrum analyzer 125
Synchronous power spectrum 189–90

T

Tape recorder 130
Telephone hoods 288
Telephone speech communication 81
Temporary threshold shift 75–77,137
Thermal insulation 334,339
TICO CV/CA 457
TICO CV/LF 459
TICO CV/M 458
Time compression analysis. 125
Time domain 152
Time domain analysis. 160
Timing gear 215
Torsional dampers 63
Torsional exciter units 180
Torsional vibration 62,63,411
Tracking filter based equipment. 186
Traffic noise . . . . *See* Road traffic noise
Traffic Noise Index (TNI) 30,268–69
Transfer function 161,176
Transient testing 174
Transmissibility . 64,65,415,420,421,423,438,439
Transmission coefficient . . . 315,316,318,322,346
Transport and Road Research Laboratory .270,272
Treaty of Rome. 464
Turbomachinery noise 277
Turbulent flow 238,247–48,252,253
Tyre noise 270–72

U

Ultrasonic frequencies 100

V

Velocity of sound 7,47
Velocity pick-ups 156
Ventilation systems 284,348
Vermiculite 344
Vibration
   damaging effects 100
   effects of infrasound 97
   effects on people 97–101
   forced 47,55,98
   human figure model 98
   principles of 47
   random 47,56,419–21
   response to 98
   steady state 52
   threshold of 404
   threshold of alarm 97
   threshold of feeling 97
   threshold of perception 97
   torsional 62,63,411
   vertical 99
Vibration analysis
   of rotating machinery 166
   synchronous 165
   See also Dynamic analysis; Modal analysis
Vibration direction determination 148
Vibration exposure criteria 101
Vibration frequency 150
Vibration frequency analysis 147,151
Vibration generator 177
Vibration-induced white finger 101
Vibration isolation 415–21
   basic methods of 423
   basic principle of 422
   efficiency of 416–17
   performance 423
   See also Anti-vibration mounts; Damping;
     Machine vibration; Resilient mounting of
     structures
Vibration level recorders 150
Vibration measurement 147–52
   calibration 151
   hand-held pick-ups 149
   readout 149
   simple instrumentation 147
Vibration modes 418
Vibration monitoring . . . . . *See* Machinery health
   monitoring
Vibration pick-up system 148
Vibration testing 177–80
Vibration transducers 154–59
Vibrator attachment points 175–76
Vortex field generation 273
Vortex noise 200,247,274
Vinyl foam 344
Vinyl products 342–44,357–59
Vinyl sheet 342,357,359

W

Walls 287
Water cooled engines 218
Waveforms 50,170
Wavelength effects 8,42,43,200,319,382,392
Weighting scales . . . . 18—20,24,25,75,127
White finger 101
White noise 59,81,420
Wind effects . . . . .12—13
Wind forces 368
Windage noise . . . . 199—200

Windows 350,369,373
Wood fibre boards 344
Wood joist floors 365,366
Work area noise levels 257
Work environment 82
Workshop noise . . . . *See* Factory noise

Z

Zoom transforms . . . . .174—75

Index to Advertisers

Acoustic and Vibration Technology	182A
Antiphon Ltd	400A
Atkinson Dynamics	182B
Atmospheric Control Engineers Ltd	312B
A.V.A. Ltd.	406B
Avomet Ltd	374B
BBN Instruments Corp	148B
Bestobell Protection	406A
Bruel & Kjaer (UK) Ltd	Bookmark
Burgess Industrial Silencing Ltd	202A
Bury Cooper Whitehead Ltd	346A
Castle-Microair Ltd	154A
W. Christie & Grey Ltd	322A
Cirrus Research Ltd	346B
Clark Door Ltd	374A
C.M.T. Wells Kelo Ltd	422B
Cobra Acoustics Ltd	322B
Computer Engineering Ltd	122A
Condition Monitoring Ltd	182A/202B/246B/405
A. Davies & Sons Ltd	422A
Dansk Industri Syndikat A/S	160A
Ewald Dorken AG	316B/C
Frank Driver Generators	374B
Econocruise Ltd	178B
Endevco U.K. Ltd	154B
Entran Ltd	VI/148A
Envirodoor Markus Ltd	282B
Environmental Equipments Ltd	144B
Ferguson & Timpson Ltd	422A
GenRad Ltd	170A/178C
Gracey & Associates	178B
Hakuto Opto/Electronics Div.	148B
Harrison & Jones (Acoustic Div)	246B
Heafield Industries Ltd	374B
Hodgson & Hodgson Ltd	374A

INDEX TO ADVERTISERS

Illbruck International	316A
IMI Norgren Enots Ltd	182A
Industrial Acoustics	Bookmark
Instron Environmental Ltd	178D
Kay-Metzeler Ltd	178B
Ling Dynamic Systems Ltd	178A
Lord Corporation (UK) Ltd.	406B
Lord Corporation Allforce Acoustics	346A
Malaysian Rubber Research & Development Boare	400B
Rich. Muller (UK) Ltd.	346A
Nicolet Instruments Ltd	122B/160B
PCB Piezotronics Inc	170A
Porvair Ltd, Industrial Div.	202B
Rion Co Ltd.	170B
Scientific Atlanta Ltd.	VIII
Selectaglaze Ltd	282A
Sound Solutions Ltd	374A
Stop-Choc Ltd	422A
Supra Chemicals & Paints Ltd	322B
Tantalic Acoustical Engineering Ltd	202A
TICO Manufacturing Co Ltd	178C
Trelleborg Ltd	406B
Trubros Acoustics Ltd	422B
Vibro-Meter AG	246A
James Walker & Co Ltd	422B
Weir Pumps Ltc.	346B
Weiss Technik Ltd	178E